WA 1305712 X
D1325613

Klaus Ehrlenspiel · Alfons Kiewert · Udo Lindemann

Cost-Efficient Design

Klaus Ehrlenspiel · Alfons Kiewert · Udo Lindemann

Cost-Efficient Design

With 320 Figures

Professor em. Dr.-Ing. Klaus Ehrlenspiel
Kemnatenstr. 47A
80639 München, Germany
ehrlenspiel@mytum.de

Dr.-Ing. Alfons Kiewert
Technische Universität München
Institut Mechatronik
Lehrstuhl für Produktentwicklung
Boltzmannstr. 15
85748 Garching, Germany
sekretariat@pe.mw.tum.de

Professor Dr.-Ing. Udo Lindemann
Technische Universität München
Institut Mechatronik
Lehrstuhl für Produktentwicklung
Boltzmannstr. 15
85748 Garching, Germany
udo.lindemann@pe.mw.tum.de

Editor:
Professor em. Dr. Mahendra S. Hundal
Via di Fabbrica 31
50024 Mercatale VP, Italy
hundal@cems.uvm.edu

Library of Congress Control Number: 2006933998

ISBN-10 3-540-34647-3 Springer Berlin Heidelberg New York
ISBN-13 978-3-540-34647-0 Springer Berlin Heidelberg New York

Springer is a part of Springer Science+Business Media

springer.com

Typesetting and Production: LE-TEX, Jelonek, Schmidt & Vöckler GbR, Leipzig
Cover Design: Erich Kirchner, Heidelberg
Printed on acid-free paper SPIN: 10933084 62/3100 YL – 5 4 3 2 1 0

Editor's Preface

This translation has been developed from the fifth edition (2005) of the Authors' German text, "Kostengünstig Entwickeln und Konstruieren". The discussion on the methods of cost reduction and cost-driven design draw heavily on the "Systematic Method of Design", the development of which is described in German language literature of the past several decades. An English language edition [Pah05][1] of a classic text is available. In this approach, the design process is divided into the phases of task clarification, concept development, preliminary design (embodiment) and the detailed layout. Design forms the core of the coordinated process of modern product development, known under the rubric of Concurrent Engineering. This book stresses the collaboration of the different departments in a company, which is essential for the development of reliable and low-cost products.

The first edition was a best seller for a number of years. Eventually, as the fields developed, a second edition came out in 1998. The development of methods (a major portion of that from the research in the authors' Institute,) as well as the demand for the book, led to the third, fourth and fifth editions appearing in quick succession. It was the first book that, with a scientific approach, showed the connection between design, production processes and the product costs. This provided the basis for cost-driven product design.

The overriding goal in preparing this translation was to convey the ideas of the authors as closely as possible.

German references

The references in the book are primarily to German literature. No attempt has been made to provide references to English language literature on the subject, as those are abundantly available to the English language readers from other sources.

There are references to DIN (Deutsche Industrie Normen – German industry standards) and VDI (Verein Deutscher Ingenieure – German Society of Engineers) guidelines. Knowledge of either is not necessary in following the book. They have been retained from the German edition as the original source material. Many of these are available in English or have been adopted as ISO standards.

Notation

The book is intended for an international audience. Although the currency used is the US$, the $ sign is placed after the number. As a compromise, numbers are shown as follows: 12 345.67 instead of 12,345.67 as is customary in the USA, or

---

[1] Expressions in square brackets denote References at the end of the book.

12.345,67 as used in Europe. The citations of **Figure** numbers are in **bold** where the pertinent figure is discussed in detail. Likewise, in the Index the page numbers are in **bold** where the corresponding keyword is discussed at length. Important sentences have also been printed in **bold**. In the appendix we satisfied a wish from the readers to have a consolidation of the most important checklists and rules for practical cost reduction work. We hope this will be helpful. These guidelines for cost reduction have a grey narrow band at the end of a page.

On a personal note, I met Prof. Ehrlenspiel for the first time in 1986. The year before, in 1985, he had published the first edition of the present book. During the year 1987–88, I spent my sabbatical year at the Institute of Mechanical Design (now, Product Development) at the Munich University of Technology. Since that time, I have been fortunate to be in continuous contact with him and his colleagues at the Institute. My 1997 book [Hun97] is an outcome of this association.

South Hero, Vermont and Fabbrica, Toscana 2006 *Mahendra Hundal*

Preface to the First English edition

The authors are pleased that after several years of fruitful co-operation with Professor Mahendra Hundal, the English translation of the 5th Edition of the German book is finally here.

We thank him for the empathy and intuitive understanding, as well as his tenacity, in carrying out the translation and editing the final product. Special thanks are also due to Dr-Ing. M. Mörtl, who, as a successor to Dr.-Ing. A. Kiewert, worked on corrections to the manuscript.

We are pleased too, as always, for good cooperation with Springer-Verlag, which realized the translation project in co-operation with ASME. In this regard, we make a particular mention of Dr. D. Merkle.

Munich, September 2006 *Klaus Ehrlenspiel • Alfons Kiewert • Udo Lindemann*

Preface to 5th Edition

Reducing costs is a continuing task, and the reason for this fifth edition of the book.

Contents and breadth of the book were developed at the Munich University of Technology, and were the basis for the course “Cost Management in the Machine Industry” and for a VDI seminar entitled “Technical Designers Lower Costs”. It is also used for courses at other universities and technical colleges.

The primary changes from the fourth edition are additions to the literature and correction of errors that were discovered.

Lowering costs is important in all countries. Therefore, the translation of the book into English has been an ongoing task, being carried out by Professor Hundal of the University of Vermont, USA. It is soon to be published by Springer Verlag, New York, in co-operation with the ASME.

Munich, April 2005 *Klaus Ehrlenspiel • Alfons Kiewert • Udo Lindemann*

Preface to the 4th German edition

The pressure of costs is continually increasingly. Target-cost based design and development are of current interest and more necessary than ever before. The cost reduction of –30% described in Figure 2.3-2 often taken as a "standard precondition", next to an increase in performance in the case of new designs and product revisions!

From the feedback on this book, from industry training courses and industry contacts we know that the book helps in the process. So, every few years a new edition comes out. We are happy, of course!

This time much of the contents were re-worked and new material added.

We have always known from the practice how important the pursuit of costs is during a project. Therefore we have introduced two more systems for this purpose in Sect. 4.8.3.2.

The prolongation of the service life of products benefits both the user and the manufacturer. Therefore Chap. 5 on "Influencing the lifecycle costs", was enhanced by Sect. 5.4.

Then there is the always-current topic of variant management! The demand for customer-specific products works diametrically opposite to the wish for low costs. Variant management must help in finding a compromise. The contents of Sect. 7.12 were revised and expanded. Two new examples for modular design were added: Modular design application in Porsche sports cars and for tractors (Prof. Dr.-Ing. Renius).

The costs and prices were converted to the Euro. And this, despite the fact that all numerical data in the book are only "relative values". On one hand the numbers must not show the original industry values in the examples, on the other hand it is not possible to keep the cost information continuously up-to-date. The matter is only of the examples from practice, which anyway always contain company-specific cost information (see Figure 7.13-2).

Further, a table of contents in English was inserted as a first step toward a complete English translation of the book. We thank Prof. Hundal of the University of Vermont for that. This helps foreign readers get an orientation to the subject and introduces important specialist terms to German readers.

Finally, this took care of the new spelling, adding to the literature citations and the correction of mistakes.

During this extensive revision, Dr.-Ing. A. Kiewert has carried the main burden. Mr. H. Nyncke M.A. has carried out the computerization of the manuscript over several months. To them, many thanks for their commitment. Springer Verlag has supported us in well-proved manner.

Our wish for you as readers is, that you get a better "cost view" with the book and tackle the "cost lowering" problems.

Munich, May 2002 *Klaus Ehrlenspiel • Alfons Kiewert • Udo Lindemann*

Preface to the 3rd German edition

The book apparently meets a strong need in practice and teaching. Therefore the 3rd edition is here after a good year.

With regard to the contents, the sections on evaluation and variant management were greatly enhanced. In particular we satisfied a wish from the readers concerning a consolidation of the most important checklists and rules for practical cost reduction work. Also, it is hoped that the clearly discernible appendix on the "guidelines for cost reduction" will be helpful for that.

Of course the literature was added to, and the mistakes found were corrected. In cost reduction there is a consistent improvement potential.

We, together with the publisher, appeal to you to implement this also!

Munich, June 1999 *Klaus Ehrlenspiel • Alfons Kiewert • Udo Lindemann*

Preface to the 2nd German edition

We are publishing this book because we have found that with the thinking and methods described here, 20–30% of the manufacturing costs of the products can often be lowered in the practice – quite apart from the overhead costs and lifecycle costs.

That appears to us to be an intelligent and additional possibility for the stabilization of the much talked about "position Germany". At least, in addition to the simple reduction in personnel, as the sign of "Lean Production".

The book was re-worked completely, vis-a-vis the **1st Edition of 1985**. This is true in particular for Chap. 2 through 6. The experiences from many industry projects and from approximately 90 industry training courses taught in and with the industry were utilized for this. (Sect. 7.13 presents, for example, the results of an almost 20-year cooperation with 8 to 15 companies of the Gear Drives Research Association FVA, under the heading "Cost Benchmarking").

Further, the increasing knowledge in methodical design and development was taken into account: Adapting universal methods to the specific concrete problem, integrated product design with increasing specialization, and emphasis on early development phases, since the essential decisions are made here.

Our aim was to consider the modern production and assembly processes. Generally, there is hardly any relevant literature about their time and cost properties. In addition a lot of things were elaborated upon, and then again dropped for space reasons – sacrificed to the editor's red pencil. Our experience showed that the production technology and parts suppliers' markets are developing so rapidly that for each concrete case the appropriate applicable knowledge must be procured individually. The book can provide only the basic stimuli.

Our **thanks** go, first of all, to all the colleagues and employees of the Institute of Mechanical Design for the contents-related work, in particular Dipl.-Ing. M. Moertl, Dipl.-Ing. J. Wulf, and Dipl.-Ing. U. Phleps.

The scientific support personnel, Dipl.-Ing. C. Geng, Mr Dipl.-Geogr. M. Krämer, Dipl.-Ing. Dipl.-Wirtsch.-Ing. M. Reichart, Ms E. Carbajo and Ms. C. Stubenrauch spent a long time transferring text and figures to the computer.

We also want to thank our sponsors. Many projects that were supported by the DFG, the FVA and the BMFT, have added to the knowledge gained.

Of course, in the same way we have also learned a lot from industry, from the companies and their employees.

The Springer Verlag is to be thanked for the careful realization of the book and good cooperation. In this regard we want to especially mention Dr. Merkle.

Munich, May 1998 *Klaus Ehrlenspiel • Alfons Kiewert • Udo Lindemann*

We would be pleased with your suggestions and critical review.

Please address the communications to:

Technische Universität München
Institut Mechatronik
Lehrstuhl für Produktentwicklung
Boltzmannstrasse 15
85748 Garching
Germany

Tel.: + 49 89 289 – 151 31
Fax: + 49 89 289 – 151 44

sekretariat@pe.mw.tum.de
http://www.pe.mw.tum.de

Contents

Nomenclature

Page numbers where definitions are given or the symbol's primary appearance, are in parentheses. In the book, **capital letters** (***MC*, *MtC*, *PCs*, *PCe***) are used for absolute quantities and **lower-case letters** (***mc*, *mtc*, *pcs*, *pce***) for quantities as percentages (usually with reference to *MC*).

Symbol	Description
φ	Size ratio (re: size range) (s. Chap. 7.12.5.3)
φ_L	Linear size ratio (Fig. 7.12-23)
Subscript 0	Basic (initial) embodiment (s. Chap. 7.12.5.3a)
Subscript 1	Succeeding embodiment (s. Chap. 7.12.5.3a)
AOC	Administrative overhead costs (Fig. 8.4-2)
AOCAS	Administration overhead costs surcharge rate (Fig. 8.3-3)
AsC	Assembly costs (s. Chap. 7.13.7b)
ASC	Administration and sales costs (s. Chap. 8.4.3b)
ASOC	Administration and sales overhead costs (Fig. 8.4-2)
ASOCS	Administration and Sales overhead costs surcharge rate (Fig. 8.4-4)
CAS	Cost accounting sheet (Fig. 8.3-3)
C_{fix}	Fixed costs (Fig. 8.5-2)
C_{one}	One-time costs (s. Chap. 7.5.1)
C_t	Total Costs (Fig. 8.5-2)
C_V	Specific material cost (s. Chap. 7.9.2.3)
C_V^*	Relative material cost (on volume basis) (s. Chap. 7.9.2.3)
C_{var}	Variable costs (Fig. 8.5-2)
DC	Direct costs (Fig. 8.3-3)
DDC	Design and development costs (Fig. 8.4-2)
DDOC	Design and development overhead costs (s. Chap. 8.4.3)
DDOCS	Design and development overhead costs surcharge rate (Fig. 8.4-4)
DfC	Differential cost with SHC, to a basic variant without SHC (Fig. 9.3-3)
Fix	fix (Fig. 8.5-2)
IC	Product introduction (launch) cost (Malus) (s. Chap. 7.5.1)
IC_p	Introduction cost per product = IC/N_{tot} (s. Chap. 7.5.1)
L_n	Number of lots (s. Chap. 7.5)
LCC	Life-cycle costs (Fig. 5.1-3)
m	Modulus (gear) (Fig. 7.13-4)
MC	Manufacturing cost (Fig. 8.4-2)
MC_w	Manufacturing cost per unit weight (Weight-cost ratio) [$/kg] (s. Chap. 9.3.2.1)
mcs	Ratio of production cost from set-up times (to manufacturing cost) *mcs* = *PCs/MC* (s. Chap. 7.7.1)
mtc	Ratio of material cost (to manufacturing cost) *mtc* = *MtC/MC* (s. Chap. 7.7.1)

MtC	Material cost (inclusive of mass-dependent costs) (Fig. 8.4-2)
MtDC	Material direct cost (Fig. 8.4-2)
MtOC	Material overhead cost (Fig. 8.4-2)
MtOCS	Material overhead cost surcharge rate (s. Chap. 8.4.2)
MU	Monetary units (e. g. $) (s. Chap. 9.3.4.2)
N	Lot size (s. Chap. 7.5)
N_{tot}	Total manufactured quantity (s. Chap. 7.5)
OC	Overhead costs (Fig. 8.4-1)
OCS	Overhead cost surcharge rate (Fig. 8.4-1)
P	Power (Fig. 7.12-24)
PC	Production cost (Fig. 8.4-2)
pce	Ratio of production cost from individual times (to manufacturing cost) *pce* = *PCe/MC* (s. Chap. 7.7.1)
PCe	Production cost from total time units (s. Chap. 7.5.2)
PCs	Production cost from set-up times (Set-up cost) (s. Chap. 7.7.1)
PDC	Production direct cost (s. Chap. 8.4.2)
PLC	Production labor costs (s. Chap. 8.4.2; Fig. 8.4-2)
plc	Labor wage rate (s. Chap. 8.4.2)
POC	Production overhead cost (Fig. 8.4-2)
POCS	Production overhead cost surcharge rate (Fig. 8.3-3)
S_{u}	Ultimate strength (s. Chap. 7.9.2.1)
SDC	Sales direct costs (Fig. 8.4-9)
SHC	Shaft-hub connection (Fig. 9.3-3)
SOC	Sales overhead costs (Fig. 8.4-2)
SOCS	Sales overhead costs surcharge rate (Fig. 8.3-3)
SPDC	Special production direct costs (Fig. 8.4-9)
SSDC	Special sales/marketing direct costs (Fig. 8.4-2)
T	Torque (Fig. 7.3-2)
TC	Total (factory) cost (Fig. 8.4-2)
t_{e}	Total production time per piece (Fig. 7.6-2)
t_{E}	Piece-proportional, non-reducible part of the first time t_1 (s. Chap. 7.5.1b)
$1\text{-}t_{\text{E}}$	Time portion reducible by quantities (s. Chap. 7.5.1b)
t_i	Indirect production (idle) time (Fig. 7.6-2)
t_{m}	Direct machine (production) time (Fig. 7.6-2)
t_{n}	Piece time for the n*th* run (Fig. 7.6-2)
t_{re}	Recovery time: Time that is required for operator to rest/recover (Fig. 7.6-2)
t_{s}	Set-up time: Basic set-up time, set-up recovery time, set-up ***extra*** time (Fig. 7.6-2)
t_{x}	Extra time: Additional time that is required due to human involvement, over and above the scheduled time for a job (Fig. 7.6-2)
Var	Variable (Fig. 8.5-2)
W	Weight (Fig. 7.12-23)
yr	Year (Fig. 4.6-4)

MATERIALS:

GCI	Gray (cast) iron
MCI	Malleable (cast) iron
NCI	Nodular (cast) iron
CS	Cast steel

1 Introduction

1.1 Cost reduction – an issue in product development

This book shows how from the beginning, starting with product development the costs of a product may be lowered, or kept below a required limit.

The book analyzes more than the customary organizational measures, for example the close cooperation of product development and manufacturing, as recommended in value analysis. Methods, data and examples are shown that are directly related to product costs, including the correlation of technology and costs. Typical questions in this regard are: How to find a basic low-cost solution? What is the cost structure of the given product or of similar products? Which components of the product and what kinds of costs must be specially kept in mind? What to begin with most effectively? How to lower the costs of materials, and the production and assembly costs of the product? How to employ Target Costing for that purpose?

It makes much more sense to lower the costs right at the beginning of product development, rather than afterwards by the usual steps of personnel reduction when the costs are found to be too high.

We have dealt with these methods in our research for over 30 years. The first edition of our book, in 1985, brought out important knowledge and methods that were acquired from, and used in practice, e. g., designing to a cost goal, long before Target Costing became widely known. In the intervening time we have gained more knowledge. We know what the needs in practice are and where the weaknesses lie.

What is often missing is the motivation and courage to get rid of old practices and adapt to new ideas and ways of thinking and working. Product developers should be quite clear about the fact that they have the greatest influence on product costs. That is why they need to master their own technology, as well as the knowledge of product costs. And they need methods that will enable them to attain not only the physical objectives, but also the cost targets.

An important prerequisite for this to come about is a close cooperation of product development, production, cost control, marketing and procurement, in the sense of bringing down the "walls" between the departments (cf. Fig. 3.2-2). **Cost reduction is a team task.** Departmental egotism and refusal to share knowledge drive up costs!

If this is truly taken to heart, our experience indicates that a reduction in manufacturing costs of 10–30% is possible. Even more can be achieved by applying new concepts (Sect. 4.8.2).

Cost management today is a **necessary addition** in the development of **innovative and high performance products**, about which customers would be enthusiastic, and which fulfill the market requirements. This should be said at the beginning of a book on cost reduction. **No enterprise can survive only by reducing costs**, or by making products too costly by overengineering.

1.2 Aims of the book

The book is aimed primarily at product designers and developers. They have a great deal of influence in the early phases of product development. That is where the important decisions are made – but they cannot do it alone. Therefore this book is also written for those willing to cooperate – from production, cost control, marketing and procurement.

We recommend it to instructors, and most importantly, the students, so they may take these methods and mental attitudes into the practical world.

We intend to establish the following **learning objectives**:

- How can we develop products to predefined **cost targets** (Target Costing)? How do we establish cost targets? How do we stick to these targets?
- What types of collaboration, organization, which methods and remedies have stood the test of time in this respect? In short: How to set up **cost management** for product development?
- Which are the **most important parameters** for the product developers **that affect costs**, and what is the best way to apply them?
- Which **costing concepts** and **types of cost calculations** are important for the product developers? What should be their basic knowledge of industrial management?
- What knowledge has been gained through research and practice? What is the current **state of knowledge**?
- How can methodical design be coupled with cost-driven design? Thus, how can **innovative** and **low-cost** products be developed in **one** development process?

1.3 Structure of the book

No book that deals with how costs may be affected can be written without clarifying the term right in the beginning (**Chap. 2**). What does one understand by the term cost and **which costs are of significance for the product developer**?

Since design and development, together with product planning, have the greatest influence on the product costs, it is therefore important that the underlying reasons and the conclusions that follow, should be explained to the whole organization.

Chapter 3 shows what we understand by cost management and what the consequences are for product development work and the cooperative activities with other company divisions and with the suppliers.

A focal point of the book, describing **organizational possibilities as well as methods and tools for cost management**, is contained in **Chap. 4**. We show how they function and how they may be introduced and put to use. It is obvious that here consideration must be given to differing company sizes, product complexities and production quantities.

Since a company wants to supply user-friendly products, the costs that the user will incur with the product should also be minimized. **Life cycle costs** are a measure of user-related costs, and they are treated in **Chap. 5** from the user's viewpoint. This supplements the previous discussion from the manufacturer's viewpoint, which dealt with **manufacturing costs**.

The control of **total factory costs** is dealt with in **Chap. 6**, particularly the reduction of overhead costs through product development. As a field that has been neglected thus far, it is particularly important due to the increasing proportion of overhead costs.

Chapter 7 goes into the details of possibilities of lowering of the **manufacturing costs**. Here alternative production and assembly processes and materials are addressed so that they may be mastered. Also discussed is the management of variants, with size ranges and modular designs and other possibilities for controlling the increasing amount of the variety of variants.

The basic concepts and the advantages and disadvantages of the common procedures of **cost calculation** are shown in **Chap. 8**. The important make-or-buy decisions can be made on this basis.

Since no cost objectives can be held to in design without **concurrent cost calculations**, **Chap. 9** presents the procedures used in practice, as well as new and effective methods.

Short **examples from the practice** are shown in the book repeatedly. How the costs in a company may be reduced in a concrete fashion, is shown in **Chap. 10** with the help of two detailed examples. In addition, the various methods of concurrent calculation are explained with the example of a simple product.

The present book shows a broad palette of possibilities for a company to lower its costs in a deliberate procedure. In this regard it is not a matter of a literal application of the methods presented. They must be chosen to suit a given situation and then applied – difficult as that may be.

➔ **Methods shown here have been proven in practice.**

➔ **We wish you success!**

1.4 For an easier use of the book

Cost reduction is a complex problem. Accordingly, there are many different views, viewpoints and starting points.

The readers will find that the quickest method to start in the right direction is with **Sects. 4.5** and **4.8**, as well as the examples given in **Sects. 4.7**, **10.1** and **10.2**.

Then, if they feel comfortable with the theme, "guidelines for cost reduction" given in the Appendix provide a summary for the day-to-day work.

In addition, at the beginning of **Sect. 4.6** there is an overview in tabular form of the **measures for cost management**, including section and figure references, and in **Sect. 7.10.4** a summary of **rules for cost-driven product development**.

Finally, we have used "boxed" text to indicate important points as follows:

➔ Important guidelines and rules for the practical use are highlighted in the text by placing these in such "boxes".

The cost data are given in US$. We must point out that the **cost data** given in this book for parts, materials, etc., cannot be used for other purposes. They are always **company-specific** (Sect. 7.13, Cost Benchmarking) and **time dependent**, since the boundary conditions (wages, prices, etc.) change with time. Wherever cost figures are given in examples, the underlying company data have been concealed. Also, in case of materials we have used the old terminology; the Material No. (e. g., 1.0570 for St 52-3) was not used.

The **significance** of each respective **concept** is given in the Subject Index on page numbers in **heavy** type.

2 Cost Responsibility of the Product Developers

This chapter will explain some cost concepts relevant to product development. Then, the influence the company departments participating in product development have on how costs originate in a company will be investigated. This will show that product development has the greatest significance in company cost management. However, for the long-term success of a company, innovation should not be neglected for the sake of costs.

2.1 What are costs?

In our society, the very existence of human beings is tied to a continual outlay of money. Food, clothing, and shelter are all are basic needs that will always need to be satisfied anew through expending financial resources. For this reason, each of us is familiar, from our own experience, with the problem of how costs originate.

In economics, **costs** (see Sect. 8.1) are generally defined as consumption of goods for company's output, expressed in terms of money. In this context, goods may be material, energy and company facilities, as well as labor, information, or the utilization of capital and the rights of others, e. g., intellectual property rights. For a company's output, creation of goods includes **providing products** or **rendering services**. Thus, we always aim for so-called **added value**, which means that the result of the goods used has more value than the sum of the incurred costs.

In a company, the costs and the possibilities of reducing the costs can be viewed from a variety of perspectives. From the aspect of efficient product development, the costs that are of interest here are those incurred by the products. In this regard, a classification of costs (as shown in **Fig. 2.1-1**) is useful. The basic costs that originate with the product itself are the **manufacturing costs** (*MC*, cf. Chap. 7), that is, those costs that can be assigned directly to the manufacturing process. Essentially, these consist of the material and production costs for the product. In addition, there are costs that cannot be directly assigned to the product manufacture (e. g., **administrative costs**). They are combined with the manufacturing costs to form the company's **total costs** (*TC*, cf. Chap. 6). The total costs, in turn, contribute to the **lifecycle costs** (*LCC*, cf. Chap. 5) and are reflected in the product sales price.

The lifecycle costs are costs that accrue to the product user, and are the sum of all costs associated with purchase, use, and disposal of the product (cf. Fig. 5.1-5) They may be roughly classified into the following types of costs:

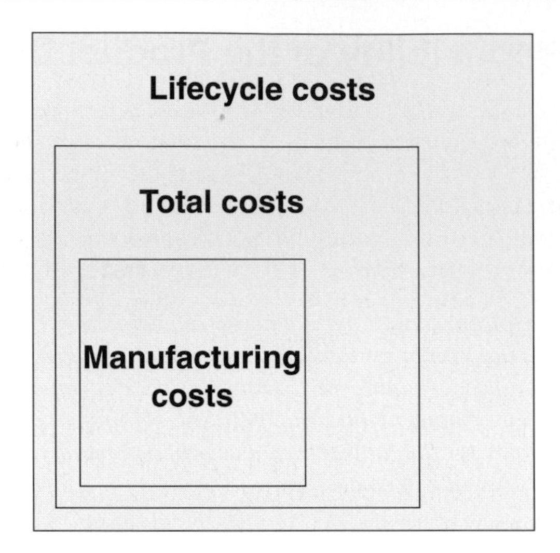

Fig. 2.1-1. Classification of costs

- **Initial costs**, which consist essentially of the initial price of the product. The resale price of the product at the end of its useful life could be subtracted from the initial cost.
- **One-time costs**, for example, costs for transport, installation, startup, personnel training, and disposal.[1]
- **Operating costs**, for example, ongoing costs for energy, operating materials, and their disposal, as well as wages for service personnel.
- **Maintenance costs** for servicing, inspection, and repair.
- **Other costs**, for example, capital costs, taxes, insurance, and breakdown costs.

For the user, lifecycle cost is the criterion by which a product's economic efficiency can be measured. This strictly economic view of a product's cost/benefit ratio is increasingly important in the field of capital goods. The lifecycle costs are a vital selling point that can be used to provide a contractual guarantee to the customer. In the consumer goods sector, other factors often play an important role in making purchase decisions (e. g., simply the price). It is less common to evaluate such products strictly according to their expected lifecycle costs, although this aspect is also important (Fig. 5.2-1).

There is a fundamental **contradiction** between the interests of the **product user** and those of the **manufacturer**. The manufacturer's primary interest is the maximization and assurance of the company's **profit**. Put simply, profit is the difference between the product's selling price and the total cost of product realization. That is why the manufacturer strives to reduce the company's total costs as much as possible, by developing cost-effective products and rationalizing company-internal processes. However, beyond the total costs, the manufacturer is also interested in

[1] According to the Take-Back Law in Germany, these costs may have to be covered by the manufacturer.

the product's lifecycle costs. In this way, the product's market competitiveness (i. e., the customer's interest) is clearly improved, as long as the company is not forced by regulatory requirements e. g. to address product disposal.

There are a number of ways to classify the total costs at the company level, which are significant in cost management. There is the fundamental classification according to **cost types** (cf. Sect. 8.3.1), for example, material costs, personnel costs, or capital costs. Cost accounting further breaks down costs into **direct costs** and **overhead costs** (cf. Sect. 8.3.2) as well as **fixed costs** and **variable costs** (cf. Sect. 8.5). As shown in **Fig. 2.1-2**, these are not "other" costs, just different ways of looking at the total costs in the company.

Direct costs are costs that can be assigned directly to so-called cost units (cf. Sect. 8.3.3). By **cost units** we mean the company's individual products or service activities. Typical direct costs are production material costs or production wages (direct labor). In contrast to these, the term **overhead costs** includes those costs that cannot be thus assigned. Examples of such costs are administrative costs, officers' salaries, CAD costs, heating costs, etc., which cannot be assigned to a specific cost unit.

Furthermore, costs may be designated as fixed or variable, according to whether they are dependent on the company's workload or production quantities. Examples of variable costs are material direct costs or direct labor, which accrue only when production takes place. Fixed costs such as rent, depreciation, or salaries are, as a rulc, independent of the company's workload.

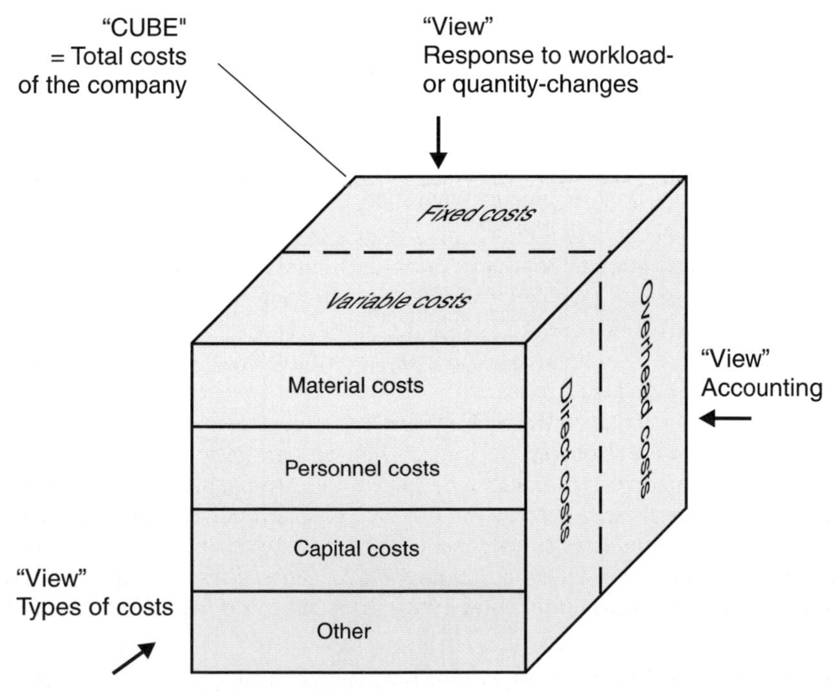

Fig. 2.1-2. Various views of the total costs of a company

2.2 Who affects costs in a company?

The aim of a company's activities is to increase and ensure its profit, (i. e., its return on capital). Since profits equal the earnings less the costs, there are basically three approaches to maximizing profit, as shown in **Fig. 2.2-1**. These are generally used in parallel.

The first approach is to **increase the company's earnings**. Strategies to achieve this include offering market-driven, "better" products, shorter supply times, and improving sales and customer service. **In particular, companies in countries with high competition exist by offering innovative products** [Gau00]. This must be emphasized at the very beginning of a book on cost reduction.

The second approach is to **reduce the total costs**. This can be the result of, for example, **rationalizing the product realization process**. This concept refers to all the measures that make the company operations more efficient, thereby reducing the costs of manufacturing a product or of providing a service. Other measures, such as computerization, production automation, reducing personnel costs, providing faster order fulfillment, or reducing inventories can also help in this regard.

Parallel to that, the company must follow the strategy of **developing cost-effective products**. This includes developing product concepts that are economical

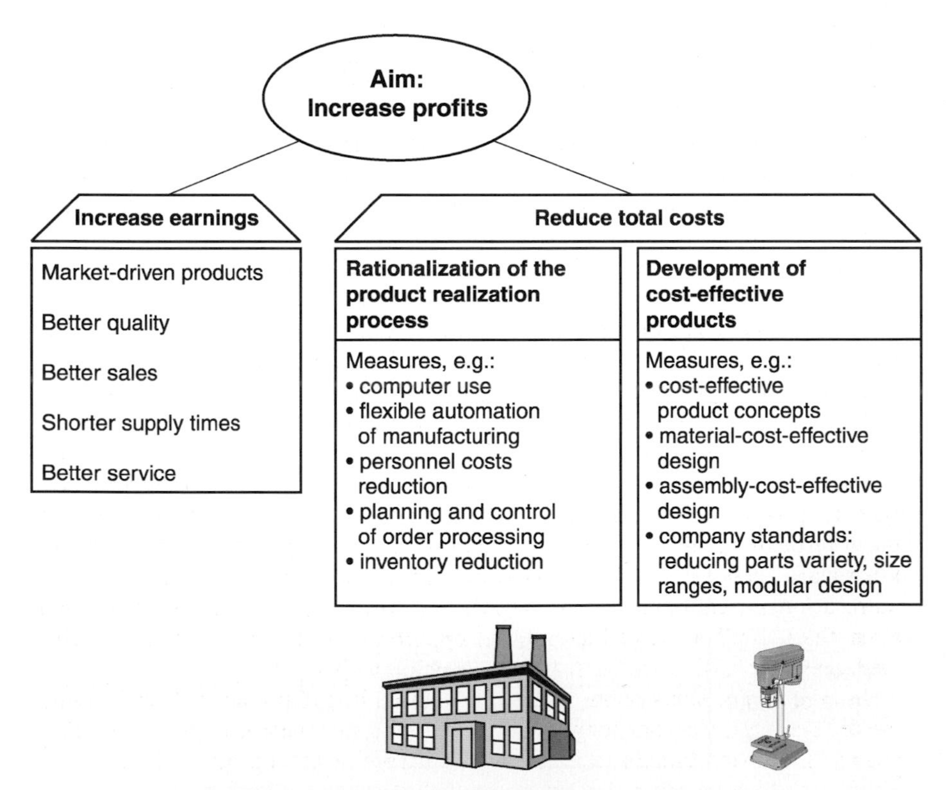

Fig. 2.2-1. Alternatives for increasing a company's profits

from the ground up; reducing the variety of parts; developing shape designs that contribute to economical production, assembly and materials; and company-internal standardization. In contrast to the rationalization measures described above, the development of cost-effective products requires cost reduction steps that work only in the mid- to long-term.

How do these three strategies fit into company operations to ensure higher profits? To answer this question, we look next at the five most important phases (Fig. 2.2-2) of the product lifecycle:

- **Decisions** about the product are made during project planning or product planning, and then forwarded to either the customer (on basis of a delivered offer) or to a responsible person within the company. In the making of these decisions, there are basic differences.

  In the case of **project planning** originating from a customer request, a special product or plant concept is usually initiated. What is important here is a very precise interpretation of the customer requirements, as well as being able to present possible solution alternatives along with their associated cost effects. The customer's technical requirements (e. g., function or performance, weight, size, etc.) are addressed by approximate computations and the design process, and the costs are estimated.

  Project planning results in a bid or an offer. In spite of what might be very short time spent on project planning, there must be evidence that the cost target can be held to before the bid is released. Therefore, only known elements (functions, parts, and their costs) are generally used in project planning. Thus, when the order comes in, it can be worked on as a variant or an adaptive design.

  In contrast to the above, **product planning** pertains to an internal decision on whether a new product or plant should be developed for a number of customers. Customer inquiries are anticipated, from which profitable orders may result in the future. In this regard, the questions are:
  - What types of problems will lead to customer inquiries in the future?
  - What types of solutions will result in additional profitable orders?

  Product planning is long-term focused, is strategically aimed, and is the result of an internal decision process, frequently for a new or an adaptive design. The degree of novelty in this decision, regarding the market as well as the product and the technology, raises the entrepreneurial risk.

  Product planning starts with facts about the company and its milieu. The first step is to determine which areas are to be searched for product ideas, and analyze the potential of the enterprise. Then, the product ideas are developed and taken through a multiple-step evaluation process. Strategic product decisions in this regard cannot generally be made right away, nor strictly planned. They require a longer, continual preparation during which the organization's internal strengths as well as available external opportunities are identified and evaluated, and the possibilities for action are considered [Ger02].

  In making decisions during product planning, it is important to bear in mind the different interdependent systems, the markets, customer groups, distribution channels, demand structures, legal restrictions, manufacturing processes, etc. These should be combined, along with the company objectives, into a logical entity [Mai01]. Knowledge of the features of the various systems and their interdependence is a prerequisite for product planning. The resulting decisions must be portable and possible objectives must be attainable.

- **Product development** includes all the phases which, following the decision to proceed with development, lead to the start of regular production. In the development process, the product properties must be defined such that its use by the customers and its production meet the criteria set during project planning. The aim of product development is to compile the documents pertaining to production and use.
- In **production**, the actual product takes shape, largely following the guidance from product development.
- The product is bought by the customer for its **utilization**. Project planning, product development, and production are, therefore, oriented toward the benefits the product will provide the potential purchasers [VDM97]. This implies an orientation to user benefit!
- The **disposal** of the product at the end of its use completes its lifecycle.

Costs arise during individual lifecycle phases that add up to the lifecycle costs over the product life (cf. Fig. 5.1-3). These can vary widely according to the type and use of the product. The problems associated with this will be discussed in detail in Chap. 5. At this point, the even more important question is: Exactly when during the product lifecycle are the decisions made regarding a product's special attributes, thus adding more to the costs?

Figure 2.2-2 shows schematically the possibilities of influencing costs, and how the costs increase with the successive life phases of a product. These two curves run in opposite directions! In the beginning phases, in which we have the most influence, the least is known about the future costs. It is obvious that during project and product planning the costs of a vaguely defined product are known only very roughly, whereas the possibilities of influencing these are the greatest. At the beginning of the product development process alternative paths can be chosen. At its end the lifecycle costs of the product are largely set, even if they are still not known. In the phases of production, use and disposal, still another cost optimization can be carried out. This optimization is of the individual processes, based on the development outcome. If an automobile motor is developed then there is little leeway as far as manufacturing and operating costs are concerned. By a clever choice of production processes, or with especially cautious driving, one could still save on costs. The largest part of the lifecycle costs, however, can be changed little, since they can be influenced only at the beginning of the lifecycle. This is equally true as far as advantage by innovation or the alignment of the product toward customer-use is concerned. These statements hold for new designs. In case of adaptive and variant designs, the possibilities of influencing costs are fewer (cf. Sect. 4.8.2).

We cannot emphasize enough the importance of the early life phases for the product's success. Any mistakes made here can be corrected later only with great effort, if at all. Thus we can realize the importance of **early cost identification** (cf. Chap. 9). The following should be brought to the attention of traditional product developers, who tend to be rather technical.

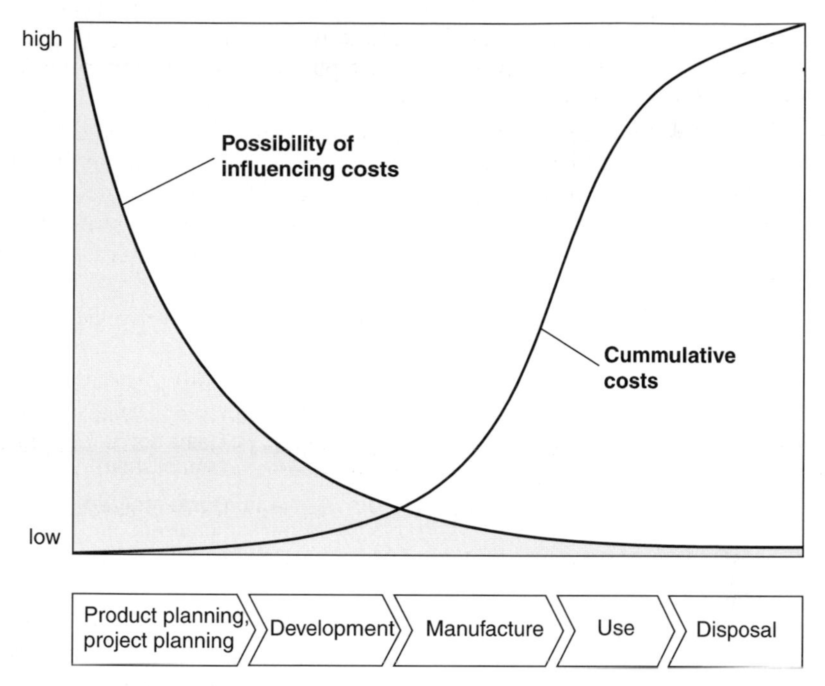

Fig. 2.2-2. Possibilities of influencing and establishing costs over the lifecycle of a product: The "dilemma of product development" (an example of a new design)

- **Each technical commitment also** represents **an establishment of costs.**
- This establishment of costs must be checked **concurrently in the product development process** (the cost numbers should be available when a technical decision is made). Otherwise, the product can become too expensive, which will require time- and cost-intensive changes (cf. Fig. 4.2-3).

From practical experience, the "**Rule of Ten**" was formulated, which conveys the idea of the exponential growth of costs over the life phases of the product. The later the changes are made, the more expensive they get. For example, a technical change that costs 1 $ at the task clarification stage will cost 10 $ during designing, 100 $ at the production planning stage, 1 000 $ if made during production, and 10 000 $ after shipping!

Product planning, project planning, and product development are highly significant for the later life phases of a product. Indeed, who makes the basic decisions during these activities? **Who really determines the future costs of a product?** We emphasize that the responsibility lies within the management, marketing, product development, and production functions of a company during the planning and development of a new product.

The **management** of the Company determines the company policy. They are obviously also responsible for deciding on the basic direction of the company's

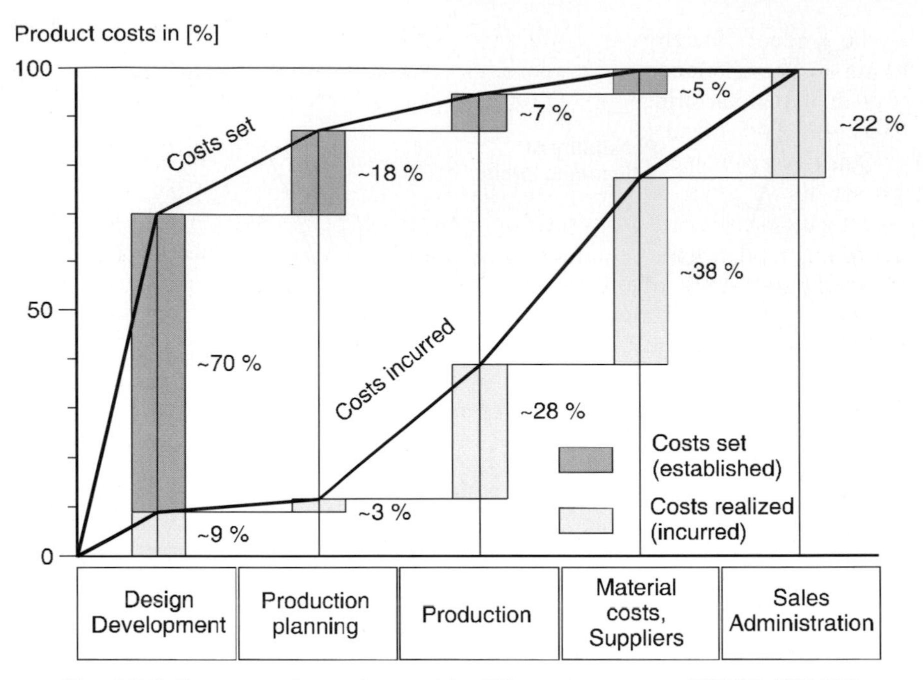

Fig. 2.2-3. Costs set and costs incurred in different departments [VDI87, VDM95]

product line. By deciding on a given product or a product line, they also determine its essential costs, even if they are not yet known.

Technically competent **sales** and **marketing** departments are links between the market, the customer, and the company. Thus, their job is to analyze the wishes of potential customers and, on basis of this knowledge, initiate the planning process for new products. Including sales and marketing, and thus the customer requirements, in product planning, project planning, and product development is a key to the product's success. These departments also gather cost objectives from the market and, thus, provide concrete cost limits in the product development process.

The **product development** department is the "unit" – the body responsible – for the project planning as well as the development processes. Here, the information coming from different areas is converted into a marketable product or product concept. One of the most important tasks of product development is to bridge the gap between the customer's dream product, the planned technology changes, and what is technically and economically feasible – a tough job, indeed! When the design is complete, the manufacturing, operating, and disposal costs are largely fixed. Thus, it can be said that product development has the greatest direct quantitative influence on a product's manufacturing and lifecycle costs (cf. **Fig. 2.2-3**).

A company's largest costs arise in **production** and **purchasing**. Therefore, these merit the greatest attention in the rationalization process. Production and purchasing should be integrated into product planning, project planning, and development of new products right at the beginning because optimizing part production and assembly processes is easiest and most effective during the shape design

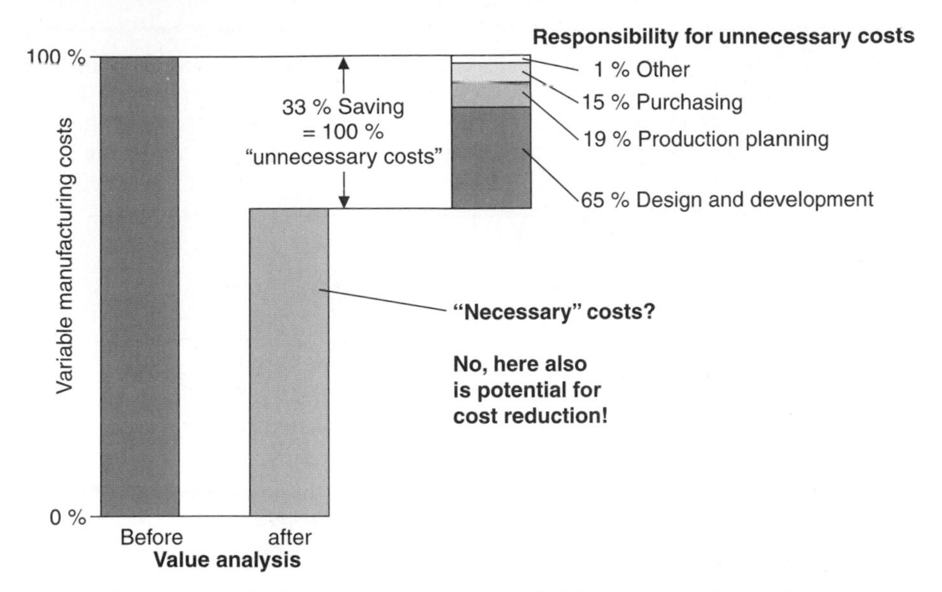

Fig. 2.3-2. Average manufacturing cost reduction of 135 items by value analysis; responsibility for "unnecessary" costs [Kie79]

are no different from other technical parameters in terms of being a "final insurmountable boundary". This is illustrated next with the example of several generations of the transmission in a construction machine.

Figure 2.3-3 shows the evolution of part quantities and costs of a transmission with an hydraulic torque converter over a period of three decades. Over that time, the company was able to reduce the number of parts in the transmission, as well as its manufacturing costs, by approximately 70% (cf. Sect. 7.11.1). The technical development seems to be a market-driven (supply and demand) evolutionary process when looked over this long time span. Technical and cost-related improvements result from continually increasing knowledge; the important thing is that we should aim for this, not be driven to it.

Basically, product development is bounded on one hand by the customer's interest in getting the highest utility and, on the other hand, by the manufacturer's interest of obtaining the highest possible return on the invested capital. In addition, it is affected by a number of boundary conditions that change with time. The time-variability of almost all important parameters leads to the fact that for a cost-efficient product design there is neither an absolute nor a stable relative optimum. **The product developers therefore find themselves in the situation where they need to find an optimum but do not know where it is. They only know that it is continually changing.**

Thus isolated measures for the design of an economical product do little to improve the general cost picture, and then only for the short term.

➔ Effective measures for increasing cost efficiency must be combined into a continually planned and guided product and process optimization. Integrated cost management may be considered the organizational shell within which the collaboration of a number of individual measures is coordinated.

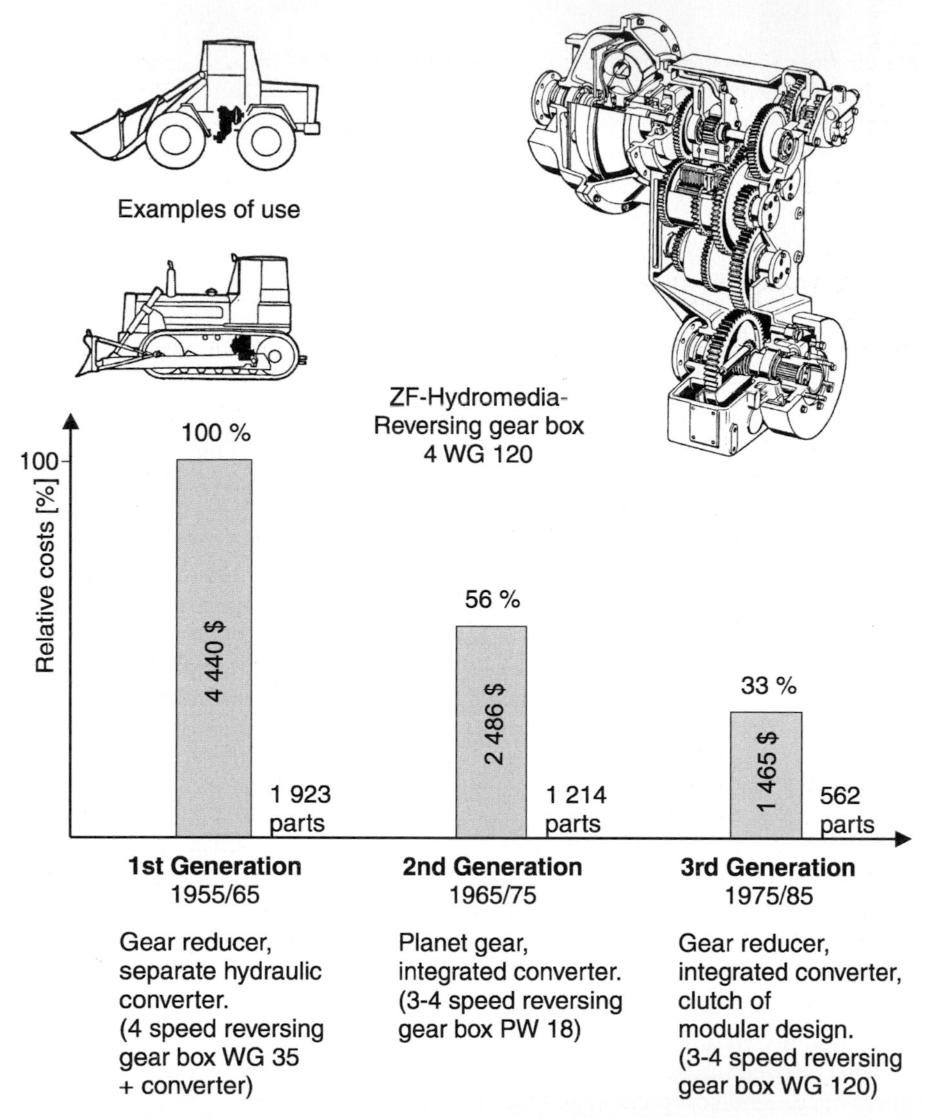

Fig. 2.3-3. Chronological evolution of costs and power throughput for transmissions with hydraulic torque converters (from ZF Co., inflation considered)

3 Cost Management for Product Development

The first section of this chapter will present the objectives of cost management in the field of product development. This is followed by a discussion of target-cost oriented procedure as the main idea of modern cost planning in product development. In the second section some of the problems are addressed, which might appear now or in the future, as cost management is adapted to industrial practice.

3.1 What is cost management?

Independent of any special perspective of product development, cost management may be defined as follows:

"Cost management is targeted and systematic steering of costs. The aim is to influence the costs of products, processes and resources by using concrete measures, such that an appropriate company success is achieved, and its competitive strengths are improved for a long time to come". [Fra97]

In product development and manufacture, a company applies a number of different processes that will be treated in more detail later in this book. These processes are feasible only with use of resources (energy, material, labor, etc.) that are basically tied to the incurring of costs. The development and manufacture of automobiles, for instance, requires raw materials, energy, and production facilities, as well as personnel with various qualifications and training. All personnel in a company should be designing the ongoing processes for the long term, in such a way that a **competitive product is realized with the use of a minimum amount of resources.** For the sake of cost management in the company, it is necessary to create an environment that enables each employee to take this on as a personal responsibility. Management, in its leadership role, must address three types of functions relating to cost: 1) the **strategic procedure**, 2) the **operational procedure**, and 3) **shaping the company's boundary conditions**. **Figure 3.1-1** represents the interrelation of these three elements of entrepreneurial action, which will be discussed in more detail below.

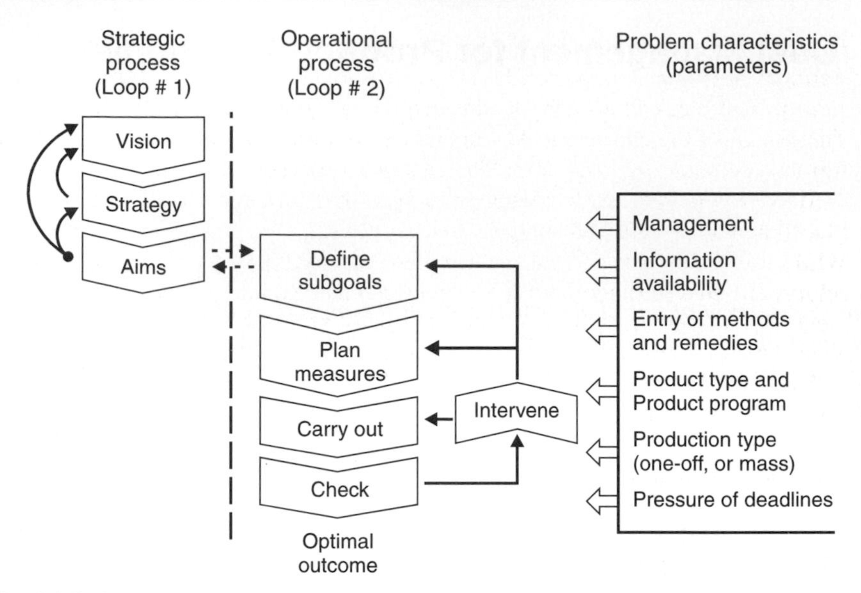

Fig. 3.1-1. Cost management is similar to the process of general management: Both the strategic process and the operational process must be organized as closed-loop control systems

The **strategic procedure** must answer three questions:

- **Where do we want to go?** A **vision** of the future.
  The foundation of strategic planning must be a vision of the company's long-term view and its future standing in the marketplace.
- **How do we get there?** A **strategy** for the future.
  The long-term measures point the way to realizing the vision; they shape the environment for the strategic actions.
- **Which operative aims** must we reach along the path? These are the milestones on the path to the future. The long-term strategy will be transparent if clearly defined subgoals can be derived from it. These subgoals determine the effort that must be expended in the company.

The **subgoals** that have been set in strategic planning form the basis of a company's **operational process**. The operational process can be divided into three fields of activity:

- **Plan:** The course the process will take (by which the given subgoals will be reached) must be ascertained within the confines of the available resources. The process must be started by using suitable measures.
- **Implement and check effectiveness:** After the process is started, its progress must be monitored with appropriate control mechanisms. If the results deviate from the goal, that should be recognized early.
- **Intervene:** In case of serious deviations from the defined goal, the process must be modified by suitable intervention.

In Fig. 3.1-1, the **closed-loop control character** (cf. Fig. 4.2-2) of strategic and operational planning is indicated by the return arrows. Iterations of this management process are often necessary to successively approach the objective.

The design of the entrepreneurial boundary conditions is characterized by leadership and personal development. The planned processes can run smoothly and cost-effectively only when a suitable employee structure enables all employees to be placed and motivated according to their abilities.

While this holds true for the management of any type of enterprise, **product development cost management** has **special needs**. Product development's strong influence on cost generation in various company processes was addressed in the previous chapter. Cost management must therefore be brought to bear in the interest of a cost-effective process organization. To this end, the product development processes must meet the following three basic requirements: (1) they must be market-driven; (2) they must be cost-driven; and (3) they must result from cost-efficient development processes.

3.1.1 Developing market-driven products

The need to develop market- or customer-driven products that provide interesting and inspiring products to customers may at first seem simplistic. In reality, this is the most basic prerequisite for any successful company action. Products must be sold at a profit and in sufficient numbers; the needed customers can be won over if they see attractive or at least sufficient utility or benefit based on the product price. From the customer's viewpoint, utility can have emotional elements (customer enthusiasm) as well as economic parameters. A **product that people are enthusiastic about** can command a higher price, and in many situations this can be a decisive competitive advantage over other similar products.

The product development process is always focused on an assumed **future market situation**. Therefore, it always progresses under **continually changing boundary conditions**. The marketing departments of every company must anticipate, as accurately as possible, the needs of potential customers. **Design and development** should be brought into **this goal-defining process early**, for three reasons:

- It deals with a **common ground** that exists between the **fundamental technical possibilities**, the talents and facilities of the company, and the wishes of a customer focus group at the start of project planning or the planning of a new product. The product palette must consider not only the market, but also the available resources.
- A careful assessment must be made of the customer's **technical understanding** as well as the **technical status** of the competition, one's own company, and of suppliers, in order to develop a realizable scenario for the future.
- **Customer wishes** are often **diffuse**, and they must be reduced, in a **technical sense**, to their **essential core**. This will ensure that there are no superfluous requests that might cause technical or economic problems later, which the customer is not really willing to pay for.

3.1.2 Developing cost-driven products

Developing low-cost products must be based on specifications free of all nonessential demands.

➔ Products are cost-effective if they make profit (i. e., if their total cost is significantly lower than the price attainable in the market).

During the development of low-cost products, product developers must constantly focus on the technical function of the product and the processes involved in its manufacture.

➔ A good product development team not only comes up with products that satisfy the customer, but also specifies manufacturing processes that are suited to the company.

A decisive factor in this regard is the **information culture** within the company. The many interdependent factors can be considered only if information (including cost information) can flow freely internally, without bureaucratic hindrances (Fig. 4.8-5). Cost management in product development creates a developmental environment in which free flow of information can exist (Fig. 3.2-1).

3.1.3 Realizing cost-efficient product development processes

In looking at cost reduction, we should not lose sight of the fact that product development itself can incur significant costs (3–25% of total costs, average 9% *TC*; cf. Fig. 2.2-3). This is especially true for customer-specific products that are manufactured in single units. Considering the widespread increasing cost pressure, the product development process should also be cost-efficient (cf. Sect. 6.2). Thus, **the product development time** must often **be significantly shortened**. Time and costs can be reduced through an efficient design of processes. A second and perhaps more important reason to shorten the development time is the fast pace of life today. The time factor is a decisive competitive advantage in constantly changing markets. Whoever can offer an innovative product first can often realize a large profit; however, there can also be higher risks.

In order to realize the three central requirements of the product development process, as described above, many different measures must be implemented (Fig. 2.2-1). These can range from developing fundamentally new technical products and improving computer use, all the way to a thorough organizational restructuring of the company. Therefore, this continuum of measures can only be realized in its totality in an integrating total concept. This is described, for example, in *Integrierte Produktentwicklung* [Ehr06]. The methodology of cost management in product development will be presented in Chap. 4, with this comprehensive structure as the background.

3.2 Problems of cost management in product development

In practice, the success of a product development process depends directly on **function fulfillment** and holding to the **cost** and **time goals** (the "magic triangle" of quality, time, and cost). For resolution of goal conflicts, see Sect. 4.5.3.3. If the product development process runs into problems, it implies that one or more of these parameters could not be held within the limitations. Frankenberger [Fran97] and Badke-Schaub [Bad04] has studied several actual product development processes in industry and analyzed the collaboration mechanisms that had a positive or negative influence on the results. These interrelationships are complex and will be presented here in a simplified form (cf. **Fig. 3.2-1**).

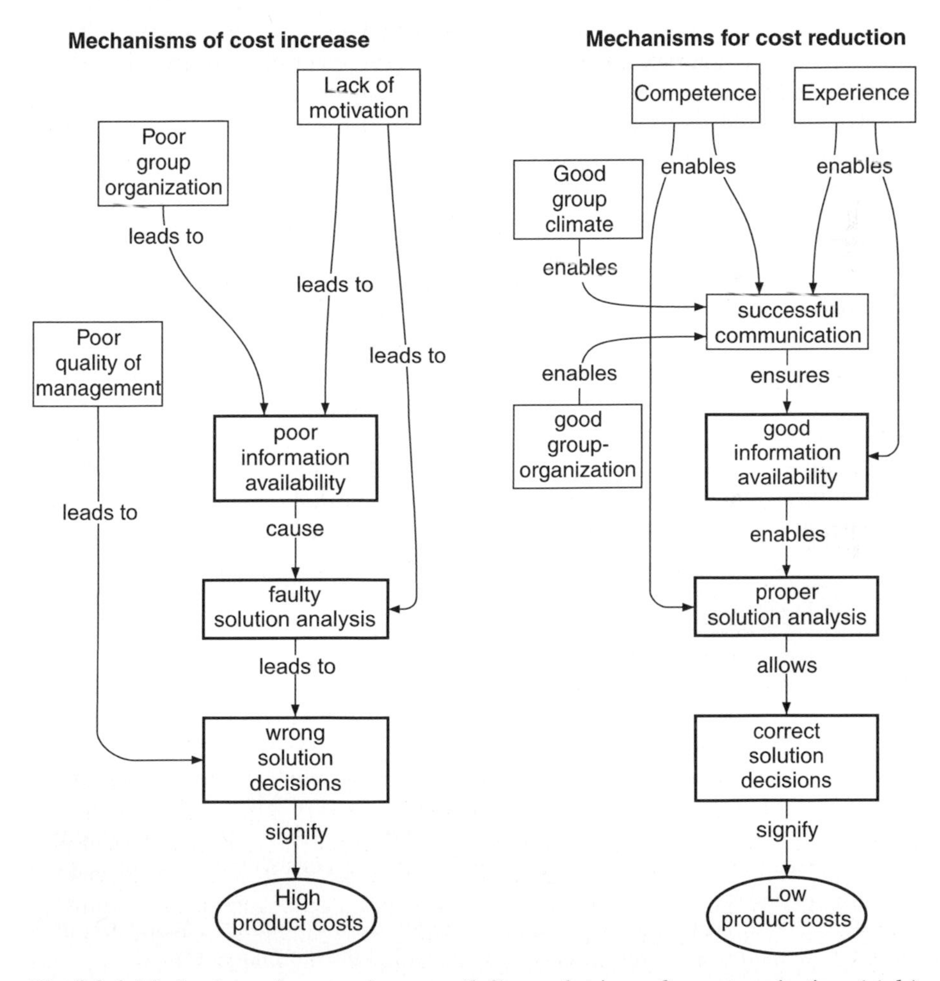

Fig. 3.2-1. Mechanisms for cost increase (*left*), mechanisms for cost reduction (*right*) [Fran97]

Low product costs (Fig. 3.2-1, right) can always be attributed to the **correct solution decisions** during product development. However, such decisions can be made only on the basis of a **valid analysis of the solution variants**. In developing valid analyses, **employee competence** plays an essential role. **Good and rapid information availability** within the company is exceptionally important, and this depends on **employee experience** as well as the **communication** between the employees. In addition, analysis of these connections showed that the quantity and quality of such communication are influenced very strongly by **each individual's experience**.

Failure mechanisms can be contrasted to these linked success mechanisms in the same manner (Fig. 3.2-1, left). **High product costs** can always be attributed to **wrong solution decisions. Poor group leadership can be fatal** to product development because it can create undesirable **informal hierarchies**. If individuals use such hierarchies for **exercise of power**, it frequently interferes with the **decision-making quality**. During product development, wrong decisions are often made due to defective solution analysis, which is often based directly on the lack of employee motivation or poor availability of information. This, in turn, depends greatly on **individual motivation** to seek the needed information. An unsuitable group organization interferes directly with **information availability**, and also has a negative effect on **employee motivation**. An unsuitable group organization interferes directly with **information availability**, and also has a negative effect on **employee motivation**.

Frankenberger's investigation confirms the overriding importance of **punctual information availability** during the development of low-cost products, which makes it possible to **recognize costs early** in the process (cf. Sect. 4.6.1 and Chap. 9, Sect. 9.2). His investigation points out the mechanisms that support or prevent information availability in the company. Thus, it is apparent that the difficulties of introducing cost management practices arise less from technical problems than from problems in the company's organizational and social structure. The central **problem areas of management, information availability**, and **use of methods and measures** will be addressed below.

3.2.1 Management

According to Frankenberger, poor management sets mechanisms in motion that negatively affect product costs in the development processes. On the other hand, his analysis of cases of successful, cost-efficient product development indicates that the quality of management is not an explicit performance-influencing factor. This reveals a very important management phenomenon: Management comes to people's attention only if difficulties show up. **On the other hand, when company processes progress optimally, management appears to work in an imperceptible, invisible way.**

Management has the greatest influence on the development of social structures within a company, and on what is commonly called the company culture.

- **Personnel management**
  Companies thrive on the quality and motivation of their employees! Good personnel management fills every position (function) with suitable employees. This suitability of function “carriers” will be described fully in Sect. 4.3.
  Many companies must manage, at least in the short term, with the existing staff. However, to best utilize the abilitiesof the available employees, reorganization and personnel development and education are often necessary.

➔ Among the most important requirements of good personnel management is to provide **inspiration** to all employees from time to time. For this purpose, professional development opportunities should be pointed out to them, as well as the promotion and qualification of employees through education, and regular changes in their range of tasks.

- **Configuring the organization**
  In the past few years, many companies in the machine industry have undergone reorganization. The classical line structure has been replaced by a product-related organizational structure (e. g., sectional organization, profit centers, segmentation) or a matrix organization. This trend points to a change in the general understanding of what is an efficient organization. Formerly, the practice was to organize the company with extensive hierarchical, Taylor-type separation of the functional areas and employee specialization. This showed however, that the informal cross connections, that are absolutely necessary for the development and production of complex technical systems, were hindered rather than encouraged through such an organizational structure. Also a strict spatial separation of the departments for marketing, product development and production makes communication difficult between the employees, who are all working on the same task. This often results in the loss of individual employees' sense of responsibility for the product and for the smooth course of company functions. Both problems also complicate effective cost management in design and development.
  A stronger **product-oriented form of organization** (e. g., a section or matrix organization) can enhance the necessary in-house information networks. In such organizations, employees from different specialties are assigned to a specific product development team and are physically co-located. Integratingindividuals into a team usually induces responsibility and motivation for the shared success of the product. Management must always be willing to develop or transform an organizational structure adapted to the requirements of its own company. However, the most polished organizational structure is not worth anything as long as it exists only on paper. Achieving its total acceptance by the majority of the employees often proves to be the more difficult part of the management task. Introducing a section- or matrix-type organization has the risk of at least a partial loss of employees' professional homes.
  Organizational measures associated with comprehensive cost management must not only must designed to improve the horizontal networking in a company, but must also consider the mental differences between the different hierarchy

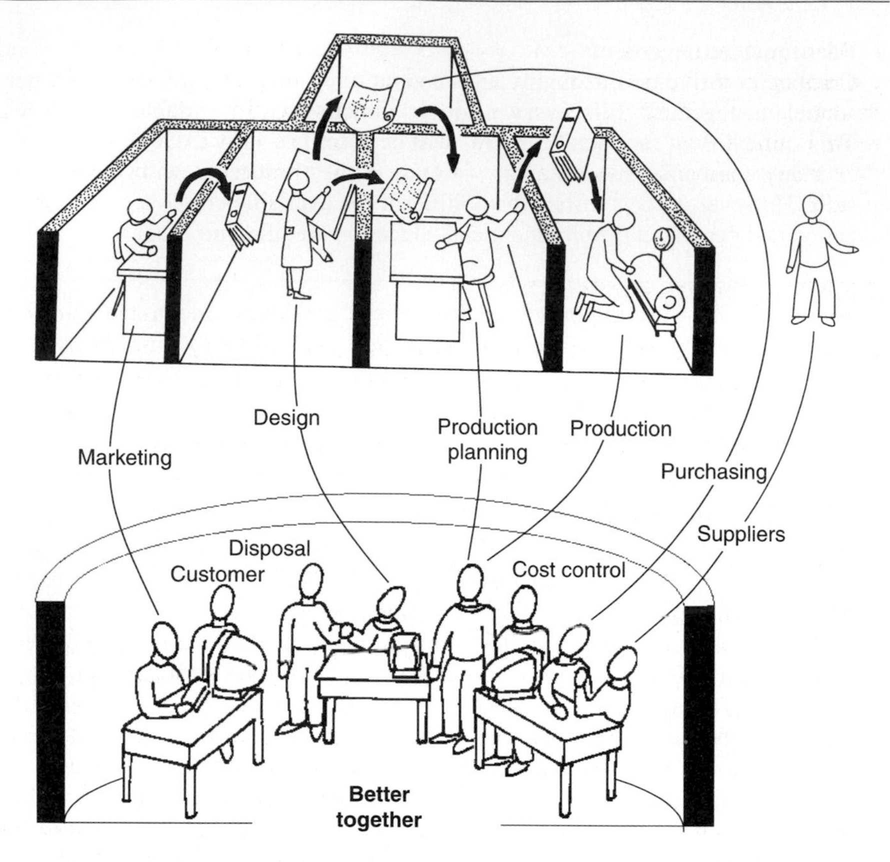

Fig. 3.2-2. Teamwork overcomes mental walls between departments [Ehr87b]

levels (cf. **Fig. 3.2-2**). While cost management focuses on effective cost control, the specialized and detailed knowledge indispensable to it belongs at the professional employee's level.

> ➔ Making cost problems a challenge for every individual employee is an important goal of the organizational measures implemented in cost management.

- **Planning employee capacity**
  It costs to lower costs! Employees' time adds to the costs of design and realizationof cost controlmeasures in design and development, regardless of whether it is for preparing cost information or searching alternative design solutions.

> ➔ Effective cost management presupposes that this additional working time is planned for, and is in fact available to the relevant employees.

Management must understand that a design department that is always under a **deadline constraint** [Ehr06] hardly has time to develop any low-cost solutions. In that case, finishing the drawings and documents has the highest priority. The same is true for an overloaded production planning or cost calculation group – they cannot provide advice early enough if they cannot finish their own jobs [Ehr93].

3.2.2 Information availability

Frankenberger [Fran97] also demonstrated that poor availability of information creates a bottleneck in the development of low-cost products. It is particularly necessary to combat the often-widespread "cost ignorance" in a company. This can be accomplished in three ways: (1) through **improving communication** with the knowledge sources (e. g., through production and cost advice, cf. Sect. 4.6.1); (2) through definitive **preparation and presentation of cost information** (e. g., with software) in the product development process; and (3) through goal-oriented **training of the product development engineers** (Sect. 4.8.3.4).

- **Communication**
  Ignorance of product development costs concerns detailed cost knowledge as well as comprehending the overall picture. In most companies, the personnel from cost control, production, and procurement (who have the greatest competence regarding costs) are involved only at the margins of the product development process, if at all (cf. Fig. 2.2-3). On the other hand, the actual product development engineers are often uncertain about costs, since costs were kept secret in many companies until a few years ago. Good product developers know their own weaknesses and can generally make up for their individual knowledge deficits through close cooperation with other departments. When necessary, they turn to competent contact persons in marketing, purchasing, production planning, or cost accounting, or to suppliers. Correctly appraising one's own knowledge as well as making personal contacts within and outside the company sometimes overtaxes employees (and managers) who are under time pressure; this can have a considerable impact on product development. There have been instances in many larger companies where millions of dollars could have been saved if people had asked appropriate questions of the correct source, and not refrained due to lack of time, or conceitedness or shyness.

> ➔ Developing an **atmosphere of general openness and cooperation** in a suitably organized company is important step for promoting the development of low-cost products.

Only in this way can the walls between different departments be broken down, enabling constructive, well-communicating social networks to develop. Figure 3.2-2 shows how every department can see itself as an exclusive entity instead of optimizing products through interdepartmental cooperation.

- **Problems of cost information**
  In target costing, as it is consistently implemented in the product development process (cf. Sect. 4.4.3), the currently foreseeable product costs and the established target costs must be continuously compared to actual and anticipated performance.

> ➔ Effective control of the target costs requires that at any time during the product development process, we can determine how much a product would cost at that particular planning state (Sect. 9.1.2, Performing design-concurrent cost calculations).

Only in this way can deviations from the projected target costs be found in time, allowing countermeasures to be taken. The difficulty is in reliably estimating additional costs, particularly in the early stages of product development (cf. Fig. 2.2-2). A number of computer-aided **methods for early cost recognition** have been developed that can be adapted to diverse product groups and production structures (e. g. unit and series production; s. Sect. 3.3.1). Chapter 9 shows a critical evaluation of the most important procedures and describes their correct application.

Company-internal cost information must be provided for all the methods described above. **Cost data cannot usually be carried over from one company to another**, since manufacturing facilities, costing methods, and cost structures vary among companies. Consequently, early cost recognition procedures are portable only in their basic concepts, but not in the technical details (cf. Fig. 7.13-2).

3.2.3 Applying methods and tools

Although many effective cost management methods have been developed over the years, only a few have actually been adopted in general industrial practice. A serious knowledge deficit exists with regard to these available methods (Sect. 4.8.3.4) and, generally, only the methods that are in fashion are utilized. Their introduction is often random and lacking a full understanding of the context in which they should be applied. Thus, failures are seemingly preprogrammed, which, in turn, increases aversion to methodical procedures.

Analogous to production resources are the **information processing resources** – the methods that must be implemented, but not by chance or intuition. The commitment must be planned and people trained for it.

> ➔ Methods must be selected in a company-specific way and adapted to the respective situations.
>
> ➔ The users must have appropriate support to familiarize themselves with the respective methods. Methods are learned by training and practice!

Position **Field**	Questions about cost knowledge	Questions about cost goals	Questions about methods used
	Overall: company-related		
Company head **Technology**	e.g. • At what workload is the break-even point reached?	e.g. • Cost goal for the company? ...For the product program?	e.g. Reasons for • Target Costing? • Reengineering?
	Related to product program and product		
Department head **Product development**	e.g. • Total product cost structure? • Weaknesses of job order costing?	e.g. • Cost goal for the product?	e.g. • Distribution of cost goals? • Evaluation of suppliers?
	Related to assemblies and parts		
Specialist **Design**	e.g. • Which is the most cost-effective screw?	e.g. • Cost goal for assembly?	e.g. • How to calculate manufacturing costs?

Fig. 3.2-3. Task-specific classification of the components of cost management

As in every learning process, there will be mistakes and setbacks that should be taken into account. The methods, the help, and the data that a detail designer needs are different from those the chief of product development requires. **Different tasks call for different methods**, as shown by the example in **Fig. 3.2-3** and discussed in Sect. 4.8.3.4. Similar to production planning, there should also be methods or information planning that answer the following questions: Who needs what information? Who must know which help tools and methods, and which ones to apply where?

3.3 Adapting cost management

We have already stated that the cost management methods shown here must always be adapted to the given situation, as needed (see also Sects. 4.8.2 and 6.2). Companies differ greatly with regard to the **product type**, the **product program** (Sect. 3.3.1), and the **production methods** (Sect. 3.3.2). Other parameters that have a significant effect on the cost management development and realization are the goal, the extent, and the type (Sect. 3.3.3) of the object under consideration.

3.3.1 Type of product and product program

- **Simple or complex products**
  The complexity of the products that a company manufactures has considerable influence on the basic cost management procedures. While simple products consisting of few parts often can be improved by an individual designer working

with production, purchasing, and manufacturing cost calculations, a considerable staff expenditure is required to optimize more complex products. With many supposedly simple products made in large lot sizes, the **complexity** is often **concealed in the manufacturing process**. In such cases it is necessary to shed light on the often very opaque connections and exchanges between design and production. This is especially important for companies with a variant-rich product program where, with design decisions, cost implications must always be considered through the ascertained number of variants (Sect. 7.12).

- **Consumer goods or capital goods**
  Depending on whether a company classifies its products as **capital goods** or as **consumer goods**, there are usually differences in product planning and determining cost objectives. In the capital goods sector, potential customers compare the technical and economic aspects of the different items, but decisions to buy consumer goods are often influenced by emotion.
- **Individual customers or anonymous customers**
  To determine the planned sale price and the ultimate cost objective, the customer must absolutely be included in the product planning process. It makes a significant difference whether a company offers products specifically to a given **single customer** (design-to-order) or to an **anonymous target group**. Thus, a specialty machine manufacturer would work very closely with the future client during the project planning phase (Sect. 5.2) to determine the cost objectives along with task clarification. During product planning for an anonymous group of customers (e. g., as in the automotive industry), determining cost objectives is considerably more difficult if competitors with the same product functionality are not already in the market. In such cases, market analysis identifies a spectrum of hypothetical customers and their needs, after which further product development can be carried out. Hypotheses derived from market analysis may or may not be valid, and so the risk of developing a product that fails in the anonymous customer market is considerably higher (Sect. 4.5.1).
- **End-user products or OEM products**
  The differences between products sold to end-users versus those to be used in other products (OEM products) are as follows:
  End-user products
  Manufacturers of end-user products must deal directly in the market. While they have considerable freedom in product design, they may also carry a higher risk.
  OEM products
  The market for typical OEM products includes car parts, computer parts, fittings, standardized parts, etc. and has dynamics different from the end user market. The competition is generally high and the goals (including the possible net profit) from the customer (e. g., the automobile manufacturer) are very strict and detailed. Thus, suppliers usually have little freedom in product design. In addition, the suppliers are frequently small to medium-sized companies who often have little time to develop their products and even less time for a multi-loop product optimization process. However, there are also cases where suppliers have innovatively improved the automobile manufacturer's product design and realized good profits.

- **Innovative or technically mature products**
  The technical maturity of the products also has an effect on cost management.
  Innovative Products
  Innovative, relatively new products such as cell phones, computers, etc., are subject to very short development cycles. Here the priority is for short development times and quick application of new technologies. Great cost reductions are often possible due to potential subsequent development and new technologies becoming available.
  Technically mature products
  Technically mature products (e. g., industrial transmissions) have longer development cycles. Cost reductions are possible mainly through optimizing details.
- **Size of the company** (large company versus small or medium-sized company)
  Large companies often have cost management advantages because they have more capital power, better development and manufacturing methods, stronger market power, etc. However, disadvantages might exist due to inertia, organizational hierarchies, and departmental thinking. **Small to medium-sized** companies have the advantage of less division of labor and more familiarity among the employees. Thus, cooperation is potentially better and easier to organize. Integration happens among employees since they are active in multiple functions. On the other hand, employees in small to medium-sized companies are often pressed for time, making it difficult to manage projects systematically in the early development phases.

3.3.2 Types of production

- **Single-unit versus limited-lot and series production**
  Depending on whether we are dealing with **single-unit** or **series production**, the strategies for cost reduction are fundamentally different. While in single-unit production the design effort for cost reduction can very quickly exceed the potential cost savings, such efforts can be profitable in mass production, with its corresponding large quantities (Sect. 4.8.2, Fig. 4.8-1). The designers of single-unit products can often develop low-cost solutions using intuition, supported by brief consultations and applying rules of thumb (Sect. 7.10.4). On the other hand, the development of series products in larger quantities usually requires precise analytical techniques.
- **Extent of the production capability (outsourcing portion)**
  When product development or production is done outside the company, special cost management measures must be implemented. The usual procedure of sending a tender with detailed drawings and a request for bids to a large number of potential suppliers is not always the most effective way to get bids for low-cost products. It is better to **provide the suppliers with the cost management methods** used in your own company for designing and manufacturing. Naturally, this works only when there are just one or a few suppliers and a relationship of trust has been built up. The requirements list, including the cost objectives, should therefore be developed collaboratively. The production capabilities and

cost structures of the suppliers should be known, and cost calculations should be disclosed. Of course, an adequate profit must be agreed to with the supplier. Such a procedure has become common with car parts suppliers, among others (cf. Sect. 7.10.2).

3.3.3 Aim and scope of cost management

This book basically assumes a target cost-oriented product development process. Therefore, it should be obvious that this further assumes continual control of a realistic cost objective, the same as for technical requirements (e. g., performance, strength, etc.). However, we know from experience that this seemingly self-evident fact is certainly not realized in all companies. The two extremes are:

- Costs are either not considered at all during product development, or, at best, secondarily.
- The development of the new product is under extreme cost pressure, since the company is already a disaster case.

For the first situation, we have already given numerous product development cost reduction suggestions. We recommend that you work between the two extremes and introduce product development that is oriented to a cost target. Since routine cost target-oriented product development is still so rarely used in practice, even though pressure from the competition increases and the boundary conditions continuously change, development projects are coming under extreme cost pressure. Here is some additional advice in this regard:

- Involve and motivate your customers and suppliers in as many fields as possible.
- Develop strong company management support.
- Tighten the organization with clear tasks, responsibilities, and regular meetings.
- As far as possible, do not look at only one product and its development. Rather, consider the whole product program and the affected product realization processes.
- Do not neglect the work on old tasks.

Another perspective, alluded to earlier, is the scope or the significance of the product considered in cost management. The scope may be considered in different ways:

- As a portion of the total sales (is the entire product program or is only a small part considered?)
- Costs of the product: 1 000 $ or 10 000 000 $?
- The actual product (the whole product, or only an assembly or a part?)

If the scope is somewhat limited, the product developer can implement cost management (see Sect. 4.8.2) alone or collaboratively (e. g., with process planning and purchasing). If the scope is large, then cost management is a job for a project manager and should be supported by all the involved parties.

4 Methodology and Organization of Cost Management for Product Development

This chapter presents the methodology and organization of cost management from the perspective of product development. First, the basic elements of cost management are related to the comprehensive methodology of integrated product development. This comprises the essential points of the product lifecycle, organization, methods, and tools. Within this context, the practical realization of cost management is discussed. The chapter closes with a short overview of the best-known alternative cost management techniques.

4.1 Elements of cost management

If a company encounters a financial loss situation, the motto usually becomes "Down with the costs, whatever it costs!" The economic crises of the last few years produced a diverse range of instruments with which the costs in a company could be lowered on a large scale. Experience has taught us, however, that such shrink cures do not take a company quickly into a new phase of prosperity.

Short-term cost management is comparable to first aid at an accident scene. Through these measures the accident victim can survive and be transported to a hospital. It would not occur to an emergency physician to leave the patient to his or her fate after only those first aid measures. Likewise, while short-term measures can provide some breathing room for broken enterprises, the responsible parties must permanently safeguard the company's economic success through **mid- and long-term cost management**.

Short-term cost management implies radical lowering of costs through:

- Reducing personnel costs
- Closing manufacturing facilities, and outsourcing
- Selling parts of the company
- Streamlining the product program

Frequently, cuts take place across the board. That can be dangerous if, at the same time, the opportunities for mid- and long-term cost control are seriously harmed. **With the development function paralyzed due to such cuts, a company's product range would hardly fit into changed market conditions!** In this book, we will concentrate on mid- and long-term cost management in product development, since the developments in this field have demonstrated huge opportunities for influencing the process.

Effective product development cost management effective must consider all aspects of the product lifecycle, from initial planning through the disposal of a product. That means that the organizational issues related product development must be given the same attention as technical problems. Therefore, as Ehrlenspiel [Ehr06] described it, product development cost management must be based on an integrated methodology of product development. The elements of an integrated product development (IPD) methodology will be described below; **Fig. 4.1-1** shows their relationships. The actual techniques of cost management associated most closely with these different elements will be dealt with in subsequent chapters.

The **product lifecycle** is at the core of every product development and is thus central to IPD methodology. The different processes that form the total product lifecycle will be discussed in Sect. 4.2. We will also see what conclusions may be drawn from that in the interest of a maximum economic benefit of the company and the action plan (see also Chap. 5). Since **people** are the central figures in all product lifecycle processes, we shall give them special attention, particularly with regard to the processes associated with product realization, in Sect. 4.3. The key question is how optimal work conditions can be developed by formation of appropriate company and teamwork **organization**.

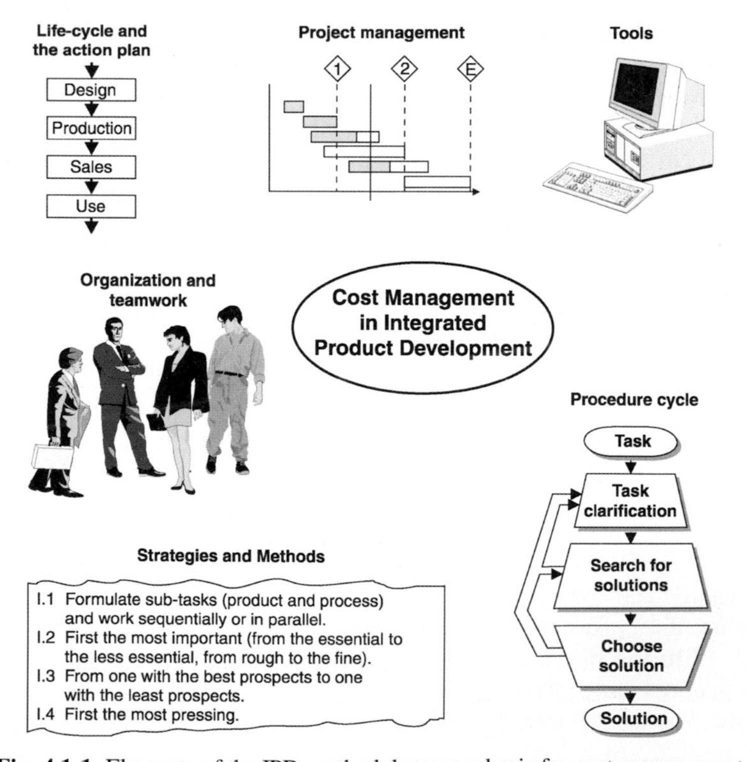

Fig. 4.1-1. Elements of the IPD-methodology as a basis for cost management

Section 4.4 will show the **Procedure cycle**, universal **strategies** that control the target-cost oriented procedure during technical development, and the **action plan** derived from both of these. We can speak of successful cost management only if it manages to realize the preset economic goals through the technique of **Target Costing**.

The deployment of a number of **tools** is combined with the action plan (Sect. 4.6), that are symbolized by the PC in Fig. 4.1-1. However, these are not all necessarily tied to a computer.

The bar graph in Fig. 4.1-1 shows that **project management** controls the overall course of events in product development in IPD methodology, including cost origination and cost control. It forms a basis of practical cost management implementation.

These are the supporting pillars of the IPD-methodology. This forms basis of the cost management procedure in the company. However, this is not the only conceivable viewpoint. Section 4.9 presents a number of competing and supplementary methods [Cla91, Cla92, Bin97, Hun97, Mon89, Wel98].

4.2 Processes in the product lifecycle

Seen from a high vantage point, a product's lifecycle appears as a network of connected and interwoven subprocesses. **Figure 4.2-1** schematically shows (without claim to completeness) the chronological sequence of the different lifecycle subprocesses, including their overlaps. The thin black arrows indicate the influence that decisions made during product development have on the subordinate subprocesses. Such a graphical display cannot fully show the necessarily close interlacing of the development process with each of the remaining subprocesses,

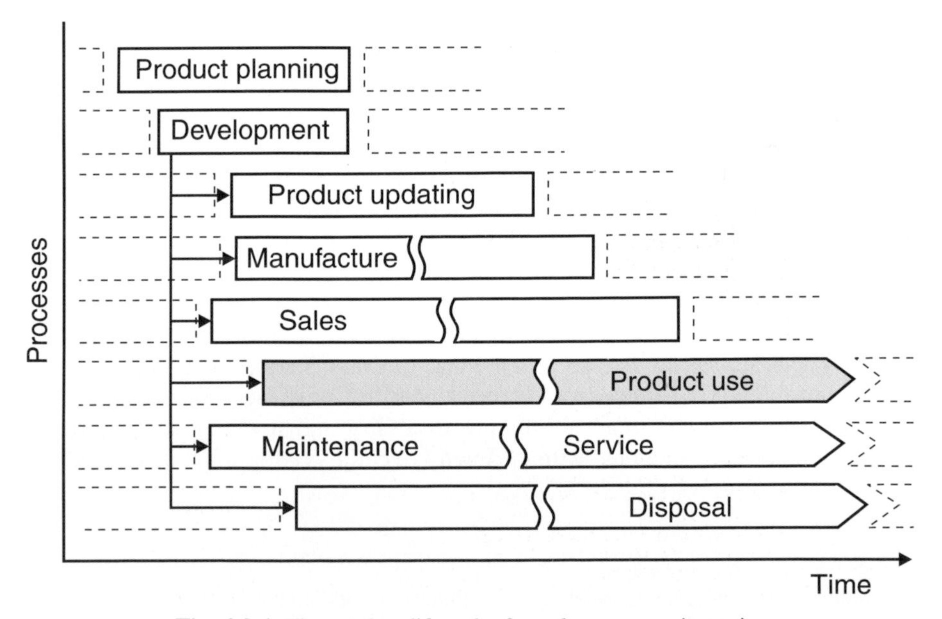

Fig. 4.2-1. The product lifecycle, from the process viewpoint

which is a prerequisite for successful and cost-efficient product development. In fact, in industrial practice not enough emphasis is given to the idea that the developers must **design not only the product, but also the whole process chain**, of which they themselves form a part. Their goal should always be to optimize these processes to maximize company profit.

The idea of **Simultaneous Engineering** (SE) was developed to support this process. The benefits of SE are **time saving** during the product realization; **cost savings** with regard to the total (factory) costs and the lifecycle costs; and product **quality improvement** from the customer's viewpoint.

Teamwork and organization within SE teams will be discussed more fully in Sects. 4.3.1 and 4.3.2. Here, we shall delineate the connections among the different product lifecycle subprocesses. A greater emphasis is given to these during product development through a procedure using Simultaneous Engineering. The generalized lifecycle model in Fig. 4.2-1 does not contain all the important aspects for all different types of industrial products. **It is thus essential that each product developer create a unique model for the total lifecycle of the product to be developed.** Incorporating this comprehensive viewpoint into the new product development process makes it easier to realize all of the project goals (cf. Fig. 5.1-3).

Cooperation of company functions

Effective cooperation of company functions results from making **connections** among product planning, production, cost control, marketing, sales, assembly, procurement, use, service, product care[3], and disposal. This is how subprocesses are linked to the overall product development process.

- **Product planning**
  Product planning defines the sequence of product development tasks. The product developers must be included at the beginning of this decision-making process, since they are best able to assess the **realizability and potential problems of a new project from a technical viewpoint**. However, **innovative product planning stimuli that come from the development department** are even more important. Experience has shown that, indeed, such initiatives do not always come from the product end users.
- **Parts production, assembly, procurement and cost control**
  An **intimate collaboration among development, production, assembly, procurement, and cost control is a key to project success**, but many companies fail to achieve it. The problems in coordinating product development and manufacturing are so diverse [Ehr93] that in Chap. 7 only some important aspects of these relationships will be dealt with in detail. Nevertheless, it can be summarized that the demands for processability and low cost origination in production can be met only with the intensive collaboration of all concerned.

[3] Product care in this context refers to the maintenance, updating and similar efforts devoted to the product by the responsible person(s) in the company.

- **Marketing and sales**
  Marketing and sales listen to the customer and must therefore be closely tied to the development process. This gives a company the flexibility to quickly react to changes in the market, and reduces the danger of developing beyond the customer!
- **Utilization**
  Providing optimal benefits to the customer is, of course, at the core of the development process. With capital goods, this is often tantamount to an increase in the customer's profit [VDM97]. Therefore, the developer needs direct contact with the customer so as not to bog down in high-cost overengineering (Sect. 5.2). Marketing and sales do not suffice as the only sources of customer information.
- **Service**
  Comprehensive product service (technical support and maintenance) is becoming increasingly more important to customers. The development and associated groups can provide information on all essential customer problems.
- **Product care**
  Routinely revising and updating stock products helps keep current products attractive right up to the introduction of successor products. Such continuous changes must be taken into account in the development of manufacturing and procurement processes. In order to keep product revision costs low, such necessary changes must be preplanned during product development (cf. Sect. 5.5).
- **Disposal**
  What happens to a product at the end of its lifecycle is becoming increasingly important, due to new legal considerations. Therefore, low-cost, eco-friendly disposal must also be considered at the product development stage (cf. Sect. 7.14).

In general, processes do not run reliably by themselves. In an analogy to feedback control systems, **product development can be understood as a closed-loop control system (Fig. 4.2-2)**. It is a matter of balancing several controlled variables with the required values of the control variable, under the influence of disturbances. The process to be controlled (e. g., a simultaneous engineering team project) is the controlled system in this case. The task of the project manager, as the controller, is to monitor the actual state of the process against the desired state – the input variable, which is defined by the project goals. If the controller senses a deviation between the actual and desired states, an intervention is needed in the process to bring about a closer agreement. Such a deviation from the planned task execution could represent, for example, a perceptible deviation from the cost target established during the development process. In this case, the agreement of the provisional cost calculation with the cost target must be restored by changes in the present development stage (cf. Fig. 3.1-1, Fig. 4.4-2).

Figure 4.2-3 shows the timeline of such a control process in which design, production planning, resource and investment planning, as well as manufacturing are involved. Here, throughputs follow one another as partial control loops in a disordered sequence. **It becomes clear that a parallel cycling of all partial control loops reduces the development time considerably through simultaneous coordination within the team.** Organizing the **processes as short, closed-loop control**

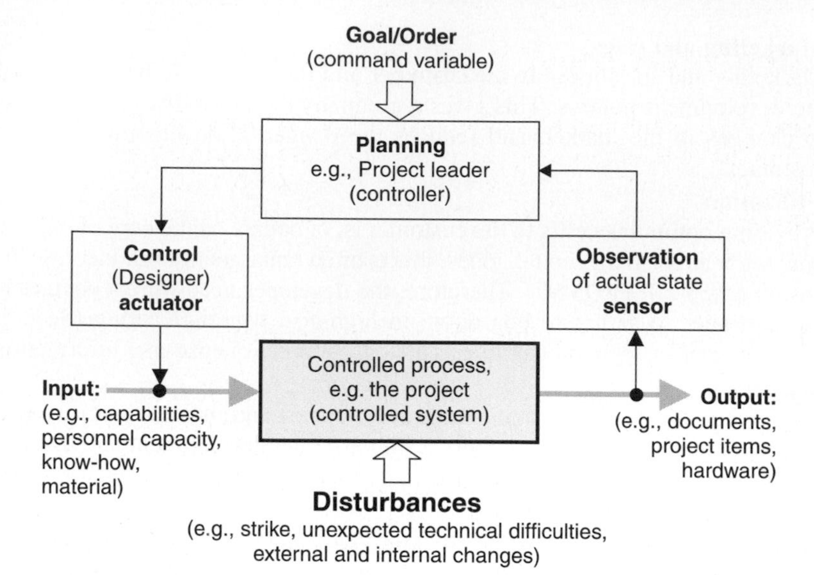

Fig. 4.2-2. Product development as a control system [Wit84]

systems is therefore especially important. Otherwise, poor coordination between individual processes (e. g., development and utilization processes) can result in a large difference between the actual and desired states that might not be caught in a timely manner. This could result in a product failure in the market – the product can become too expensive and thus unmarketable (cf. Fig. 4.4-2).

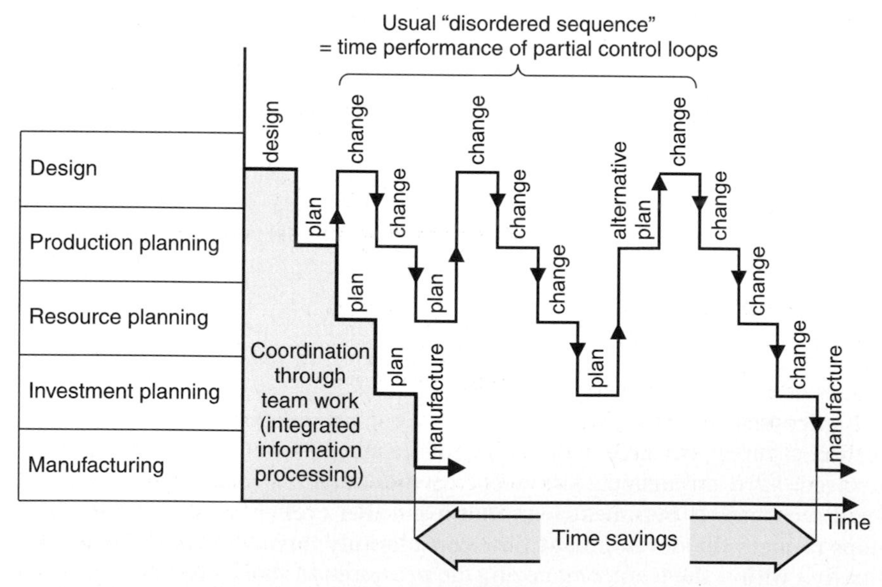

Fig. 4.2-3. Coordinated information processing saves repeated changes [Wit84]

4.3 The human being in the organization

Demands on a company's employees and its organizational structure will be discussed in the following sections. The need for close cooperation among product development and other departments places demands on individuals and on the cooperation within the team (Sect. 4.3.1), as well as on the company organization (Sect. 4.3.2).

4.3.1 The individual and the work within the team

In every company, people must work with other people at all times. Each of us has had the experience that cooperation with others can occasionally be very difficult. How does it happen that we can communicate with one colleague in a few words, while the cooperation with another always tries one's patience?

Because smooth cooperation among employees can provide an invaluable competitive advantage, many companies attempt to actively organize the form of internal and external cooperation. Accordingly, companies often use **personality models** that are oriented to employee **attitude** and **development**. Personality models can function as a guideline for employee selection, help newly hired employees understand the company culture, and foster employees' desire for continual personal training and development. This learning process systematically enhances new employees' social, job, and leadership competence.

Of course, many **external factors** affect the individual development of employees in an enterprise. The company's **Work climate** is extremely important. It can significantly influence whether an employee considers his or her job as a personal challenge that can be **actively managed**, rather than simply an annoying duty. Another factor that affects employees' motivation is the **professional perspective** that is afforded them. As professional employees complete assigned tasks, they sometimes experience a let-down or decrease in motivation unless they are given a new job assignment at the appropriate time (timely job rotation).

Teamwork

Development and production of technical systems is driven today, at all levels, by intensive cooperation in teams. A **team** here implies a group of people who are **goal-oriented**, usually working against a **deadline**, and working **jointly** on the solution of a **problem** [Bor96].

- **Advantages of teamwork**
  As opposed to one person working on a job, **interdisciplinary teamwork** is distinguished by a **higher quantity** and **quality** of ideas and opinions, as well as greater **knowledge** and a broader **judgment basis**. Due to the breadth of different experiences in the cooperative activity, effective mechanisms can be developed in the team to **avoid errors** and achieve **consensus**. Team decisions usually gain a greater acceptance than individual decisions, since they are supported and implemented by several persons. From the viewpoint of

cost control, one frequent result of teamwork is a significant speeding up of company processes. However, since teamwork often introduces a relatively high personnel load, there should be a balanced approach between teamwork and single-person activities.

- **Problems of teamwork**
  Teamwork can sometimes be difficult due to problems that can occur during typical collaboration in a group. In teamwork, articulate, quick-witted people often have an advantage in that more **deliberate** and **thoughtful** team members may be **pushed aside**. If conflicts of this kind are not resolved by mutual agreement, real teamwork will not come about. Another problem that can occur in a group is group **self-censorship** for the preservation of **harmony** ("group thinking"). This can result in good solutions being suppressed, risky decisions being invariably avoided, and wrong decisions not being corrected. Such group-dependent indecision will result if the team does not establish awareness of **responsibility** for the **common goals**.

These problems derive from shortcomings in employees' **social competence**. In fact, introducing teamwork is no guarantee of immediate productivity miracles. It will take time for a culture of teamwork to develop in a company. This process can be supported by well-planned training by which employees are prepared for work in interdisciplinary teams.

Rules for effective teamwork:

➔ Teams need a clear **goal** and a **capacity** and **cost planning schedule**, with corresponding task distribution.

➔ A team should contain no more than **seven** or **eight members**; larger groups work less effectively.

➔ A team should have a **patron** or **sponsor**, alternatively, a **guiding group** or steering committee within management.

➔ For more complex tasks that require more participants, the members should be divided into a **core team** and an **expanded team** Those not in the core team are invited only on an as-needed basis.

➔ Adapt the team composition to the work progress.

➔ The **team leader** should be considered a **moderator** ("*primus inter pares*") rather than a supervisor.

➔ Defend team decisions to the outside, even against resistance.

➔ Team decisions must be **recorded**, perhaps in minutes of team meetings.

4.3.2 Integrative forms of organization

Until now, we have considered the problems of intra-company cooperation only at the levels of individual employees and small groups. However, in this section we direct our attention toward a **company-wide organization** for **teamwork**. The three best-known organization types in industrial firms are:

- **Line organization**
- **Project organization**
- **Matrix organization**

It is not easy to compare these types of organizations in terms of efficiency. Subdividing large organizations into **product-related units** (e. g., profit centers, cost centers, or segments) is helpful for employee cooperation and motivation. But beyond that, the so-called **"subsidiarity principle"** should be incorporated in every organizational structure. This principle states that in **hierarchical organizations, decisions should always** be **made** at the **lowest possible decision level**. Higher levels are activated only if a functional group is overburdened and needs help. In this way, decisions are made where the greatest job competence is available, and the upper management levels are thus not burdened [Pic96, War92, Wil94].

Simultaneous Engineering

An important instrument for the improvement of interdisciplinary collaboration in the enterprise, which has found wide circulation in the recent years, is Simultaneous Engineering[4] [Eve95, Cre90]. Simultaneous Engineering is an integrating procedure from which we understand the **focused, interdisciplinary collaborative** and **parallel working** of **product, manufacturing, procurement** and **marketing development** in which the **entire product lifecycle is** considered (Fig. 6.2-2). The organization is supported and enhanced by hands-on project management [Cre90, Kan93, Pfe93].

Simultaneous engineering pursues three goals:

- **Save time during product development and realization** [VDM98]
- **Decrease in total or factory costs**
- **Improve quality improvement from the customer's viewpoint**

Rules for implementing Simultaneous Engineering:

➔ **"Do it right" – Early identification of properties** for product, production, use and disposal:

- More capacity for **task clarification** (e. g., by application of Quality Function Deployment).

[4] In the English-speaking countries the term "Concurrent Engineering" (CE) is more common.

- **Integration** of experience – the **interdisciplinary SE team** allows the experience from similar products or processes to be applied at the early stages in the process of development.
- Early application of **calculation, simulation** and **cost estimation** (Early identification of properties).
- Early execution of **experiments** (e. g., manual experiments, mockups, rapid prototyping, etc.).
- Implementation of intermediate revisions, **release discussions**, reviews at agreed-to milestones (e. g., concept, realization and serial release, Fig. 4.4-2).

➔ Running jobs in **parallel**:
- Jobs that usually run sequentially (e. g., development of product and production system) are run as close to parallel as possible, through **good communication** and making provisional assumptions.
- Core team sets **groups working** in parallel in the realization phase and takes care of the necessary networking.

➔ **Speeding up** the work:
- **Reducing meetings documentation** through good verbal communication. However, good documentation on the results and the product is essential too!
- **Effective tools** (e. g., for information search, CAD/CAM, EDM/PDM systems, databases).

➔ **Project organization**
- Remove team members from the line for a limited time.
- The **core team** and expanded team form the entire team.
- Have a cooperative but strong and motivating **team leader**.
- Strictly adhere to the project **flowchart** with milestones.

Definite measures for improving the collaboration of product development with other departments must be developed for each company-specific situation (Fig. 6.2-2). This helps in lowering the costs. An example of such a measure is the introduction of **manufacturing** and **cost advising,** as described in Sect. 4.6.1.

4.4 Methods of cost management in product development

In a company organization, those subtasks broken down for small groups or individuals to focus on are generally too complex to be worked on in unstructured form. For this reason, a methodical procedure is necessary. The **procedure cycle**, with its various strategies, is a general-purpose tool applicable to all types of problems (see also Sect. 4.9.2).

4.4.1 Solving problems with the procedure cycle

The **procedure cycle** shown in **Fig. 4.4-1** is a **scheme for problem solution**. It arises from the psychology of thought and systems engineering, and corresponds to the control system concept in Fig. 4.2-2 [Ehr06]. This has been confirmed by empirical investigations (as will be shown in Sect. 7.4 [Dyl91]). As a universal action model, the cycle allows a **structured attack on difficult and complex problems**. According to this model, every problem solution may be broken into three principal steps: **task clarification, solution search**, and **solution selection**.

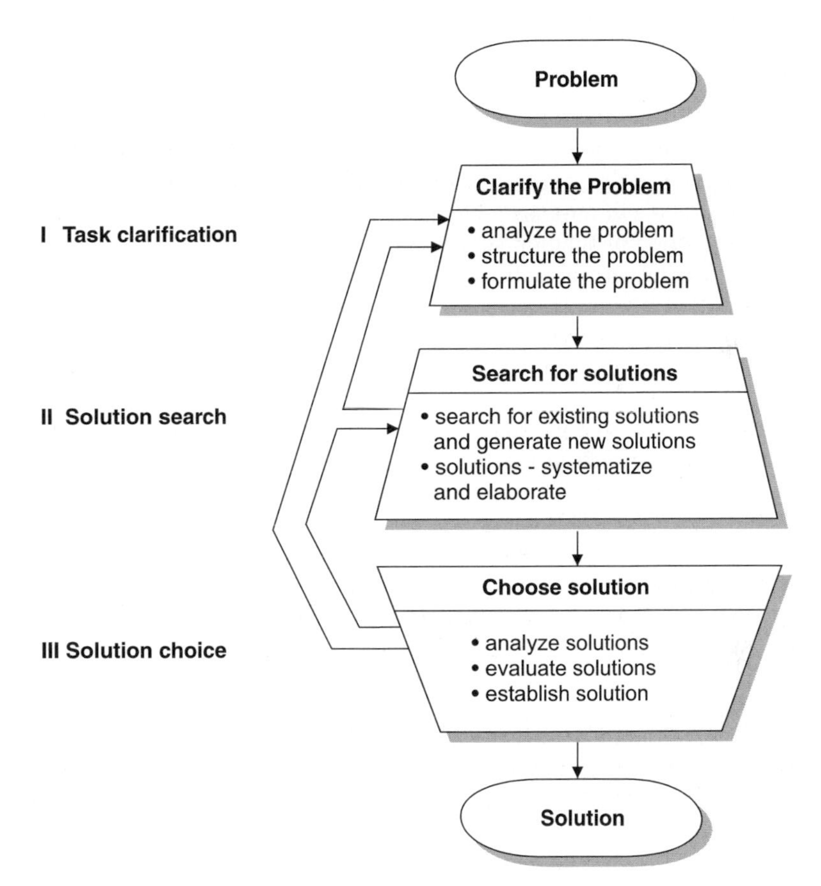

Fig. 4.4-1. The procedure cycle as a general problem solution method

- **Task clarification**
  The operational step "clarify the task" can be subdivided, in turn, into three individual tasks: analyzing the task, structuring the task, and formulating the task. In **analyzing the task**, the ultimate goals are systematically clarified. In a product development task, that means understanding the clients' ideas and objectives

(e. g., in particular, their expectations on the price). At the same time, it is necessary to grasp all relevant boundary conditions and to identify the advantages and disadvantages of existing solutions. The clients' assumptions and the situation analysis enable goal definition, from which the requirements of a solution to the problem can be derived.

Structuring the task breaks up the entire problem into manageable subproblems that can be delineated from each other. This is perhaps the most important step in coping even with complex problems. One or more people can then act upon the defined subproblems according to their logical dependence and importance.

Formulating the task summarizes the result of task clarification in a document, for example, a requirements list. This document forms the basis for the following steps in the procedure cycle. Task clarification also includes planning actions (schedules, participants, etc.).

- **Solution search**
  The goal of the solution search operational step is the **creation of a solution field** from which a solution meeting all of the requirements can be selected. The explicit separation of solution search and solution selection reduces the danger of walking exclusively on well-trodden paths during the problem solution.

  Many problem-oriented techniques exist for the search or synthesis of solutions. The most important methods for development and design of low-cost products are fully described in Sect. 4.5.2, and in Chap. 7. Synthesis techniques can be divided into three categories: (1) the **intuitive solution search**, (2) the **solution search supported by creativity techniques**, and (3) the **solution search aided by systematic methods**. These techniques are adapted to the problem situation (task type, time frame, and available knowledge). In theory, for every task there might actually be too many desired solutions, so it becomes a matter of finding the best low-cost solutions in the shortest possible time. Therefore, every solution search should use systematic and ordered methods to guarantee that no promising solution options are overlooked [Ull92].
- **Solution choice**
  This operational step can also be split into three individual steps: analyze solutions, evaluate solutions, and establish the solution.

  While **analyzing the solutions**, the particular qualities of the different solutions must be determined to have a basis for a rational evaluation of all alternatives at your disposal. For the individual subsolutions, you should also examine their compatibility with the overall system.

  In **evaluating the solutions**, the existing properties of the solutions are compared with the requirements to be met (desired properties). In this way, advantages and disadvantages become transparent as to the risks and success probabilities of individual solution variants.

  Establishing the solution is based on the evaluation of the alternate solutions. Depending on how strongly the evaluation of the solutions can be formalized, the definitive decision either results directly from the evaluation procedure or needs an additional intuitive consideration.

It should be noted that only in rare cases would the solution of a problem follow a linear course, as depicted by the procedure cycle. The arrows in Fig. 4.4-1 show how it is possible to **retrace** the operational steps, if needed. Thus, during solution search, for example, it might become evident that the product requirements are not clarified sufficiently, which initiates another iteration of the "clarify the task" step.

The procedure cycle can be adapted to different problems by bringing in suitable concepts and measures [e. g., holding target costs (Fig. 4.5-7) or lifecycle target costs (Fig. 5.3-1)].

4.4.2 Strategic organization of the procedure

With complex problems, the control of the process during problem solution becomes more involved. Nesting procedure cycles inside each other leads to the formation of more extensive process structures for which graphic planning and representation techniques exist. Among these are **action plans** (e. g., according to [VDI93]; see Figs. 4.4-2 and 4.8-2) and **methods of project management** that more strongly stress the chronological course of a process (cf. Sect. 4.8.3).

How, then, can the structuring of complex problem solution processes occur and how can the designers find their bearings? Faced with such a task, all sensible human beings intuitively make use of certain basic methods, regardless of whether they are planning a vacation trip or designing a concrete mixer. These methods are called **strategies**.

Examples:

- **"From the logic of the problem"**
  Usually, the most favorable procedure for its solution comes directly from the problem. Restrictive boundary conditions often force a specific strategy, and make alternatives unacceptable.
- **"From the urgent to the less urgent"**
  The procedure adjusts itself to the time pressures of the problem.
- **"From the important to the less important"**
  What is important and what is not must be determined in the task clarification. This is coordinated with the rest of the strategy.
- **"From the abstract to the concrete"**
  It can be helpful to start from an abstract problem consideration and go on to the concrete problem, particularly with new and unusual problems.
- **"From the simple to the complicated"**
  It can make sense to begin with the simple sub-aspects of a problem and to work toward the more complicated ones.
- **"From better prospects to less favorable prospects"**
  This strategy is driven by the probability of individual process steps being successfully concluded. When steps that have better prospects are set in the beginning, the steps with poorer prospects can usually be avoided and resources may thus be saved.

Target costing

Target costing is a **means of cost control in product development processes**. It was developed in the 1960s in Japan, from industrial practice [Tan89, Mon89, Sak89a, Sak89b, Cla92, Hor93, Seid93, Ehr93b, Fra93, Pee93, Sak94, Bug96, Stö97, Bul98, Stö99], and is a successor to value analysis [Mil87] (Sect. 4.9.2). There were similar approaches in Europe (e. g., [Kes54, Lor76, Ehr85]) and the United States [Les64], where this method of cost management has been intensely studied since the 1980s. Since then, it has been talked about by everyone and has been applied in many companies.

In target costing, the properties of a product to be developed are determined from market analyses and, simultaneously, a hypothetical sales price is established that conforms to market conditions. If we subtract the planned profit from this price, we get the so-called **target cost** that provides the financial framework for the development process (Fig. 4.5-3, right). This approach fundamentally changes the relationship between the manufacturer and the customer; it changed from a **seller's market** to a **buyer's market**. In the past, when manufacturers developed a new product they would ask:

"How much will this product cost?"

They could actually determine their price based on that in the market. Today, manufacturers are forced to ask the inverse question **at the start** of product development:

"How much is the customer willing to pay for the product?"

By using different methods, which are reviewed in Sect. 4.5.1.4, the total target cost can be split up, based on the product's individual functions, assemblies, or components. These are called the **partial target costs**. Ideally, holding to the established partial target costs is watched closely during the entire ensuing development process. The basic idea is to control the **development process** with regard to product costs in the **management cycle**: *plan – check – intervene* (cf. Sect. 3.1, Fig. 3.1-1).

The differences between traditional and target cost-oriented product development become clear in **Fig. 4.4-2**. In traditional product development, a development task is often carried forward to the complete manufacturing documentation (drawings, parts lists, etc.), before a preliminary calculation of the product cost is made by the production planning and cost calculation departments. Changes at this point usually result in large expenditures of money and time and result in long timelines and ineffective feedback (Fig. 4.4-2, left). Often, the product development group gets no feedback on the actual costs incurred. In contrast, in target costing the results are checked for compliance to the total cost target during all stages of product development (Fig. 4.4-2, right). Observed deviations can be fixed immediately, and the iterative loops become as short as possible ("short feedbacks"). Thus, the development processes can also become shorter while keeping within the planned product costs. This has positive effects on development costs (reduced costs of changes) and the company's market presence (Fig. 4.2-3, Fig. 6.2-2).

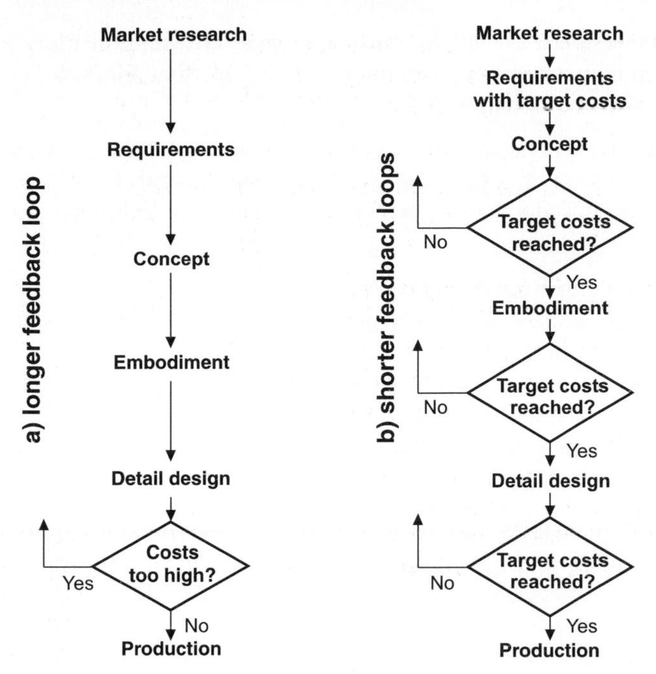

Fig. 4.4-2. Comparison of traditional (*left*) and target-cost oriented (*right*) procedure in product development while implementing the concretization steps

The significant long-term success of target costing is due to three main reasons:

- **Determining a sales price very early in the development process forces a precise analysis of the market situation and market trends.** That enables a company to demand the development of customer- and market-driven products, right at the beginning. By coupling the product's individual functions to specific target costs, product properties that are uneconomical from the customer's viewpoint can be recognized early and weighted lower, or removed from the requirements.
- **With consistent, concurrent control of the preplanned partial target costs, possible deviations can be identified and corrected at the earliest possible moment.** With changes made during the development process, the central insight obtained in Chap. 2 applies [i. e., the longer the decision to be revised is postponed, more expensive the changes are (the "Rule of Ten", cf. Sect. 2.2)]. In this way, the emphasis lies, as a matter of course, in the early stages of product development.
- Traditional low-cost design was oriented largely toward the product itself. Here it was a matter of setting up the defaulted product with the least expenditure on material, people, and machines. Design and production planning remained narrowly focused, mainly on production operations: Must the feather key groove be made with an end mill, or can a face-milling machine also be used? A company's different products were often considered completely separately from each other, so that overriding problem areas (e. g., logistics or customer service)

were not considered at all. **In contrast, target costing considers all product development processes as connected to the product lifecycle, according to their relevance and importance.**

The advantages of a target-cost-oriented procedure during cost management have been discussed thus far. Four important task statements, which must be resolved with the introduction of the cost management in industrial practice, result from this approach:

- **How do we determine total target costs?** (cf. Sect. 4.5.1.3)
- **How do we get the partial target costs from the total target costs?** (cf. Sect. 4.5.1.4)
- **How do we check the compliance of target costs during product realization?** (cf. Sect. 4.5.3.1 and Chap. 9)
- **How do we steer the development process to minimize goal conflicts as far as possible?** (cf. Sect. 4.8.3)

To complete these tasks, a series of **specific organizational methods** is available, which usually need to be **adapted to company-specific situations**. It is a matter of establishing realistic cost objectives and recognizing additional costs as they arise during the development process, as early as possible. For this reason, departments that follow in the logical course of the development process (e. g., production and assembly planning, and costing) are included in the decision-making at the beginning. This can be achieved through teamwork, simultaneous engineering, cost consultation, etc. **Parallel task processing must take the place of sequential work.**

4.5 Integration of methods for target-cost oriented development

Sections 4.1 through 4.4 introduced the basic cost management modules. These will now be integrated into a **target-cost oriented development methodology**, according to the design methodology point of view. The course of a development task is based both on the work phases (defined by project management; cf. Sect. 4.8.3.1) as well as on the concretization steps defined by the design methodology (i. e., concept, embodiment and detail design; cf. Fig. 4.4-2). All work steps can be dealt with per the procedure cycle (Fig. 4.4-1), and will therefore be handled accordingly in the following sections. An example is given in Sect. 4.7.

4.5.1 Task clarification: Requirements clarification, target costs establishment and their distribution

In the development process, the first step of the procedure cycle consists of precisely clarifying the demands on the new product (Fig. 4.4-1). This includes, for example, the maximum permissible manufacturing cost. Target costing determination and distribution of target costs is especially important [Tan89, Seid93, Frö98, Stö99].

4.5.1.1 Clarify the requirements

Experience has shown that lack of clearly understanding the goals and demands on a product is the chief cause of project delays and endless changes, along the lines shown in Fig. 4.2-3 and Fig. 4.4-2 (left). This is a matter of interface, or getting over the intellectual wall between the manufacturer and the known or unnamed customer (Sect. 3.3.1). If we as designers or planners do not know what we want, or where potential problems lie, or how we should spend money, we should not be surprised later by failures and expensive changes!

A **requirements list** should be developed, divided into demands (fixed demands, minimum requirements with numerical values) and wishes (qualitative information). This is a starting document for a project, and must be compiled across departments and modified repeatedly in the course of the project. Accordingly, it should contain details about the origin of its information and, as is now common in product documentation, a change process (document revision control) must be set up (for target conflicts, see Sect. 4.5.3).

A requirements list can begin with a known customer's specifications, or marketing and sales data. It should contain not only technical requirements but, as shown in **Fig. 4.5-1**, also target costs (Sect. 4.5.1.3) and requirements from ergonomics, industrial safety, and industrial design. Besides these external requirements, it should also contain the company's internal organizational requirements such as schedules, development budgets, and personnel capacity, which must be clarified and established. The entire lifecycle of the product, from the start of development through its disposal, should be discussed by an interdisciplinary team and any pertinent information should be included in the requirements list. It is important to include customers and all known suppliers in the requirements list; see Fig. 3.2-2.

It might appear that a requirements list is necessary only for a large project, but that is not the case. For smaller tasks (e. g., the design of appliances or tools), the early documentation of performance, interface dimensions, and schedules in a short list with the client's initials can also prove useful.

4.5.1.2 Function analysis

Starting with the requirements list, the functions that the product must fulfill are defined using function analysis. Function analysis thus stands at the threshold between task clarification and the solution search. It allows the developer to examine in detail the purpose of the system to be designed, before thinking about concrete solutions. The literature (e. g., [Aki94, Ehr95, Pah05]) offers a number of suggestions for the formulation of functions and function structures. In many cases it is sufficient to formulate the basic functions of the considered technical system as verb and noun combinations (e. g., "amplify force", "separate materials", etc.) and to visualize these with the aid of a **simple or hierarchically organized list** (Fig. 4.7-2). Only if this procedure does not prove to be practical, as, for example, where there are many relationships between the different function carriers of a system, should people think about using one of the above-mentioned tools.

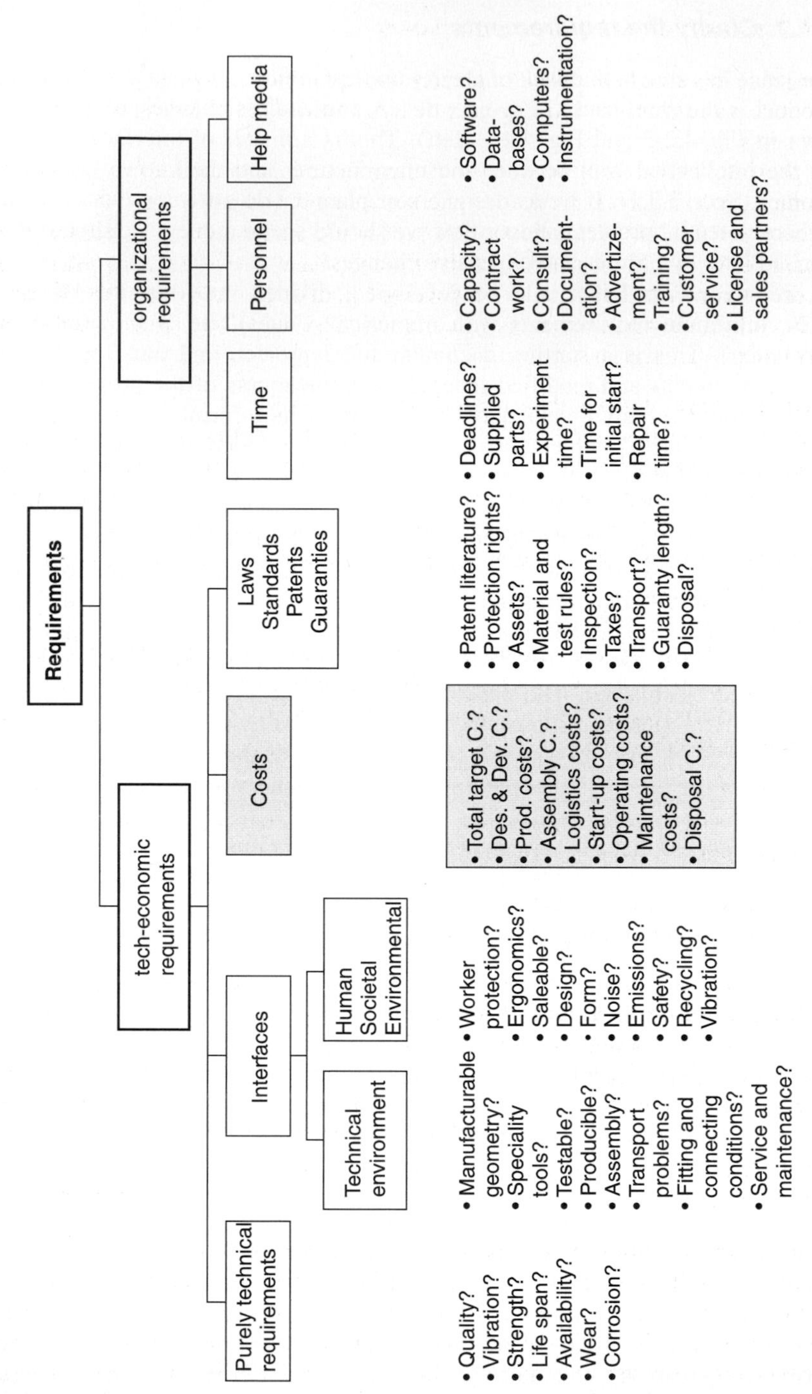

Fig. 4.5-1. Types of requirements with some lifecycle-oriented questions (checklist for the preparation of the requirements list)

4.5.1.3 Establish the total target costs

During task clarification, **target costs for the product** must also be established. The entire process of product development that follows is based upon these preset target costs.

A problem frequently encountered during the establishment of target costs is the time elapsed between the date of market analysis and the beginning of the sale of a new product [Bug95, Ehr96, Fra93, Hor93, Lak93, Wes 02]. **Only if the forecast set up at the beginning of the development process still applies at the date of the market launch can the planned profits be realized with the product!** In order to reduce, as far as possible, the risk of failure in the market, two things must be guaranteed:

- **Reliable target cost must be established**
  It is impossible to make totally reliable forecasts about the market for the date of the planned product introduction. However, through careful analysis of existing trends and a cautious extrapolation, the risk of a wrong forecast can be minimized. If the total target costs are established, they must also be reliably estimated as to whether they are in fact attainable for the company under the given circumstances.
- **Short development times**
  The best way to ensure reliable market forecasts is to keep the development time for a new product as short as possible [Smi98]. The shorter this time is, the lower the probability of events taking place that might contradict the earlier assumptions. Methods to shorten development times are discussed in Chap. 6 (cf. Fig. 6.2-2).

The starting points for the establishment of the total target costs are the two crucial questions in target costing:

- What does the customer want?
- How much is the customer willing to pay for that?

This information alone, however, is not enough to formulate realistic target costs because the customer would always like to have the best solution and pay the least amount, or nothing. Frequently the necessary costs for a "property" or a demand cannot be quantified. For example, can automobile customers specify how much they are willing to pay for a car, or that the car will satisfy all licensing regulations, or that the transmission will never leak, etc.? The apparently simple question "How much is the customer willing to pay?" must therefore be considered from other points of view (**Fig. 4.5-2**), as follows.

- **How much is a potential customer willing to pay for our product?**
  According to the philosophy of target costing, the potential customers are at the center of target cost estimation because they are the ones who are expected to buy the product in sufficient numbers. A product will be bought if it in some manner promises the customer benefit and/or profit, or if it is somehow exciting. The evaluation of a possible user, and thus the customer, can be split into objective and subjective customer benefits.

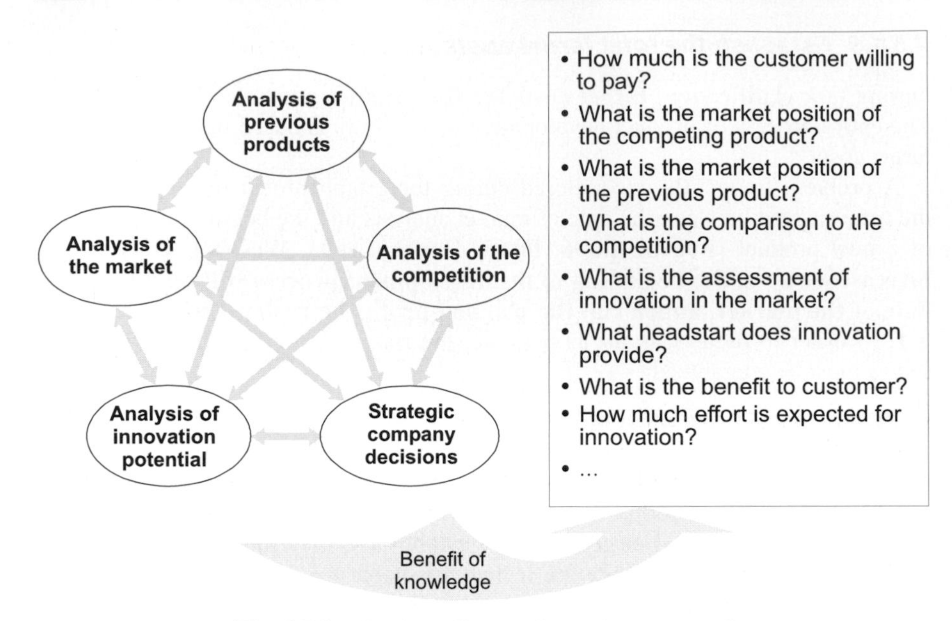

Fig. 4.5-2. What has effect on the total target costs?

The **objective customer benefit** means, strictly speaking, the product's efficiency. This results from the product's **functionality, purchase cost**, and eventual **operating costs**. Other evaluation criteria, such as the reliability or the service involved with the product, naturally also enter into this.

On the other hand, the **subjective customer benefit** implies the rather vague evaluation criteria that may be attributed to the public's taste or the current spirit of times. This includes the product's **appearance** or the **logo image** [Cla91]. Additional inducements offer product properties that Seidenschwarz designated as **"excitement features"** [Seid97]. They distinguish the product from its mass of rivals and give it an individual character.

Subjective benefits can also be identified with many market analysis methods (after [Koh99]). These include personal customer surveys (expensive, and best used with a concurrent checklist); questionnaires (usually of little benefit); polling experts within the company (low expenditure, but has the danger of company blinders); evaluations by external experts (institutes, technical societies); and portfolio analysis (Fig. 4.6-8). Additional marketing aids include conjoint analysis and the analytical hierarchy process [Aak95, Arn95, Gre88, Mal93]).

- **Analysis of the market ("market into the company")**
  With market into the company, the total target costs are derived from the market. The price attainable in the market ("target price") is determined through market analysis. Market analysis is also used for determining product features and their value estimate from the viewpoint of the customer. After subtracting the planned profit ("target profit margin"), what remains are the costs allowed by the market ("allowable costs").

- **How well do previous products perform and how much do they cost?**
Of all of our planned product's competitors, we know our own previous product the best. We should use our detailed knowledge about this product's costs and the cost structure, in particular, for checking the realizability of the target costs derived from the market, under our own company's development and production conditions. Target costs that have been agreed to at the beginning will show their effect only then, if they have in fact been held to also during the market launch. Analysis of previous products enables us to judge which target costs can be attained for a specific product under the conditions present in the company (and at its suppliers). Such cost information should available in structured form, and that structure should be oriented to the product shape, the product functions, and the various cost categories, among others.
- **Analysis of the competitors ("out of the competitor")**
Even if the potential customer of our product is put in the center of the target cost estimation, it would be frivolous to ignore possible future competing products in the market. This is because a prospective customer's evaluation of our product will depend strongly on what functional and price alternatives the then-current competing products offer. Finally, it is every company's nightmare to be confronted, at the end of an expensive development process, with an innovative competing product which, at one stroke, totally destroys the market chances of their own product. A comparison of rival products with one's own products (previous product, concept of a new product, etc.), considering the market requirements, provides a number of valuable hints.

  As a rule, we do not know what products rivals will launch. During the target cost establishment, we are thus forced to **estimate the customer benefit by means of the available competing products** and thus the **presumed innovation potential**. At the same time, as far as feasible, the competitors' manufacturing costs should also be estimated. It is understood of course that while applying these techniques is important, the results are uncertain.
- **Analysis of customer benefit / cost-effectiveness analysis ("out of customer benefit / cost-effective analysis")**
The economic benefit that a customer can generate through the purchase of the product provides information about a justifiable investment level and, with that, about a reasonable price for the product. This procedure is particularly suited to products with new functionality, as well as those with optional extra features. While this approach is less well suited to consumer goods, it is frequently used with capital goods where the considerations of reliability, maintenance, and operating costs, etc. can also be included for pricing the investment.
- **Analysis of innovation potential ("out of innovation")**
Innovations have great influence on how target costs are established. On one hand, it is important to clarify how the market judges the innovation and how much of a head start it brings against the competition. With a positive assessment, there is potential for better prices. On the other hand, technology developments in the suppliers' market and by the product manufacturer can lead to cost-reducing innovations in products and/or processes. These include new materials, sensors, electronics, manufacturing processes, calculation methods, use

of flexible software for controls, etc. These innovation potentials from technology must be incorporated in the process of finding target costs.

- **Consideration of strategic decisions**
  Strategic company decisions influence the determination of target costs. These can be, for example, defined target costs that allow strategic entry into a specific market segment. These can be restrictions on product realization, such as preferred suppliers, in-house production due to low facility workload, acceptance obligations at subsidiary companies, etc.

The answers to these questions can supply ideas and define limiting conditions for a first, quick solution search. With this search, the product's possible cost reductions can be estimated under the conditions existing at the company. The demands on the new product, which have already been identified, form the basis of this early solution search. They also show that with restricting cost objectives it is not sufficient to consider only the development of a product; the entire process chain must be examined and modified as needed. Based on all available information about the possible cost targets, and considering the company situation, target costs are established by the management and/or with the agreed-upon parties responsible for the implementation (e. g., the project manager).

Most designed products draw upon similar earlier products. In such cases, it stands to reason that the calculations based on the previous product should be used for estimating the costs of the new product (cf. **Fig. 4.5-3**, left). Such calculations include the purchased parts and the probable production processes. The total costs and the sales price are determined from the manufacturing costs by using the known surcharges (Fig. 8.4-2). The target market price must be based on the competing products and, therefore, might be lower (cf. Fig. 4.5-3, right). Besides, the

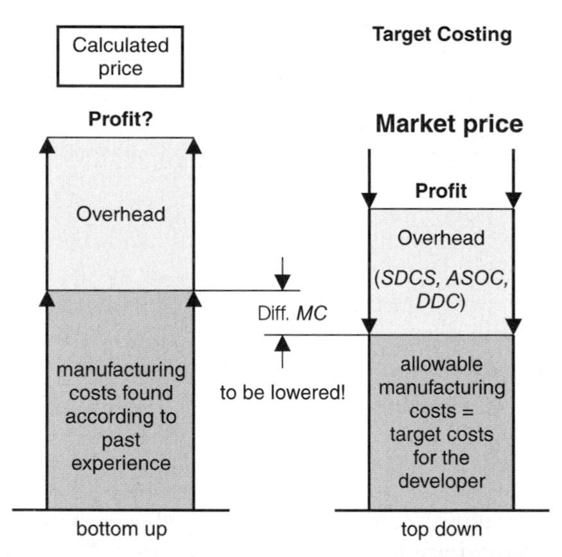

Fig. 4.5-3. The market price determines the cost target (not the expected manufacturing costs, but the market price!)

conventional **"bottom-up"** pricing, the decisive target manufacturing costs for the developer must also be calculated **"top-down"**.

At the same time, the planned profit and the estimated overhead costs are subtracted from the market price. The initially calculated manufacturing costs must be reduced during the development process by the cost difference (ΔMC). For this purpose, it makes sense to divide this difference into the product functions or components.

Before the total target cost and the partial target costs are determined by mutual understanding, following points should be critically considered:

- Why is our product so expensive (e. g., smaller quantities than the competition, other performance features, etc. cf. Sect. 7.5)?
- Why can the competition offer so much more, with a lower cost (e. g., larger quantities, manufacturing in a low wage country, etc.)?
- How much are the customers ready to pay, for what (e. g., which functions do they really need, etc.)?
- What do the customers need (e. g., higher performance), and how much would they pay for that?

The answers to these questions can provide ideas and limiting conditions in a **first, quick solution search**. With that, we can estimate the realizable **cost reduction potential** for the product, for the **conditions existing at the company**. In this regard, the techniques described in Sect. 4.5.2.4 for solution search, with the help of creativity techniques, are particularly useful. The previously identified demands on the new product form the basis of this preferred solution search. The process, as it evolves, depends strongly on the result of the following types of assessment:

Case 1: The estimate shows a cost reduction potential that enables a sales price that will be clearly below the competition's price. Then **the established sales price is used as a target price**.
Case 2: The estimate shows a sales price which is in the range of the competition's price. Similar to the first case, **here also a target price will be determined that is clearly lower than the competitor's**. It may be expected that the difference between the recognizable cost reduction potential and the established sales price can still be eliminated during the course of product development (see Sect. 10.1).
Case 3: The estimated sales price is still clearly above the competition's prices without any further cost reduction potential being recognizable. If, in spite of that, a target price that is reasonable from the viewpoint of the market situation is stipulated, the risk is very high that the targets cannot be held to during the development process. Under these circumstances, **the planned product development must be reconsidered**. In this case, it is recommended that a project study be carried out for an intensive solutions search to safeguard the product.

If the target price for the product being designed is firm, the **total target cost for the product manufacture** can be established from that by **subtracting the planned profit and the overhead costs** (Fig. 4.5-3).

4.5.1.4 Dividing the total target costs into partial target costs

A product that different developers or development groups work on, usually consists of several components (assemblies), and so the overall target must be split into sub-targets that are binding on the individual developers.

Prerequisites

- Start with a similar **forerunner product** whose cost structure is known.
- The **product developers** working with the partial target costs must be associated with the **process of cost distribution** from the beginning. Only in this way will they be motivated to attain these goals and will be able to address the potentials for cost reduction.
- Practice has shown that the established **partial target costs cannot be unalterable values** during the development process. It often happens that some components have a higher cost reduction potential than was earlier estimated, and others have less potential. The partial cost targets must then be mutually balanced.
- Even if the starting point for partial target costs cannot be directly assigned to assemblies per customer preferences, requirements, and functions, they should be **established for assemblies**, if possible (cost deployment) [Rös96, Sau86, Tan89, Frö94, Fre98, Stö99]. Only the partial target costs of assemblies can later be unambiguously controlled during the concurrent calculation.
- **Customer benefit** must always be the starting point for establishing partial target costs ("How much is the customer ready to pay for this function, quality, or unit"). However, every product has certain components that the **customer will recognize as being essential** (e. g., wheels for a vehicle, a housing for a gearbox). For these, the customer could not balk at an acceptable portion of the purchase price. On the other hand, there is often a customer benefit that costs relatively little but for which a customer is willing to pay well (e. g., interesting colors for a car).
- The **target cost distribution** must be defined by the **interdisciplinary team** to compile all relevant experiences and ideas. A visualization of the contributions is strongly advised.

Procedure

In spite of the scientific debate [Seid93, Rös97] about target cost division, we propose a pragmatic procedure, based on the above conditions, that is in line with Figs. 4.5-3 and 4.5-4. Distributing the total target costs can serve to formally secure the methods finally specified, but remains primarily **a determination agreed to in the collective discussions** (Sects. 10.1.4; I.2b).

- 100% of the total target costs are allocated according to **customer-relevant functions and qualities**, as well as to known **previous components** or assemblies (**Fig. 4.5-4**). As an enhancement, we can also attempt to allocate the distribution according to estimated **competitors' component costs**. That shows three cost structures which are very different, yet provide a first direction.

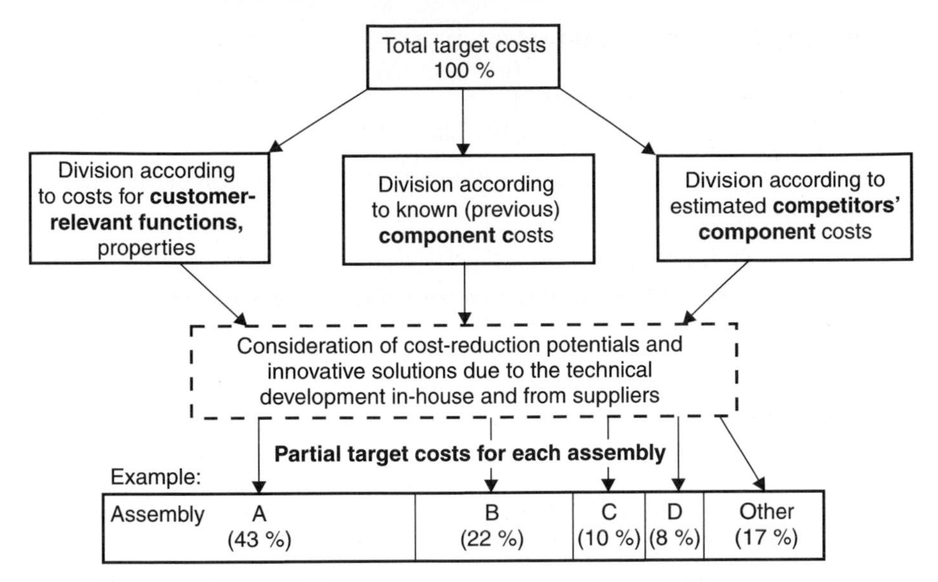

Fig. 4.5-4. Example of the separation of total target costs into partial target costs of components or assemblies A through D of a future product

- The cost structures are then modified by the innovation and cost reduction potentials are usually identified in team discussions. For example, the team might decide that a device can be controlled more flexibly and at a lower cost by software instead of electromechanical means, or that a certain supplier has particularly promising subassemblies, or that a lower-cost material will suffice.
- Finally, all findings are concentrated in the new partial target cost structure that is henceforth binding on all parties. With that, the respective responsible parties are also established.

There are many ways to establish and divide a new product's target costs. These are assigned in the following listing according to the elements of target cost establishment described at the beginning of this chapter.

- **Customer's** aspect (cf. [Aak95, Arn95, Bau93, Gre88, Mal93, Rie93, Sei93]):
  - Contract discussion (cf. Sect. 10.1)
  - Customer survey
  - Conjoint analysis, analytical hierarchy process [Arn95]
- **Competitor's** aspect:
  - Market analysis [Hei91]
  - Portfolio analysis (cf. Sect. 4.6.6)
  - Benchmarking (cf. Sect. 7.13) [Cam94, Mer94, Pie95, Kre97, Bro98]
  - Product clinic [Wil98, Wil99]
- **Previous products'** aspect:
  - Cost structures (cf. Sect. 4.6.2)
  - Pareto analysis (cf. Sect. 4.6.2)

- **Technological development** aspect:
  - Design methodology [And81, Bre93, Ehr06, Hub92, Kol94, Pah05, Roo95, Rot94, Ulr95, VDI93, Wal97]
  - Innovation methods [Alt84, Kap96, Lin93, Ter97]

The examples in Sects. 4.7 and 10.1 demonstrate how to determine and allocate a product's target costs.

4.5.2 Solution search: How are low-cost solution approaches developed?

After **clarifying the requirements** and **establishing** the associated **target costs**, a technical concept must be developed that will realize the product development objectives. First, note down and arrange the solutions that emerge immediately. For that purpose, it has proved useful to look at what energy types to use and thus evaluate mechanical, electrical, magnetic, or hydraulic options. The product developer should find the way by applying the fundamental procedures of design methodology, as explained fully in [Ehr06], for example. After the requirements have been clarified, the sequence of design methodology operational steps are explained briefly below. (Note that instead of the chronological term "phase", we use "concretization step".)

The means and measures from **Steps I (task clarification)** and **III (solution selection)** of the **procedure cycle** (Fig. 4.4-1) are applicable for **every phase** of **the development**. It is different for the solution search in Step II, where there are specific helpful measures, depending on the concretization step, which are presented here only in part (for the concretization steps, see Fig. 4.8-2).

For the **solution search** in the **concretization steps**, the following are applicable:

- **Conceptualization:** A systematic solution search evaluates physical effects (Sect. 4.5.2.2), or shape variations (Sect. 4.5.2.3) in which the functionally active surfaces and movements are influential (direct shape variation: Fig. 4.5-6).
- **Embodiment:** Solutions can also be found by indirect shape variation—variation of the material properties (Sect. 7.9) or the production, joining, or assembly processes (Sect. 7.11).

  Moreover, the design types can be changed. Examples are light and small design (Sect. 7.9.2.2); production-specific designs (e. g., cast, sheet metal, and welded designs) (Sect. 7.11.2, 7.11.3, 7.11.5.3, 7.13.4); integral or differential designs (Sect. 7.12.4.3); design in size ranges (Sect. 7.12.5); and modular design (Sect. 7.12.6).
- **Detail design:** Since embodiment leads smoothly into detail design, the final concrete decisions about product details come about at this step, but the above-mentioned methods of finding solutions are still partly valid. More specialized factors are part standardization (Sect. 7.12.4.1), dimensional tolerances, and roughness (Sect. 7.11.6).

In addition, the **types of design** play a role [Ehr06; Pah05]:

- A **new design** contains all of the concretization steps.
- An **adaptive design** retains the given concept but changes the sizes, shapes, location of the assemblies, or the material or manufacturing processes. Principles of secondary functions can be new.
- For **variant designs**, mainly the geometrical decisions are changed. Concepts, materials and manufacturing processes are kept intact.

The search for available solutions (Sect. 4.5.2.1), the use of creativity techniques (Sect. 4.5.2.4), and use of a morphological matrix (Sect. 4.5.2.5) are applicable for all of the concretization steps.

4.5.2.1 Search for available solutions

We do not need to break our heads over what others have already thought about – there is no need to reinvent the wheel!

Existing solutions to technical problems can be found in our **own company**, in the **market**, in **competing products**, in **patents**, in **catalogs** [VDI82], in checklists (e. g., Fig. 4.6-7), and in the technical literature [Bir92, Bir93, Büt95, Ehr06, Rei93, Sche97]. These can often be realized reliably, quickly, and at a low cost.

4.5.2.2 Solution search by using physical effects

Most problems, with which the mechanical engineer is confronted, are solved by the **technical application of physical principles**. Therefore, a basic clarification of the physical solution options should be a component of the methodical procedure [Rod91].

An example of developing technical systems by modifying the physical effect associations is the evolution of computer printers during the last twenty years (cf. **Fig. 4.5-5**). The original dot-matrix print head evolved into the thermal transfer process, then to the ink-jet print head, and finally to laser printing, as at present. The physical mode of operation of these devices was continually modified. The modification of the principle of physical action was a decisive factor in developing lower-cost designs for these devices.

The steep dive in prices for ink-jet printers can be ascribed to the fact that the piezoelectric drop generation was superceded by thermal bubble-jet technology. This shows how a change of the physical principle (concept change) can result in a huge cost reduction (Sects. 4.8.2 and 7.3).

The search for basic physical solution options for technical problems can be methodically supported through summaries of physical effects in so-called **checklists** or **design catalogs** [Ard88, Ehr95, Kol94, Pah97, Rot94]. Furthermore, there are computer-aided solution systems that suggest options from analysis of patent literature [Lin98a]. Fig. 4.7-3 shows the application of physical effects by an example.

	Principal solutions for printers		**Physical effects**
Dot-matrix printer	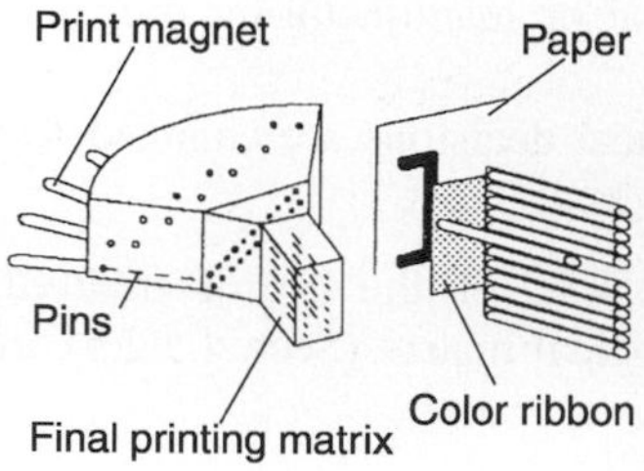 Print magnet Paper Pins Color ribbon Final printing matrix	In a 24 pin printer, two rows of opposed pins (final printing matrix) produce an unbroken continuous line. The pins are pushed out by an electro-magnet and return by spring action. With this technique pictures and signs are also possible.	• Electro-magnetic attraction • Impact • Adhesion
Ink-jet printer (thermal)	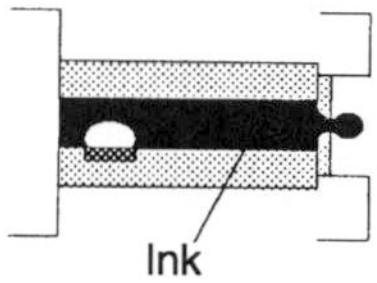 Ink	Principle of the ink jet printer: By heating with a heating element a steam bubble forms which expands explosively and shoots the ink out through a nozzle.	• Gay-Lussac • Elastic deformation press. propagation • Bernoulli • Adhesion (capillary effect)
Ink-jet printer (piezo-electric)	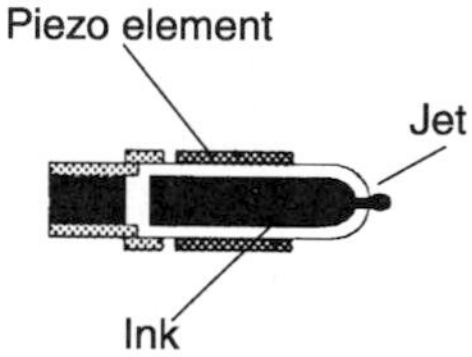 Piezo element Jet Ink	Here is another version: By applying an electric voltage a piezoelectric element constricts around the ink container and thus pushes out the ink. Advantage: The print heads last longer.	• Piezo effect • Pressure propagation • Bernoulli • Adhesion (capillary effect)
Thermal transfer printer	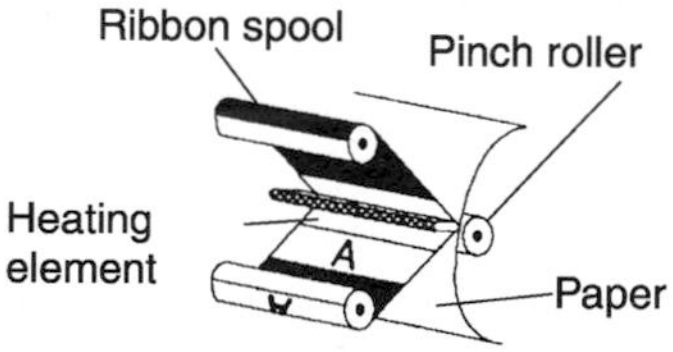 Ribbon spool Pinch roller Heating element Paper	In a thermal transfer printer the color is also transferred to the paper by heat, and in fact from three different color films. The heat "burns" the color into the paper.	• Change of physical state • Plastic deformation
Laser printer	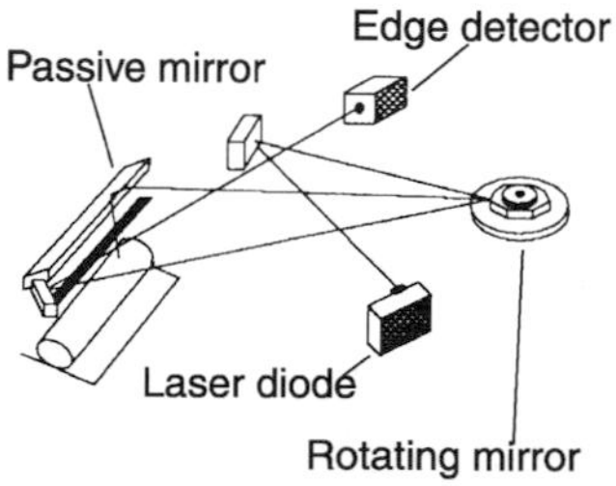 Edge detector Passive mirror Laser diode Rotating mirror	In a laser printer the laser beam is passed through a rotating mirror. Through the passive reflector it reaches on the copier drum, that it pointwise unloads or does not unload.	• Radiant heat • Thermal effect • Electrostatic attraction • Adhesion

Fig. 4.5-5. Some basic solutions for printers, along with their physical effects [Rep94]

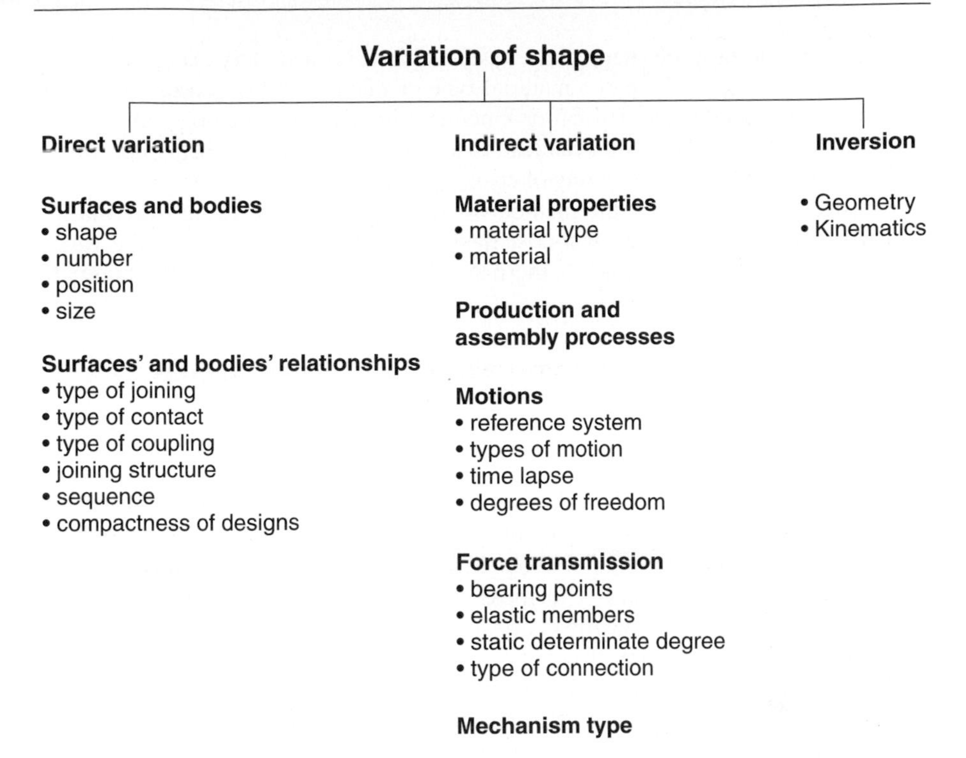

Fig. 4.5-6. Overview of shape variational features (illustrative sketches and examples are given in [Ehr06])

A similar beginning, with which physically different solution elements in the software **TRIZ** are offered, found far spreading in the industry [Her00]. The procedure is based on the preliminary work of Russian patent engineer Altschuller, which analyzed an immense number of patent specifications according to its solution principles [Alt84].

4.5.2.3 Variation of shape

Shape variation is a basic technique in which the designer systematically develops the product's shape [VDI98] (Sect. 7.4). A primitives' geometry can be the starting point for shape variation, as it often arises with new designs from the application of the physical action principle. A second possibility is to start with an existing solution with the goal of improving upon it, or to adapt it to a modified task statement [Geu96] (Figs. 4.7-4 and 4.7-5).

In systematic, goal-directed work, the designer uses design methodology [Ehr06] to classify the features according to which variation could be implemented. **Figure 4.5-6** shows an overview of possible variational features. Here, distinctions are made among **direct variational features**, **indirect variational features**, and the so-called **inversion**. Under direct variation are summarized all features which refer to the direct modification of the shape in the considered systems. In indirect variation, by contrast, the shape of the systems is not immediately changed. This is the case, for example, if a part's material type or manufacturing process is varied. Inversion refers to the geometrical or kinematic inversion of the operating principle. In extreme cases, inversion can lead to a whole series of parts being discarded, which usually leads to the lowering of costs.

Of special importance is the method of shape variation for **design-for-manufacturing (DFM)** and thus a **low-cost design** of components. Beyond the variational features shown in Fig. 4.5-6, there are many design rules for DFM; these are discussed in full in Sect. 7.10.4.

4.5.2.4 Solution search with the help of creativity techniques

To stimulate the creativity of the participants of an interdisciplinary team can be very rewarding. There are different techniques for this [Gam96, Osb57].

- After a problem has been clarified and formulated, the goal of **brainstorming** is for the group to produce a large number of possible solutions. Brainstorming can be carried out in groups of about **5–15 persons** who are, in the ideal case, from **different disciplines**. The ideas generated by the participants must be visualized (e. g., on boards, flip charts, etc.) so that the group may be stimulated to further idea generation. It is very important during a brainstorming session that the suggested ideas are not criticized. While this often results in a few improbable or unfeasible ideas, it is the only way the frcc thought process can produce promising innovative solutions. The evaluation of a brainstorming session occurs later. A variant of brainstorming, which is characteristic particularly for larger groups, is **brainwriting**, where the participants write their ideas down on cards that are then discussed in the group [Osb57].
- **Synectics** is a modification of brainstorming, in which the group uses analogies from non-technical fields (e. g., biology, art, etc.) to stimulate the solution of a technical problem. After the group is familiarized with the problem, its members attempt together to make comparisons or analogies to other fields. Upon the analysis of a stated analogy, it is compared to the posed problem. From that, a new idea can then be developed into a problem solution [Gor61].
- In the **6-3-5 method**, **six** persons in a group each develop **three** solution proposals and take turns to elaborate on these **five** times. As during brainstorming, they begin with a common formulation of the problem before each person begins an individual solution search. Everyone has approximately five minutes to outline and to explain, with keywords, their three different solution proposals. Then, each form is passed on to the next person. This procedure is repeated until every participant has worked on each form. Specialists should evaluate the result of the session [Roh69].
- The **gallery method** is particularly well suited to a group solution search in **shape design** problems. After the problem is explained, every participant makes sketches of possible solutions. These solutions are displayed so that everyone can look at them, mentally process them, and discuss them with the others. The solutions then can be further worked on in common or individually, and made accessible to each member [Hel78].

The results that emerge from creativity techniques should be integrated in same manner as existing solutions, into a systematical processing of the solution space in the form of a morphological matrix (cf. Sect. 4.5.2.5).

4.5.2.5 Concept development with the morphological matrix

Several solutions for a sub-problem are usually found, and these subsolutions must be integrated into functioning overall concepts that can be evaluated.

The **morphological matrix**, initially developed by Zwicky [Zwi66], is a well-proven means of combining subsolutions into an overall solution. It is a matrix, in which the subsolutions are displayed for each of the subfunctions to be fulfilled by the system (cf. Fig. 4.7-7, top). The morphological matrix offers an excellent overview of the available subsolutions and in this way can suggest further variations in the field of physics or in shapes. Therefore, it should be set up right at the start of the solution search to use for concurrent documentation of the design process [Ket71].

How can an optimal machine concept be developed with the aid of a morphological matrix? In theory, it should be possible to hypothesize all of the conceivable combinations of subsolutions and subfunctions, and simply evaluate them. However, this would probably take much too long. Instead, the individual subsolutions can be combined based on the strategies that were introduced in Sect. 4.4.2. For example, for designing low-cost products the strategy **"from the simple to the complicated"** should be considered: **The concept should be put together from the simplest subsolutions that can also be easily combined.** In general, this means that physically, geometrically, and kinematically compatible subsolutions are to be interfaced with each other. This method can prevent the undesirable and costly **overengineering** of a system.

4.5.3 Solution selection: How can the best solution be selected?

From the several proposals for a solution, the best must be selected – a decision must be made. We should prepare methodically for important decisions. The **properties of the alternative solutions** must first be analyzed, which then enables the **solutions** to be **evaluated** with regard to the **defined decision criteria**, thus facilitating a decision.

4.5.3.1 Analysis of product properties

Analyzing the properties of alternative solutions vis-a-vis the requirements (Sect. 4.5.1.1) should be done dispassionately. This is a matter of comparing the **present properties** (the found solutions) with the **desired properties** (the requirements, Fig. 4.5-1). But if this evaluation is done in haste, certain so-called killer phrases such as "That has never worked" or "That is too expensive" often come into the discussion, thus immediately sweeping aside some promising solutions.

We can use four types of analysis methods for technical properties (e. g., interface properties), as well as for costs (corresponding to Fig. 4.5-1):

- Deliberations and discussions with competent persons
- Calculation, optimization, reference number comparison
- Simulation
- Experimentation, rapid prototyping, practical realization of products and processes (during cost analysis the results are usually production times)

Since these are rather well known analysis methods for technical qualities and are not the core subject of this book, here is some information pertinent to **cost analysis**.

- It is essential to **know costs** and lower cost alternatives **before** the decision in favor of a solution is made. For this, **relative statements** from specialists such as production planners, cost consultants, industrial engineers, etc. can provide information like "This solution is about 40% lower in cost than the previous one". The process of cost estimation is discussed in more detail in Sect. 9.2. However, at times an absolute cost value must also be calculated or estimated so that you can verify whether the proposed cost objective is attainable. Quick calculation procedures can be used for this purpose, as described in Sect. 9.3. With in-house production, a production planner's **rough preliminary calculation** can be quite useful, based on memos for the probable job routing with estimated times for manufacturing operations and costs. Also, suppliers can be asked about costs – bids usually come in the fastest!
- **Cost statements** can be highly **uncertain**, according to Fig. 4.8-3 and Fig. 9.1-1, particularly in the early stages of development. This is often offensive to technical people, but it must be considered. It is better to take care of the costs relatively early (the developer's responsibility!) rather than proceed with only technical aims. Exceeding the cost target often renders useless a considerable part of the previous work. Furthermore, there are still the costs of additional changes to be reckoned with (see Fig. 4.2-3). For a discussion of attainable cost accuracy, see Sect. 9.3.7.
- A relatively recent but more dependable cost analysis method is the **preliminary calculation**. In case of single-piece production, nothing much changes after that for schedule reasons. Manufacturing and materials management can only rarely compensate for developmental cost overruns. In **mass production** with its correspondingly extensive product development and higher accuracy requirements of cost calculation, the procedure will run iteratively between product development and preliminary calculations, and will converge to the cost target.

In mass production, production and assembly processes are often also redeveloped and optimized according to cost criteria, in parallel with product development.

4.5.3.2 Evaluation and decision

Evaluation is a company's **assignment of value and weighting** certain product properties but, in fact, the **customers** (users, the market) have the final say. What value they attribute to the product is manifested during use after the sale, unfortunately too late for consideration during the decision-making process. We must consider how we evaluate, usually without direct customer participation, still staying as close to the customer as possible. Since the manufacturers must also let their interests be known, an integrated evaluation becomes difficult.

To reduce these difficulties, here is a series of some **formal evaluation methods**, along with references for further details. We must not let ourselves be deceived by the formalism, however, and trust the numbers too much. These methods only help the evaluation team and the chief decision-maker examine the evaluation closely. Similar to "distributive estimation" (Sect. 9.2), it helps them judge criteria (type of property) more precisely from the customer's viewpoint. Thus, **preparing for a decision** is a learning process, enabling comprehensive judgments on future product implementation. In view of the variety, weighting, and contradiction of properties, and the uncertain customer viewpoint, a necessarily subjective **decision** must be made.

- Using the following **procedure during evaluation** (point evaluation) can reduce the uncertainty:
  1. Compile the evaluation **criteria** (properties), in writing, from viewpoint of the customers, the markets, the manufacturer, and of society. As far as possible, evaluate the technology and the costs together in the team, always referring to the product or project requirements (Fig. 4.5-1).
  2. **Weight** these criteria as needed (use multipliers for a point evaluation).
  3. Analyze the solution alternatives with regard to the criteria, do a **point allocation** (e. g., 4–10 per criterion), and multiply by the weights (see also Fig. 10.1-5).
  4. **Critically compare** the solution alternatives according to their point sum; determine whether a good solution has weak points that can still be reduced.
  5. Decide.
- Some **evaluation methods** [Ehr06, Pah05, Bre93]:
  - Advantage / disadvantage comparison (in the team)
  - Selection list (see Fig. 4.7-6)
  - Simple point evaluation (see above)
  - Weighted point evaluation (see above; see also Fig. 10.1-5)
  - Technical-economical evaluation according to VDI 2225 [VDI97]
  - Economic value analysis [Zan70]

4.5.3.3 Summary of the methodical procedure

The above description shows rather briefly which methods and measures we can employ during the development of a technically satisfactory product, particularly from the cost viewpoint. See the examples in Sect. 4.6 and Chap. 10 for more detailed concrete realization.

Target conflicts complicate the picture. For example, high performance and throughput are required for a typical machine but if at the same time the manufacturing costs must be reduced, a compromise must be found. From industrial practice, an

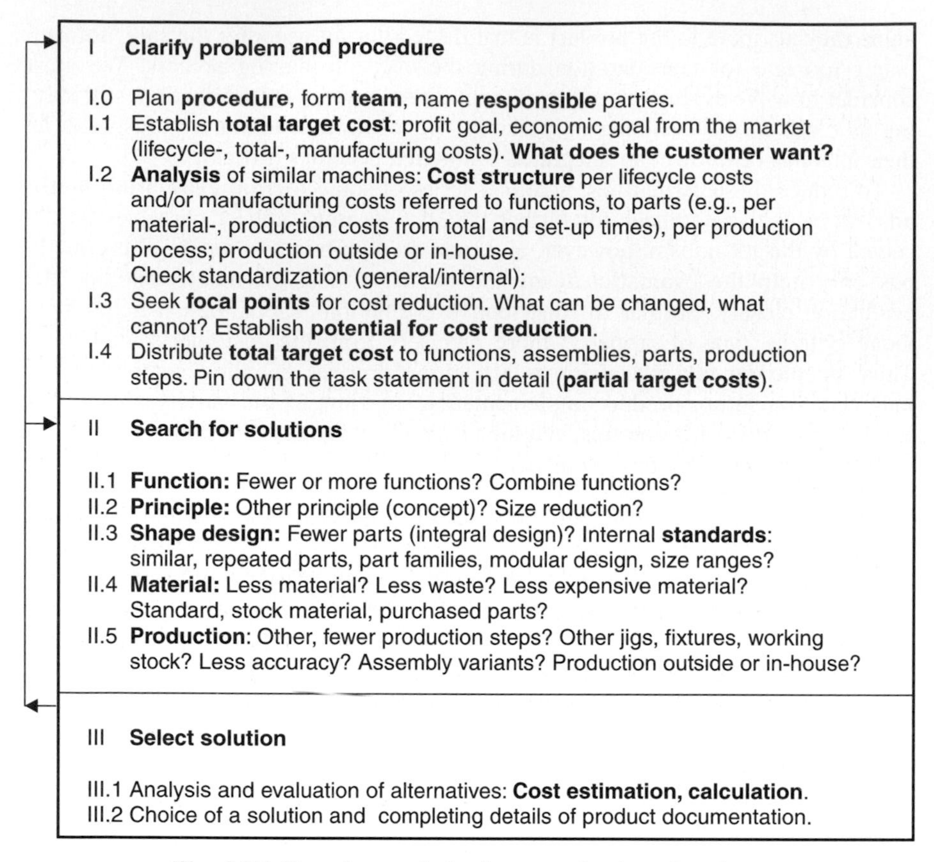

Fig. 4.5-7. Procedure cycle for the cost reduction of products

effective strategy is to make the target conflict known and designate responsible parties for each of the goals [Eil98].

To a practitioner unfamiliar with design methodology, the variety of new terms might seem overwhelming. These methods and benefits are like tools in a toolbox. This person must first become acquainted with these tools and find out how they work, and determine the tasks and situations in which they can be of help. Not all of the tools can be simply applied one after the other!

We know that this flexible "breaking down" can be very difficult and frustrating in a specific case, but it can also lead to happy "aha" experiences. The procedure cycle enhanced for **constructive cost reduction** has proved itself frequently in practice, as shown in **Fig. 4.5-7**. An expanded version is contained in the **Appendix** under **"Guidelines for Cost Reduction"**. The procedure cycle adaptedto lifecycle-based product cost reduction is shown in Fig. 5.3-1.

Referring to Fig. 4.5-7:

- In **Principal Step I** (task clarification), the technical requirements and the total target costs must be established after team building and action planning (Sects. 4.5.1.1 and 4.5.1.3). These should be divided up according to functions, assemblies, and possibly parts or costs for production operations (Sect. 4.5.1.4). Then, search for **focal points** and **potentials for cost reduction** by performing a cost structure analysis of any previous products or similar products and of competing products. After this, the task and the procedure should be carried forward in detail by determining **partial target costs** (Fig. 4.5-4).
- **Principal Step II** is the **search for different solutions** or subsolutions (Sect. 4.5.2). The search for different solutions is necessary since we cannot be sure that the first solution found is the one with the lowest cost. In Principal Step II (with the Sub-steps II.1 through II.5), suggestions for cost reduction are introduced; see, for example, those derived from the cost analyses of 135 products from 42 companies (cf. Fig. 2.3-2, [Ehr78]). More ideas are shown in Fig. 4.6-7.
- The costs of the compiled solutions are determined and/or estimated in **Principal Step III** (cf. Sect. 4.5.3, Chap. 9), most preferably in parallel to the solutions search or, at the latest, immediately afterward. Thus, the lowest cost solution can be selected. If the cost objective is not reached, new clues must be derived from the calculation regarding cost focal points and alternative possibilities. Another solutions search can be performed, as in Principal Step II. The new cost structures are useful for that purpose. Occasionally, in Principal Step I the requirements must be clarified again and coordinated with the customer (Figs. 4.4-1 and 4.4-2).
- Product documentation (drawings, parts lists, etc.) is then developed.

Figure 4.5-7 summarizes the procedure and these hints in a rather condensed form. It must be further refined and adapted in the details, particularly during solution search. This is done, for example, with the basic rules for a low production-cost design in Sect. 7.10.4, to special rules for individual components (Sect. 7.13) and production processes (Sect. 7.11).

4.6 Resources and means for supporting cost management

This book shows methods and measures for cost management and low-cost design. An overview of the helpful measures contained in this book with references to the corresponding chapters is shown in Figs. 4.6-1a and b. In addition, important specific aids are especially elaborated upon in the following sections.

As **Figs. 4.6-1a** and **b** show, there is a whole series of **organization plans, methods and tools** for cost management. Which of these to use and how they are implemented in practice must be tested on an individual basis. They must always be **adapted** to the given situation (Fig. 6.2-3). Furthermore, the employees might need to be **trained** in the application of these aids.

	Name	Refer to chapters and sections
Methods	Benchmarking	Sect. 7.13
	Design to Cost, … to Manufacturing, … to Assembly	Sect. 4.9.1
	Kaizen	Sect. 4.9.1
	Complexity reduction	Sect. 6.2
	Continuous improvement (CI)	Sect. 4.9.1
	Cost management	Especially, Chaps. 4 and 10
	Cost target • Establishment • Divisions	 Sects. 4.5.1.3; 10.1 Sects. 4.5.1.4; 10.1
	Outsourcing	Sects. 6.2.3; 7.10
	Project management • Project cost search • Trend diagram • Bar chart	Sect. 4.8
	Quality and costs	Sect. 7.11.8
	Quality management QFD	Sect. 4.9.1
	Target costing	Sects. 4.4.3; 4.5.1.3; 4.5.1.4; 10.1; 10.2
	Management of variants • Use of repeating parts • Search for similar parts • Feature table • Size ranges and modular design	Sect. 7.12
	Procedure cycle	Sect. 4.4.1
	Value analysis • Value configuration • Value improvement	Sect. 4.9.2
Org. Meas.	Production and cost consultation	Sect. 4.6.1
	Profit center, cost center	Sect. 4.3.2
	Simultaneous Engineering SE	Sects. 4.3; 4.6.1; 4.8.3.1
Tool	Pareto analysis	Sect. 4.6.2; Fig. 4.6-4
	Evaluation procedure	Sects. 4.5.3; 4.5.3.2; Fig. 4.7-6
	Checklists	Sect. 4.6.5
	Conjoint analysis	Sect. 4.5.1.4
	Cost deployment	Sect. 4.5.1.4
	Function costs	Sect. 4.6.2
	Break-even quantities Procedure comparison	Sect. 7.11.1, Fig. 7.11-5; Sect. 10.3
	Competition analyses	Sect. 7.13

Fig. 4.6-1a. Overview of the measures for cost management. Organized according to methods, organizational measures, tools (left column)

	Name	Refer to chapters and sections
Tool	Development Concurrent Calculation, Short Calculation • Estimate • Difference cost calculation • Search calculation, similar part calculation • Weight cost calculation • Material cost method • Design equations • Parametric calculation • Cost growth laws • Computer-aided calculation	Chap. 9 Sect. 9.2 Sect. 9.1.4 Sect. 9.3.1 Sect. 9.3.2.1 Sect. 9.3.2.2 Sect. 9.3.3 Sect. 9.3 Sect. 9.3.5 Sect. 9.4
	Design catalogs	Sect. 4.5.2.1
	Cost structures	Sect. 4.6.2
	Market analyses	Sect. 4.5.1.4
	Morphological matrix	Sect. 4.5.2.5
	Portfolio analysis	Sect. 4.6.6
	Process cost accounting	Sect. 8.4.6
	Rules, trend information	Sect. 4.6.4
	Relative costs	Sect. 4.6.3
	Search for similar parts	Sects. 4.5.2.1; 7.12.3; 7.12.4
	Preliminary / follow-up calculation	Chap. 8

Fig. 4.6-1b. Overview of the measures for cost management organized according to methods, organizational measures and tools

➔ The most **important aids for cost management** are:
- Expert knowledge and knowledge of costs
- Collaboration, organization and motivation
- Systematic procedure

➔ The aids must be adapted to each specific case and to each specific company!

4.6.1 Advising on production and cost in design

One very quick and effective measure for low-cost design is the advising of Engineers working on low-cost designs can benefit greatly from input from colleagues chosen specifically for that purpose. These can be from production planning or, where appropriate, from the costing office, materials management, or value analysis (**"production and cost consultants"**) **Fig. 4.6-2**. An increasing number of

Organizational alternatives

- Advisor is **nominated** and is requested by product development to participate, **as needed**.
- Advisor comes on **agreed-to fixed schedules** to certain development groups, stays otherwise, e.g., in production planning.
- Advisor's **office** is in **product development** and works in production planning only as needed.

Results

- **Reduction of effort** and **time required** for making changes by design, for production and test planning, for work-stock planning, NC-programming, inventory and purchasing;
- **Quality improvement** (design for manufacturing);
- **Omission of additional cost reduction measures** with corresponding efforts for making changes.

Fig. 4.6-2. Organization plans and results of the production and cost advising

companies choose this path. Good design engineers have always consulted with colleagues experienced in production technology [Deb98, Lin93a].

Such a **consultant has the task** of setting up comparison calculations and making cost estimates directly in the design office. The consultant can also control adherence to the cost target jointly with the design engineer and, along with purchasing, can procure bids for purchased parts or outsourcing. The consultant advises about available jigs, fixtures, and tools and can point out resources' bottlenecks. The consultant must have very highly developed professional and interpersonal skills, since this position requires an understanding of the way people think, and special knowledge of product development, production, and industrial management. The consultant must also be able to function as a service person without appearing to be a "know-it-all" [Dun82, Ehr85, Reh87]. It is advisable that the consultants should remain assigned to their home departments (production, industrial management, etc.) so as not to lose their specialized knowledge.

4.6.2 Cost structures

Cost structure is the breakdown of costs into different parts (Fig. 4.6-5). Absolute or relative divisions may be used. Cost structures can be formed according to all cost categories (e. g., lifecycle costs, manufacturing costs, etc.) and according to different points of view (**Fig. 4.6-3**).

Cost structures are intended not only for distributing the cost target (Sect. 4.5.1.4), but are generally of help in finding the essential cost focal points. They save us from "not seeing the forest for the trees". For example, we could be

Organization of cost structures according to

- Costs for requirements (similar to Figure 7.2-1)
- Costs for product properties or features (e.g., costs for low service requirement and low noise; Figure 4.5-1)
- Portions of life cycle costs (Figure 5.1-5)
- Costs for product functions (e.g., mixing, driving, emptying: "function costs") (Figures 4.6-6; 10.1-5)
- Parts classification (Pareto analysis) (Figure 4.6-4)
- Costs for assemblies and parts (Figure 4.6-5)
- Material and production costs (Figures 7.7-3; 7.7-4; 7.13-15)
- Production costs referred to individual processes (Figures 7.7-2; 10.2-2)
- Production costs from set-up and total times (Figures 7.7-4; 7.12-15)
- Fixed and variable costs (Figure 8.5-1)

Fig. 4.6-3. Possibilities for forming cost structures

occupied with the costs of a few dollars for screws while inadvertently ignoring several thousand dollars of materials that could be affected.

As an example, the cost structure of a turbine transmission (Fig. 4.6-5) shows that the housing, gear, and the pinion shaft alone constitute 75% of the manufacturing costs of all the parts (the "A" parts, in the sense of Pareto analysis.) It is therefore more important to strive at first for a low-cost design of the housing, than to be concerned with the shaft seals or the oil piping.

Then, observe what a large portion of the material cost is in the gear and the gear shaft (approximately 45% of the manufacturing costs). From that might come the idea to use case-hardened steel (16 MnCr 5) instead nitrided steel (31 CrMoV 9) for the gear because it costs about the half as much (Fig. 7.9-10).

Cost structures of the most important products should be available at least to the lead designers. In this way, products become "cost transparent". During designing, cost structures are as useful as components and process cycles for recognizing specific clues for cost reduction. They provide the designer with robust and understandable cost information on a long-term basis [DIN87a, Seid93]. They are valid not just for one product, but for a whole series of similar products. Even if the costs change with time, their ratios to each other remain constant over long periods [Ehr85, VDI87, Ehr80b]. By means of cost structures, the cost effects of technical changes can also be estimated more easily and more accurately during design (Sect. 9.2). By comparing cost structures of similar products, differences are recognized that give hints about too-expensive components.

Pareto analysis

Pareto analysis (**Fig. 4.6-4**) is a type of cost structure. With it, parts of the total entity (e. g., the machine parts of a product) are arranged according to one property, such as costs, mass, sales, reliability, etc.), giving rise to three classes: A, B, and C. Class A has the largest amount of the property, class B a middle amount, and class C the smallest. The subdivision occurs by free assessment. The purpose of a Pareto analysis is to find focal points for a specific purpose, such as separating the essential from the nonessential.

Standard values for the classification, according to experience from industrial practice, are:

5% of the parts make up	75% of the costs of the total
20% of the parts make up	20% of the costs of the total
75% of the parts make up	5% of the costs of the total

In individual cases, there may be deviations from these classifications. The simplified 80/20 rule (also known as the Pareto rule) can also be applied (i. e., 80% of the costs arise from 20% of the parts). Figure 4.6-4 shows a typical Pareto analysis of the products of a company according to sales and profit. We recognize from the sales analysis that only three products, **a**, **b**, and **c**, comprise the largest portion of the sales. From the profit analysis, we realize that the profits from the products **c**, **e**, and **b** dominate the picture.

Figure 4.6-5 shows a similar Pareto classification. We can see that only about 16% of the parts ("A" parts: housing, gear, and pinion shaft) comprise approximately 75% of the manufacturing costs. Thus, cost-cutting measures would begin

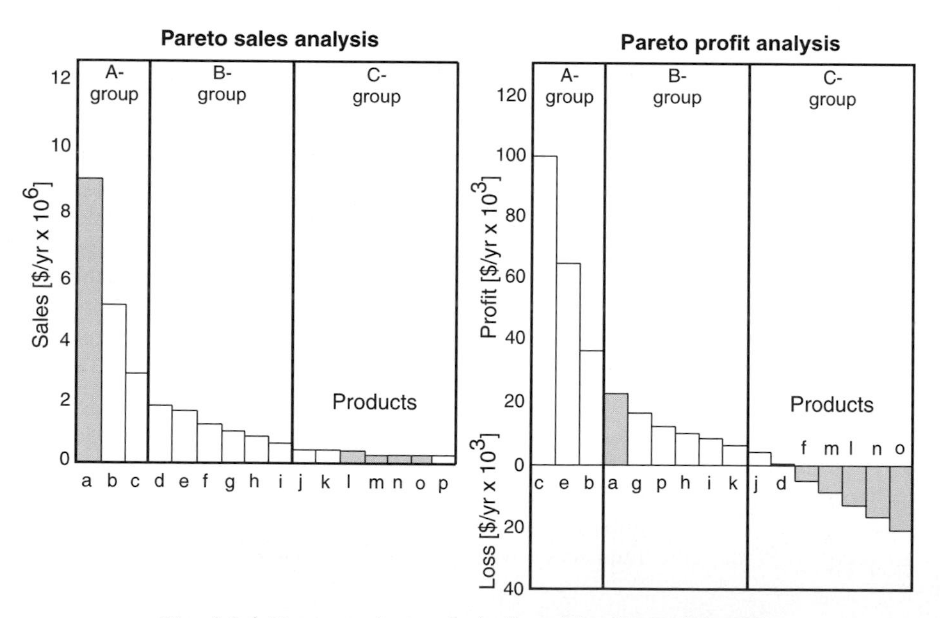

Fig. 4.6-4. Pareto product analysis (from Meerkamm [Mee89])

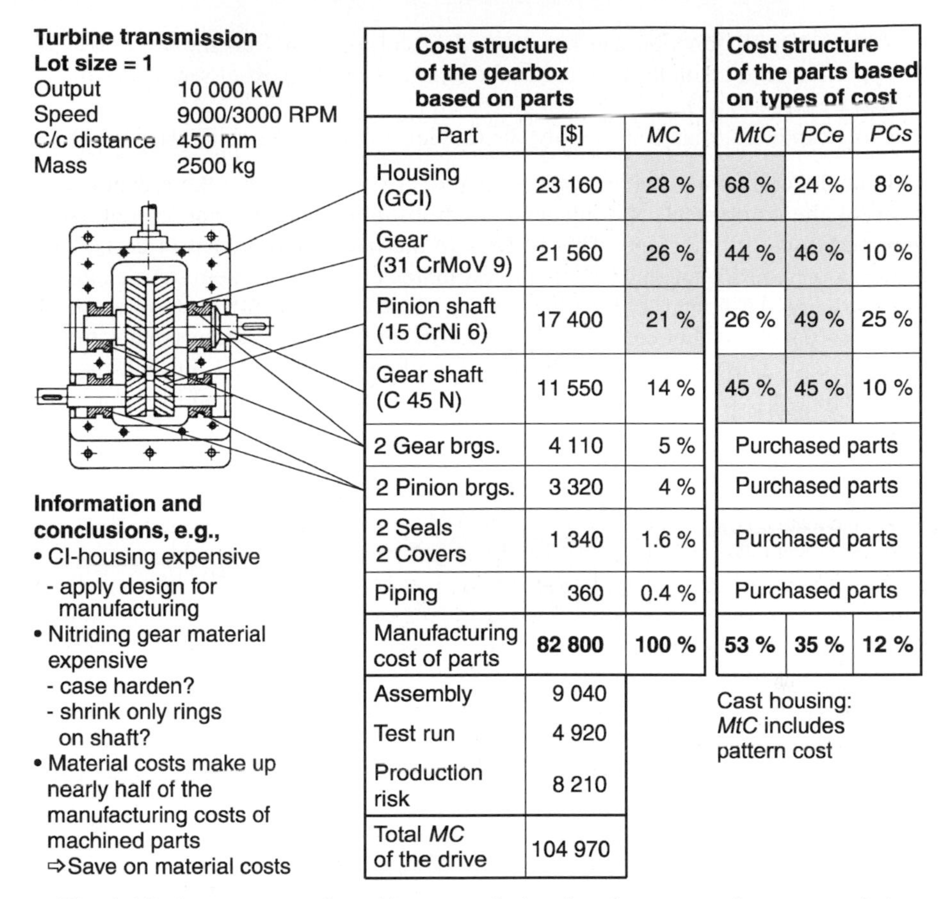

Cost structure of the gearbox based on parts			Cost structure of the parts based on types of cost		
Part	[$]	*MC*	*MtC*	*PCe*	*PCs*
Housing (GCI)	23 160	28 %	68 %	24 %	8 %
Gear (31 CrMoV 9)	21 560	26 %	44 %	46 %	10 %
Pinion shaft (15 CrNi 6)	17 400	21 %	26 %	49 %	25 %
Gear shaft (C 45 N)	11 550	14 %	45 %	45 %	10 %
2 Gear brgs.	4 110	5 %	Purchased parts		
2 Pinion brgs.	3 320	4 %	Purchased parts		
2 Seals 2 Covers	1 340	1.6 %	Purchased parts		
Piping	360	0.4 %	Purchased parts		
Manufacturing cost of parts	**82 800**	**100 %**	**53 %**	**35 %**	**12 %**
Assembly	9 040				
Test run	4 920				
Production risk	8 210				
Total *MC* of the drive	104 970				

Fig. 4.6-5. Cost structure of a turbine transmission (based on parts and cost categories)

with these parts. Approximately 56% of the parts are "C" parts and therefore are of lesser importance, comprising approximately 2% of the part manufacturing costs. Section 9.3.7 will demonstrate how Pareto classification affects the cost computation accuracy (see also Fig. 10.1-12).

Function costs

Function costs are the costs necessary to realize a function [VDI87]. The purposes include:

- Solution variants can be compared concerning their costs per subfunction, not just for components and assemblies (Fig. 7.11-44). This also facilitates the comparison of competing products (Sect. 7.13).
- Subsolutions with the most favorable function costs can be systematically combined.

- The costs of a function can be assessed in comparison with the value that is apportioned to a similar function in the market (see the cost of secondary functions, Sect. 4.5.1.4).
- Costs can be estimated at a highly abstract level (i. e., while conceptualizing).

Function costs are based on function analysis (Sect. 4.5.1.2) of the product. The costs of the components sharing function fulfillment (the function carriers) are assigned to the subfunctions. This is not a to-the-penny assignment of costs, but, rather, a method for making reasonable estimates of important items. An example is shown **Fig. 4.6-6**. Parts of the transmission shown in Fig. 4.6-5 are here assigned to five subfunctions. For components that fulfill several subfunctions (shafts, housing), the costs must be divided up and functionally justified here, using an estimate. The calculation can be carried out very easily with a spreadsheet and the numbers

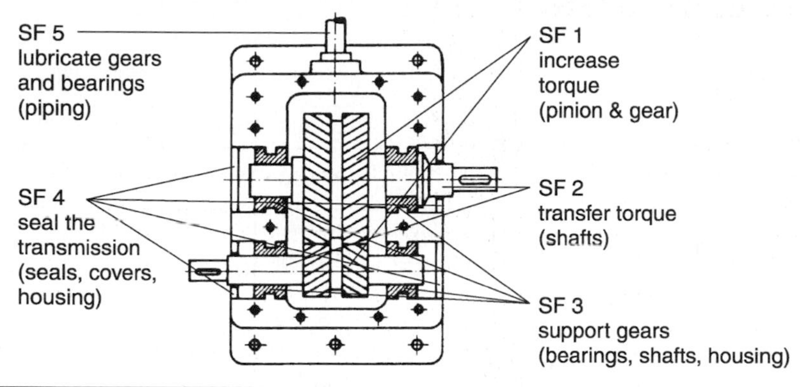

		Subfunctions									
		Portions					Costs [$]				
		SF1	SF2	SF3	SF4	SF5	SF1	SF2	SF3	SF4	SF5
Part	Costs [$]	Magnify torque	Transfer torque	Supprt. gears	Seal trans.	Lubr. trans.	Magnify torque	Transfer torque	Supprt. gears	Seal trans.	Lubr. trans.
Cast housing	23 160			60 %	40 %				13 896	9 264	
Gear	21 560	100 %					21 560				
Pinion; shaft	17 400	70 %	15 %	15 %			12 180	2 610	2 610		
Gear shaft	11 550		50 %	50 %				5 775	5 775		
2 Gear brgs.	4 110			100 %					4 110		
2 Pinion brgs.	3 320			100 %					3 320		
2 Seals 2 Covers	1 340				100 %					1 340	
Piping	360					100 %					360
	82 800						**33 740**	**8 385**	**29 711**	**10 604**	**360**
							40.7 %	**10.1 %**	**35.9 %**	**12.8 %**	**0.4 %**

Fig. 4.6-6. Function structure of a turbine transmission

can be varied. The parts are shown in percentages in the left portion of the table; these are multiplied in the right portion by the part costs of Column 2 and the function costs added up on the right at the bottom. We see that increasing the torque – most important subfunction SF1 – implemented by the gears incurred only 40.7% of the manufacturing costs. Supporting the gears and shafts – subfunction SF3 – which the customer does not really value, costs nearly the same, 35.9%. Cost-cutting measures should therefore be applied here. One should consider whether a bearing's function could be taken over by a coupled machine (this is integrated design, often implemented in practice).

Function costs are an essential approach of value analysis [Mil87, VDI95]. They can also be combined with the QFD method [Aka92, Dan96] that is based on the consideration of functions.

It should be observed that **function costs are determined on basis of available concrete solutions** (Fig. 7.11-44). They **cannot be** stated **for functions** for which **no solutions** and **costs** exist.

4.6.3 Relative costs

Relative costs have been used for a long time as aid in low-cost design [Les64]. The VDI Guideline 2225, Sheet 2 [VDI77] was published in 1964, the first catalog of relative costs for materials (Fig. 7.9-10). Before 1970, some German and Swiss companies formed a Relative Costs Working Group [Bus83]. In DIN 32990 and DIN 32992, the terms processing and presentation of relative costs are standardized [DIN87b, DIN89a–b, DIN93]. In [DIN87b] the results of a research association program for the processing of fundamentals of relative costs [Bei82a, Ehr83, Spu82] are summarized. A definitive investigation by RKW [RKW90] stated that, other than the fundamentals, it is not reasonable to produce generally binding documents for cost information. They must unfortunately be generated anew specifically for each company (Sec 7.11.5.5b). In Japanese target costing literature, these are called Cost Tables [Gle96, Sch98a, Seid93, Tan89].

Relative costs are developed when the costs of products in size ranges, materials, manufacturing process, etc. are related to some basic cost.

Figure 7.9-10 shows an excerpt from VDI 2225, Sheet 2 [VDI77]. A catalog for relative costs of material is given in that guideline. The cost of round steel stock, 35–100 mm diameter (USt 37-2, DIN 1013), 1 000 kg lot size, are employed here as the basis or reference. The costs of the other materials are divided by the cost of the basis material. Relative material costs (on volume basis, C_V^*) are shown. If the costs of other materials change proportionally to the costs of the basis material, the relative costs remain unchanged. Therefore, there is the advantage that it is **rarely necessary to update** the data. Furthermore, the **numbers** remain relatively constant over the long term (e. g., in the **range of 0.5–30**). Nevertheless, the material relative costs must be generated in-house and checked or updated annually [Ehr80b].

Figure 7.11-28 shows relative costs of work produced by drilling and boring machines at the Voith Company. We see that the operating costs increase progressively with changing dimensions and tighter tolerances. Here the **problematic nature** of **all numerical aids for low-cost design** is again evident. This data is valid only for the Voith Company's products and manufacturing processes (single-unit production of large water turbines [Bus83]). It cannot simply be applied in other companies (e. g., those, which make small products, or use series production, etc.). For accuracy, see also Sect. 9.3.7.

Relative costs have the following **advantages**:

- The relative cost numbers are **easy to distinguish**, since they have few digits.
- Relative costs **change less with time** than do absolute costs, especially when the reference object is chosen such that the costs of the solution variants change in same manner.
- With relative cost information, there are **fewer problems regarding secrecy**. Thus, much cost information in this book is to be understood to consist only of relative values. All company data presented here have been altered and certain limiting conditions are omitted. This is why readers cannot directly apply this data to their own circumstances.
- If relative costs are integrated **into the factory standards**, a **preliminary selection** of expected low-cost solutions may be achieved.

> ➔ Relative costs help guide the solution search and enable a quick solution selection.
>
> ➔ Relative costs are specific to the company and field.
>
> ➔ For cost management (attaining and verifying the cost objective) current cost data of the company must be used.

4.6.4 Rules

We are usually not conscious of rules, but they encompass the majority of our experiences, determine our actions, and influence goal setting. Because rules are formed from constructive experiences, successes, failures, and complaints, they are the most helpful measures used in designing. Rules are not always valid, however, since they associate only a few parameters from a complex relationship.

For low-cost design and design for manufacturing (DFM), rules for different manufacturing processes are given and described in Chap. 7. They must be modified, however, for a **company-specific use**.

Rules are valid only under specific limiting conditions, which are usually not clear. This is shown in Fig. 7.3-2 for a comparison of costs of spur gear sets, for a speed ratio of 1:10. When large torques are transferred (large spur gear sets), the costs of materials dominate so that the multiple-path gear assemblies containing more parts become more economical in spite of the higher number of parts. On the other hand, for smaller torques, the two-stage gear set is more economical because the production costs are dominant, and fewer parts are more economical. Furthermore, the **validity period** must be considered. Through new manufacturing processes (e. g., high-speed milling machines), the DFM rules for the design of milled parts also change. The application of rules must therefore always be critically checked.

In spite of these restrictions, the need for and benefit of rules must be stressed. During single-unit production, not every "small" design decision, (e. g., the establishment of every dimension or form) can be calculated exactly. It must however be decided upon quickly, based on the product developer's knowledge (Sect. 4.8.2, Fig. 4.8-1, Case A, Sect. 7.10.4).

4.6.5 Checklists

Checklists are another form of rule representation (**Fig. 4.6-7**). In checklists, general experiences, steps, etc., can be compiled, so that nothing is omitted. These lists can support market analysis and, at the beginning of the design work, help stimulate thinking. They can also be used systematically at the end to check the design (Figs. 4.5-1 and 10.1-2). All the methods described here achieve the best effect when they are attuned to the specific needs (products, materials, production facilities, etc.) of the given company.

Figure 4.6-7 shows a checklist [Hei93a] that details points II.1, II.4, and II.5 from Fig. 4.5-7. Thus, designs can be analyzed at the beginning of the design work

Function

Are the functions of the part/assembly clear?
Is the function realization unambiguous, simple and safe?
Can functions be integrated into another part?
Can the functions be separated into several parts?
Are the material and production expense justified for the function realization?

Material

Are lower-cost raw material or purchased parts obtainable?
Can another lower-cost material be utilized?
Can standard parts (modular design) be used?
Can the unfinished part be made from another semi-finished material?
Can wastage/scrap be reduced by form design?
Can the unfinished part be cast, forged, sintered or made from sheet-metal?
Can the raw- or semi-finished material be pre-processed by the supplier?

Production

Is there in-house expertise for the production technology?
Is the part within the company's spectrum of parts?
Must the part be produced in-house?
Are the production times justified?
Is the sequence of work operations optimal?
Is the production cheaper on other machines?
Are other processes feasible for material removal, surface treatment, joining and assembly?
Do all finished surfaces serve for function fulfilment?
Must all active surfaces be finished?
Are lower surface quality and looser tolerances acceptable?
Can different dimensions be made more uniform?

Fig. 4.6-7. Checklist for cost reduction regarding function, material, production [Hei93a]

and after the preparation of the drawings, according to the potentials for cost reduction. Obviously, these points can also be used with the solution search, or the analysis of available designs (Sect. 4.5.1).

> ➔ The most important checklist is the requirements list!

4.6.6 Portfolio analysis

Portfolio analysis is used for discovering action possibilities in early stages of product development, or during the task clarification, if exact data or "soft" (numerically indefinite) numbers are still not available. **Figure 4.6-8** shows a portfolio analysis for cost cutting measures in a product program for products "**a**" through "**f**", from Fig. 4.6-4. The cost pressure to be expected from the competition, about the expected volume of turnover in each case, is applied in three classes – low, medium, and high. We recognize immediately that cost-cutting measures are most urgently needed for product "**a**".

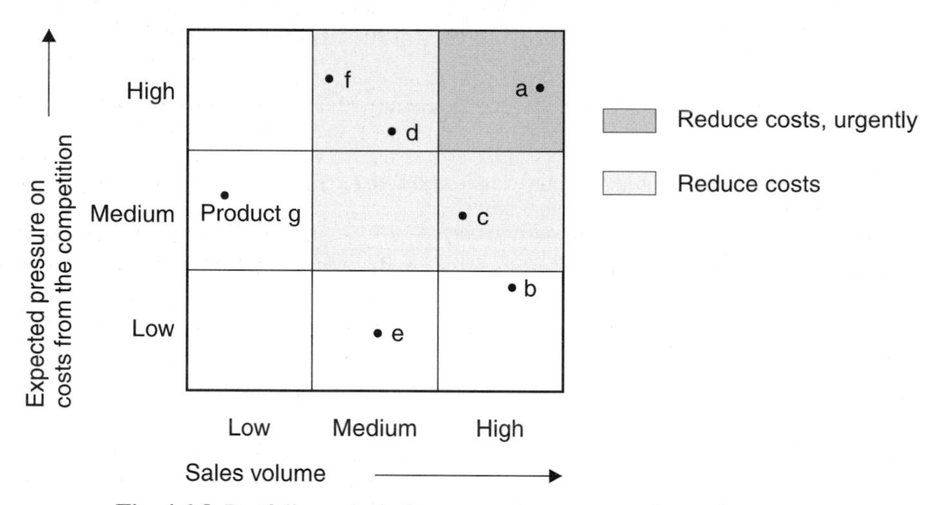

Fig. 4.6-8. Portfolio analysis for cost cutting measures for products a to g

4.7 An example of the methodical procedure: A marking laser

4.7.1 Task clarification

Here we will continue the discussion begun in Sect. 4.5, using an example from industrial practice to elaborate on the basic methodical procedure. We want to clearly demonstrate how to gain systematically **economic advantages** by a flexible application of **target costing** and **design methodology** techniques. However,

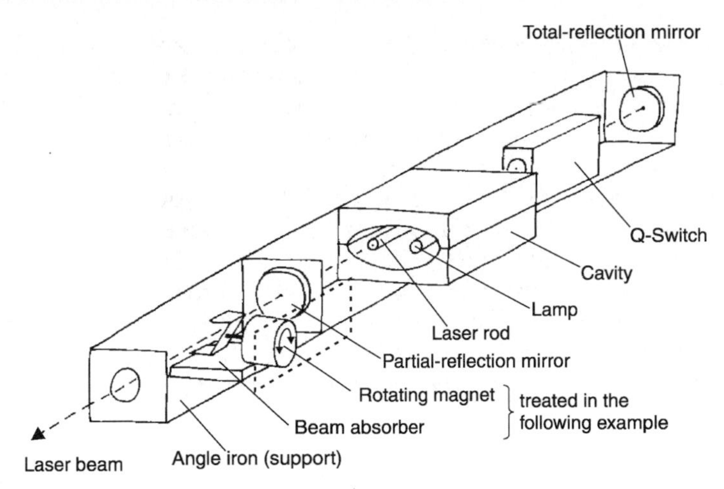

Fig. 4.7-1. Basic construction of the resonator of a marking laser

the design process must not be forced into a straitjacket that hinders rather than helps product development. This means that it is left up to the designers to individually decide on which portions of the methods they use and which methodical steps might be skipped. This presupposes a detailed knowledge and experience in methodical working, although this is unfortunately not the case.

Our **example** deals with an industrially produced NdYAG solid-state laser with approximately 80 W output, which is commonly used for lettering and marking work. **Figure 4.7-1** shows the basic construction of the laser resonator. The manufacturing cost of the device needs to be lowered. The optical laser lamp supplies energy to the optical laser rod, in which the actual laser beam is generated. Both components are put into the so-called cavity that is chilled with water to dissipate the large amount of heat generated. Between the two mirrors, coherent laser light is produced that passes through the partial-reflecting mirror and then out of the resonator. With the aid of the Q-switch, the laser beam can be switched on and off at high frequency. In addition to the Q-switch, the shutter and beam absorber form a safety system that can interrupt the laser beam. All optical components are mounted on an angle iron that ensures their alignment on the optical axis. The whole system is encased in an integral plastic housing (not shown in Fig. 4.7-1) that carries both the water and power supplies.

With this cost reduction project, the technical requirements were established (the requirements list is not shown here) and the total target costs were determined for the system shown in Fig. 4.7-1. The need for a 23% reduction of the laser manufacturing costs resulted from an analysis of the wishes of potential customers, from the competing products, and the company's previous product (from which the total target costs were derived. The target costs were divided into partial target costs based mainly on the cost reduction potential of the individual assemblies recognized during the analysis of the company's previous product.

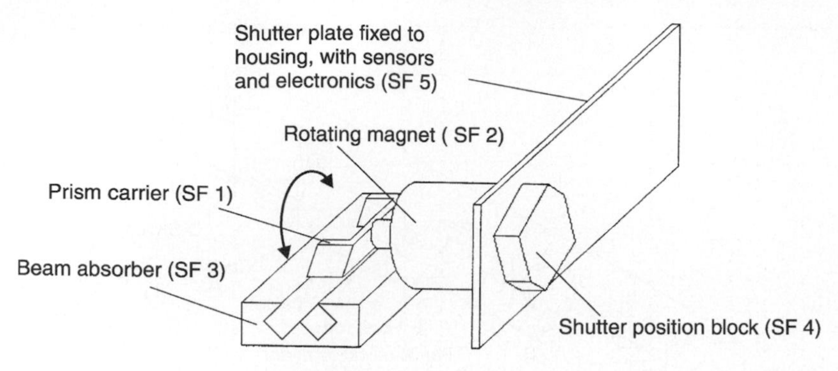

Part	Subfunction	Manufacturing cost	
Prism carrier	SF 1: Control laser beam	25 $	15 %
Rotating magnet	SF 2: Position change of deflection fixture	50 $	30 %
Beam absorber	SF 3: Receive beam energy	35 $	21 %
Shutter position block	SF 4: Locate position sensors	37 $	22 %
Plate with sensors and electronics	SF 5: Check position of rotating magnet (shock-free)	20 $	12 %
Total		167 $	100 %

Fig. 4.7-2. Construction, subfunctions and cost structure of the shutter unit

We want to concentrate further on the shutter subsystem with beam absorber [Web97]. For the subsystem shutter, a **25% cost reduction** (partial target cost) was assumed. The shutter is a safety system in the resonator that enables control of the laser beam at high frequency, and it also catches the beam in the beam absorber. **Figure 4.7-2** shows the basic construction of the shutter unit and the distribution of the actual costs to the individual subfunctions analyzed by the team.

4.7.2 Solution search

In order to attain the established **partial target costs** of the laser shutter, this unit had to be totally reworked. After the requirements list was defined (which differed only slightly from that of the previous product, except for the target costs), a **function analysis of the existing system** was carried out. The **subfunctions** to be fulfilled were assigned to the components of the system, arranged in a list.

The subfunctions identified through function analysis formed the basis for the search for **physical effects** for their fulfillment. For the subfunctions "**SF1: control laser beam**" and "**SF3: receive beam energy**", no genuine alternatives to produce the desired physical effects were found (i. e., reflection and absorption). On the other hand, for the subfunction "**SF5: position check**", a whole series of physical alternatives were discovered; these are depicted in **Fig. 4.7-3**.

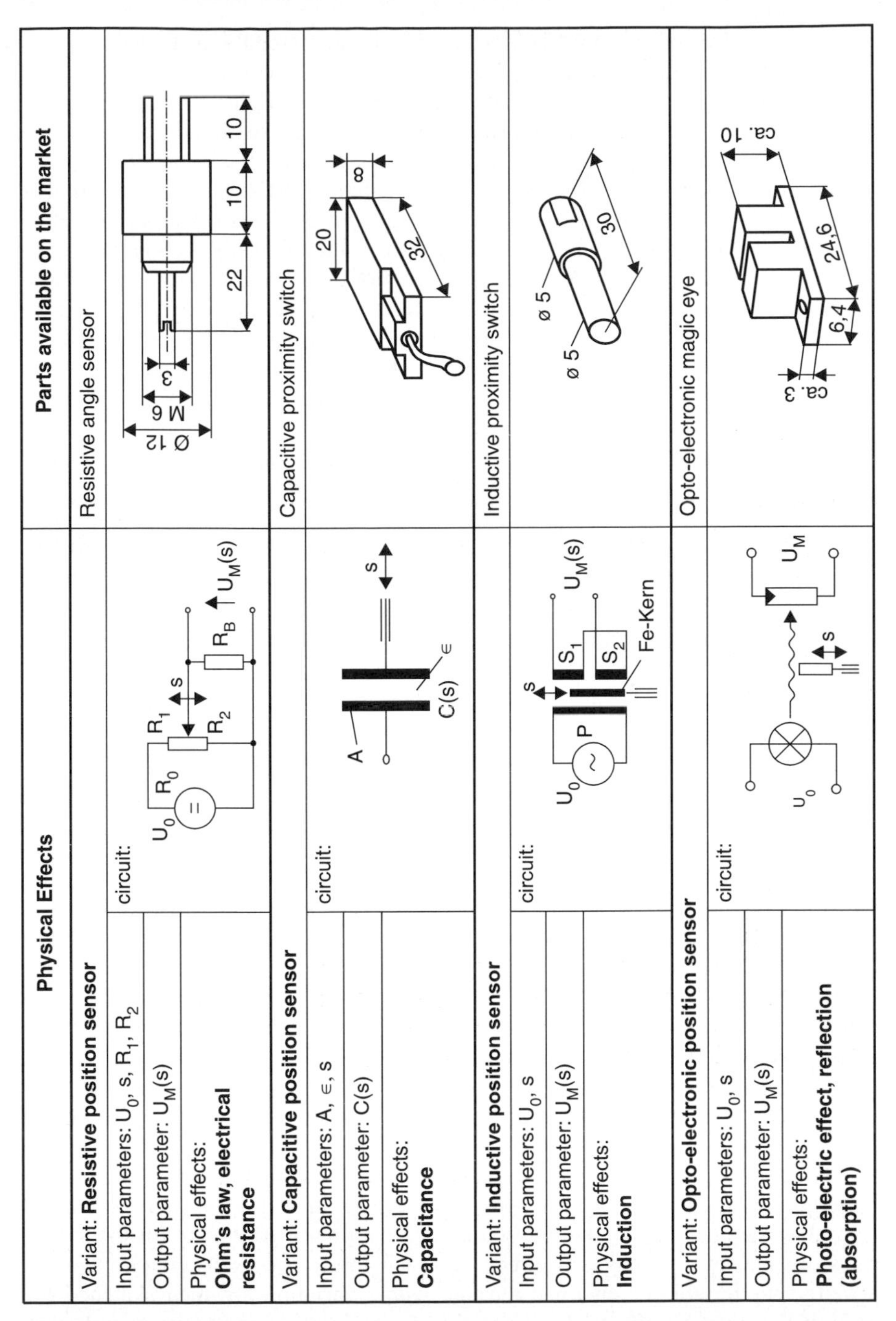

Physical Effects	Parts available on the market
Variant: **Resistive position sensor**	Resistive angle sensor
Input parameters: U_0, s, R_1, R_2 Output parameter: $U_M(s)$ Physical effects: **Ohm's law, electrical resistance**	
Variant: **Capacitive position sensor**	Capacitive proximity switch
Input parameters: A, ∈, s Output parameter: C(s) Physical effects: **Capacitance**	
Variant: **Inductive position sensor**	Inductive proximity switch
Input parameters: U_0, s Output parameter: $U_M(s)$ Physical effects: **Induction**	
Variant: **Opto-electronic position sensor**	Opto-electronic magic eye
Input parameters: U_0, s Output parameter: $U_M(s)$ Physical effects: **Photo-electric effect, reflection (absorption)**	

Fig. 4.7-3. Physical effects and commercial solutions for the fulfillment of the subfunction SF5 'position check'

After the physical possibilities were clarified commercially available solutions were searched, since electronic components such as sensors must be bought. A component that fulfills the required subfunctions (Fig. 4.7-3, right) for every physical effect was soon found. It turned out that all of the solutions found and combined made the shutter position block superfluous, which comprised 22% of the manufacturing costs. Upon further processing, the subfunction "**SF4: locate position sensors**" was also eliminated due to this function integration. The evaluation and selection of one of these solution options are dealt with in the following section.

Without basic changes to the overall concept of the laser resonator, and without introducing changes not desired by the customer, the subfunction "**SF3: receive beam energy**" could up to now be realized only through the absorption of the optical output. Since the water-cooled beam absorber plate comprised 21% of the manufacturing costs of the assembly, an especially intensive search for cost reduction through variation of shape was undertaken (Fig. 4.5-6).

Figures 4.7-4 and **4.7-5** show the initial solution in which the laser beam is caught in a beam trap formed from two grooves offset at 90 degrees to each other, and is reflected repeatedly until its entire energy is dissipated. Two diagonally bored holes, which must meet at the top, cool the absorber plate. This is difficult to produce and is therefore expensive. In the first shape variation step, **Variant 1**, the plate was made thicker (variational feature: size) and the number of holes increased (variational feature: number) to avoid the need for diagonal bores. In **Variant 2**, the grooves acting as a beam trap were replaced by a single bore that is easy to produce (variational feature: form, and manufacturing process); the laser beam travels around in this bore until its complete absorption. Finally, in **Variant 3** the location of the hole and the cooling channels relative to each other was changed (variational feature: location), which permitted a more compact design and better heat dissipation.

Besides a pneumatic cylinder for the efficient fulfillment of the subfunction "**SF2: position change of deflection fixture**", the physical effect chain consisting of an electromagnet and an elastic spring was the only remaining choice. A quick search for purchasable solutions produced a simple solenoid-spring combination that was lower in cost, by a factor of five, than the rotary magnet that had been previously used.

A solenoid was used for the realization of "moving the deflection device" but, since it cannot rotate, it had to be moved axially into the beam path of the laser. During the solution search for the subfunction "**SF1: control laser beam**", design of the existing prism carrier was changed. The change was to convert the rotational movement into translational (variational feature: type of motion) and to simultaneously simplify its manufacturing (cf. Figs. 4.7-2 and 4.7-7).

All **compiled solutions** were assigned to the corresponding subfunctions in a **morphological matrix**. The morphological matrix in Fig. 4.7-7 is distinguished from the usual matrices in that the solutions are represented, except for the lowest rows, on the level of physical effects. In each lower row, subsolutions that are actually purchasable or obtained through shape variation are then sketched for each of the physical effects. Based on this synopsis, the lowest-cost overall concept could now be developed by evaluating the different subsolutions for the shutter assembly. The procedure is shown in the following.

Initial variants of the beam absorber plate

Weak points:
- Cooling channels improperly shaped: uncertain intersection point; clamping fixture necessary; questionable if anodizing layer present on the surface to be drilled; drill size too small, danger of tool breakage.
- Number of operations too large for changing tools.
- Whole geometry too complicated.

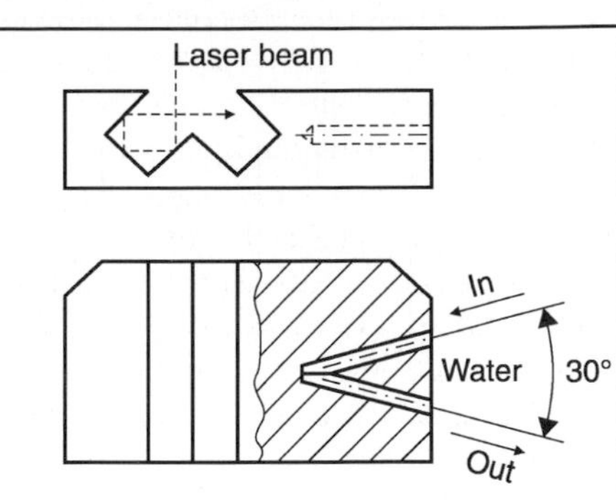

Variant 1

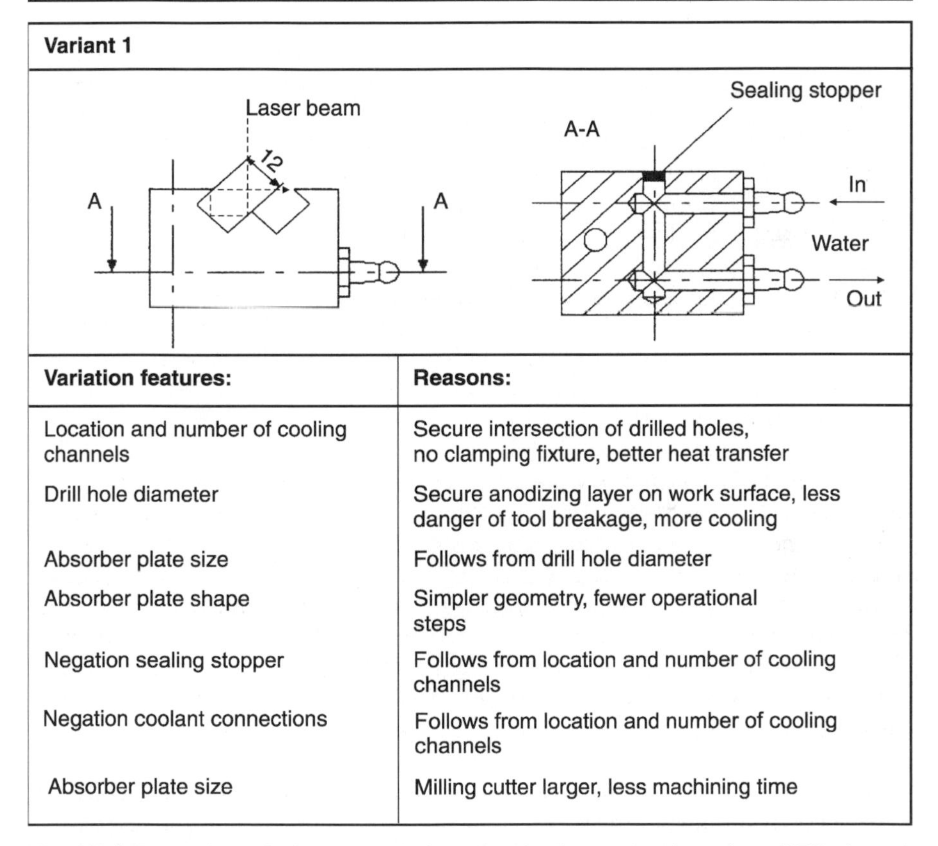

Variation features:	Reasons:
Location and number of cooling channels	Secure intersection of drilled holes, no clamping fixture, better heat transfer
Drill hole diameter	Secure anodizing layer on work surface, less danger of tool breakage, more cooling
Absorber plate size	Follows from drill hole diameter
Absorber plate shape	Simpler geometry, fewer operational steps
Negation sealing stopper	Follows from location and number of cooling channels
Negation coolant connections	Follows from location and number of cooling channels
Absorber plate size	Milling cutter larger, less machining time

Fig. 4.7-4. Processing of alternate solutions for the beam absorber plate (SF3) through variation of shape

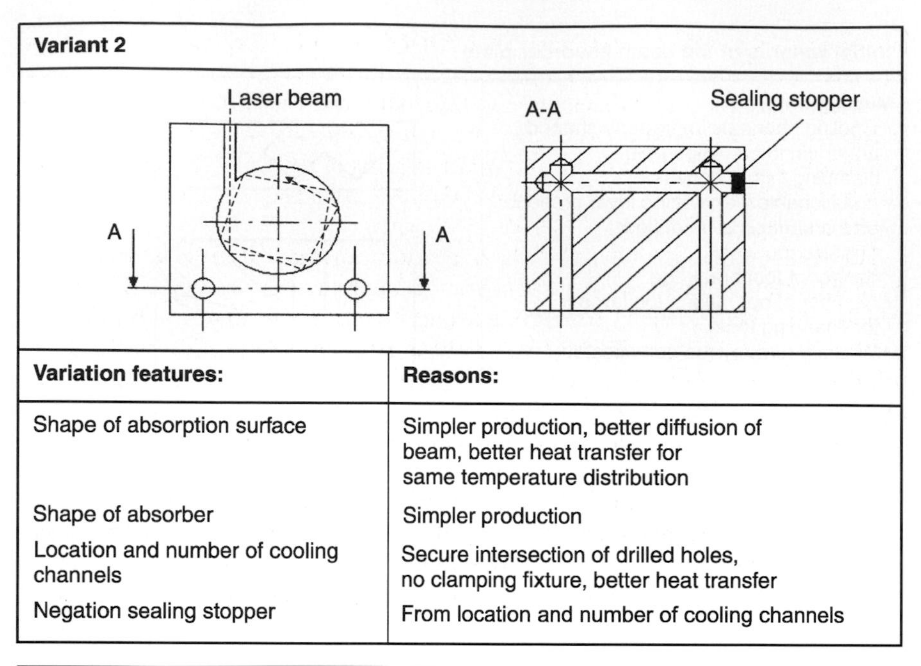

Variation features:	Reasons:
Shape of absorption surface	Simpler production, better diffusion of beam, better heat transfer for same temperature distribution
Shape of absorber	Simpler production
Location and number of cooling channels	Secure intersection of drilled holes, no clamping fixture, better heat transfer
Negation sealing stopper	From location and number of cooling channels

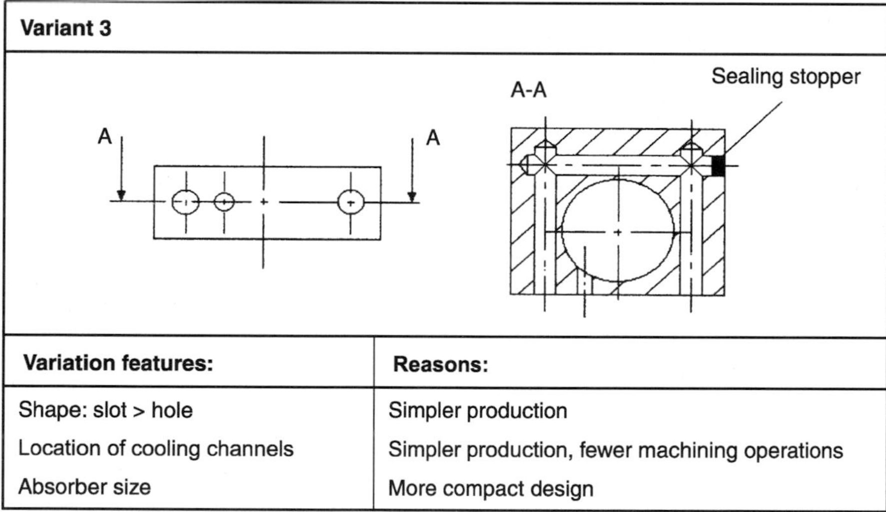

Variation features:	Reasons:
Shape: slot > hole	Simpler production
Location of cooling channels	Simpler production, fewer machining operations
Absorber size	More compact design

Fig. 4.7-5. Processing of alternate solutions for the beam absorber plate (SF 3) through variation of shape

4.7.3 Solution selection

Section 4.7.2 described how different subsolutions for the fulfillment of the subfunctions SF 1 through SF 5 were compiled methodically, which were to be realized in the shutter unit. The solutions for every subfunction were then **analyzed** and **evaluated** to select the **subsolutions most favorable** with regard to **function** and **cost**. Only after that could a new **concept** for the entire assembly be developed. In this case, this procedure was possible since the different subsolutions could be combined without problems. The compatibility of individual subsolutions with each other must often be tested first in systems in which the interaction of the different subfunctions is more difficult. Then the possible solution combinations can be analyzed and evaluated as overall concepts.

For realizing the subfunction "**SF 2: position change of deflection fixture**", three different solutions were chosen from the morphological matrix (Fig. 4.7-7, middle): the rotary magnet from the existing design, a solenoid, and a miniature compressed-air cylinder. In the first **analysis step**, the **properties of the three solutions** were determined from the manufacturer catalogs, referring to established criteria in **Fig. 4.7-6**. Soon it became clear that the solenoid with a gap would represent the lowest-cost solution. The computational simulation of the time performance of the three solutions (how fast can the prism carrier be moved into the beam?) was delayed due to the expected computational expense, until there was a provisional evaluation with the aid of a **selection list**. In this list, a subsolution's satisfactory fulfillment of a criterion was indicated with a "+", and non-fulfillment with a "–". Missing information was indicated with a questionmark.

Analysis of the properties of different variants:

Evaluation criteria	Solenoid	Miniature cylinder	Rotating magnet
Price, $	12.50	22.50	60.00
Energy type	electrical energy	compressed air	electrical energy
Size, mm	22 x 25 x 32	ø 8 x 35	ø 25 x 25
Time response	?	?	?
Life span	very good	good	very good
Assembly properties	very good	good	very good
Disturbance safety	very good	good	very good

Evaluation of the properties of different variants:

Evaluation criteria	Solenoid	Miniature cylinder	Rotating magnet
Price	+	+	-
Energy type	+	-	+
Size, mm	+	+	+
Time response	?	?	?
Life span	+	+	+
Assembly properties	+	+	+
Disturbance safety	+	+	+
Overall value	?	-	-

Fig. 4.7-6. Evaluation of solutions for the subfunction 2 "position change of deflection fixture", with a selection list

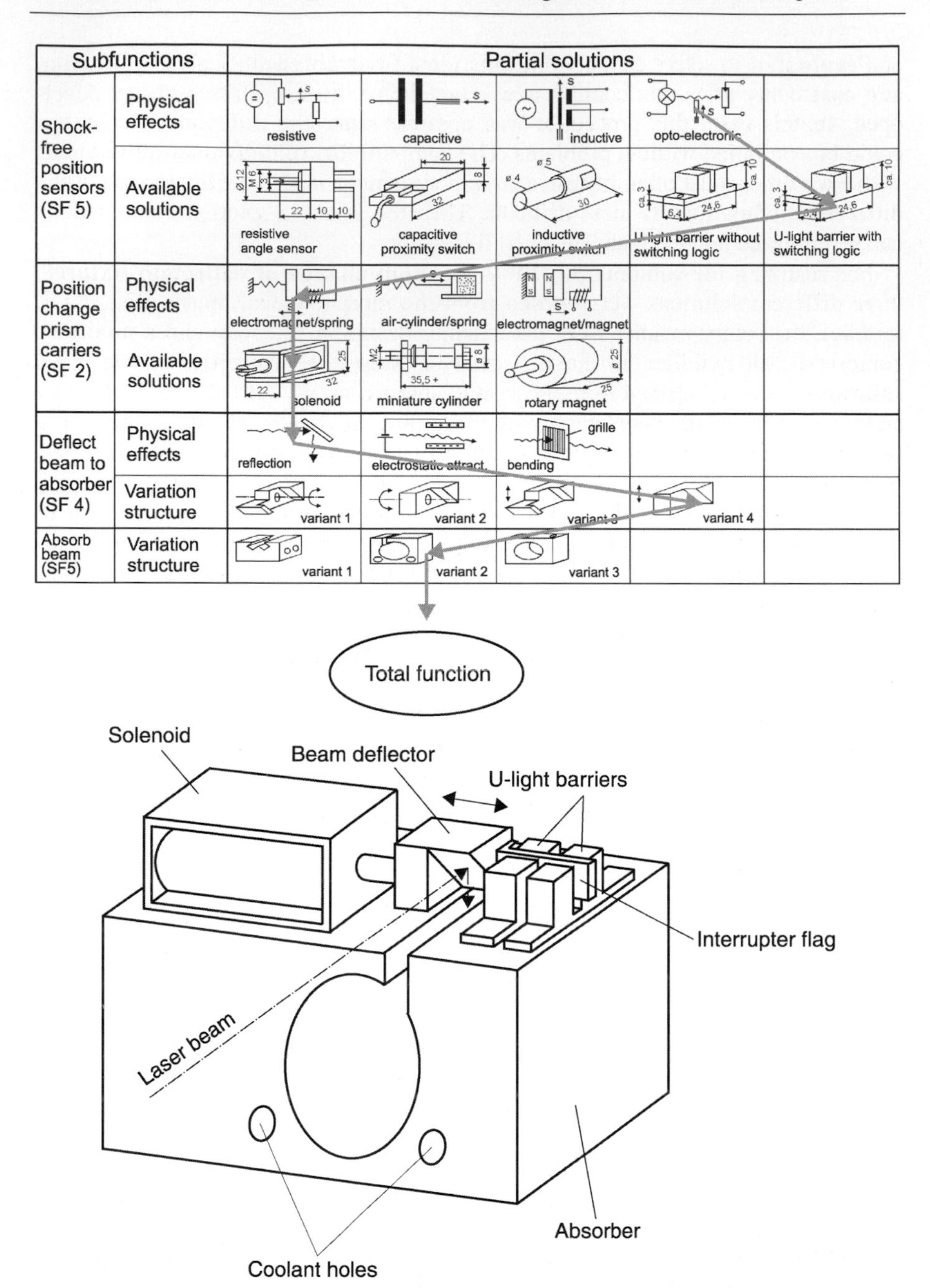

Fig. 4.7-7. Concept development from the morphological matrix

If a subsolution could not fulfill an evaluation criterion, it was considered totally inadequate. The miniature cylinder and the rotary magnet were ruled out as solutions for cost reasons, regardless of the unanswered question about their time performance. On the other hand, the suitability of the solenoid depended exclusively on this question. Analysis and **calculation** of the stroke with prism carrier mounted verified that the maximum 10 ms for the interruption process demanded in the requirements list would be met.

In selecting a solution for realization of the subfunction "**SF 5: position check**" it became clear that not all criteria had the same importance within the overall system. For this reason the position sensors were selected with the aid of use-value analysis, not shown here. The magic eye with integrated switching logic emerged as the most favorable solution. In this case, the expense for the fulfillment of the subfunction "**SF 4: locate position sensors**" was reduced to a simple mounting option. The expensive shutter position block component could probably drop out of the picture.

The prism carrier to fulfill the subfunction "**SF 1: control laser beam**" was optimized for manufacturability by shape variation. A thorough analysis of the qualities of competing variants was not considered necessary due to the component's simplicity.

For the same reason, a strict evaluation was omitted during the selection of the subsolution for the subfunction "**SF 3: receive beam energy**". Here, the reliable function of the favored variant was only checked with the aid of a simple experiment (cf. [Ehr06]).

After selecting the most favorable subsolution for every defined subfunction, the basic elements of the concept were established. The combination leading to a concept (**Fig. 4.7-7**, bottom) did not present any more problems. A preliminary rough **cost estimate** based on this concept showed that the cost of the entire unit would be lowered by approximately 30%.

Even if this value could not be held to after detailed design and more precise cost calculations of the concept, the targeted cost reduction of 25% for the shutter assembly could still be slightly exceeded.

4.8 The practice of cost management

4.8.1 Introduction to cost management

Almost everyone in a company has concerns in the introduction of a complex thought and action system due to inner, personal attitude or aim, organization, methods and aids. It is not a measure that can be rationally controlled, as, for example, the purchase and installation of a networked computer system would be. **It deals with the people involved**, whose behavior can be rationally understood only to a limited extent. That is why it is often difficult to use objective facts and arguments to convince people to accept something new. Human beings are often driven by unconscious feelings and self-esteem issues.

Thus, so as not to suffer any disaster during the introduction of cost management, certain **management practices** must be considered:

➔ The goal, planning, and execution of the introductory project must be worked on with the **participants** and **involved personnel**. Only thus can they be motivated to organizational learning. Otherwise, internal blocks (a "without me" position) can result.

➔ Managers must convincingly **live up to** the management measures that affect them. This is more important than might first appear. In management, objects and human being are combined; on the objective side are rationality, structure, and theory, while on the subjective side are confidence, feelings, and visions. [Dae95].

➔ The **introduction** itself should be implemented as an **actual managed project**. An external moderator can be assigned to the team as a neutral person, supporting situation analysis as well as processing and implementing the measures [Amb97].

➔ In the beginning, it is advisable to tackle a smaller, manageable project in a product area that promises success. An example is a cost reduction project for a product that is used for determining which organizational measures would be practical and which methods and data are missing. This **introductory project** can be followed by a more systematic weak-point analysis or situation analysis, from which further approaches to the introductory project are derived.

Therefore, the introductory project can have the following **phases**:

- **Project start** from the client (e. g., the management): Setting goals, procedures, expected results, time plan, formation of the project team, choosing the project manager. Concentration on a pilot product area.
- **A model introductory project** of a product, a precursor for situation analysis.
- **Weak-point or situation analysis** by means of the introductory project (organization, methods, aids, data, task progress, etc.) for the purpose of task clarification for the introduction of cost management.
- **Solution search** concerned with measures in organization, methods, aids, data, task progress. **Implementation plan**: Responsible persons, time plan, cost schedule?
- **Carrying through** the measures.
- A model trial on a **second cost reduction project**. What are the findings from that? What can be improved or optimized?
- **Decision by the management** to make cost management as the standard working method – also in other product lines. What must still be changed? What can be stressed, what can be neglected?

Experience shows that it is important to not interrupt the project after the first stage of enthusiasm, since the introduction effort is considerable and the benefit is not quite discernible yet. **Organizational learning is usually a long-term process**.

Furthermore, such a **project** must be **implemented in an agreeable way**. In accordance with Fig. 6.2-3, not just **one** item is to be stressed. Figure 6.2-3 shows qualitatively the scale of the modification, for four modification steps. The farther away from the center, the stronger the modification was. In the "deficient introductory example", only the use of software was stressed. All other important fields (education, organization, methods, etc.) remained underdeveloped. So there will be no benefit and no acceptance!

4.8.2 How much effort is justified for cost reduction?

Experienced designers and product developers who have worked intensively with product manufacturing costs know where and when they must pay particular attention, what they can ignore, and whom they should ask for advice. **They design for low cost from the very beginning**, (e. g., according to the rules in Sect. 7.10.4). That does not mean, however, that they ignore information and advice from the cost management group.

All designers ask themselves **which direction justifies special efforts for cost reduction** – concurrent cost calculation, team building, or the development of alternatives [Kie88]?

Here is a very **simple** example of this line of questioning:

- **How much time** should a design engineer, whose time is worth $ 1/minute (60 $/hour) spend **thinking** about reducing a material cost by 10%? For an appliance of 10 kg mass, 1 kg of steel (costing about 0.5 $) would be saved. The **answer is simple**: So that the thinking does not become more expensive than the resulting cost reduction, the design engineer may think only for (0.5 $ / 1 $) × 60 = 30 seconds!
- This is too little! **Low costs must be kept in mind right from the beginning**. The designer must know how to design, of course, and should draw upon experience from typical previous cases. It is best in this case if the designing is fast, since design time constitutes the largest cost element in new product development.

The **justifiable expenditure depends** on the **cost per piece**, the **quantity to be produced**, and the **attainable cost reduction**. This will be reviewed in detail by using the three cases, A, B, and C, shown in **Fig. 4.8-1**. The left column corresponds to a conventional type of design, without special effort for cost reduction. In the right column, 20% **more** design (development) time has been added in all cases to obtain the necessary information in order to **reduce costs by 30%**. In all of the latter cases, the 30% cost reduction is achievable and the hourly design costs are 60 $.

	Conventional design without special effort, re. costs	**Special attention paid to low-cost design** with 20 % more design time and a 30 % cost reduction
Case A: Small appliance (10 kg), made in lot size = 1		
• Design costs (hourly rate 60 $/h)	5 h ≙ 300 $	5 h + 20 % = 6 h ≙ 360 $
• Manufacturing costs	50 $	50 $ - 30 % = 35 $
• Manufacturing and design costs	**350 $**	**395 $**
		more expensive by 45 $ = +13 %
Case B: Large, custom-made transmission, made in lot size = 1		
• Design costs (hourly rate 60 $/h)	300 h ≙ 18 000 $	300 h + 20 % = 360 h ≙ 21600 $
• Manufacturing costs	50 000 $	50 000 $ - 30 % = 35 000 $
• Manufacturing and design costs	**68 000 $**	**56 600 $**
		saved: 11 400 $ = -17 %
Case C: Small assembly, mass-produced (10 kg), lot size = 10 000		
• Design costs (hourly rate 60 $/h)	200 h ≙ 12 000 $	200 h + 20% = 240 h ≙ 14 400 $
• Manufacturing costs	500 000 $	500 000 $ - 30 % = 350 000 $
• Manufacturing and design costs	**512 000 $**	**364 400 $**
		saved: 147 600 $ = -28,8 %

Fig. 4.8-1. Additional development effort is useful with large, expensive products in single and series production

- **Case A**: The above **small appliance** of 10 kg mass gives incurs 50 $ in manufacturing costs (*MC*) and a design time of five hours. The sum of manufacturing costs (*MC*) and design and development costs (*DDC*) amount to 350 $ in the conventional case on the left; on the right, with 20% more design time, the sum is 395 $. Therefore, there is no saving and, in fact, the appliance becomes more expensive by 45 $ (13%), in spite of a manufacturing cost reduction of around 30%. The inferences were already discussed above.
- In **Case B**, a **large custom-made transmission** is to be designed for single-unit production. It needs 300 hours of design time and will incur manufacturing costs of 50 000 $. The possible cost reduction is 30%, as in Case A, but in this case will yield 15 000 $ in savings. Here, the 20% additional design time (60 hours) is justified, since a net of 11 400 $ (17%) will be saved. Thus, with high costs the effort in using methods and aids for cost reduction is justifiable. The design time for the same product type (complexity) is virtually the same, regardless of whether it is a large, expensive object or a small one with lower manufacturing costs (see Sect. 8.4.4).

- **Case C** has similarly high total costs, but for different reasons. As for Case A, it is a small unit (10 kg mass) but it is made in lots of 10 000 units, giving rise to total manufacturing costs of 500 000 $. The cost saving of 147 600 $ is enormous; a decrease of manufacturing costs of only 10% brings in a total savings of 47 600 $ (about 9%).

Summary:

With **high-cost** (expensive, large) products in single-unit production or with low-cost products made in large production quantities, special efforts for cost management are justified (see similar information at the end of Sect. 7.13.8).

When dealing with **low-cost** items such as small appliances and devices made in single units, the most cost-effective way is to design quickly with knowledge available from similar designs. There must be company-specific rules and people must be trained in their use. **Generally applicable rules** that can be used to lower costs, even without concurrent design calculation, are shown (A, B1, and B2) in Sect. 7.10.4. Furthermore, as these examples show, the probable cost reduction attainable should also be considered, as described below.

Empirical values for reductions in manufacturing costs

The following should help in developing a rough estimate of the justifiable cost reduction effort, particularly for single-unit and limited-lot production. The **costs**, in the sense of what has been said above, must be known. Previous cost-cutting measures must be kept in mind.

- **About 10%** in manufacturing cost reduction can often be achieved by **slimming down** the product (i. e., material cost reduction, better suppliers, inspection of the setup in manufacturing, better manufacturing and assembly conditions, and standardization of parts, materials and joining processes). An example is shown in Sect. 7.13.8.
- **About 20 to 30%** in manufacturing cost reduction might result from an **intensive target costing project**. This might be a matter of an adaptive design using the measures described above. The basic product solution (concept) is maintained, but a rework of the concepts for secondary functions may be necessary. Teamwork, as in simultaneous engineering, is obviously required (see examples in Sects. 10.1 and 10.2).
- **About 40%** in **manufacturing cost reduction**, and **more** – often a required goal – requires that the quality of the use of methods by especially qualified personnel must improve significantly. **New concepts** for the product must be developed by a methodical design procedure (Sect. 7.3, Sect. 10.1.4, Sub-Sect. I3). Close, interdisciplinary cooperation and support by management is a prerequisite (Sect. 4.3.2, Fig. 6.2-2). The **risks**, including technical risks, **are higher**. Such **large cost reductions** cannot usually be achieved through development measures alone. Far reaching changes in all the activities in the product area must also often be made, such as outsourcing, company restructuring, etc.

4.8.3 Implementing cost management

Since product costs usually affect all areas or departments in the company, **cost management** must be introduced **across all disciplines** and put into practice in all daily work. That presents difficulties in many companies, where narrow, specialized fields and departmental thinking prevails (Fig. 3.2-2) [Ehr93, Hei95, Gra97a, Hor97, VDI98, VDI99, Stö99].

To achieve a high degree of acceptance and a favorable cost-benefit ratio, cost management must:

- be built up simply and transparently
- support an interdisciplinary teamwork, oriented toward project progress
- advance employee proficiency and motivation

This succeeds only if the interdisciplinary work methodology is coupled with a strong planning and control system for schedules, product and project costs, as well as capacity in product development [Bur93, Der95, Sau86].

4.8.3.1 Interdisciplinary work methodology

Corresponding to VDI Guideline 2221 [VDI93], the design and development process can be organized across disciplines into the **concretization steps** shown in **Fig. 4.8-2**, which were in part subdivided into further **work phases**. The diamond-shaped decision steps enable the precise control of the total development process (due to monitoring the interim results) relatively early in the task execution (function, quality, product and process costs, time). See also Fig. 4.4-2.

It is clear that to integrate the different disciplines (e. g., mechanical, electrical/electronic engineering, process engineering, and software), the concretization steps and work phases must be extended and detailed. The following rules have proved effective for this purpose [Der95, VDI98]:

➔ **Define work phases wisely**

- Define few work phases, which are clearly separated so that the interim results can be monitored.
- Separate the work phases so that an unambiguous assignment of personnel capacities can occur (e. g., development engineer versus detail designer responsibilities).

➔ **Formulate work phases to be understandable**

- As far as possible, make the work phases alike for all disciplines and yet understandable to all. Formulate standard work phases to reduce interface problems.
- Formulate work phases so that they can be employed for all types of designs (new, adaptive, or variant designs, see Sect. 4.5.2).
- Do not formulate production phases to be either too abstract or too concrete (intelligibility versus universality).

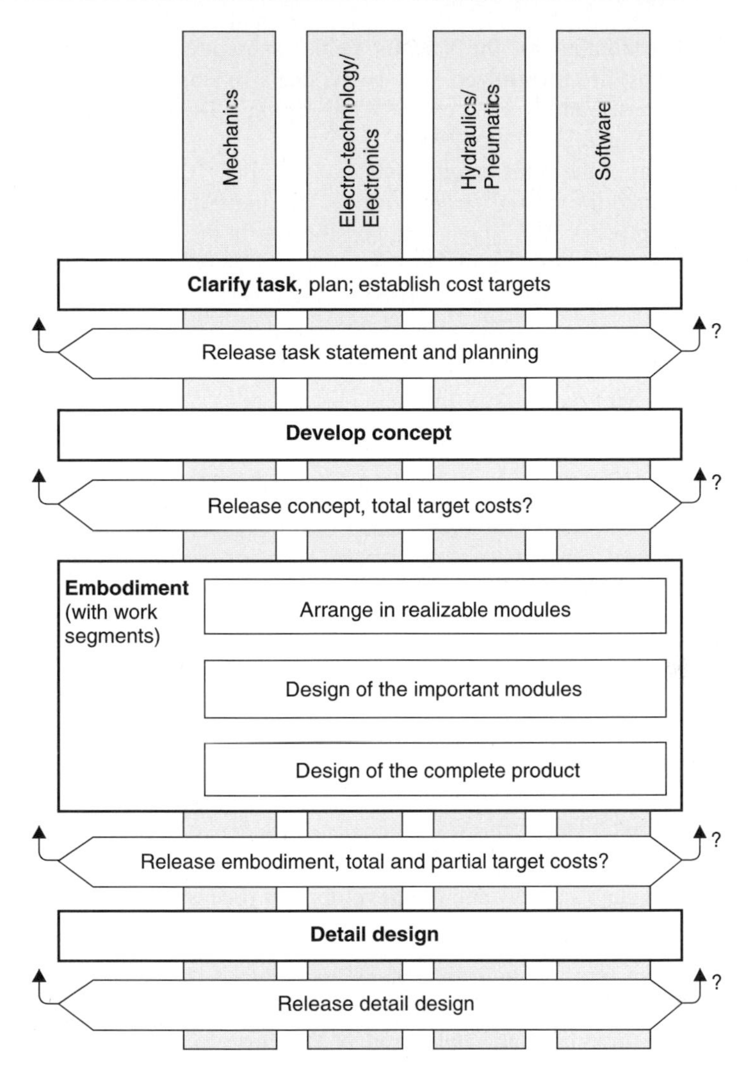

Fig. 4.8-2. Work methodology for interdisciplinary collaboration in product design and development, proved in practice [Der95] (see Sect. 4.8.3.2a)

Our **experiences from the practice** of this methodology can be summarized as follows:

- The implementation of **simultaneous engineering** is facilitated because the sequence is obligatory for everyone. This is achieved especially by the **coordination of the individual areas**. Ensure timely recognition of the feedback effects of solutions in one field upon solutions of other fields (for example, the choice of the type of spindle can influences the drive motor type, which in turn affects the control software).

- By early coordination of the various fields, subsequent, expensive **changes** (iteration loops) are **minimized**. The early, decisive stages of development are especially stressed and worked on across disciplines. **Development time** is thus **reduced** (see Fig. 6.2-2) [VDM98].
- The attainment of the **total target costs** (see also Fig. 10.1-12) of the product is facilitated, since the partial target costs of the individual fields are discussed earlier in the process. More favorable possibilities in the range of solutions may be recognized (e. g., finding out in time that a mechanical operation is too inflexible and also expensive, as opposed to a software solution with appropriate electronics).
- For the **employees,** this results in
  - increased "outside the box" thinking (i. e., an understanding for other fields, their problems and restrictions)
  - an increased feeling of responsibility for the entire product
  - a desirable learning opportunity through mutual exchange of knowledge.

4.8.3.2 Planning, operation and control of cost reduction projects

Why are special planning, operation and control necessary?

Practical experience in projects for cost reduction of complex products over long periods shows the following:

- Target costs and required characteristics of the product to be developed are often lost sight of over details. One does not see the forest for the trees. This happens primarily because the work of several departments must be coordinated. Figure 4.8-2 shows the needed coordination.

That the product stays in view and the development processes are kept under control, the following are necessary:

- A tight **project management** that plans and controls the activities and schedules, as described in Sects. 4.2 and 4.3 (Fig. 4.8-4)
- A planned development and definition of the **target costs**, their division according to functions, assemblies and parts of the product to be developed. Furthermore, a running comparison of target costs with the accruing costs (**comparison of target to actual costs**; see example in Sect. 10.1).
- A **constantly updated cost calculation** and **estimation** for the parts, assemblies and products in question, in order to achieve the target costs according to plan.
- A **concurrent documentation**, in simple cases this can be produced with a spreadsheet (Sects. 9.1.2; 10.1; Fig. 10.1-8). In projects that are more extensive and in the long term, special programs are needed.
- A **computer-aided system** must be introduced that is adapted specifically to the company for schedules, costs, and product properties and for the product structure. All participants must adhere to this, which is kept current by the departments involved under direction of the project leader. This in effect is what was said in regard to Fig. 4.8-2.

This presents in essence the following **problems**:

- The cost estimation and calculation in the early phases of the product development is difficult. Which production planner or costing person dares estimate costs based only on sketches, production ideas and assembly suggestions, which upon realization of the product, are found to be far from the reality? Cost control and production planning people are used to making calculations based upon reliable technical data. Here, at the beginning of a project however, the data are unreliable (**Fig. 4.8-3**). At the start, the designers also have the same technical uncertainty with new solution ideas. All those involved must simply trust and use their experience for early cost recognition, otherwise no progress is possible. Figures 9.1-1, 9.3-9 and 9.3-10 provide hints of possible errors and for error reduction. As the process of product development proceeds, the information available increases, estimates become more firm and finally we have accurate actual costs.
- Usually the biggest problem is one of **mutual confidence** and **cooperation** among the departments of design, production, assembly, purchasing, cost control and costing (Fig. 3.2-2).
- Further problems result from the fact that the departments in larger enterprises may be distributed over **several locations**. It is necessary to pursue not only the product, but also the parts and tool procurement. Additionally the software used by the part suppliers can be different [Ren97].

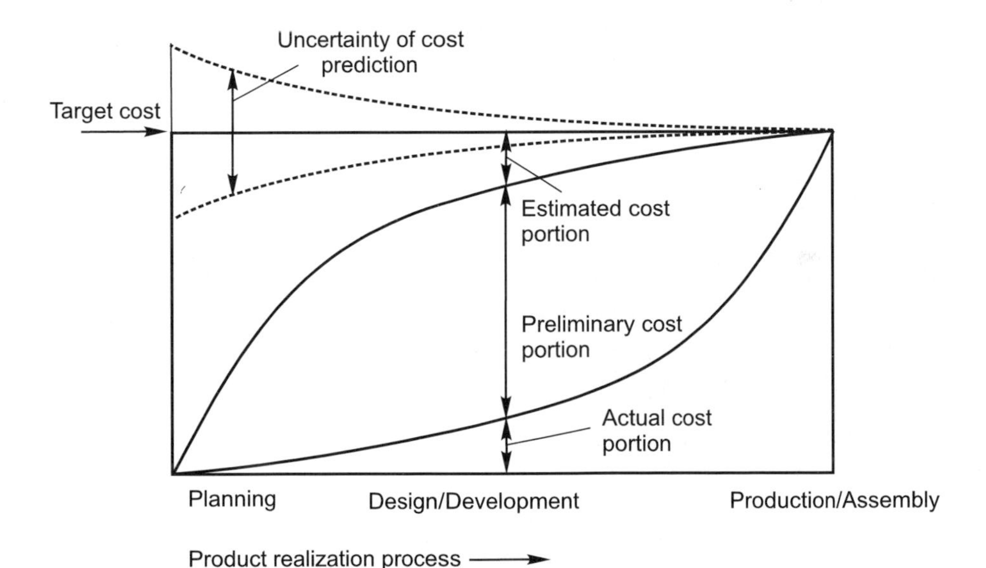

Fig. 4.8-3. In the course of the product creation process, the product costs contain calculated costs, estimated costs and actual costs already accrued. Initial uncertainty turns to confidence [Stö99]

Examples of systems realized in practice

We describe here three systems realized in practice. They were developed in different places for different enterprises and thus for different requirements. Their software base is different and there are additional systems on the market, as described by Renner, et al [Ren97]. The aim here is not to recommend a given system, but rather to demonstrate which possibilities and problems are involved.

First, we present an overview:

a) First, the simpler **Integrated Planning and Control System (IPCS)** that ties job methodology with concurrent cost calculation, according to Fig. 4.8-2 [Der95, VDI98].
b) The **Systematic Control Procedure (SCP)** for cost reduction projects, corresponding to the procedure cycle (Fig. 4.4-1) linked to a detailed, uniform computer-aided cost calculation and documentation. A database (Concept Database CDB) was tested for that purpose. This allows the solution ideas to be stored with their costs, their savings potential, their market life, in such a way that it can be used again for other cost reduction projects of similar type [Lin01]. The concept database will not be described here since it is not an actual planning instrument.
c) The **Target Costing Management System (TCMS)**, which arose in the course of several target costing projects and supports the continuous monitoring of the attainment of target costs [Stö99].

In the following, the three systems are described more in detail.

a) Integrated Planning and Control System (IPCS)

In general, **cost calculation** is used in companies **independent** of **schedule** and **capacity planning systems**.

In cost calculation, the product costs are accounted for, but very seldom the costs of the design and development process. Instead, the working hours for certain pre-assigned activity classes (subprocesses) are recorded in the scheduling systems. These are seldom represented as process costs.

In addition, it often happens that with scheduling systems the classification into subprojects is not sufficiently product-oriented, e. g., according to assemblies (modules in Fig. 4.8-2). Furthermore, the differentiation does not follow the concretization stages (e. g., conceptualizing, embodiment, and layout), so that product-orient recorded times and costs cannot be compared at the beginning on estimated expenditure for subprojects. Thus, a comparison of desired and actual values is only possible for the overall project, but not for individual subprojects.

In order to get **dependable systems** for **cost calculation** and for the **schedule** and **capacity planning**, the two systems were unified [Der95, VDI98].

The product structure as shown in Fig. 4.8-2 and the work methodology described above are the starting points. For the **schedule** and **capacity planning**, the division into subprojects results from the product structure and from the work

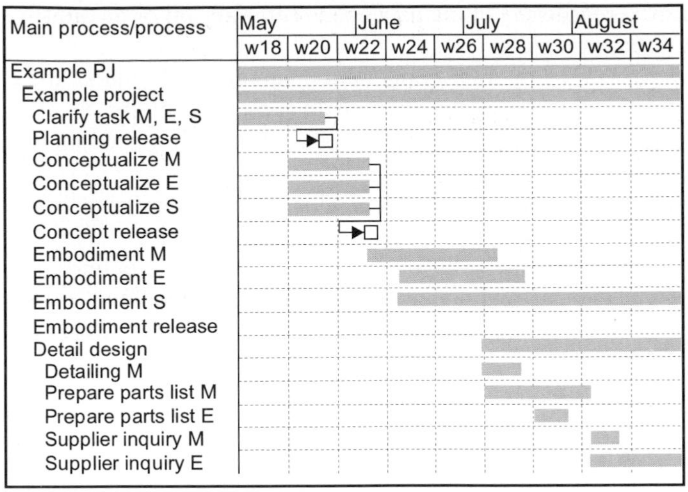

Main process/process | May | June | July | August
w18 w20 w22 w24 w26 w28 w30 w32 w34

Example PJ
Example project
Clarify task M, E, S
Planning release
Conceptualize M
Conceptualize E
Conceptualize S
Concept release
Embodiment M
Embodiment E
Embodiment S
Embodiment release
Detail design
Detailing M
Prepare parts list M
Prepare parts list E
Supplier inquiry M
Supplier inquiry E

Fig. 4.8-4. Interdisciplinary schedules (M=mechanical, E=electronics/electrical, S=software [Der95])

methodology, the division into subprocesses. A subproject within an overall project consists of, e. g., the entire order processing of a machine within a plant, or an assembly within a machine. A job number can be assigned to the subprojects. Subprocesses within a (sub) project are the concretization steps described in the work methodology, and according to specificity, the work phases. For the **cost calculation** the cost units are derived from the product structure with corresponding work number, and from the work methodology the subprocesses for the reporting back of the man-hours and process costs. So both systems are (for process times and process costs) optimally coordinated with each other.

For an efficient implementation of the planning, operation and control system, a continuous **computer support** is recommended. Thus clear deadline schedules or capacities can be planned, e. g., with simple standard programs for project engineering (**Fig. 4.8-4**, see also [Stu97]). With that, capacity bottlenecks can be recognized early and can be eliminated. Further, it is possible to assess processed orders from an economical viewpoint by the link-up of these systems with the cost accounting system. The findings resulting from that form the basis for planning new projects.

Experiences gained thus far show that the planning, operation and control system described here can bring following advantages:

- The acculturation of the employees promotes **acceptance** and improves the quality of the feedback.
- Knowledge from experience, such as the expenditure for the standard work phases, develops with time and increases the **planning confidence** for new projects.

- The uniform registering of the data opens **comparison** or **evaluating possibilities** for different projects.
- The standardization on a generally held planning level reduces the **care effort** considerably and increases the clarity with complex projects.
- The system is **up-to-date**. A regular updating is more important than an exaggerated planning accuracy.
- The system supplies important information for basic decisions in particular for the project progress reports.
- With the use of suitable software the data are available online, so that a short reaction time is possible, the input and evaluation expenditure is very small.
- The coarse planning assumed as standard, referred to the project level, can be made finer for use at department level.

Project control in a graphic mode regarding the planned and realized times, project and product costs, can also be carried out with the **trend diagrams** shown in **Fig. 4.8-5**.

The creation of the figures is very simple: In the abscissa the desired times DT_i or the desired costs. DC_i are put on to the individual project milestones P_1 to P_i In the ordinate the corresponding actual values AT_i or AC_i appear. If the project runs exactly according to plan, both values meet on the 45° lines. Above that the project runs too slowly, or is too expensive.

With the example of the concrete mixer (Sect. 10.1, Fig. 10.1-8) **Fig. 4.8-6** shows how the target costs of the product developed during the project with the project milestones (meetings). The data from the cost search table in Fig. 10.1-8 has been enhanced and represented graphically.

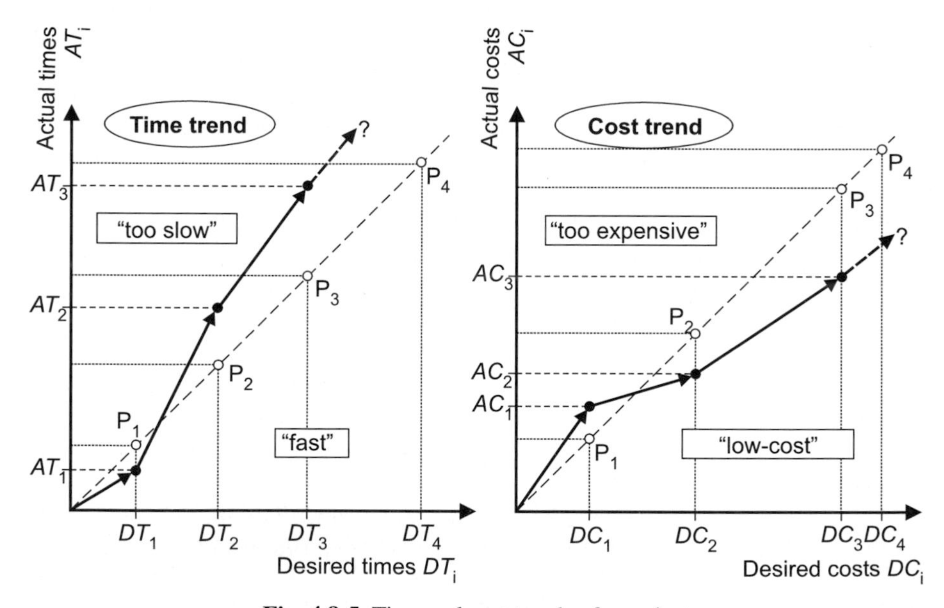

Fig. 4.8-5. Time and cost trends of a project

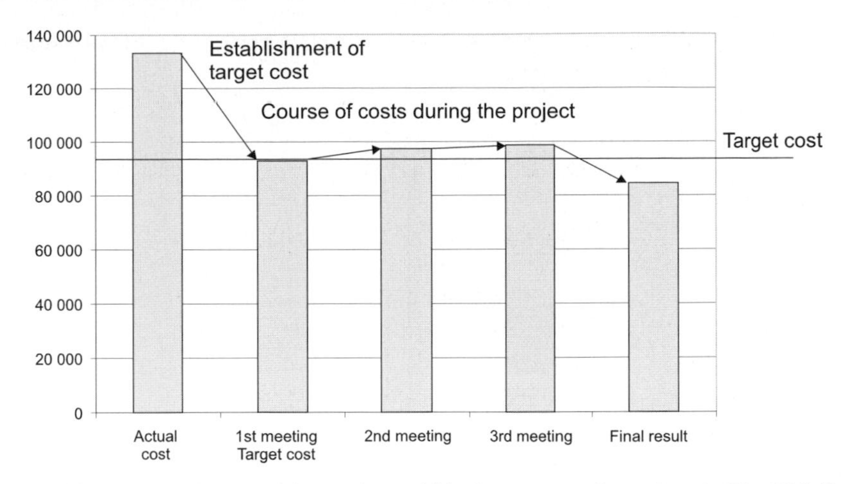

Fig. 4.8-6 Trend of costs of the product within the context of a project (s. Fig. 10.1-8)

b) Systematic Checking Procedure (SCP) for cost reduction projects

In the following part, a methodology for a checking procedure will be introduced. This was developed in the Institute for Product Development at the Munich University of Technology in cooperation with several companies. Scientists and graduate students from the Institute worked on industry projects for durations of half a year or more. The task in each case was to reduce the costs by around 25 to 30% while maintaining or improving the quality. From an extensive product range the best-selling systems were considered in more detail and potentials for optimization generated by using the procedure described in the following. In one of the projects for cost reduction in heavy equipment industry, a problem showed up first: to estimate potentials for cost reduction with assemblies and their production or purchase. The questions were: "With what shall we begin? What can we disregard?" Furthermore the urgent need of an adequate cost calculation in the enterprise became clear [Lin01].

Building the systematic checking procedure

In order to improve teamwork, and to enable a checking of the process by the project leader, all gathered information must be connected together systematically. For this purpose, the data must be present in a suitable, usually digital form. An efficient procedure was developed for this purpose as shown in **Fig. 4.8-7**.

The following shows the process of proceeding from an idea regarding cost reduction, to formulating an explicit statement, according to Fig. 4.8-7:

Cost information is needed when selecting assemblies and components. The information would be examined closely during the cost-cutting process. As a rule, a Pareto analysis is carried out for this purpose (1). The information on the manufacturing costs come from material costs (*MtC*) and production costs (*PC*) from system level with the assemblies (2 above) as well as up to the subsystem level (e. g., the parts) (2 below). With the help of suitable methods of idea-creation, (3)

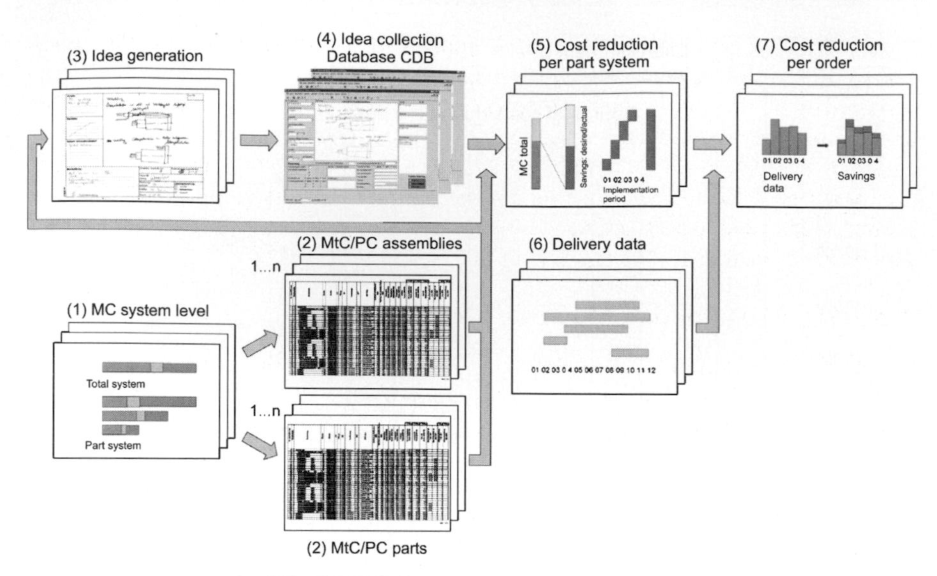

Fig. 4.8-7. Systematic checking procedure

solutions are generated and combined for the parts to be searched. These should lower the manufacturing costs and/or improve the value. These ideas are incorporated in a database (4) (see above, Concept Data Base). Further work concretizes these in such a way that their costs can be estimated. The ideas are then evaluated regarding their potential for savings (5). In addition, the cost information of the subsystem from (1) and (2) is necessary. These are joined with the forecast of future tasks (supply operation), that is, with the quantity and the supply data of the individual systems (6). Then it can be determined when an idea can become effective and what are the overall savings for the system (7). The consideration of process costs is also possible with this systematic method. One-time costs, such as alteration and tool costs, must also be considered.

In this way, cost-cutting measures become transparent and can always be clearly represented. What was really saved, however, is found only later after production and a follow-up calculation!

In the project has also shown what the **deficiencies** are in cost **calculation**. The data are usually not present in a form suitable for a cost reduction project. A calculation for individual parts or assemblies and for functions is particularly difficult and requires much expenditure of time. The way out – to estimate the costs often fails because of the lack of confidence and experience of those doing the calculations (Fig. 4.8-3). The cost structures for individual parts are also somewhat different and thus not comparable. Frequently, the costs of individual elements of a product cannot be determined at all or only with substantial effort.

A consideration of the **function costs** of a product often also becomes difficult, if the same parts can have different functions or the product cannot be divided meaningfully into individual customer-relevant functions.

In order to guarantee the care and continuity of the **systematic checking procedure** it is necessary to designate **a responsible person**. This person checks the correctness of the inputs (actual costs), steers information and the results. This person also provides for the logical coupling of the solution ideas, the distribution on the individual products and the tabulating of possible savings and ensures continuity in the process. The information needed for the cost reduction project is summarized. Actual cost savings as well as the potential for savings can in this way be made visible at all times for different assurance stages.

Problems of estimating the possible outcomes

In other projects, the estimation of the potential for cost reduction of a change also showed up as a substantial problem. It should be made plain whether a rough or an advanced idea is worth pursuing or should to be abandoned, that is, whether an economic advantage can be achieved. Technically necessary changes are thus excluded from this view. For a precise cost evaluation, the design of the changes must be present, which itself costs in capacity and/or money. An early evaluation of the possible potential for savings can thus be very inaccurate, on the other hand a late evaluation is likely to increase costs which cannot be recovered from the expected savings. Large changes can be estimated only with difficulty in regard to the development effort of the product. Small changes show only a small potential that can be rapidly wiped out by the resulting expense.

Against this backdrop, a detailed cost calculation is therefore indispensable. To these must be added the costs of materials, labor costs, development costs, tooling costs (or total investment costs), administrative costs of making changes, as well as costs to the management of the additional variants. In doing so, it should be noted that the fixed costs of the previous solution must be carried by the new solution, e. g., the old tool costs that are to be charged to the total quantity. This double burden through extra charges complicates the efforts for cost savings, in particular with smaller quantities. On the other hand at times more quantities are sold because of the cost reduction, if the savings are passed on to the customers. However, even with a positive measure the question arises as to where, in fact, the costs are saved. A concrete saving is achieved only if long-term changes in the enterprise structure are also initiated such as staff reduction, sale of machines, higher turnover, etc. The same problem shows up also with the financial evaluation of product variants: an additional or left-out variant will be evaluated with series products at up to 15 000 $. These costs do not appear directly, but only over a company-wide, long-term context over all variants in process cost accounting.

c) Target Costing Management System (TCMS)

For several companies a computer-aided system was provided, which uniformly supports the target costing process (as shown in Figs. 4.4-1 and 4.5-7) [Stö99]. It was tested in practice with many different products. Examples of such products were the following: at Siemens in Berlin and Erlangen, gas-isolated high voltage switching systems (single and small-batch manufacturing) and at Pretema Company in Pforzheim, punched parts for electrical and precision machines as well as

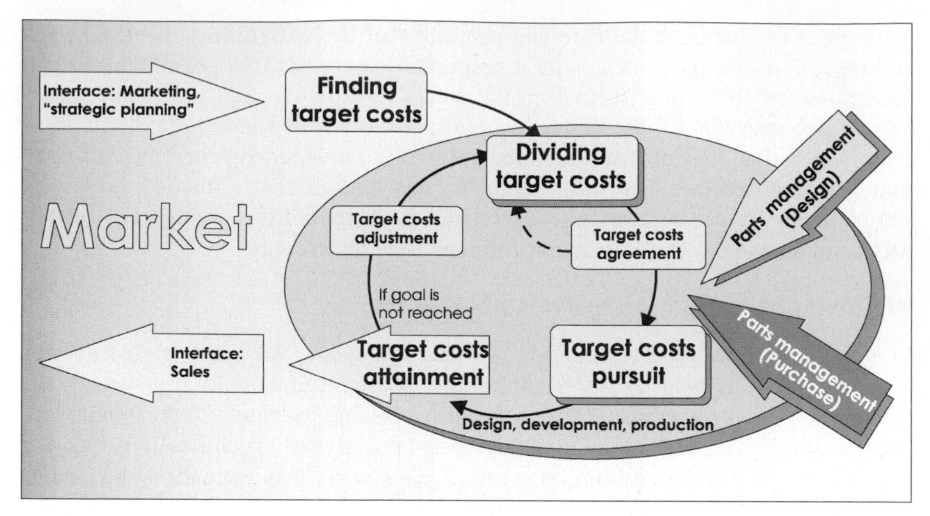

Fig. 4.8-8. Target Costing Management System (TCMS)

the associated punching tools. Thereby the punched parts represent mass production, the tools the single-lot production.

The so-called target costing management system (TCMS) is described in the following by the example of high voltage switching systems, as shown in **Fig. 4.8-8**. We start with the requirements of the technical divisions (left above). With partial feedback, a computer-aided target cost identification follows target cost splitting, search and attainment. The target costing process is constantly supported from project engineering, to design and development all the way to production planning and cost calculation. It must be adapted to the conditions in the company.

The system was developed at the Institute for Product Development based on Microsoft Access software. It will be described here by using an example of switching gear design.

Target cost identification takes place on the basis of the costs of similar precursor products, coordinated with project engineering and/or sales from customer contacts and the probable prices of the competition. The total target costs are thus given. Strategic decisions of sale are kept in mind.

This is followed by the **division** of the **target costs**. The offered product can be split up relatively easily into partial solutions from earlier experience, since it is usually not entirely new. Then, the total target costs are split into partial target costs according to product functions, which can be assigned individual partial solutions (**Fig. 4.8-9**). Thus, partial target costs can be developed, stepping from the total product level, to function, assembly and down to the part level. Standard costs from one's own products, purchased parts, competitor products and information from the market, can flow into the target cost distribution.

In the next step, the design begins under the guidelines of the set target costs. The TCMS informs the designer about product and subject attributes, schedules, fixed target and resulting actual costs, the available time on the current order and/or about previous orders for comparison (**Fig. 4.8-10**).

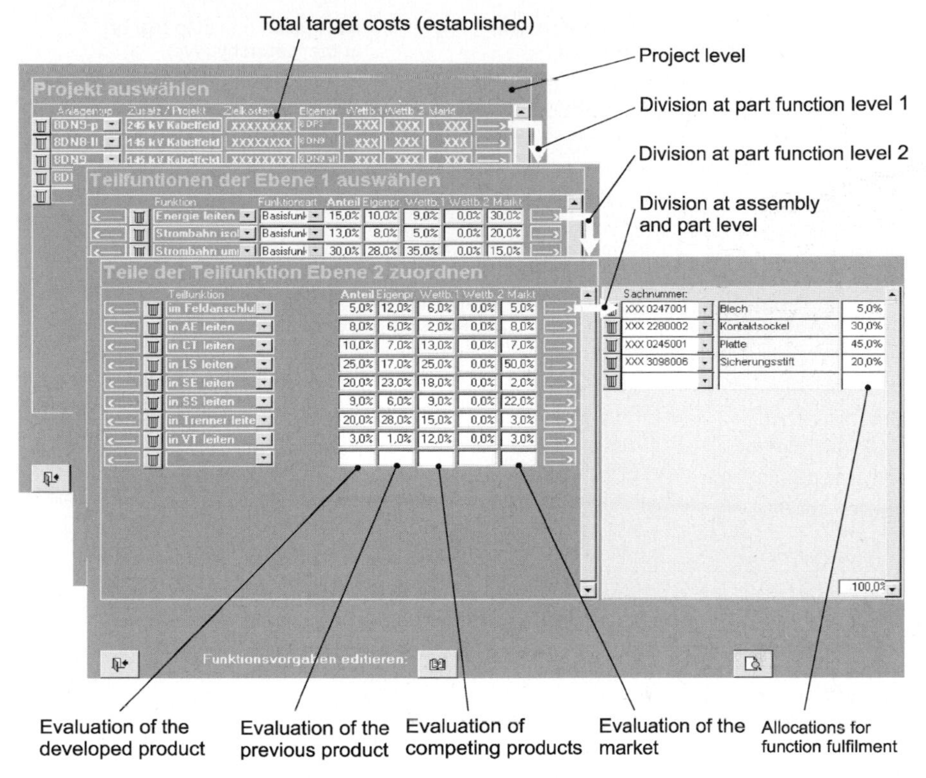

Fig. 4.8-9. Masks of the TCMS for the target cost division on functions, assemblies and components

Apart from these practicalities, many other possibilities are available to the designer such as establishing optimization possibilities.

The actual ensuing costs are administered in this application by sales. The cost data is determined thereby by inquiries to suppliers and to own production. Because in the present case each company unit is its own cost center, in-house production likewise represents actual costs, as do the offers received from the suppliers. Just like the designers, the people from sales can also document cost reduction possibilities in the system.

For the determining the ensuing costs and checking the target cost attainment, a further module is provided in TCMS for **target cost pursuit**.

This module shows the product target costs and the costs that arise as a comparison of nominal and actual values, according to the state of the existing design. It also yields data concerning the probability of achieving the target costs. Possible checking measures are also considered. According to Fig. 4.8-3, thereby a process goes from one of a large cost uncertainty to one of full certainty.

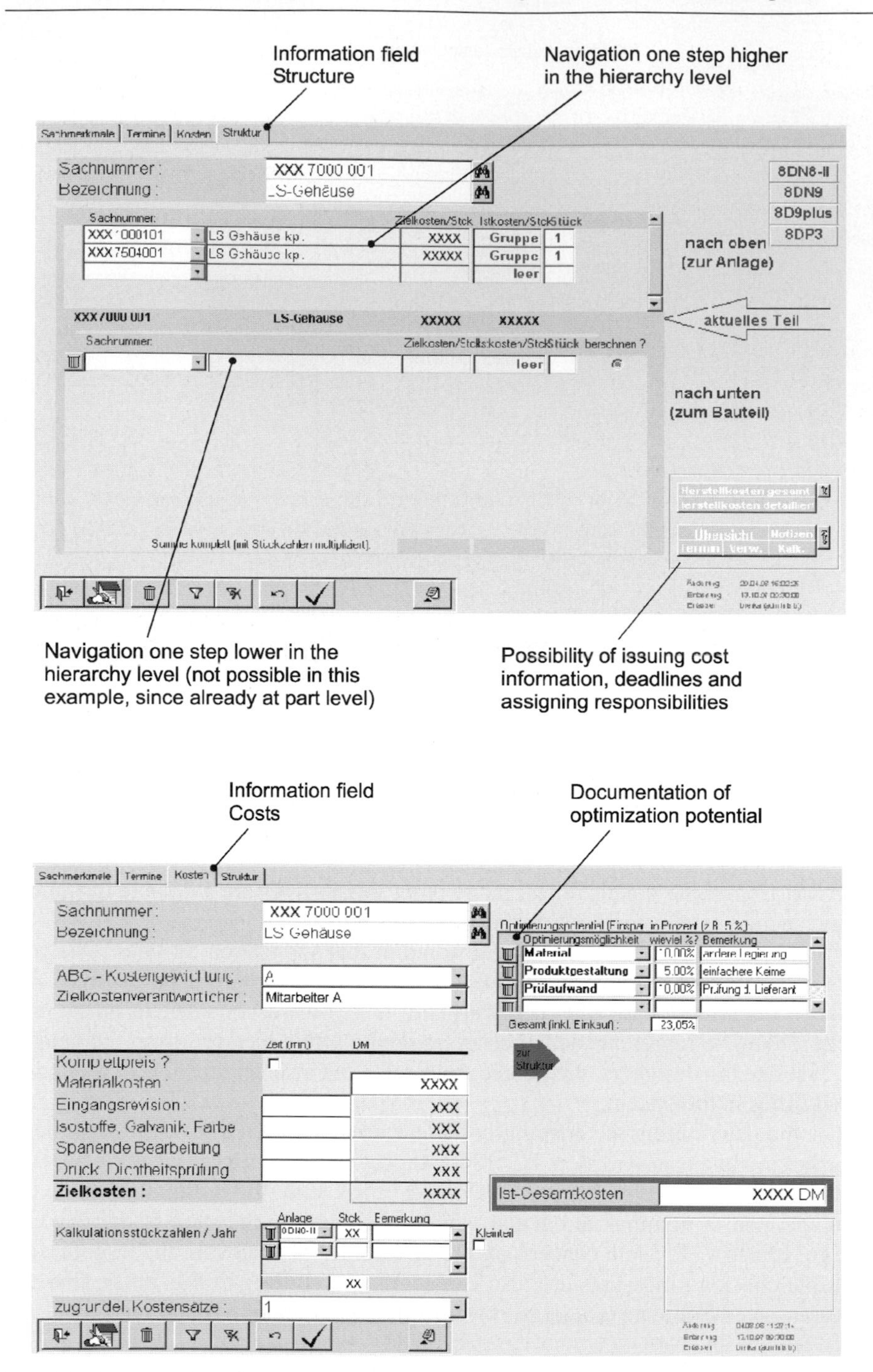

Fig. 4.8-10. Examples of masks of the TCMS for parts management in design for the shape structure and the costs of the product

Then, according to Fig. 4.8-7, we run into the last phase of **target costs attainment**. If it turns out here that the given target costs cannot be attained with the former measures, a "molding" of the target cost is carried out. This involves a repeat of the preceding procedure, with more discussion over where savings are still feasible in the product. Pertinent questions in this regard are for which characteristics is the customer really willing to pay? Are there other possibilities in production and sales that were ignored (e. g., entire production in low wage countries, more direct sales)?

4.8.3.3 Company-internal prerequisites

Simply working on cost reduction is not enough. Costs arise in many departments and are a collective problem. Therefore, costs must be lowered in a coordinated fashion, across disciplines.

The most important prerequisite of those shown in **Fig. 4.8-11** is that all measures are desired, introduced, implemented, and supervised by the **company management**. If management does not stand behind the process, the efforts of individual groups will bog down and the work will be for nothing. An **example** from practice: The design department develops a cost information system (short calculation formulae, computer-aided bid estimation s. Sect. 9.3) but the system becomes useless due to lack of updating. This can happen when, for example, a department head in production planning or costing refuses to share the necessary data or if there are changes in the executive board. Therefore, such information

A) Industrial & Production Engineering Prerequisites

- "Must" assumptions
  - Prices for materials must be known, up-to-date and documented for easy accessibility.
  - Prices for parts must be known, up-to-date and documented for easy accessibility.
- "Can" assumptions
  - Time values for important production sequences;
  - Hourly rates for important machines [$/h];
  - Make work plans, cost calculations for products accessible on a PC: Updated times and costs.

In short: Manufacturing costs and starting prices must be released and may not be kept secret.

B) Organizational Prerequisites

- Cost reduction "costs" time and qualified personnel in many departments. The management must want it, initiate it and keep a check on it: Ready and willing to invest!
- Colleagues in Design & Development, Production, Costing ... must work with the new methods and tools ... and acquire them. Therefore they must be drawn into the information and decision process at the right time.

Fig. 4.8-11. Examples for operational and organizational prerequisites for low-cost design

systems should be so embedded in the day-to-day operations that their maintenance becomes mandatory.

Extensive measures for cost reduction should begin only if management is prepared for the necessary **investment for personnel, time, and financial resources** (Sect. 4.8.1). It is clear that cost reduction requires expenditure – cost data must be procured, processed, and used constructively (Sect. 4.8.2). Furthermore, it must be made clear here that financial data, at least the manufacturing costs, must be made available; it is as essential as technical data. Secretiveness in this respect harms the enterprise. How are costs to be lowered in designing if nothing is known about the costs? However, the cost data must be handled confidentially, that is, definitely only inside the company.

4.8.3.4 Information and continuing education

Traditionally design engineers often have **little knowledge of costs** because their training is mainly technical, and because of the lack of business information flow. This deficit must be eliminated by additional educational measures (in-house and external), since we cannot affect the costs if we do not know anything about them (Sect. 3.2.3).

Knowledge requirements differ according to the extent of responsibility: A detail designer should know the cost categories that influence manufacturing costs and how they may be affected in shape design. A design group leader, on the other hand, should also understand the various types of cost accounting (full and direct costing, fixed, proportional costs, etc.), as well as profitability and lifecycle costs. The group leader should to be informed about essential parameters such as dimensions, lot size, number of parts, etc., which influence costs. This individual should also know about the department costs and their allocation to the products (Fig. 3.2-3).

Additional training in cost reduction is most effective when provided in-house, organized by production planning or costing (see Introductory Project, Sect. 4.8.1). Cost reduction is best motivated by the success of one's own product. Cost analysis of specific products with regard to their origin and calculation (according to components, cost categories and production operations) is most important. Trusting collaboration is a prerequisite. Participants should be employees from all affected departments (i. e., from purchasing, project planning, cost analysis, etc.). In addition, classes may be offered by external sources.

4.9 Other well-known cost management methods

In addition to those described above, there are many other cost management methods that have been developed either directly for cost reduction or to indirectly influence cost reduction.

Due to space considerations, these are addressed below only briefly, along with references. Only value analysis is discussed at length in Sect. 4.9.2, since it has been used for cost reduction for a long time.

4.9.1 Overview

- **Benchmarking** is a method of learning about performance features through comparison with world-class products, in order to become even better. Cost benchmarking will be addressed in Sect. 7.13.
- **Design to Cost (DTC), Design for Manufacturing (DFM), and Design for Assembly (DFA)** have arisen from information exchanged between universities and industry. These are methods, tools, collections of rules, codes, and programs that are especially useful for cost reduction (DFC) [Dom85, Mic89], and for improved manufacturability (DFM) [And91, Boo94] and assembly (DFA) [Boo89]. They have become well known, or are in use, particularly in English-speaking countries.
- **Quality management** usually also favorably influences costs, because in the process of product realization, quality, time, and costs are closely coupled [Pfe93, Rei96]. According to W. Edwards Deming [Dem86], improving the quality of product realization processes should be addressed first. Time and cost reductions follow from that. Simply decreasing costs could lead to loss of quality and, thus, customer dissatisfaction [Sch98]. Quality-oriented methods include **Total Quality Management** (**TQM**) per the DIN EN ISO 9000ff standards [DIN94], **Kaizen** or **Continuous Improvement** (**CI**) [Ima93], **Quality Function Deployment** (**QFD**, a formalized method to uniformly acquire customer preferences for planning enterprise processes [Bos91, Dan96, Ehr06, Kin94b]), and **Failure Mode and Effect Analysis** (**FMEA**, a formalized, analytical method of systematic recording and avoidance of potential mistakes [Rei96, Sche93]). QFD [End00, Lei01] and FMEA can be extended with function costs with regard to target costing. It should be noted that all of these methods have the goal of developing better and thus lower-cost or products. All of these methods can be enhanced with cost considerations. We recommend that the reader experiment freely, without "method blinkers".

4.9.2 Value analysis

Value analysis is a method developed by L. D. Miles [Mil87] for the solution of complex problems. It supports a work schedule (**Fig. 4.9-1**) on system elements (management and behavior) and was adopted as a German standard, DIN 69 910 [DIN87d][5]. Value analysis can apply to products, procedures, services, information content, and processes.

The **purpose** of **value analysis** is to increase the value of something, not just to lower its costs, but also to improve its worth – its benefit, function, or performance, for example [VDI95, EUR95, Sti98]. In about 60% of the cases, value analysis is used for cost reduction of products that already exist; they are redesigned using value analysis. This approach is called **value improvement**. By comparing the

[5] The standard DIN 69 910 of 1987 was withdrawn in 1996, since an EN composition is under preparation. As an alternative, the "VDI Guideline 2800, Value Analysis" of July 1997 is valid.

Basic step	Sub-steps
	(Processing intensity and, if necessary, also the sequence of the sub-steps in each basic step, are project dependent)
1 Prepare project	1.1 Name moderator 1.2 Accept order, establish tentative goal, with conditions 1.3 Set individual goals 1.4 Limit the scope of investigation 1.5 Establish project organization 1.6 Plan course of the project
2 Analyze object situation	2.1 Obtain information on the object and its environment 2.2 Obtain cost information 2.3 Establish functions 2.4 Ascertain solution-dependent plans 2.5 Allocate costs to the functions
3 Determine desired state	3.1 Evaluate information 3.2 Establish desired functions 3.3 Ascertain solution-dependent instructions 3.4 Allocate cost targets to desired functions
4 Develop solution ideas	4.1 Gather available ideas 4.2 Use idea generation techniques
5 Establish solutions	5.1 Establish evaluation criteria 5.2 Evaluate solution ideas 5.3 Condense and present ideas for solution principles 5.4 Evaluate solution principles 5.5 Elaborate on the solutions 5.6 Evaluate solutions 5.7 Provide decision pattern 5.8 Reach decisions
6 Realize the solutions	6.1 Plan details of realization 6.2 Introduce realization 6.3 Monitor the realization 6.4 Finish the project

Fig. 4.9-1. Work schedule for value analysis, per DIN 69 910 [DIN87d]

before-and-after costs, a proof of success is possible. Unfortunately, great effort is often required to making such changes and sometimes this reduces the design engineers' motivation. After all, they designed the original product whose cost is to be lowered by value analysis under considerable time pressure. To minimize the costs and time required for changes, **value configuration** is increasingly preferred, in which the method is used during product development, before the product exists [Bro68a, Bro68b, Bro89, Kre81, VDI95, ZWA92].

Value analysis is a systematic procedure that utilizes a work schedule (Fig. 4.9-1). It contains the steps of the procedure cycle shown in Fig. 4.4-1:

- Basic step 1 "Prepare project" corresponds to "structure task"
- Basic step 2 "Analyze object situation" corresponds to "analyze task"
- Basic step 3 "Determine desired state" corresponds to "formulate task"

- Basic step 4 "Develop solution ideas" corresponds to "solution search"
- Basic step 5 "Determine solutions" corresponds to "analyze, evaluate and establish solutions"

Additional characteristics of value analysis are:

- **Interdisciplinary teamwork**, which captures the knowledge of all affected fields and all aspects that are relevant for a problem (see Sect. 4.3.1). Instead of an isolated point of view or a decision only from the viewpoint of one discipline, the problem undergoes an integrating deliberation. The value analysis work schedule and its basic ideas (e. g., functional and cost-oriented thinking and cooperation) can be modified, and they can also be used for an individual's work [Reh87].
- **Thinking at the function level**, to move away from existing solutions and thus arrive at new solutions. The existing costs of the object are assigned to the functions so that we can recognize from these **function costs** the central points to concentrate on, as well as identify the unnecessary functions and costs (Sect. 4.5.1.2). A "cost target" is stated.

  In value analysis, a function is the action of an object. It is described very clearly by a noun and verb combination. There are main functions, subfunctions, help functions, unwanted functions, and the prestige function.
- **Involvement of the management**, from the clear awareness that value analysis can be successfully introduced only when the leadership understands, desires, and supports the necessary activities. Decades of experience have proven that this is fundamentally valid for the introduction and application of all working processes and methods. This, even though their effectiveness can be only partly verified quantitatively and often occurs so much later that the success is not immediately observable (Sect. 4.8.1).
- **Cooperative, flexible behavior** of all collaborating or affected persons. Anyone with management experience knows that the most flawless methods, statements, and plans come to nothing if they are not dealt with and implemented by motivated and qualified personnel. Hard-working, responsible employees who put the team's success above their own will create the cooperative, healthy working environment that is the sign of a successful team. Naturally, this can only occur jointly with inspiring leadership. Personnel are the most important resource of an enterprise!

Value analysis has been shown to bring about following results:

Krehl demonstrated **directly measurable successes** in cost reductions of a variety of products subjected to 800 value analyses [Kre81]. These amounted to, on average, 23% (range 5–75%) of the variable manufacturing costs. In 80% of those value analyses, the repayment period was less than one year.

Value analysis produces **qualitative successes**: better, more market-oriented products, enthusiasm for teamwork, more job satisfaction, more direct exchanges of information, and procedures that are more rational.

Value analysis has evolved over the last few years into **value management** [Kre81, VDI95], which links problem solution methods with project and management tools, as well as methods for a decided organization of the psychosocial field. It should be kept in mind that emotional relationships could decisively determine the work output.

Value analysis and **target costing** have much in common. Target costing is oriented more strongly toward the cost objectives of the customer or of the market [EUR95, Bul98, Sti98].

The concern of this book is to use input from both fields, combining it all into an integrated product development procedure [Ehr06]. This has its scientific basis in methodical design [Rod91, Pah05]. Thus, innovative and low-cost products can be developed with the desired quality and within the desired time frame.

5 Influencing the Lifecycle Costs

In this chapter, we will see how the product developer can reduce the lifecycle costs of a product, thus providing advantages to the user. Their importance is discussed in the explanations given in Sect. 2.1. It is clear that modifying lifecycle costs is possible only by keeping close contact with the customer – the user. This has great potential for development benefits in machine engineering as well as in plant engineering. Often, the user does not know the principal factors in the lifecycle cost structure but the manufacturer can alter them by changing technical parameters.

5.1 What are lifecycle costs?

Before we can discuss the composition of lifecycle costs (*LCCs*, **Fig. 5.1-1**), some important concepts must be clarified.

Every product has a lifecycle during which costs accumulate. The user's costs keep mounting, from the initial concept or order, through product development, production, to its disposal. These are generally direct costs (e. g., operating costs) or indirect (e. g., higher manufacturing costs), that he must ultimately pay for. Figure 5.1-3 shows the **product life span** of a single product, which is different from the **market life** of a product type, as shown in Fig. 5.1-4. Market life is the period during which a product, a product type or a product range is built and marketed by the manufacturer. In this chapter we will discuss only costs associated with the individual product life span.

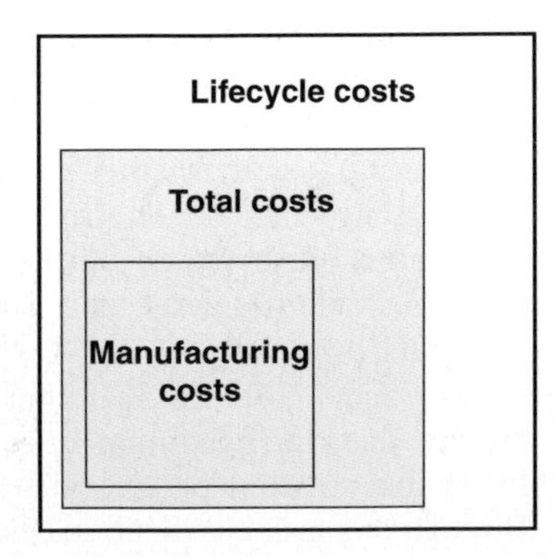

Fig. 5.1-1. Manufacturing, total and lifecycle costs

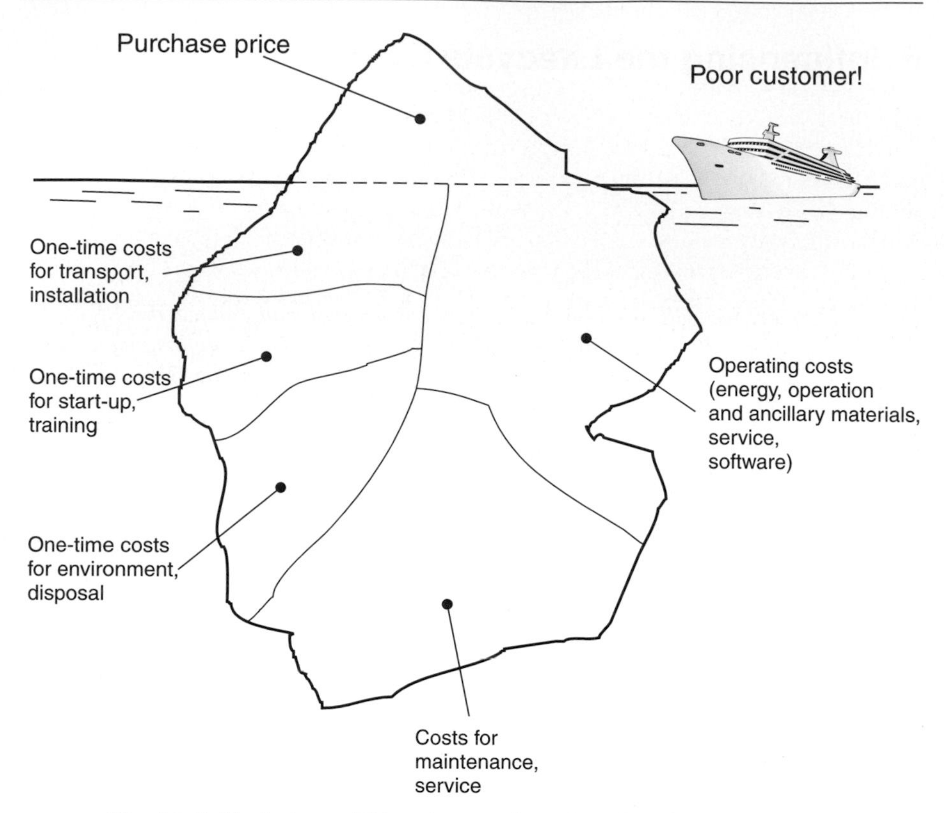

Fig. 5.1-2. The iceberg of lifecycle costs from the point of view of the user

Unfortunately, the user often does not understand the total cost picture. This can be compared to an iceberg (**Fig. 5.1-2**): the customer might be aware of only the tip – the purchase price – and know nothing of the initial costs, the learning curve, the cost of software and adapting it to new uses, or the many new versions which must be acquired. There are also environmental and disposal costs. All of this could be positively influenced if the manufacturer knew what costs would befall the user, and conveyed their importance.

For example, a **photo printing machine manufacturer** asked us for assistance in lowering his manufacturing costs because his selling prices were under competitive pressure. We asked the manufacturer whether he could help his customer (a photo printing lab) lower his costs so as to better afford to buy his machine, but he could not provide much information. It turned out that paper and chemicals constituted 70% of the *photo* manufacturing costs, and the portion that could be influenced by the machine manufacturer was only 15–20%, of which 60% was attributable to the customer's personnel costs for operating and servicing the machine. The manufacturer realized that, for maximum benefit to his customer, he should automate his machine so that only one person (instead of two) was needed to attend to the machine. Through this task clarification, the product development task was thus now very different from at the beginning: The manufacturing costs of the machine therefore became a little higher instead of lower, But the customer could save approximately 5% of the manufacturing costs for *each picture*. On the other hand, if the machine sales price had been reduced by 10%, the picture production costs would likely have been only 0.5% lower! Such user information was not easy to find, as is so often the case. Users are generally too afraid to disclose their costs, or they divulge only a part thereof. A manufacturer can become privy to such data only if he can develop a relationship of trust with his customers.

In this example, the machine manufacturer and the user were in two separate companies, independent of each other. Therefore, only **one** information barrier had to be overcome (Fig. 3.2-2). With bigger plants (e. g., chemical plants), an engineering company might come into play, only designing the plant, not building the system parts or operating it. In such cases, more barriers must be overcome. In such cases, a common product development team is helpful, or at least a task clarification team that brings together the goals and the expertise of the user, the plant engineering company, and the machine or equipment manufacturer.

Thus, it becomes clear that lifecycle costs are important not only for the creation of products but also for the development of processes that come about through an interaction of people and machines.

What do lifecycle costs consist of?

Lifecycle costs are those seen by the product user as the sum of all costs (Fig. 5.1-5), starting from the purchase and during the use (product life span) of a product (plant, machine, device, apparatus, etc.). These are:

- **Initial costs**, basically consisting of the purchase price, perhaps less the resale value (often written off, plus interest, converted to current costs)
- **One-time costs**, e. g., costs for transportation, installation, startup, personnel training, and disposal.
- **Operating costs**, e. g., ongoing costs for energy, supplies and their disposal, and wages for the operating staff.
- **Maintenance costs**, e. g., for service, inspection, and repair
- **Other costs**, e. g., capital costs, taxes, insurance and costs of breakdown.

The elements of these costs are shown in a simplified form in **Fig. 5.1-3**. Over the product's life span, the costs are assigned and added up (the user's capital expenditure is indicated here as a fixed amount, without interest). In the beginning, during the development of the product, the costs are still low.

They accrue only for ideas that are actually brought to paper (design and development costs). The costs that will increase greatly later in the lifecycle are largely already set at this point, for the product and its use (cf. Fig. 2.2-3.)

Product manufacturing costs accrue during parts production, assembly, and procurement (described more fully in Chap. 7). The design on paper is transformed into the metal and plastic of the machine. The costs from departments down the line are added to these, which result in the total cost (more fully defined in Chap. 6). When the product sale occurs, the cost jumps when the manufacturer adds profit and the risk provision (warranty cost). The users consider the purchase

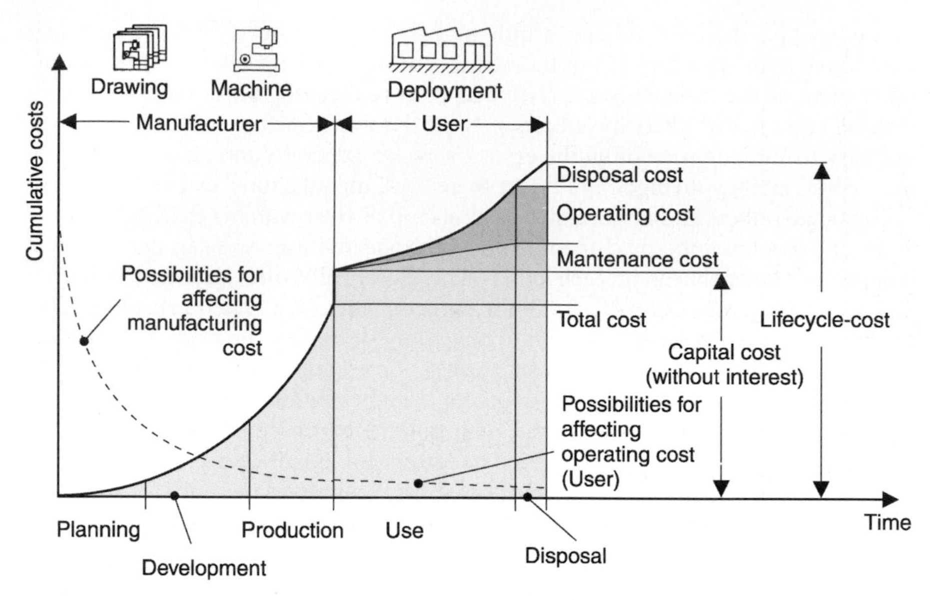

Fig. 5.1-3. Lifecycle costs during the individual product life span

price as their capital expenditure, which would also include the one-time costs to the user for transportation, installation, etc. The user's operating and maintenance costs keep increasing until the end of the product's life, and can amount to many times the capital cost. **Minimizing** these lifecycle costs accruing to **the user should be the foremost goal of a cost-conscious product developer** [Coe94, Fis94, Gro04, RTO00, VDM97, Weu99].

Figure 5.1-4 shows the **market life** of a product type, as opposed to the product life span of a single product shown in Fig. 5.1-3. A given product can have a much longer service life than claimed in the manufacturer's sales prospectus. Vintage automobiles are one such example. In Fig. 5.1-4, the costs of preliminary work are indicated for product development and introduction (*IC*, see Sect. 7.5.1). Note the breakeven point that specifies when these costs are covered by the sales and a profit is earned.

In **Fig. 5.1-5**, the terms from Fig. 5.1-3 are again clearly shown. Machines and plants are not just primary products that have lifecycle costs. We must also analyze the **secondary products** from these plants, such as the costs that arise, for example, from the production of a kilowatt of electricity, a cubic meter of drinking water, a bottle from a glass blowing plant, or, as shown above, the cost of one photographic print. These cost structures are especially informative.

Besides the components of the lifecycle costs mentioned up to now, we must consider **expanded or economic lifecycle costs**. These include cost components for the use of technical products that are borne, not directly by the immediate user, but by the community, the national economy and/or by insurance companies; an example is environmental pollution. **Disposal costs** relate, in part, to this group. In the future, they will probably increasingly be charged and carried according to the

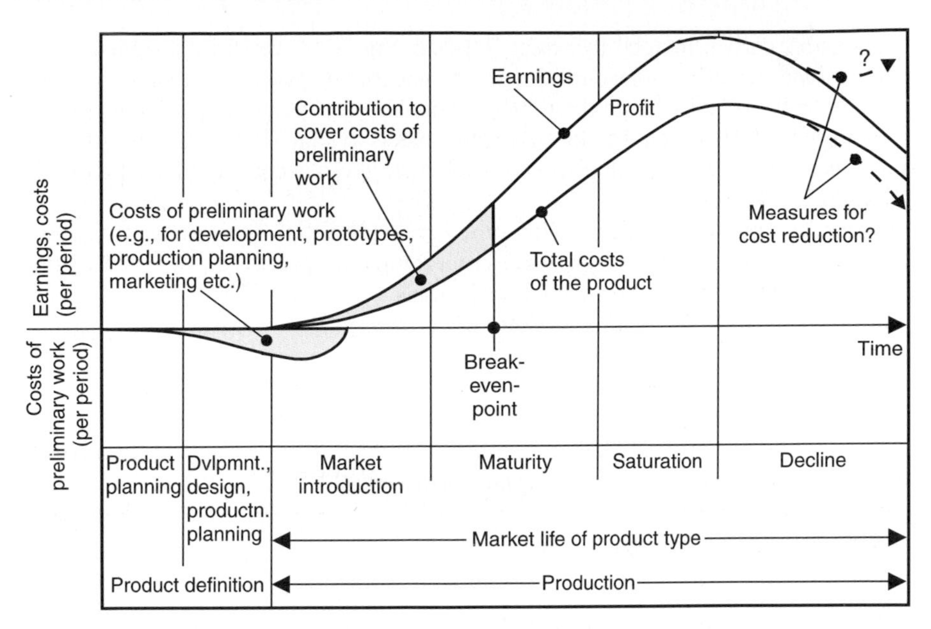

Fig. 5.1-4. The evolution of earnings and costs of a manufacturer over the market life of a product type

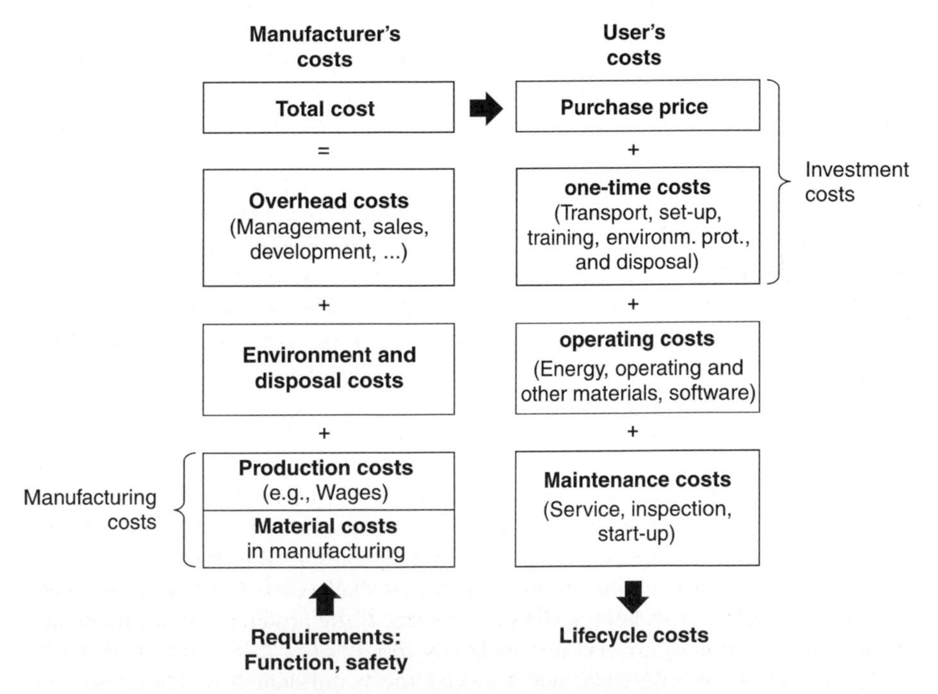

Fig. 5.1-5. Composition of lifecycle costs

"responsible party principle" (see Sect. 7.14). In Fig. 5.1-5, these are assigned to the product manufacturer, and would thus be considered as part of the total costs. Depending on the business or product line, the final user may have to pay such costs.

A uniform **calculation for the lifecycle costs** is available only in parts. All parties – manufacturer, user and disposer – calculate these according to their individual schema [Wel97].

There are **partial overlaps** with other cost concepts, such as quality costs (quality, from planning through use), environmental costs (e. g., disposal of the product), and total costs of ownership or process total costs.

It is advisable to refer the cost elements arising later in the life of the product (e. g., energy costs of alternative air conditioning equipment) relative to different procurement costs **at a specific time** (e. g., a calendar date). Only then can the different alternatives be compared in economic terms. This can be achieved by using different time values of money. We can also calculate the date of the **return on investment** (ROI) or the **breakeven point** (Fig. 8.5-2) (i. e., the period within which the initial additional purchase price costs can be made up for by lower subsequent operating costs) [Män97].

5.2 What influences the lifecycle costs?

Every type of product has a lifecycle cost structure, as shown by examples in **Fig. 5.2-1**. These are typical to a certain extent. Some parts of the lifecycle costs are contained in this cost structure. For simple devices, such as wrenches or other tools, there is only a capital expenditure; there are no operating or maintenance costs. On the other hand, for vehicles, all three cost categories are important, and the optimization task becomes complex.

The lifecycle costs of pumps in waterworks (to the extreme right in Fig. 5.2-1) include the energy costs that in turn strongly depend on the annual running time and on the pump performance [Fra82]. Thus, energy costs constitute 96% of the lifecycle costs for a 2 000 kW pumping set with an annual running time of 8 000 hours and a service life of 20 years. Consequently, the waterworks operators should consider purchasing pumps of higher initial cost, if they have better efficiency. For such performance, the pump price could be twice as high if the efficiency were only 0.2% higher, but both cases lead to the same lifecycle costs. However, since an efficiency increase of around 0.2% can be realized with a fraction of this price/cost increase, the difference would benefit the operator as genuine savings. However, very often the buyers do not value high efficiency; they pay attention only to low capital expenditure – "Only the purchase price counts". This kind of short-term thinking is widespread, but very shortsighted and essentially uneconomical.

As we see, the central points that influence lifecycle costs depend strongly not only on the product type, but also on its life span. If the product is a car, for example, the capital expenditure cost dominates for the first few kilometers that a car is driven, but over the long term, fuel becomes the paramount lifecycle cost. Also, the quality of the product and the operating mode (short-haul versus long-haul routes), and the quality of the maintenance all influence lifecycle costs. These

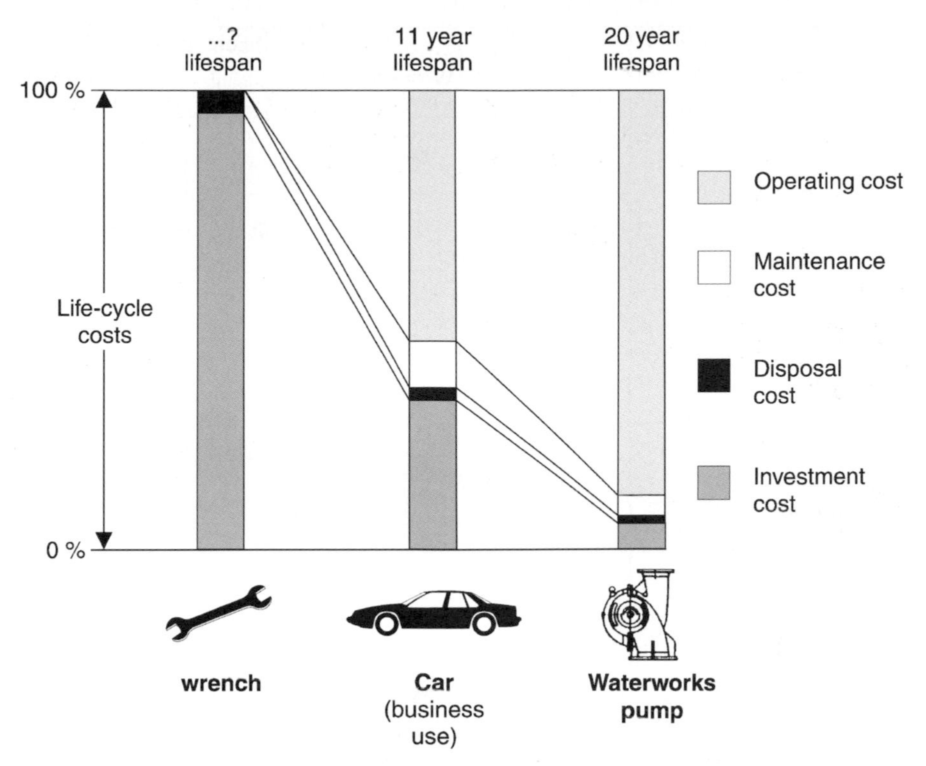

Fig. 5.2-1. Lifecycle cost structures

dependencies lead to different **cost focal points** that must be known before beginning the development of any new product, since the development process must be varied accordingly. **Figure 5.2-2** clarifies these by means of some examples.

As these focal points change over time, the conception and make-up of a product must likewise change. **Figure 5.2-3a** illustrates this with the example of a cargo ship drive (Sect. 7.3). In the 1970s, medium-speed diesel engines (Engine B) (e. g., 430 rpm) with gear reduction (430/120 rpm) were more advantageous in terms of costs, weight, and space requirements, as compared to slower diesel engines (Engine A), and thus replaced many of them. However, **Fig. 5.2-3b** demonstrates how, over time, (particularly during the energy crisis of 1973), fuel costs increased so sharply that the medium-speed diesel engine with its demand for high-quality, expensive diesel oil became too expensive in terms of lifecycle costs. Ironically, the slow-speed diesel engine burning cheap, almost tar-like fuel oils became more cost-effective. As a result, slow, two-cycle engines today power most freighters with direct-driven propellers [Puc89].

Figure 5.2-4 summarizes all the important parameters that affect lifecycle costs. In product development, these parameters become extremely important when the customer insists that the product must not only justify its purchase price in terms of functionality, but also be a good buy in terms of lifecycle costs [VDM97]. Products should thus be developed according to the customer's preferences, appropriate cost

Product	Annual use	Significant cost item		
		Write-off + interest	Energy costs	Service, personnel costs
 Car	10 000 km	●		
	40 000 km		●	
 Television	700 h	●		
 Fire engine pump	50 h	●		
Waterworks pump	8 000 h		●	
 PC (business)	1 600 h			●
PC (private)	700 h	●		

Fig. 5.2-2. Focal points of lifecycle costs (from H. J. Franke)

accounting practices, and evaluation criteria of the consumer markets. Of course, this does not mean that we might not reach another conclusion from a direct contact between the manufacturer and the customer (as in the automated photo printer example above).

It again becomes clear how crucial it is to have close contact between the manufacturer (developer) and the customers (or, even better, with direct users or operators) before and during product development (Sects. 3.3.1 and 4.2).

The influences on the lifecycle costs were pointed out above. Even if the lifecycle cost targets and requirements for the product development are clear, there is still a question of their realization. However, in general, it cannot be said which individual design measures might lead to lowering the lifecycle costs because the products and their constraints are so varied. Nevertheless, there is a methodology that promises success, as shown below.

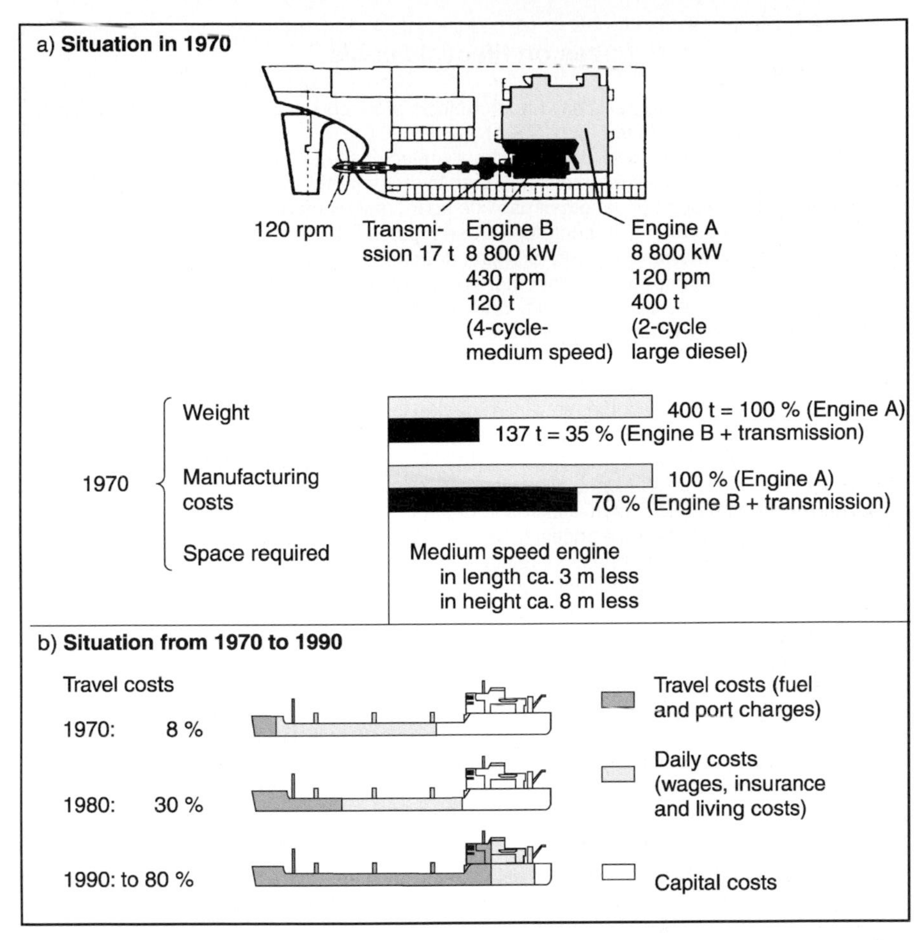

Fig. 5.2-3. Comparison of different transmission concepts in cargo ships (from BHS [Puc89])

Influences on lifecycle costs

- **Product type,** including its quality and quantities produced, e.g., as single unit or in series production (see Figure 5.2-1).
- **Design principle, e.g.,** mechanical, hydraulic, electrical, etc. - Example: mechanical transmissions are usually more efficient than hydrodynamic transmissions.
- **Product use** viewpoint of length of use, life span, environmental conditions, like dirt, corrosive substances, temperatures (see Figures 5.2-2 and 5.2-3).
- **Service and maintenance** quality of which depends on user, as well as on the manufacturer, or others, e.g., replacement parts availability, troubleshooting with remote monitoring, service availability and quality.
- **Cost structure of the user,** e.g., automation expense not justified due to low labor in a given country.
- **Costs for energy and materials used**, e.g., expensive fuels, process materials, lubricants, wear parts become relevant in design (see Figure 5.2-3).
- **Product life span and reliability,** since long lasting product with corresponding reliability are often cheaper with regard to the life-cycle costs. The opposite tendency results in short-term exploitation of fashionable variants.
- **Long-term trends**, e.g., relative increase of service and maintenance costs, vis-a-vis the purchase prices, rationalization of manufacturing, increasing pressure of competition, relative increase of energy costs.
- **Legal requirements, ordinances**, e.g., taxes on cars, oils, inspection and disposal requirements.
- **Time span**, since in general, shorter processes are effectively, cheaper ("Time is money").
- **Price policy** in a sector or with customer. The actual purchase is often important for monetary or psychological reasons. In case of car tires, e.g., car manufacturers demand low price for new parts. The supplier's profit then shows up in the spare parts business. The life span of the tires is of little concern to the user. For the car dealership the important factors for the life-cycle costs are the resale price and, related to that, also the product quality.

Fig. 5.2-4. Items influencing lifecycle costs which can be important for product development

5.3 How to develop a product to a lifecycle cost target?

The procedure cycle was introduced in Fig. 4.4-1 as a generally applicable problem solution method. The knowledge gained from practical experience is used to reduce product costs in general, as shown in Fig. 4.5-7. A further application for achieving lifecycle target costs is shown in **Fig. 5.3-1**.

Figures 5.1-5 and 5.2-4 are useful both for task clarification as well as for the solution search. For the solution search, the rules shown in **Fig. 5.3-2** should be especially helpful of product lifecycle costs (see Fig. 4.4-1).

The subject characteristics of a product (e. g., concept, shape and material) are important with regard to lifecycle costs, as are the corresponding processes affected in the product lifecycle. These processes should be analyzed, calculated, and simulated because the technical artifact are important, as well as the people concerned with them (and thus, for example, the operating method, and quality of service and maintenance.) In this regard, the product must be designed to be simple and there should be provision for appropriate training.

I Clarify the problem and procedure

I.0 Plan the **procedure.** Form the **team**. Name the **responsible** persons.
I.1 Establish the total **lifecycle costs**: Profit goal for the customer/operator, economic target from the market. **What is the customer's wish?**
I.2 **Analysis** of similar machines: Cost structure according to lifecycle costs and types of costs (Fig. 5.1-5), influences (Fig. 5.2-4), related to functions.
I.3 Search for **focal points** for cost reduction. What can be changed, what cannot? Establish **possibilities for cost reduction** with customer/operator.
I.4 Split up target costs according to types of LCC (e.g., energy and material use costs, wear costs) for functions, assemblies.
Divide the task into individual parts.

II Search for solutions (see rules, Fig. 5.3-2)

II.1 **Function:** Fewer or more functions? Function integration of processes, product modules? Function separation (e. g., special wear protection)?
II.2 **Principle:** Other principle (concept)? More automation? More software?
II.3 **Shape design:** Fewer parts (integral design)? Higher reliability? Longer life span?
II.4 **Material:** Less material? Less waste? Wear/corrosion resistant material? Material easier to dispose off?
II.5 Right **solutions** for each **individual process** of the life cycle (Fig. 5.1-5), e. g., set up, training, operation, service and maintenance, organization of training and service, disposal (see rules in Fig. 5.3-2).

III Decide on solutions

III.1 Analysis and evaluation of alternatives: **Cost estimation, calculation** (according to types of costs, see Fig. 5.1-5), testing, experiments.
III.2 Choose one solution.

Fig. 5.3-1. Lifecycle costs during the individual product life span (see Fig. 4.4-1)

Rules for reducing the lifecycle costs

1. General

Choose low-loss, reliable, life-span-optimal design principles
e. g., according to physical principle (mechanical, hydraulic, electrical), active surfaces, motions (variants, see [Ehr03]).

2. Low one-time costs

2.1 Lower **transportation costs** by mass, weight, which correspond to type of transport. Packing, weather and rust protection, which depend on the expected conditions, transportable as a "package".

2.2 **Lower set-up and training costs** by using aids, devices for setting up; instructions easy to see; easiest possible training, putting into operation (pre-requisite: put oneself in the position of the average operator). Simple, common-sense service. Means for noise and vibration isolation should be provided.

3. Low operating costs

3.1 **Save energy, reduce losses**
- Avoid energy transformation.
- Reduce friction losses (e.g., rolling motion instead of sliding motion; elastic bearings (for reciprocating motion; reduce forces transmitted through bearings).
- Reduction of flow losses and use of waste energy.
- Machines adapted for the use (e.g., speed control).

3.2 **Reduce costs of operational and auxiliary materials**
- No special lubricating oils, the user's standard oils, as far as possible. same for all units; long periods between oil changes.
- Fewest possible lubrication points, preferably permanent lubrication.
- Few and cost-effective process materials; seldom changed.
- Infrequent software changes, and then simultaneous changes as far as possible at user and other interfaces.

4. Low maintenance costs

4.1 **Inspection and service,** infrequent, and simultaneous for all units. Simple and sensible measures; easily recognizable and/or legible instructions, best with pictorial aids; tools as far as possible unnecessary or present on site. If possible, central supervision with optical and acoustic response and acknowledgment possibility. Units easy to disassemble for inspection.

4.2 **Repair costs** lowered by infrequent and simple change of low-cost repair parts, liquids, gases, without special personnel needs. Clear and unambiguous instructions on need for change. Good accessibility of parts to be changed; no special tools, no additional adjustments necessary.

5. Low disposal costs
(See Sec. 7.14.4)

Fig. 5.3-2. Rules for lowering of lifecycle costs (This is an incomplete set of examples, since the product variety is too large)

5.4 Extending service life to lower lifecycle costs

Developing products that are **upgradable** is a means for manufacturers and customers to lower product lifecycle costs. This is in addition to the known measures for reducing the individual cost elements. Products will be easier to adapt later to new requirements, which leads to a longer utilization phase. Thus, the user can distribute the one-time capital expenditure over a longer period and the product will be adaptable to changes in technology, laws, etc.

Upgrading means the benefits increase of an existing product for the product user. This is achieved by a function enhancement or function change during or at the end of the product's lifecycle [Phl99]. These function expansions are not yet *completely* designed in when the manufacturer delivers the product. However, they are included during product development to simplify a later upgrading process (looking ahead, with trend analyses, technology research, etc.). During use, an assembly or function (= module) is implemented in the original product with the upgrading process (see Example 5 in Sect. 5.5).

An upgradable product is thus upwardly compatible for functions that were not yet present during the development of the product, or which were not yet detailed. A simple, low-cost upgrade, however, requires a specific procedure in the early stages of the product planning and development. The possible triggers that can begin an upgrading project are diverse: inventions, legal guidelines, the competition's behavior, price changes, etc. These triggers (internal and external to the company)

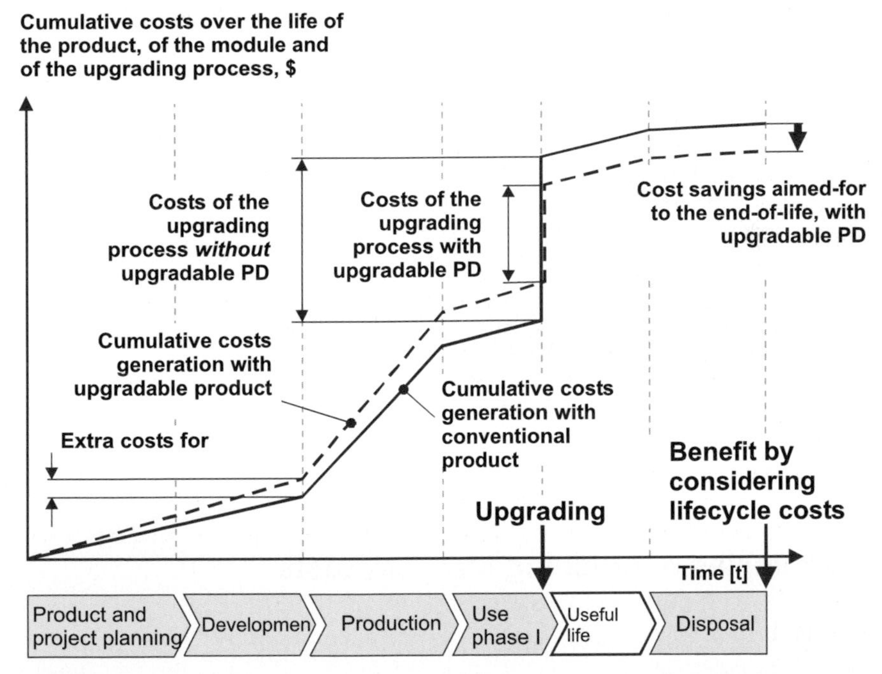

Fig. 5.4-1. Expected cost savings to the end-of-life, by use of upgradable products (PD = product development)

and their qualities must be thoroughly analyzed with trend analysis methods, scenario techniques, or similar procedures; we should try to predict their future behavior. It should then be possible to formulate an assured scenario for the product, and on this basis strategic decisions about the future direction of the company and of its products must be made early. The safeguard is used for an extensive estimation of expenditure on, and benefit from this project [Mör02].

With the planning, development, and production of an upgradable product, the manufacturer begins the preliminary work on options that the customer will need only much later (see **Fig. 5.4-1**). Because the product may be utilized longer and have a broader functional range, there are a number of **opportunities** for the manufacturer and the user:

- Closer contact between the manufacturer and the customer
- Costs for adaptations and function enhancements and changes are lowered due to the larger quantities produced
- Operating costs are lowered through introduction of new technologies
- New investments decrease
- Coordination with the customer's changed use and functional requirements
- Adapting to changed boundary conditions and influences
- Protection of valuable resources (capital, work, raw materials).

Since an entire upgrading project can take several years from the first idea through the realization of the upgrade, there are certain **risks** in this action:

- Change of the tax laws
- Prolongation of the write-off period
- Difficulties in making profitability calculations over long periods
- Interest rate changes
- Entry probability of the planned scenario
- Start of a used-product market promotes the sale of original-design machines and complicates activities for an upgrade
- Further imponderables and any unforeseeable interference (e. g., natural disasters).

By extending the service life, the ratio of operation costs to capital costs (cf. Fig. 5.2-1) changes.

Through upgrading measures, the customer has a greater than normal influence on the operating costs (i. e., the curve in Fig. 5.1-3 is raised higher). The labeling machine (Example 5 in Sect. 5.5 below) demonstrates the advantages that upgrading affords the customer.

5.5 Examples of reducing lifecycle costs

Example 1

The **concrete mixer** described in Sect. 10.1 was designed not only to a manufacturing cost target, but also to a lower wear-cost target, which is attractive to customers.

Since the wear lining of the mixing trough is most subject to wear, extensive experiments to extend this product's lifecycle were carried out in cooperation with the supplier.

Example 2

The **photo printing machine** presented in Sect. 5.1 was further automated after the operator's cost structure was established, after which the manufacturing costs for the product (photo prints) were lowered by about 5%.

Example 3

In the case of a **cigarette wrapping machine**, the issue was not just to lower the machine cost. Rather, the machine had to produce low-cost filter-tip cigarettes. The cost structure of the cigarette had to be determined; it turned out that the chief cost elements were the tobacco (about 50% of the manufacturing costs), the filter and the paper (about 45%), the investment costs (3%), and only a minor portion the maintenance of the machines (2%). Only then could reasonable constructive measures for cost reduction be considered. Since tobacco constitutes a high portion of the cost, decreasing the tobacco quantity by having dense tobacco packed only at the system interfaces (i. e., at the beginning and at the end of the cigarette and in the middle, where it is held by the fingers). Furthermore, highly reliable machines were desired, with correspondingly few downtimes for maintenance, so that production could be carried on around the clock.

As shown in the second and third examples above, it makes sense to develop products and plants such that the **user's profit** is influenced positively [VDM97]. (Principle: "Increase your own profit by increasing the customer's profit".)

Example 4

Disposal costs: See Sect. 7.14, coffee maker.

Example 5

In the research project "Optimization of the product life span" [OPL00], the success of companies in developing new machine concepts which result in extending the service life was analyzed. In the following, the example of a **labeling machine** used in bottling and packing plants is briefly described.

Prior to this project, it was usual for labelers to have an integrated design (see **Fig. 5.5-1**, left: variable rotating machine). But intensive market analysis as well as input from many customers in sales talks and in product expositions revealed that people in the beverage industry desired greater flexibility in regard to the shapes of different container types (bottles, cans, or similar objects with different heights, diameters, and forms). This led to the development of labelers with a modular concept (Fig. 5.5-1, right: modular rotating machine "Solomodule"). This allows *one* machine to perform cold-glue and self-sticking-labeling as well as hot-glue labeling onto the different containers.

By changing the labeling format, the cold-glue and/or hot-glue units and the dispensing units for self-stick labels can be easily replaced by the "plug and label method". Thus, for modular type machines, the changeover times were considerably reduced. At the same time, different units could be used in parallel, or changed over to another type of labeling after any desired period.

The machine and plant concepts presented here have the following economic advantages:

- Machine can be quickly retooled for different labeling systems.
- Units are easily accessible and transportable for maintenance and adjustment work.
- Units for other existing or future labeling systems can be retrofitted.
- Since the units can be changed relatively easily, the machine can be quickly put back into operation; thus, the breakdown costs are minimized.

In particular, the following factors contribute to longer service life of the entire bottling plant:

- The customer does not have to decide on a particular labeling process for its entire service life when purchasing a labeling machine.
- If required, the units are easy to retrofit or exchange because the electronics installed in the unit are synchronized with the rest of the plant (compatibility).
- The units can be used in parallel because their open design allows virtually unlimited access of the labeling function onto the container (see Fig. 5.5-1, right).

In this example, new product specifications were established from intensive market research and close customer contact, and a new modular machine concept was developed. The compatibility of the product and the different modules and/or units to each other was thus ensured. The customer can now purchase units independently of each other and, depending on the current production requirements, can change the units as needed [Mör02].

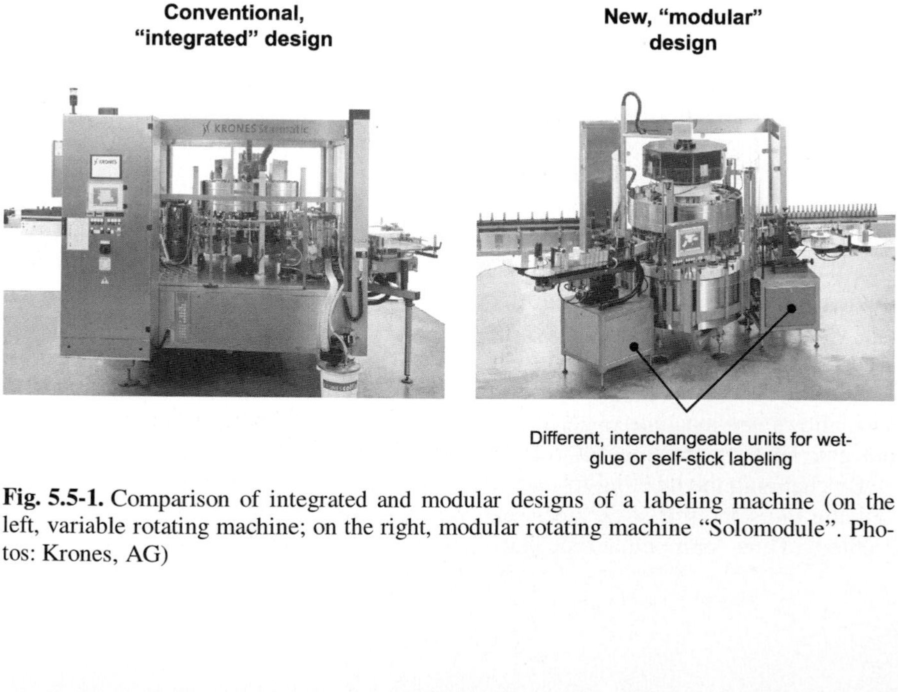

Fig. 5.5-1. Comparison of integrated and modular designs of a labeling machine (on the left, variable rotating machine; on the right, modular rotating machine "Solomodule". Photos: Krones, AG)

6 Influencing the Total Costs

Among costs, the greatest importance is attributed to lowering manufacturing costs, usually within the framework of the most cost-efficient possible product development. However, it is frequently overlooked that this is not the only way product development influences the total costs in a company. A separate chapter (Chapter 7) is dedicated to the lowering of manufacturing costs. Here we will show some important influences that product development has on the total costs of the company (see ***Fig. 6.1****). The lowering of these costs is only one aspect of the problem; it is just as important to mold the company structure so that market changes can be reacted to quickly and flexibly, and that innovation is encouraged.*

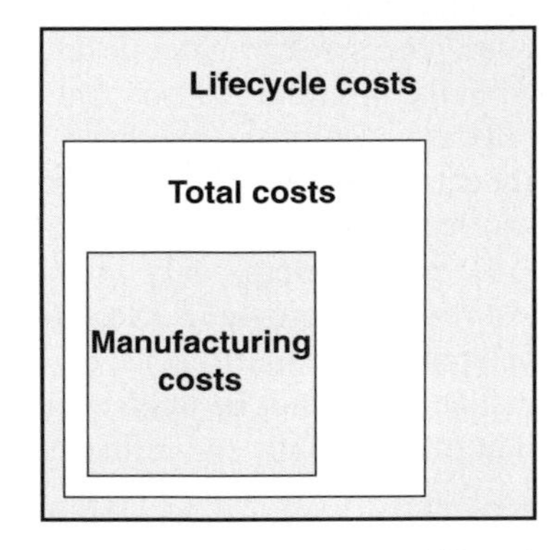

Fig. 6.1. Manufacturing, total and lifecycle costs

6.1 Total costs in the company

Figure 6.1-1 shows three representations of the cost structure of a product from different perspectives (cf. Fig. 2.1-2) in order to address the different points of view here. These are explained in detail in Chap. 8.

Total costs are the sum of the manufacturing costs, design and development costs as well as marketing and administration costs (Fig. 6.1-1, left). Direct costs are assigned to the products directly, overhead costs are distributed according to a formula (Fig. 6.1-1, middle; see also Fig. 8.4-2). Variable costs are dependent on the quantities produced currently, fixed costs cannot be changed (in the short run) in the present situation (Fig. 6.1-1, right).

Management's primary goal in cost reduction is to lowering the fixed costs. On one hand, their goal is to improve the company's earning power, since a considerable

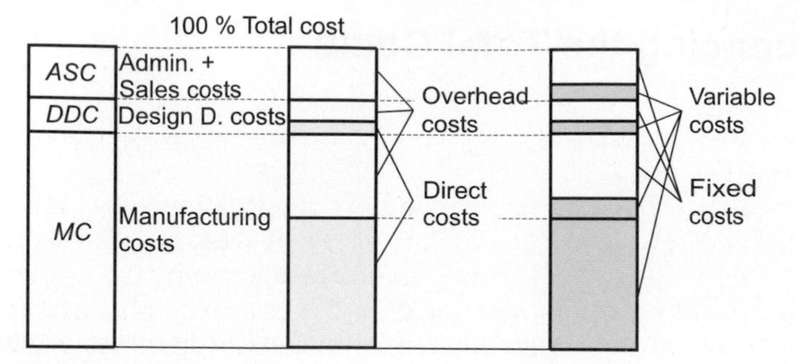

Fig. 6.1-2. Distribution of the total costs of a product (see also Fig. 7.9-1 and Fig. 8.4-2)

part of a company's reserves are in its fixed costs (frequently more than 50% of its total costs). Concurrently, the breakeven quantity with which the company can generate profits is lowered, so there is more flexibility. **By doing so, the company gains product development freedom in the market and becomes more resistant to market fluctuations (Fig. 6.1-2).**

Most important of all, the groups participating in product development bear the full cost responsibility for the process of product development. In differentiating overhead costing as applied in the machine industry, the costs of the development process itself are seen usually as a part of the overhead costs of the manufactured products. Even if the product development as was carried out as shown in Sect. 2.2, it makes up on average only about 9% of the entire costs generated in the company (cf. Fig. 2.2-3). Nevertheless, only in the rarest cases will it be spared in the company's cost reduction programs (Sect. 6.2).

In addition, product development has an influence on cost generation in a series of other company processes that should not be underestimated (cf. Sect. 2.2). First of all there is the **production process**. Only the influence of product development

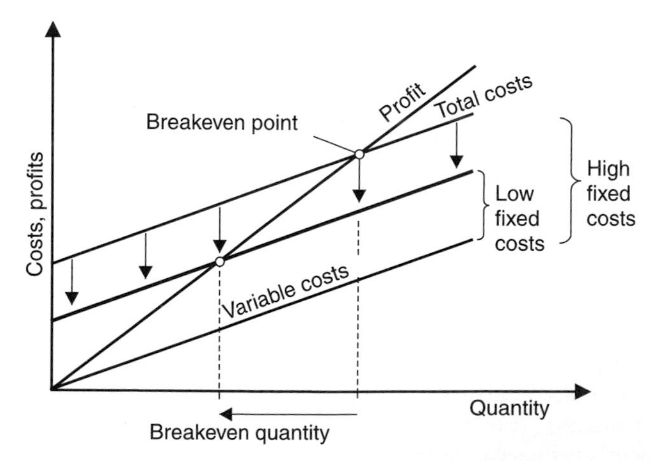

Fig. 6.1-1. Shifting of the breakeven quantity in lowering of the fixed cost portion in the company

on the overhead costs in production will be considered in this chapter. The direct influence of product development on production costs is examined in detail later in Chap. 7. The **logistics and service processes** in the company are also cost-relevant, the costs of which also depend considerably on decisions during product development (Fig. 6.3-4).

The **availability of sound and dependable information about the actual level of costs** is an important requirement for all of these notions. As mentioned above, the problem details of cause-based cost information show that the cost accounting systems are not usually suited for this task. Only the variable elements of manufacturing costs are usually determined fairly precisely; the remaining elements are only roughly estimated. Since accounting systems must satisfy a great number of different requirements, the specific requirements are considered only conditionally (cf. Sect. 8.4.3). The topics presented in this chapter are supported by **process cost accounting** (cf. Sect. 8.4.6), however its implementation in companies is more the exception than the rule due to its additional expense.

6.2 Reducing product development costs

The costs of the design and development department are identified in the operational accounting sheet of the cost center or unit, arranged according to cost categories (Fig. 8.3-3). In many companies these costs at the cost centers are considered as **the** product development costs. However, for determination of the cost-driving parameters of development, **the costs of all processes initiated that way for development and care**[7] **of the products are necessary. These processes include the diverse expenditures in marketing, production planning, purchasing, and other company functions.** The organization of product development work in projects, coupled with project budgets and project cost control, approaches the need of looking for the causes of cost build-up by following the process steps. Unfortunately in industrial practice, project work is frequently introduced without the project-based cost information being made available (cf. Sect. 4.8.3.2). Thus, the parties responsible for the project are forced to improvise [Lin92].

Against the background of everyday reality, with the many different cost accounting models that companies use and the varying requirements from the processes of product development and product care, we present the following compromise.

We must differentiate between products from **single-unit and series production**, as well as between customized and customer-independent product development. With a mix of standard and special products, extreme demands are placed on cost accounting. There is a great danger of wrongly assigning costs to standard instead of to special products; this stems from the practice of assigning surcharges (see Fig. 8.4-5 to Fig. 8.4-7). This mistake pushes many companies to price their special products too low, and the standard products in product niches to be too expensive [Ber95, Buc99, Eve92, Eve97a, Hor97, Schu89, Sak94, Sto99, VDI98a].

---

[7] Product care in this context refers to the maintenance, updating and similar efforts devoted to the product by the responsible company personnel.

The costs of product development consist of a number of different components. Whatever the models for assigning costs might be, **personnel costs** are the decisive portion of the product development costs, depending on the company's strengths in design and development. Therefore, personnel performance is a defining factor. On the expenditure side, salary levels (including the secondary costs) as well as the necessary number of employees is discussed. These costs must be seen as fixed costs in the company because in the short run it is hardly feasible to reduce personnel simply in response to market changes. Besides the pure cost considerations, the strategic importance of innovative in-house product development is also to be taken into account when thinking about product development costs.

Therefore, which essential aspects constitute product development costs and which describe cost reduction potentials, while recognizing quality and innovation capability? **Figure 6.2-1** shows a structure for dealing with this topic. The important questions are the following: **What** do we do, **how** do we do it, and **who** does it?

First, it is necessary to carefully select projects and tasks to be worked on and to **focus on projects important for the company** (cf. Sect. 6.2.2).

Because the necessary activities must be carried out with the **highest efficiency**, this is when the company must choose and form the "building and running" organization adequate for the task and the process. In this context, the question of **in-house (internal) versus outsourced (external) product development** frequently comes

Reducing product development costs Checklist		
Form focal points	**Efficiency**	**In-house or external development**
Which development work supports the goals of the company? Which development work brings the greatest advances for the company? Are the product development goals clear? Are the profitability questions sufficiently clarified? Is there additional customer benefit? Where are most of the chances and risks?	Are the responsible parties identified? Does the project have proper goals? How have the old ways of working been changed? Which limiting conditions counter motivation? Why are changes being made again? What do we learn from the completed project? ...	Do we have sufficient expertise for this development task? Do we have sufficient capacity in the time window? Is the problem clearly defined? Who advises the development partner? Is the development topic of strategic importance? ...

Fig. 6.2-1. Checklist for reducing product development costs

up. Together with the question of in-house or outsourced manufacturing, it is necessary to reach objective evaluations and decisions (cf. Sects. 6.2.3 & 7.10.3).

6.2.1 Establishing focal points of product development activities

In many companies, too **many tasks and too many projects** are regarded as urgent and necessary, and are pursued more or less simultaneously. The results thus do not satisfy expectations because the company's capacities are overwhelmed by too many projects; high-priority problems dominate the daily agenda. The company needs to develop a system for prioritizing projects, bearing in mind the requirements of the market as well as questions of profit and return on investment.

The estimated profits should be compared to the estimated efforts needed for product development – market analysis, design and development, prototypes, special production resources, etc., through training and market launch. Along with that, an adequate return on capital as well as emerging opportunities and risks must enter into the decision. The company's strategic direction should be the primary criterion in decisions about prioritizing projects. **Therefore, integrated considerations are called for, not just the isolated viewpoints of product development, marketing, or customer service.**

The profit orientation of project work must not be considered a constraint but, rather, as the condition that creates the necessary room for basic product development and innovations. The entire company, not only the product developers, must consider setting and prioritizing the goals for product development.

➔ Arrange product development projects according to priority and importance.

➔ Employ personnel focused only on one project.

6.2.2 Increasing the efficiency of product development

Personnel

Personnel qualification and motivation has the most significant influence on the efficiency of product development (cf. [Fran97, Amb97]). This is true both for the personnel in the company as well as for the personnel of associated external partners. What are the essential criteria here? On one hand, increasingly rapid changes in technology place continual demands on personnel training. Thus the initial qualifications as well as ongoing training of the employees are decisive parameters. Not only are the technical and technology-related topics of great importance here, but also the so-called "soft" factors related to collaboration; integrated and networked thinking and working; and questions of conflict resolution, etc. [Bei97].

Two essential elements of project success are the fundamental suitability of the employees and the quality of their training. There should be a **structure of qualifications** relating to the respective task areas. Also, the different specialties needed for the project or product need to be identified, such as mechanical and electrical engineering, information science, etc., and the training tracks for skilled workers, technicians and engineers should be developed. There should also be a balanced **age structure** to ensure knowledge transfer from the older to the younger employees, as well as continuously scrutinizing "old" knowledge as well as innovation in methods and tools by the junior staff. **Continuing education and training** within the company and on employees' own initiatives is becoming increasingly important for all market participants, especially for product developers (Sect. 4.8.3.4).

Motivation

Motivation is another decisive success factor (cf. [Fran97, Spr95]). Motivated employees who are focused on the total goal of a successful will be more successful in shaping the many processes of product development. Employees who are less motivated or who see only their specific limited subject, or who are more concerned about their personal sphere of influence or their professional future, will generate only average output, at best. That means that product development managers must face personnel challenges that far exceed their core areas of technical and professional expertise. However, the results can be considerable increases in performance.

> ➔ Make personnel development, management, and personnel structure the essential aims.

Organization

Related to the issues of employee qualification and motivation is the question of the right organization. This is bound to the company's goals and is based on the abilities and experience of the participating personnel as well as on the company culture and history. The "right" organization will enable many participants to effectively implement the procedures jointly.

There is no one type of organization form that satisfies all requirements and can be safely regarded as the correct one. Decisions about the form of organization must be grounded in the personnel situation, the product requirements, the history of the company, as well as the market demands. Whether the organization will be functional-, matrix-, or process-based must be decided on a case-by-case basis (Sect. 3.2.1).

The framework of the organization should include flexibility toward changing market demands, the increasing activity in decentralized development teams, as well as the increasing use of computer-based communication and data processing in product development.

> ➔ Keep the organization continually advancing.

Project management

The structure of project management is decisive for success or failure in a company. With more complex products (large number of parts, many functions, etc.) or products with high process complexity (for example, a difficult production process sequence), it is necessary to execute operation steps as in **Simultaneous Engineering**. In other words, the steps are executed in parallel to satisfy, on one hand, the requirements of the system complexity and, on the other hand, the demands of time and process sequence optimization (Sect. 4.3.2). With complex products such as cars, it is apparent that many project teams must function in parallel. The intensity and the effort of the coordinating processes must be thought out as carefully as the degree of project organization. Concerning decision competence and responsibility, the decision on hierarchies is replaced increasingly by the **subsidiarity principle:** Decisions are made where the highest professional competence is available, and complexity should be held to a minimum. Systems must be developed and used to master the requisite complexity sufficiently well.

> ➔ Make decisions where the highest professional competence and the necessary oversight are present.

Time scheduling and capacity planning

Time scheduling and capacity planning have overtaxed design and development for decades (cf. [Fra98, Pau78]), and recent investigations still show [Kle98] need for improving these activities (cf. Sect. 4.8.3.2). Dramatically shortening product development times (by as much as 50% in many industries [Smi98]), requires carrying out activities strongly in parallel, and the elimination of unnecessary work.

Investigations of product development methodology have shown repeatedly that considerable time and expense can be saved by making the correct decisions in the **early product development stages** (during clarification of the task and the first concept development). If more emphasis is placed on "up-front" efforts, fewer iterations and improvements are needed later. Shortening product development times requires investing more time and capacity in the early stages of product development than was done previously. Therefore, preplanning and early identification of critical properties of future products are needed for avoiding iterative loops (**Fig. 6.2-2**). This also enables product planners to react quickly to market changes.

In order to visualize these correlations, the following assumptions were used in Fig. 6.2-2 for product development, manufacturing, and operating costs that are based on practical experiences (other costs are not considered):

- Through better planning, more thorough use of personnel and methods, the **product development time during the integrated product realization** (thick line) compared to the conventional product realization (dashed line) **is reduced to one-half** (with new developments in single-unit production to even 40%! [VDM98]). The **product development costs remain constant** at 30% of the former manufacturing costs.

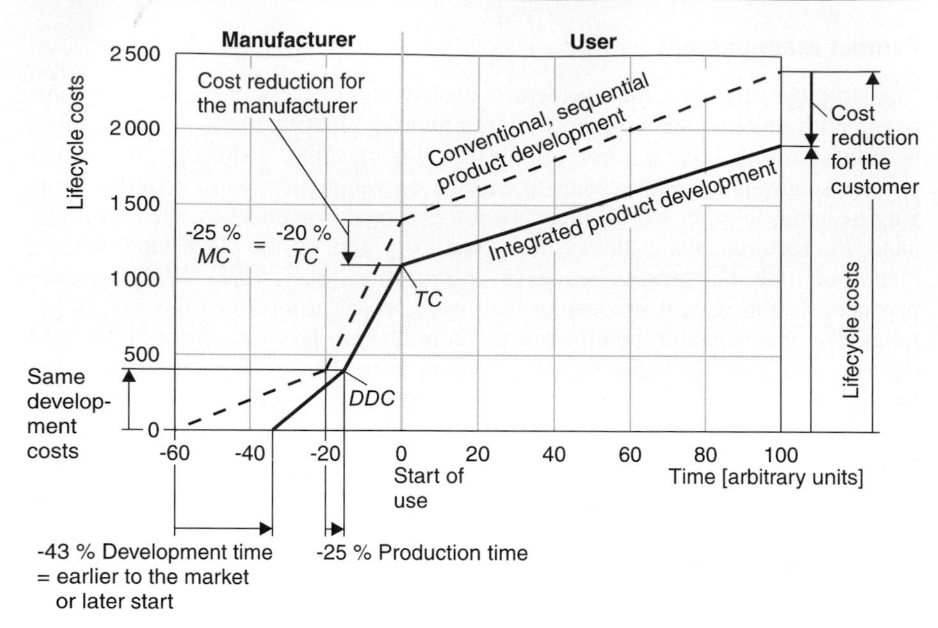

Fig. 6.2-2. Cost and time saving with more intensive product development in the early stages (see also Fig. 10.1-11)

- The **manufacturing costs are lowered** by around **25%** through integrated product development. That also results in a **production time reduction of 25%**.
- Furthermore, it is assumed that the **operating costs decrease by around 20%.**

We achieve the following advantages:

- With these hypotheses, the total **product development time** for the **manufacturer** is reduced by about **43%**. This results in either a **shorter time to market**, or the company can begin product development work later and react faster to market changes. The **total costs fall by around 20%**. It is assumed that this reduction is passed directly on to the customer in the form of a lower price.
- For the **customer** there is a reduction in price of around 20% and, depending on the operating life, a **considerable reduction in the lifecycle costs**.

> ➔ Invest in time and capability in the early product development phases.
>
> ➔ Ensure properties of products early and with low effort.

Changes

Every product development goes through iterations or optimization processes, which lead automatically to changes. Changes incur costs, use up time, and influence the resulting quality. They involve both chances and risks.

It is therefore highly desirable to avoid changes late in the product realization phases, where they cause considerable time and cost expenditure or in themselves contain great risks (see "Rule of Ten", Sect. 2.2). **Necessary changes should be purposefully introduced in early product development phases, when they require little time and create few costs, and avoided as far as possible in later stages.** This will be supported by the procedure represented in Fig. 6.2-2 and by the separation of **predevelopment** (e. g., for the clarification of technological risks) from the actual product development. If changes must be made, they should be executed effectively and efficiently. The **effects** of changes would be a valuable basis on which to assess whether such changes should be made at all, but this information is often not available (Sect. 6.3.3). It is important to know, for example, which parts and assemblies would be affected; what technical and economic effects there would be on the qualities of the entire product; which persons or organizations must be included; and how large the effort will be in terms of time and costs [Gem95, Eve97b, Con98, Kle98]. Since in many companies product care is integrated into product development, processing changes often absorbs from 20% to over 50% of the product development capacity (cf. [Con98, Lin98b]). Product changes therefore determine, in great measure, the overall efficiency of product development.

➔ Discover the causes of change and avoid them in the future.

➔ Before carrying out the changes, test the entire process chain for their effects.

Helpful measures

There are many methods for acquiring and using product development information efficiently. Computers, lists, drawings, models, and logs are examples of information carriers. For increased efficiency, there can be meetings, team sessions, letters, fax, e-mail, etc., or even the automated dispatch of documents in Workflow Management Systems. **Before using such new measures, the processes (new development, changes, releasing, etc.) should be fundamentally analyzed and optimized.** Strive for simple, unambiguous structures with few interfaces [Sch95, Kan93, Bur93, Mat98].

Examples of **conventional tools** in product development are standards, forms, checklists, reports (test, service, visits to trade shows, etc.), and use of Metaplan in meetings. In addition, the possibilities of experimental workshops should be mentioned here (Fig. 4.6-1).

➔ Increase efficiency through the use of tools for improvement of communication.

➔ Analyze and optimize the processes of development before using the new tools.

Computer support

Application of computer technology is an important means of increasing product development efficiency. CAD, FEM, company intranets, and many computer applications characterize contemporary product development work. The expected future performance improvements of computers and computer networks will expand possibilities for simulation, visualization, and providing information and communication [Mee89].

Unfortunately, many of these advantages are sometimes cancelled out by negative effects, including software version changes that require constant updating and retraining. Exchanging data with product development partners becomes difficult if everyone's software changes are not done simultaneously.

This problem becomes clearer in the case of necessary system changes by other providers or with other technologies. Consider that the change from 2D to 3D CAD systems must be accompanied by systematic personnel training. The advantages of 3D CAD are obtained only later, if at all.

Figure 6.2-3 shows, that besides the **introduction of new software**, updating personnel training and qualification, a revamping the organization must also take place, ideally with the same weighting (see also Sects. 4.8.1, 4.8.3.4).

The possibilities of macro-technology or variant generation and monitoring them in CAD or simulation systems for standardization are often exercised only to a limited extent. Here an outstanding possibility exists to automate specific processes in product development and to force compliance with agreed-upon standards and norms (cf. [Fig88, Mer96, Lin84]).

Standard interfaces and modern systems engineering solutions increasingly support **communication between different systems**. Information search and acquisition, on the other hand, still depends very much on humans. Questions to ask are: What information and of what quality is put entered and maintained on the system? Does the user know its background? Is the information also used correctly?

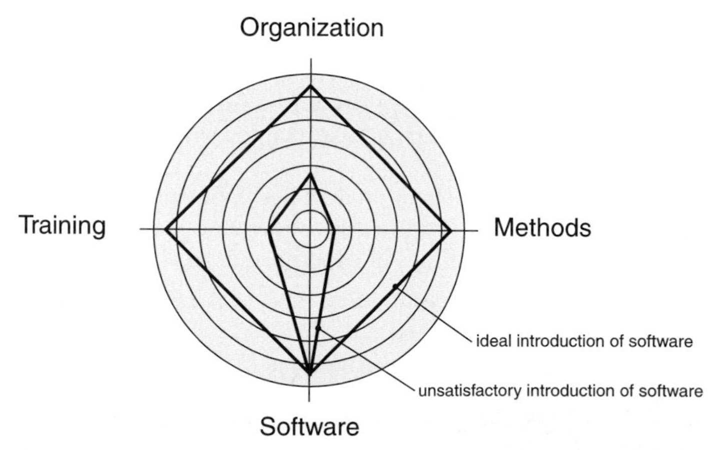

Fig. 6.2-3. Integrated introduction of modifications; example of software.

The question of documentation – its continuity, long-term availability, and usefulness – is a computer-related technical problem that is still not satisfactorily solved. Product documentation is legally required to remain available for more than ten years (40 years for car manufacturers). To get around the problems of software and hardware modifications, documentation is often "backed up" on older forms of information storage, such as paper or microfilms, or pixel images in addition to CAD data.

Selecting only the essential information required for the current task is often a great challenge (cf. [Ehr06]). In such cases, experience and the knowledge about the connections and support from other knowledge sources are necessary.

➔ During the introduction of new software, give equal weight to consideration of the applicable methods, the qualification of the personnel, and the adaptation of the organization.

Methods

While product development engineers often use as many as 20 different computer tools, they are often not truly qualified with regard to the application of design and development methods. However, if a very specific method application is considered necessary, personnel are **frequently** trained in very complex methods, for example FMEA, QFD or TRIZ (cf. [Alt84, Rei96]). Engineers often fail to understand the basics of the methodical procedure, or lack the ability to judge in the given situation which methods should be deployed, in which form, and to what depth (Sect. 3.2.3).

➔ Training is needed in methods and their application in different situations.

Knowledge management

To better utilize the available knowledge, knowledge management strives to explain the decisions through localizing knowledge and networking between knowledge people, as well as an increasing documentation of the development process. **Information** (cf. [Amb97, Fran97, Gra97]) is still **imparted predominantly through conversation**. Just the newest developments in computer-aided communications technology show the necessity of testing and using the new possibilities also in product development. **However, the deployment of all these methods based on a trust in technology alone is not critical, rather it is their use for the lasting increase in long-term efficiency.**

6.2.3 Capabilities regarding in-house product development

Personnel costs comprise the major portion of product development costs (about 60–80%). That leads to the question of the proper staffing level of, and the

strengths and weaknesses of, the in-house product development group when deciding whether to award product development work to outside sources. In combination with the issues pertaining to in-house versus external production, this question often leads to emotional rather than impartial decisions (Sect. 7.10).

Product development engineers have always relied on the development work of their partners (suppliers, design and development offices, etc.) in their own performance results. After decisions about which parts and assemblies are to be developed in-house are made, engineers must decide whether outside capacity is to be used to meet deadlines, or for financial reasons, or when supplementary external qualified personnel must be called in. Along with the question of in-house versus outside product development comes the question of **in-house** versus **external production**; this is discussed in greater detail in Sect. 7.10.3. There is no universal rule about this, and every company should develop a clear strategy and repeatedly check to what extent in-house product development is appropriate. An increase in product development outside the company can lead to a series of perspectives that can have a positive effect on company output. For example, **the fixed costs of product development** will be reduced in the company, and using external capacity will speed up the development process. One can take advantage of the particular qualifications and expertise in other companies, and build up and maintain strategic alliances with partner companies.

However, combined with that are also expenses and risks, including the possible loss of technological competence, dependence on product development partners, and increased coordination expense. Also, the product care must be appreciated as a long-term proposition [Wil96].

> ➔ Define, advance and enhance the core competencies in product development.
>
> ➔ Build up and advance product development partnerships.

6.3 Product development creates complexity in the company

As presented at the beginning of Sect. 6.2, product development activities influence not only the direct costs of product development, but also the complexity of all processes in the company. This influence can be substantial ("complex products usually engender complex processes"). The problem lies in that this complexity is not apparent in its essential nature and cannot be recognized by the usual control methods either.

Complexity in a company is determined by many things, but from a product development viewpoint, the following should be mentioned. (1) The number of different parts in the products (Sect. 7.12); (2) the number of the technologies used; (3) the number of the participating designers and development partners; and (4) the extent of networking among all these entities. Additional influences certainly come from the markets (which differ according to product), as well as from the organization of the company and other functions.

6.3.1 Costs of complexity

Complexity costs are those costs that result from the complexity of the product and the production processes. They become apparent through process cost accounting. In addition, complexity produces opportunity costs[8] that cannot be determined explicitly.

An example of opportunity costs of complexity is the commitment of valuable capacity in product development for preparation of variants and their care, which is thus unavailable for product innovation. This is similar what happens in production planning, cost control, and purchasing. A Pareto analysis may be applied here: frequently the highest amount of the sales (70–80%, is accounted for by only 20–30% of the variants designed (cf. Fig. 4.6-4).

Cautionary Note: If there are too many variants on the product side, secondary products will take the place of core products. Products that are produced at high costs due to their smaller quantities, and marketed without perceptible additional price, will work against the core products and thus lead to a reduction in earnings (Sect. 8.4.4).

If opportunity costs are disregarded, limiting the number of variants leads to considerable potential for cost reduction (cf. Fig. 6.3-3 and Sect. 7.12).

Other factors can also lead to high complexity in the company processes, such as an exaggerated in-house production capability, a high portion of outside designs, and procurement of material from a large number of suppliers (Sect. 7.12.3.1).

> ➔ Make the effects of complexity and complexity modification transparent.

6.3.2 Costs of part variety and technology complexity

Typical one-time complexity costs are the costs that arise in design and development for defining and testing new parts. Accordingly, production expenditures arise for planning, special production equipment, technology development as well as the costs of setting up series production. In addition, these one-time expenditures occur in other company functions, such as materials management, quality features, and service (cf. Sects. 7.5 and 7.12).

In the sense of the complexities described above, **administrative costs** for each additional part usually lie in the range of 1 500–2 000 $ for purchased parts and 3 000–3 500 $ or more for parts made in-house (**Fig. 6.3-1**). Administrative

[8] Opportunity costs consist of the profit contribution that must be forgone because the limited available resources cannot be used for the most favorable procedure (cf. [Ker96]).

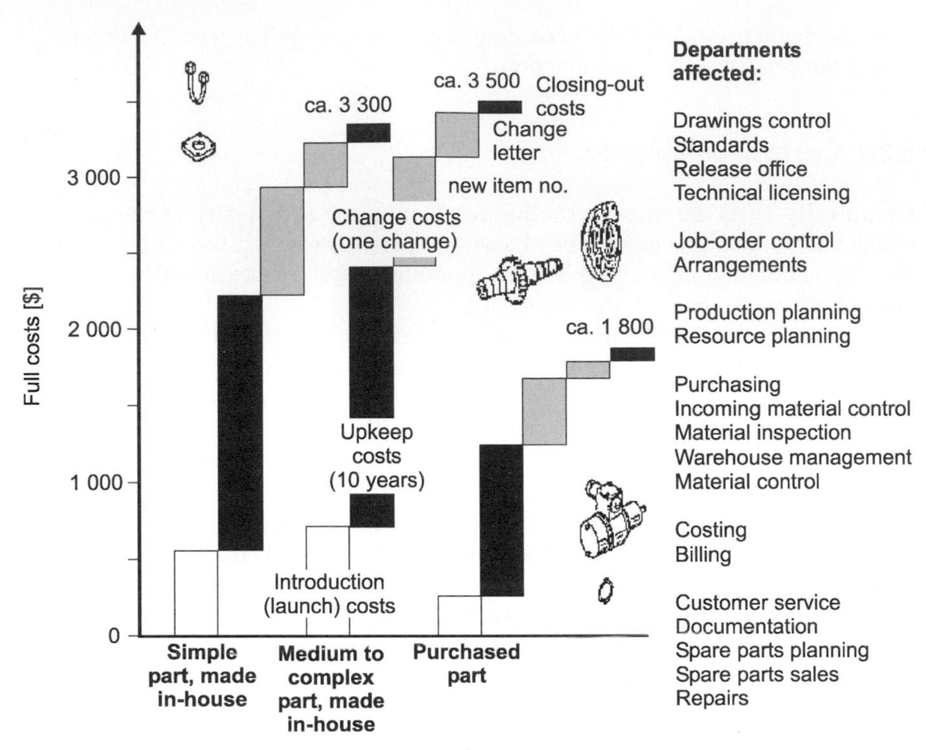

Fig. 6.3-1. Examples of administrative costs for parts (MTU diesel engines)

expenditures with the supplier are not considered here, but expenditures are included for documentation, job control, planning, production planning tasks, all logistics tasks, cost calculation and billing, as well as the entire milieu of spare parts and service. Costs for the design and development of the part are not included in these figuress.

Every part must be documented by, for example, technical drawings, parts lists, technical calculations, test reports, operating and service instructions, spare parts documentation, disposition rules, incoming goods vouchers, bills, production test records, work schedules, NC programs, tool settings, etc. At times some of these are needed in several languages. Having many parts therefore leads to needing many documents that are all in turn tied to activities.

Manufacturing processes for the parts are characterized by different production technologies. High-grade, deep-drawn body panels tend to wrinkle; case-hardened parts are subject to warping and residual stresses; and mating parts with tight tolerances complicate production control.

Thus, changes to parts result in changes in the procurement and production processes that had been running smoothly, risking disorder.

An investigation by Hichert [Hic85] shows how high the risk of chaos is when the number of the parts and thus the complexity increases in a company. **Figure 6.3-2** shows that within 10 years the number of the items in a product line more than doubled and the sales of each part therefore reduced significantly.

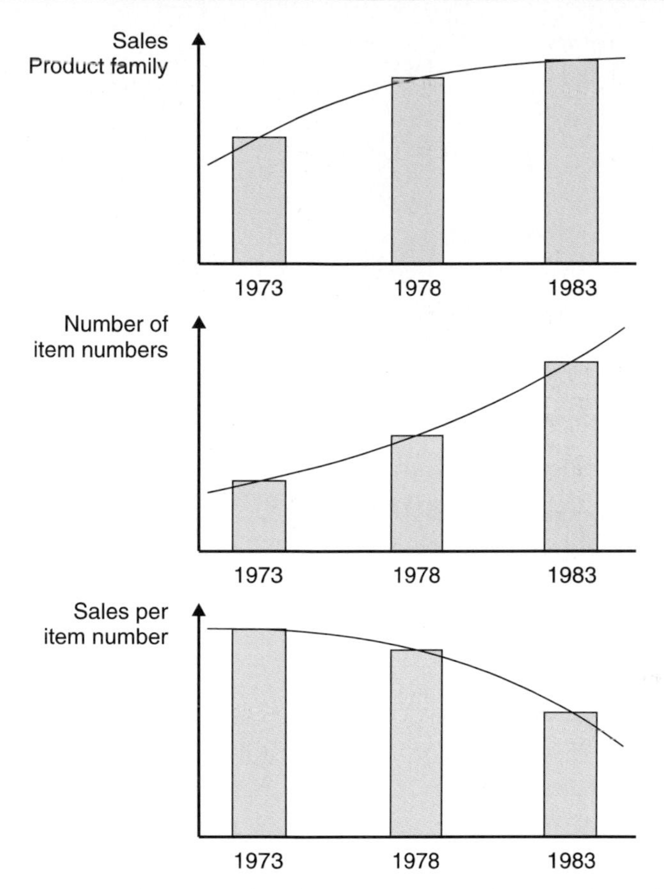

Fig. 6.3-2. The part quantity in a family of electrical products increased faster than the sales [Hic85]

> ➔ Search for existing or similar parts before deciding to develop entirely new parts.

6.3.3 Cost of product variants

Management consultants have repeatedly warned against having too many product variants. **Figure 6.3-3** shows that successful companies with a specific line of products achieve positive results with fewer products and significantly fewer assemblies and parts within their products.

Investigations in industry [Eve88] have shown that cost reductions of 10–20% of the entire total costs are possible if unnecessary variants are dispensed with. This results from savings in design and development, marketing and administration, and in a number of overhead costs that are assigned to manufacturing, production control, etc., see **Fig. 6.3-4**) [Schu89].

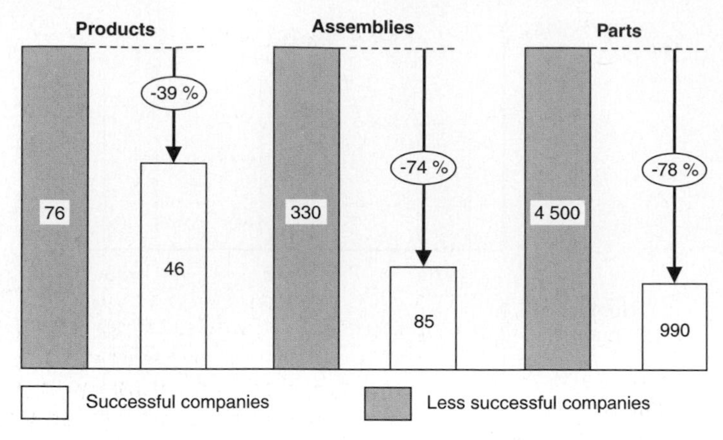

Fig. 6.3-3. Fewer variants; greater success [Hen93]

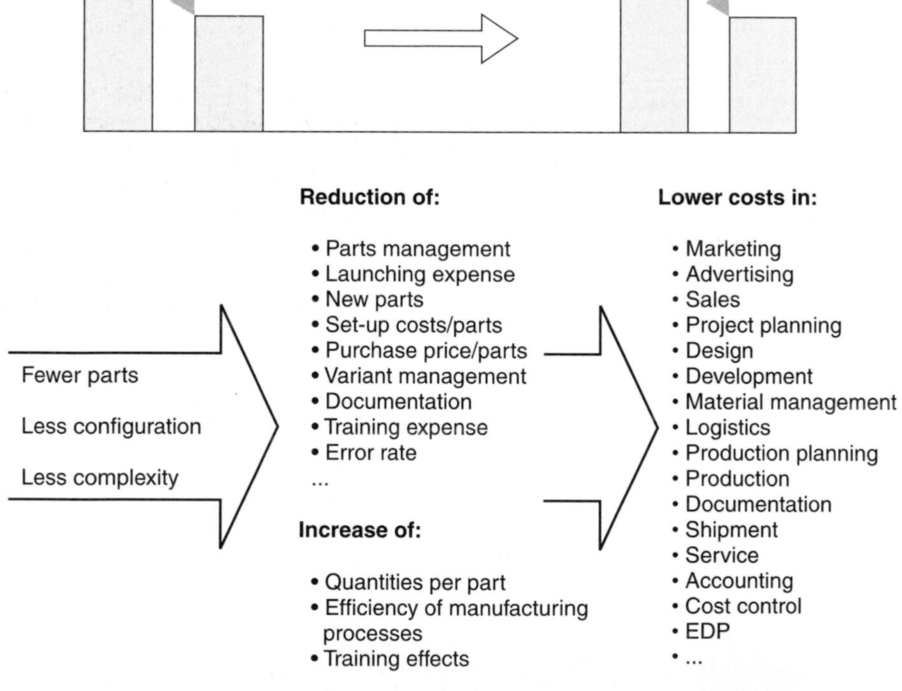

Fig. 6.3-4. Fewer variants mean cost reduction in many company functions

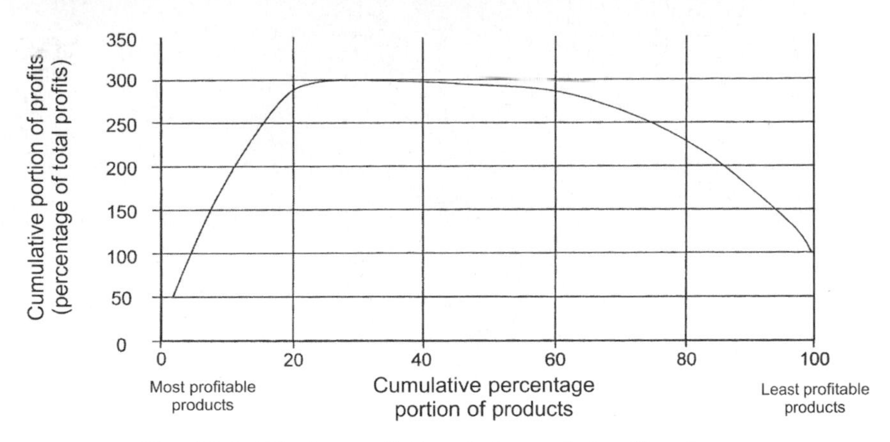

Fig. 6.3-5. "Whale curve": Cumulative profit for all products

The "whale curve" in **Fig. 6.3-5** shows the drastic effects of variant variety on a company's profits. According to Kaplan [Kap98], in this case about 20% of the products generate about 300% of the profit while the remaining 80% of the products cover their own costs, at best, or cause a loss so that in the end only the 100% profits remain. Every company must find its optimal product structure, considering the markets and their own situation. It is important to understand the often hidden cost effects of product variants. **Not every customer wish is considered a good wish from a business viewpoint**, but supplying companies often acquiesce due to competitive pressure. On the other hand, a customer preference can also be the basis for product innovation.

The goal should be to serve the market with a variant number consistent with the market needs, supported by a company-internal optimized standard, e. g., a modular system (cf. Sect. 7.12.6).

➔ Examine and reveal all of the consequences of new variants.

7 Factors that influence Manufacturing Costs and Procedures for Cost Reduction

*In cost accounting, manufacturing costs are determined by adding the material and production costs (part production and assembly). Manufacturing costs are a portion of the total costs (**Fig. 7.1-1**). Thus, the measures for cost reduction can be divided accordingly into three groups and each can be tackled in a similar way (**Fig. 7.1-2**).*

This chapter reviews a number of other factors, according to product and process features. The examples herein can stimulate in developing their own individual procedures. This, of course, must be adapted in each case to the given task.

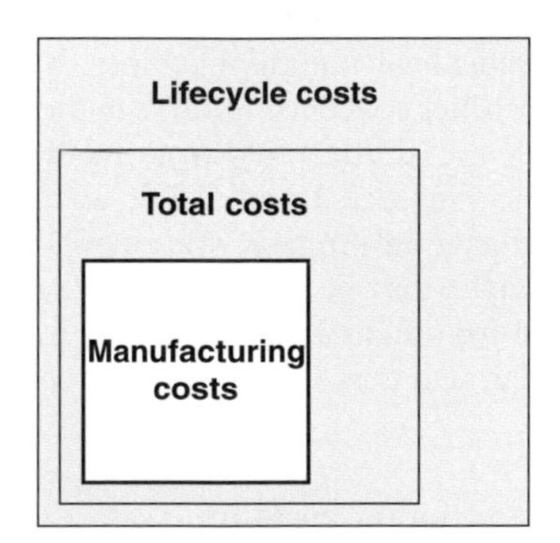

Fig. 7.1-1. Manufacturing, total, and lifecycle costs

7.1 Overview of the influences and their importance

From the profusion of factors that influence manufacturing costs, which are listed in **Fig. 7.1-3**, only a few essential ones are discussed in detail in this book. Almost all of the departments in a company have influence on costs. As the circles suggest, product development and production play the most essential part (Fig. 2.2-3). The overlapping area shows that most decisions should be made in close consultation between the production and product development departments.

It is the **concept** (i. e., the function principle) that primarily sets the costs. Examples of replacing old concepts with new, low-cost principles are: the combustion

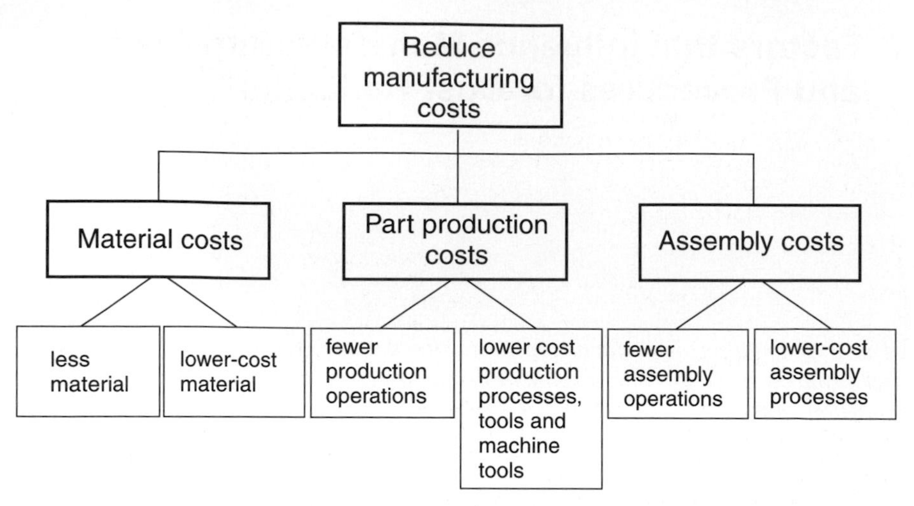

Fig. 7.1-2. Possibilities for the reduction of manufacturing costs [Kol94]

engine replacing the steam engine; ballpoint pens replacing fountain pens; and computers replacing mechanical calculating machines. It is clear that simply creating shape variants of these earlier concepts, relative to the new principles, would be ineffectual. The possibilities of influencing costs in the early stages of product development, shown in the Figs. 2.2-2 and 2.2-3, are valid only under specific conditions (Sect. 7.2). A change in the **task statement** also affects costs. A fully automatic camera, for example, may be more expensive than a non-automatic one, simply due to the demand for automation. However, upon consultation with the customer, the task statement, and thus the subsequent costs, can often be changed significantly.

Chief design factors influencing the costs are:

- **Task statement** (requirements)
- **Concept** (function principle; e. g., physical principle, with material type, number and type of the active surfaces, complexity)
- **Size** (e. g., dimensions, amount of material)
- **Quantities** produced, and thus standardization, particularly in the case of single-unit and limited-lot production
- **Production and Assembly technology**, influenced strongly by materials, quantities produced, and sizes

If these factors influencing the costs cannot be changed, other parameters become more prominent, such as tolerances, surface finish and shape details. For detail designers, these become the primary factors.

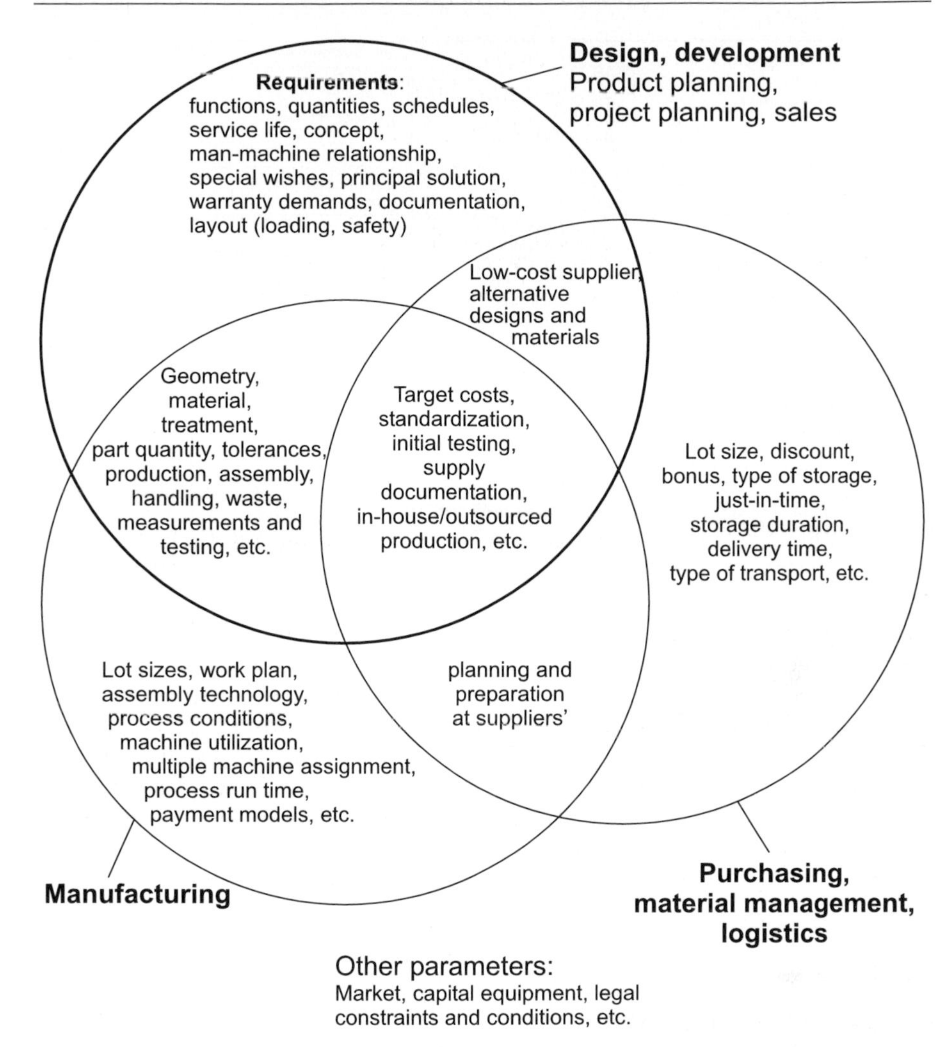

Fig. 7.1-3. Examples of parameters affecting the manufacturing costs

We can look at the influence on design not only according to certain parameters, but according to also the affected cost categories, as shown in **Fig. 7.1-4.** It is evident that overhead costs and parts of the total costs (Chap. 6) can also be changed by constructive measures.

The influences on manufacturing costs analyzed in the following sections will be elucidated by examples from the practice. These arise from our own investigations and from the literature. Since, in the latter case, the underlying data and limiting conditions are often not given, it usually is not possible to use these numerical values for one's own applications. On the other hand, trend statements or rules are often portable (Sects. 7.13.2b and 9.3.7.2, Fig. 7.13-2).

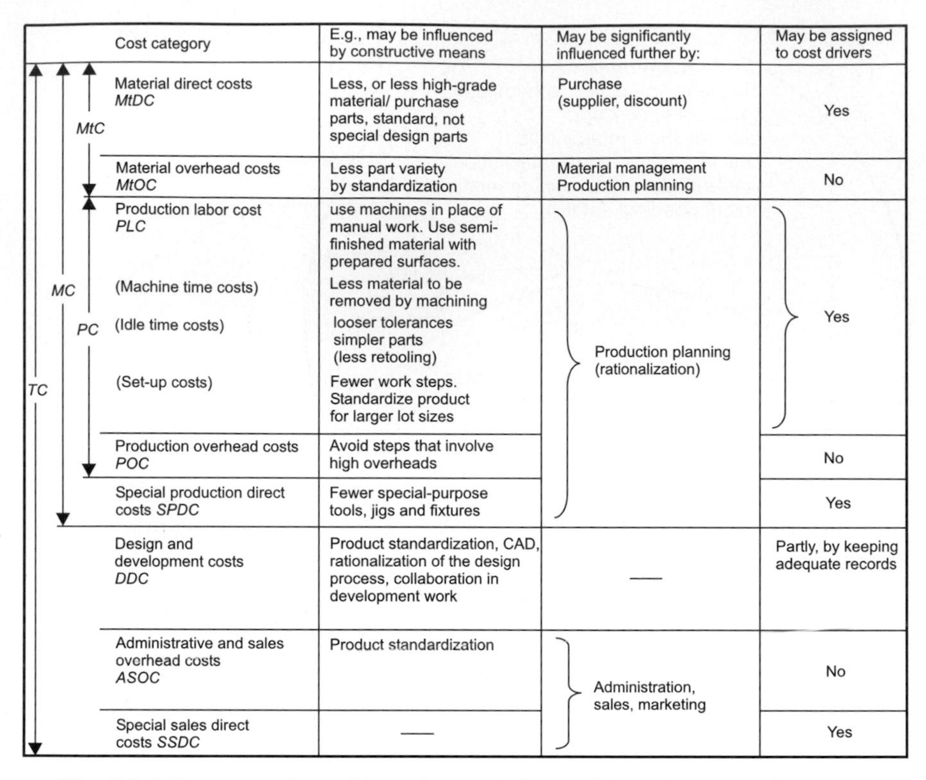

Cost category	E.g., may be influenced by constructive means	May be significantly influenced further by:	May be assigned to cost drivers
Material direct costs *MtDC*	Less, or less high-grade material/ purchase parts, standard, not special design parts	Purchase (supplier, discount)	Yes
Material overhead costs *MtOC*	Less part variety by standardization	Material management Production planning	No
Production labor cost *PLC*	use machines in place of manual work. Use semi-finished material with prepared surfaces.	Production planning (rationalization)	Yes
(Machine time costs)	Less material to be removed by machining		
(Idle time costs)	looser tolerances simpler parts (less retooling)		
(Set-up costs)	Fewer work steps. Standardize product for larger lot sizes		
Production overhead costs *POC*	Avoid steps that involve high overheads		No
Special production direct costs *SPDC*	Fewer special-purpose tools, jigs and fixtures		Yes
Design and development costs *DDC*	Product standardization, CAD, rationalization of the design process, collaboration in development work	—	Partly, by keeping adequate records
Administrative and sales overhead costs *ASOC*	Product standardization	Administration, sales, marketing	No
Special sales direct costs *SSDC*	—		Yes

Fig. 7.1-4. Cost categories and how they are influenced through constructive means

7.2 Influence of the task statement

A product and its costs are defined by the requirements in the task statement and the available solution options [Ehr77]. Every demand and its corresponding constraint should be considered not only for its technical implementation, but also for the costs arising from it. "Demands cost money". For every requirement, one should note the permissible costs, as in target costing.

That requirements give rise to costs is shown by the example of the development of an automobile (**Fig. 7.2-1**). With time, each additional requirement leads to an increase of costs. Therefore, target costing starts with functions; value analysis defines function costs and raises the question "How much does the fulfillment of each function cost"? This can be easily answered by analyzing other available products (value improvement), since the solutions for the function are known. We need only sum up the costs for the components accountable for the whole or partial function. For products not yet designed, function costs also can be estimated, if a possible solution for the function is known or the function can be implemented with purchased parts. However, it is not possible to estimate the cost effect of

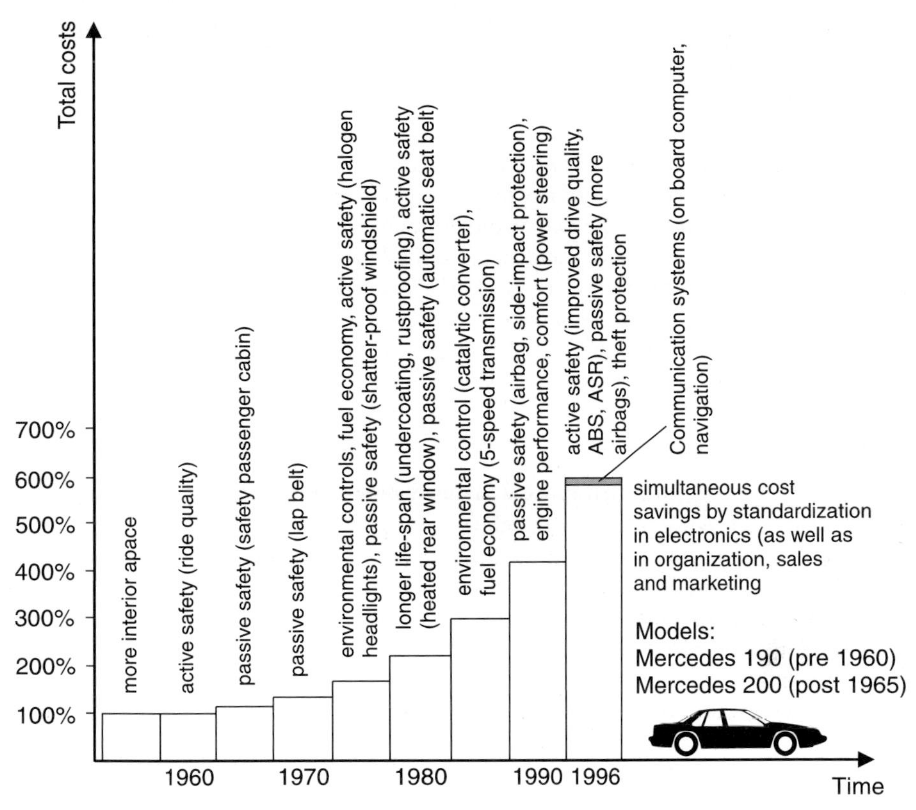

Fig. 7.2-1. Gradual increase of the requirements on an automobile with time, and the corresponding rise in costs

functional demands when no solution or even no approximate solution is known. In this case, the solutions must first be worked out and their costs estimated or calculated (Sect. 4.6.2). The same holds true for nonfunctional requirements such as safety, operability, serviceability, and recyclability. People must have sufficient experience and the necessary technical measures with known cost effects if they intend to determine the effect of parameters such as specific noise limits, or defined availability or life span.

➔ Product costs decrease if lower requirements, tolerances, warranties, acceptance conditions, prescribed standards or specifications are agreed upon.

7.3 Influence of the concept

The influence of the concept (basic solution) on the manufacturing costs is enormous (Sect. 4.8.2). In the concept, the product is defined in its essential properties, individual processes, and their order in the function structure. The types of energy and the physical effects, by which the subfunctions are implemented, are established. Thereby, and with the active movements and surfaces, essential decisions concerning product complexity and size are made (Sect. 4.5.2).

To achieve a significant cost reduction, a proven design methodology must be firmly observed at the concept stage (cf. Sect. 4.8.2). However, an inquiry [Bei74] and an empirical investigation [Dyl91], showed that less than 10% of the entire product development time is spent in design.

The cost-reducing effect of new concepts and new designs (which are fraught with risk), will be shown with examples below (see also Sect. 7.9.2.2):

- Development of a low-cost **optical laser system** (see Sect. 4.7).
- Development of the first **pocket calculators**:
  Although the basic task statement of calculating machines has not changed for many years, their costs and prices have fallen by a factor of approximately one hundred over the past ten years. Semiconductor technology introduced a completely new physical method of implementation. Greater market advances due to price reductions allowed further rationalization possibilities because of larger quantities produced. The size reduction also allowed considerable decrease in the cost of materials, which are dominant costs in mass production. Calculators could be made even smaller, since their present size accommodates the man-machine interface. The display cannot be made any smaller due to the human finger size and the legibility of the buttons.

  This example shows that a transformation from a mechanical signal to an electrical signal is possible with minimal energy and mass requirements. Such a substantial cost reduction seems conceivable only for signal transformation. With the energy and material conversion, such a size and mass reduction is generally not feasible. Nevertheless, since the introduction of the Otto engine, a decrease in the specific weight by a factor of 1 000 has come about.
- **Membrane switch**:
  Figure 7.3-1 shows how a new concept involving significantly fewer part numbers can drastically reduce the product size and costs through use of new materials and production processes. The membrane switch, suitable only for low current flows, (to the left in the figure) functions with half the number of parts compared with the electromechanical switch, and is only about 0.5 mm thick. Function integration made this possible by printing a conductor on polyester film. Its resiliency combines a push-button and a spring, and the imprinted connection replaces the contact springs (Sect. 7.12.4.3). An even greater cost reduction becomes possible if the manufactured quantities increase due to a broader range of applications (Fig. 7.5-1).

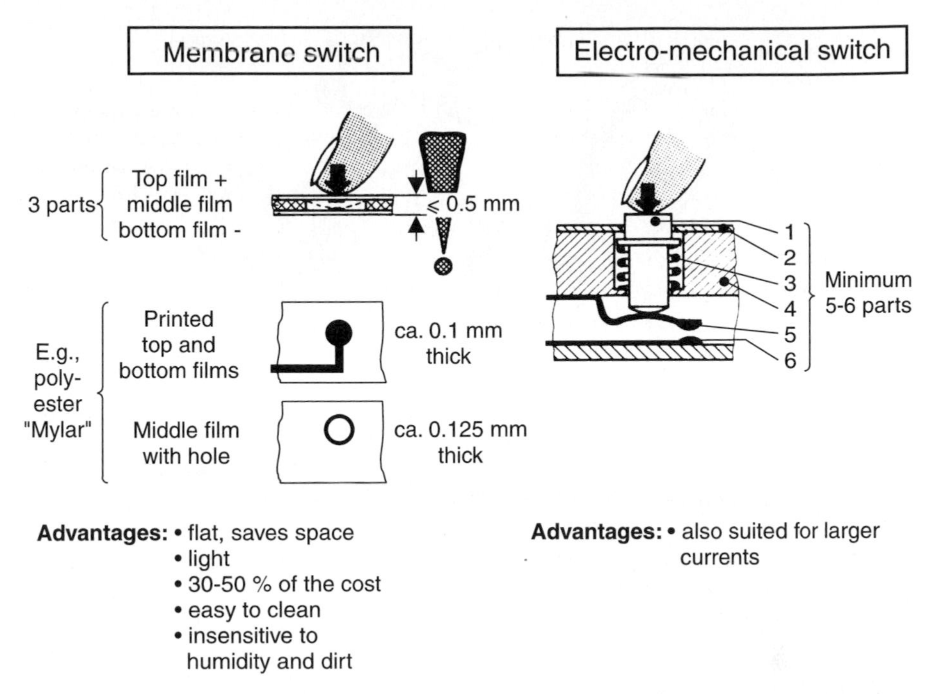

Fig. 7.3-1. Comparison of a membrane switch and an electromechanical switch. The membrane switch is small, light and economical, due to a new concept and new production processes

- **Concrete mixer**:
  The example in Sect. 10.1 will show how a machine retains the physical principle of material conversion (mixing function) while its manufacturing costs are lowered by about 30% by changes in the drive concept and the manufacturing process.

- **Four gear train concepts** for gear ratio $i = 10$:
  Figure 7.3-2 shows (not to scale) four different gear set concepts for spur gears, which realize a gear ratio of $i = 10$ [Fis83]:
  - single-stage gear sets;
  - two-stage gear sets;
  - two-path gear sets;
  - three-path gear sets.

Besides the concept, the size was varied, affected by the applied torque T_1. The spur-toothed gears are manufactured in single-unit production, ground from case-hardened steel (16 MnCr 5). The load-carrying capacity was calculated according to DIN 3990.

The manufacturing costs of the gears were computed with a computational program [Ehr82a], which contains mean production times and standard rates of German gear manufacturers. The cost comparison considers only the gears, not the shafts, bearings, housing, and assembly. The computations showed:

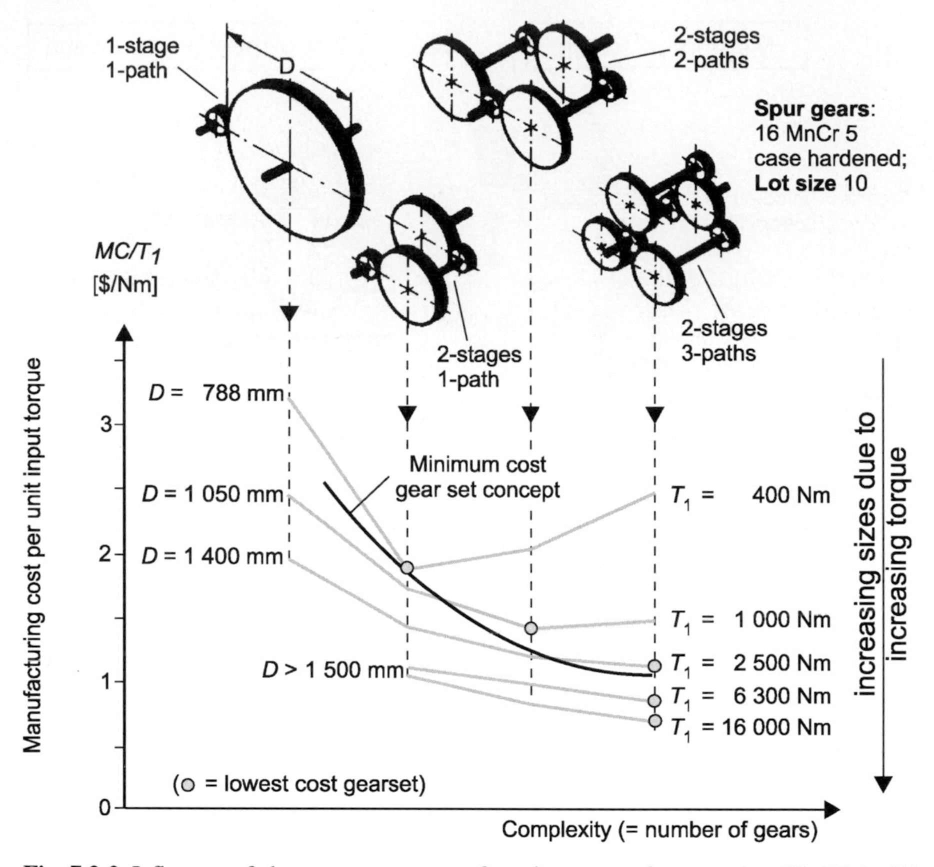

Fig. 7.3-2. Influence of the concept on manufacturing costs of gear trains [Fis83] (*valid only within the parameters shown!*) Gear-ratio $i = 10$

- The gear set weight per unit of transmitted torque (and thus the size per unit of torque) decreases strongly with increasing complexity (not presented in Fig. 7.3-2). For the three-path gear set, the weight is only 38% of the single-stage gear set. This happens due to three effects: 1) the partitioning into two stages reduces the center distance; 2) the Hertzian pressure becomes smaller in the case of lower ratios per stage (two-stage gear set), since the curvature ratios become more favorable; and 3) in multiple-path gear sets, the central gears contact more than one gear; thus, the torque diminishes for each contact to 1/2, 1/3, etc., of the single-contact value. See also Fig. 7.9-5.
- The lowest-cost concepts are different, depending on torque (size). For a small size (i. e., a low torque T_1), the two-stage gear set is most favorable. For a larger size, first the (two-stage) two-path gear set, then the three-path gear set (see curve: minimum cost gear set concept) are the best. We see that the gear set concept has at least as much influence on cost as the size, and therefore also counts as one of the main cost factors.

The reasons for this behavior are as follows. We see from the falling curves that the greater the torque, the smaller the costs per unit of torque ($/Nm). The different growth laws explain this. The torque increases with otherwise constant parameters with the third power of the size (φ_L^3). Large portions of manufacturing costs (see Sect. 7.6) increase only with $\varphi_L^{0,5}$ and φ_L^2 (Fig. 7.6-3), particularly with smaller products and single-unit production. Accordingly, the quotient $/Nm decreases with increasing size (for accuracy of rules derived from that, see Sect. 4.6.4).

On the other hand, the production costs increase with an increasing number of gears (complexity). That is caused by, among other factors, the idle times (clamping, adjusting, and measuring, etc.). Therefore, for the small torque of 400 Nm, the curve increases again with the increasing number of parts.

Why do the manufacturing costs per unit of torque become smaller as the gear sets become more "branched", as in the case of large gear sets torques in the 2 500–16 000 Nm range? This is because the manufacturing costs of large gear sets basically contain high portions of material and heat treatment costs which can amount to 50–60% of the total costs (Fig. 7.6-3 gives an idea of this). Since increased branching drastically lowers the gear set volume due to the multiple utilization of the gears, the material costs and the heat treatment costs (which are proportional to the weight) also decline (Sect. 7.13.5). The larger the gear set torque is (and thus the size), the more this influence outweighs the opposing effect of the increasing number of gears. We see how interconnected the influences on the manufacturing costs are. In addition, the cost minimum still depends on the overall speed ratio. If the ratio were $i = 5$ instead of $i = 10$, the one-stage gear set becomes most economical.

Cargo ship transmission:
In Fig. 5.2-3 we saw the comparison of two marine engine concepts, a slow-speed (120 rpm) large diesel engine connected directly to the propeller with 8 800 kW output, and a medium-speed (430 rpm) diesel engine of the same output with a planetary gear reducer. In spite of the additional gear box, the weight decreases in the latter case by around 65%, and the shipyard costs around 30%. An additional advantage is the smaller space required by the faster engine, which leaves room for more freight. We can show that other influences on the operating costs (fuel costs, maintenance, reliability, etc.) further raise this capital cost advantage, by using lifecycle cost accounting. This advantage existed before the oil crisis of 1965–74, which is why this concept was successfully implemented in about 25 freighters [Ehr73]. The basis for the size and cost reduction is that a constant driving power at the engine ($P = T\omega$) can be realized either with a large torque T_1 at a low rotation speed ω, or the other way around, by high rotational speed and a small torque. Thus, there are small, light, multiple-cylinder, and lower-cost engines.

→ **Concepts for small and lightweight design** usually result in low-cost machines. Machines with strong physical effects become small and light (for example by using mechanical and hydrostatic energy), by parallel connection of active surfaces (branching of power), and increase in speed. In general, the cost decrease is less than the decrease in weight (Fig. 5.2-3).

→ **Concepts with simple design and few parts** (e. g., function integration, integral design) are usually more economical, particularly with small products or made in large quantities.

7.4 Influence of shape

Shape or form refers to the entirety of the geometrical features of a physical product. The product's size and the number and position of its elements (surfaces, machine elements) are also related to this. Shape also refers to the tolerances in the sense of a macro- or micro-shape (i. e., the coarseness of the surfaces). It is easy to understand that the essential cost paramcters are dependent on shape. These are the size (Sect. 7.6), the material, the surface treatment (Sect. 7.9), and often the production processes (Sect. 7.11), including the assembly.

Shape is determined in the concretization steps of preliminary and detail design (see Fig. 4.4-2, Sect. 4.5.2). However, the preconditions for a product's shape are already present in the requirements and at the concept stage. Thus, establishing the shape is an essential constructive activity through all steps of product development. In Fig. 4.8-2, the design that follows the strategy "from the important to the less important" emphasizes shape; the strategy is divided into shaping the decisive modules (or parts, assemblies) and shaping the entire product [VDI98].

There really is no universal **action plan** we can recommend. A product's features, which are established during shape design, are very inter-linked, as shown in Fig. 7.11-1. How shape design is started depends very much on the task statement, the product type, what is already defined, etc. The function-related dimensions are established during design (Sect. 7.8), which are associated with material selection and the manufacturing process. After this comes the design, which takes into account the production and assembly (Fig. 7.11-2) in more detail.

Even if the procedure can only be roughly structured, the **procedure in detail** is much better known while setting the individual product features. As the designers' empirical observations have shown [Dyl91], it occurs after the **procedure cycle** (Fig. 4.4-1). At every decision pertaining to dimensions, tolerances, form, etc., the requirements must be clarified, solutions searched for, and the final solution selected. This procedure is **often not** at all **conscious** Practitioners often develop it themselves with their increasing experience, in ordinarily in the right direction. Nevertheless, the process should be rationally clarified on important items, and consciously worked through in the group or organization. Clarifying the requirements generally remains on course, or only **one** first "best" solution is foreseen.

7.5 Influence of the production quantity

The strong influence of quantity on the costs of identical products becomes clear if we think about how expensive the first cars, television sets, or electronic pocket calculators were when they were originally made in small quantities. Only a few people could afford such luxury. But the costs and prices have come down significantly, mainly through efficient, rationalized production processes and appropriate designs. What had been accessible by the privileged few is today taken for granted by all. Cost-reducing production processes are, however, only possible if correspondingly high quantities are produced. **Figure 7.5-1** schematically shows the "influence spiral" for the transition from single-unit to series production. To achieve throughput in large quantities, the product is first redesigned so it can be produced in a rationalized process. This leads to a possible price reduction due to more efficient production, which, in turn, could enlarge the market share, and so on. This scheme is essentially also valid for serial production. Redesigning for greater production quantities can come about by applying suitable production processes or through internal standardization in accordance with Fig. 7.12-4 (see also Sect. 7.12.4.1, Fig. 7.12-18).

The quantity that may be supplied to the customers is essentially market-dependent. Through product standardization, the design has a decisive influence on the quantity of similar parts or assemblies produced in-house in single-unit and limited-lot production (Sect. 7.12). With modular-designed products and those produced in varying size ranges, a compromise is reached between specific customer preferences (e. g., single-unit production) and larger quantities.

Essential for understanding the influence of quantity is the distinction between the **one-time** costs (independent of the quantity) occurring during the product development and the beginning of series production **(Fig. 7.5-2)** and the **running** production costs (costs per piece). This is similar to the differentiation into fixed and variable costs, or into direct and overhead costs (Sect. 7.12.2 and 8.3.2,

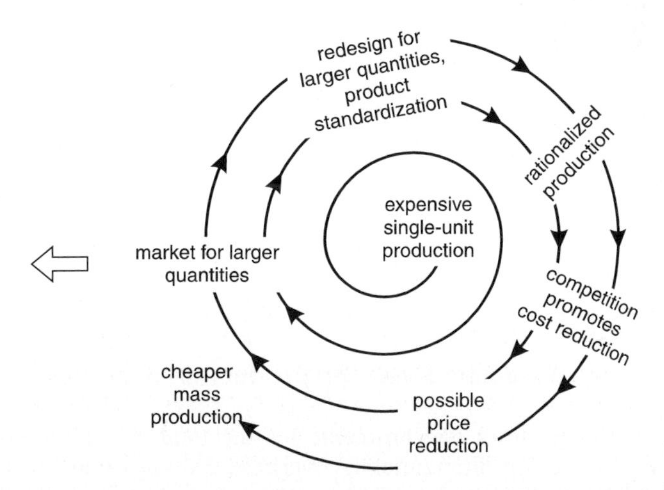

Fig. 7.5-1. The "influence spiral" for transition from single-unit to mass production

Procedures that incur one-time costs

1 Administration Marketing/Sales Project planning	• Publicity expense • Bid preparation • Order negotiations • Order confirmation • Order documentation • Order accounting
2 Design Development	• Design and drawings preparation, parts list • Calculations • Records
3 Production planning	• Production plan preparation • Production time calculation • Production control • Records
4 Production	• Setup times, learning steps for assembly, tests • Special items (jigs & fixtures, models, tools) • Internal transport facilities
5 Purchasing	• Offer procurement • Ordering steps • Scheduling • Shortfall surcharges • Billing control, payments
6 Material storage	• Entry control • Storage steps • Material provision
7 Shipment	• Shipment method • Paper work • Packaging

Fig. 7.5-2. Activities that cause one-time costs, independent of whether they are to be charged to the entire manufactured quantity N_{tot}, or to the lot size N of a product

Fig. 2.1-2). According to the procedure being considered, there are different definitions of terms.

Terms relating to the quantity produced:

- Quantity N_{tot} (total quantity produced)
- Number of lots L_n
- Lot size N ($\Sigma N = N_{tot}$), the quantity N_{tot} is frequently produced in not one but in several lots, usually in different quantities.

7.5.1 Processes associated with the production quantity

a) Costs in **(new) product development**, for the total produced quantity N_{tot}, which are charged using differentiating overhead costing usual in the machine industry (Sect. 8.4.2), are not proportional to the quantity, are the following:

- **One-time costs**: All costs which result for the product development: costs for planning, market analysis, conceptual, preliminary and detail design, calculations, tests, prototype models and specific production investments. These include also the model and tool costs (inclusive of assembly, if they are not included in the special production direct costs *SPDC*), first-time sales, marketing and advertising costs (not the special sales and marketing direct costs *SSDC*), etc. (see Sect. 7.11.2.2a, Fig. 6.3-1).
- **Terms**: Introductory costs *IC* [Ehr85], also:
  Product development costs referred to one product *DDC*p.
  Advance payments.
  Production run fixed costs.

  These costs should be converted to the total produced quantity N_{tot} **per product unit**, which is not often the case:
  Introductory costs per product $IC_P = IC/N_{tot}$

b) Manufacturing costs for production of lot size *N* of a product $\Sigma N = N_{tot}$

- **One-time costs**: One can distinguish between **direct costs** in production (setting up the machines = **production cost from setup times** (setup costs), *PCs,* and **indirect costs** in production-related fields (e. g., design and development, production planning, and possible procurement costs, including logistics costs, for the initiation of the lot order).
- **Terms**: Production costs from setup times (*PCs*, setup costs).
  One time costs C_{one} [Ehr85],
  Fixed costs, lot.
- **Calculation per product unit** with produced lot size *N*:
  Production costs from set-up times, per piece = *PCs*/*N*
  One time costs per piece = C_{one}/N

In cost accounting practice, the drop (degression) in the costs with increasing quantity is considered only infrequently. Today it is still common to assign the costs of product introduction (*IC* = marketing, project planning, design and development, production planning and materials management) in a lump sum as manufacturing overhead cost. In this way, all products, regardless of the quantity, are equally affected (Sect. 6.1, Sect. 8.4.4, Fig. 8.4-2). Only the costs for models, jigs, fixtures, and special production material (special production direct costs, *SPDCs*) are considered separately and related to the produced quantity. The decrease in production costs arises today primarily in overhead costing, only when the setup costs (*PCs*) are taken into consideration, which are spread over the lot size *N*. In this respect, it is correct to say that in overhead costing (and in the case of mixed-quantity production), the actual cost reduction as quantities increase is much larger than what is usually calculated. Process cost accounting takes more into account as to how the costs arise (Sect. 8.4.6) [Buc99, Sch96, Sto99].

7.5.2 Why costs come down with increasing production quantity

There are four specific causes for cost degression with increasing quantity:

a) Costs reduction through division of one-time costs

The product introduction (launch) costs *IC* and one-time costs C_{one} are divided by the total quantity N_{tot} or the lot size *N*. If only the production costs are obtained from set-up times *PCs*, can the manufacturing costs per piece be derived with the production costs from total time units *PCe* (Fig. 7.6-2):

$$MC_N = \frac{PCs}{N} + PCe + MtC \left[\frac{\$}{piece} \right] \quad (7.5/1)$$

The production costs associated with setup times per piece decline as a hyperbola; they are added to the production costs arising for each piece from total time units (*PCe*) and materials costs (*MtC*) (**Fig. 7.5-3**). Depending on the cost structure for the realization of a piece, the quantity-based reduction may be strong or weak. As an example, if the setup time accounts for 80% of the production costs in single-unit production, then in the production of two identical parts, each becomes 40% cheaper. If the setup time portion of the production costs is only 20%, then for the production of two parts, each becomes only 10% cheaper. We can see the influence of setup times on production costs, and the sensitivity regarding quantity-based reduction from these cost structures.

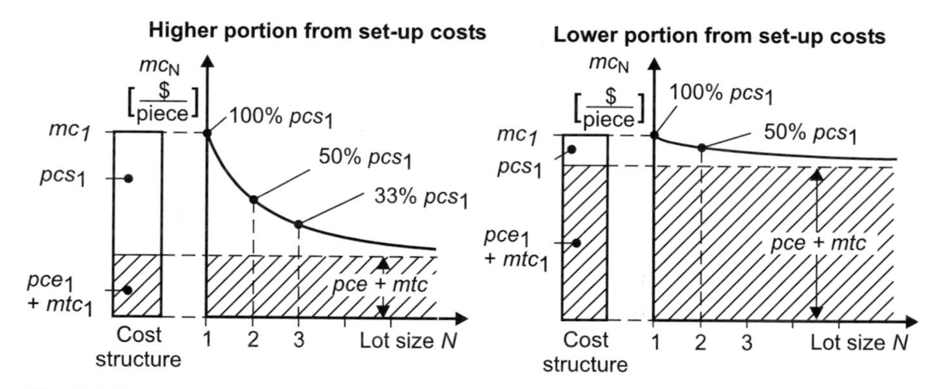

Fig. 7.5-3. Drop in manufacturing costs for products with high portion of set-up costs (*left*) and low portion of set-up costs (*right*), depending on the lot size (*see also Figs 7.7-1 to 7.7-6*)

b) Cost reduction due to training effect

We know that doing a job that we are not familiar with becomes increasingly easier from repetition, since one is trained in the intellectual and manual processes. This is valid for all activities – for designing, selling, and ordering; for preparing work schedules; or for doing assembly work or packing machine parts [Bro66a, Bro96, DeJ56]. As we see from **Fig. 7.5-4** [Bau78], the effect can be

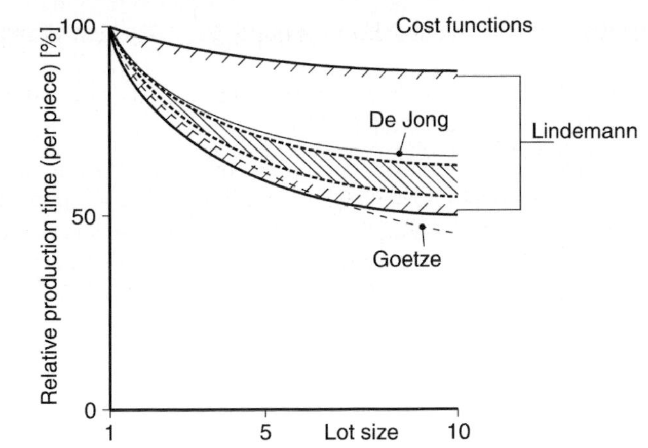

Fig. 7.5-4. Drop in production costs as function of lot size (partly due to training effect [Bau78])

considerable: In the tenth repetition of the operation, the time required is only about 60% of that needed the first time!

This influence is too seldom or insufficiently considered in cost calculations (Fig. 7.5-4). However, we know from experience that a prototype is usually more expensive than first estimated, and that after starting up the production run the costs turn out to be lower than were estimated. In this regard the terms "practice degression" and "starting curve" are used. In the literature [Bro66a, Bro96] an estimating formula is given, which is similar to Eq. (7.5/1):

$$t_n = t_1\left(\frac{1-t_E}{n^{\alpha}}+t_E\right)\left[\frac{\text{time}}{\text{piece, process}}\right] \qquad (7.5/2)$$

with t_n = Piece time for the nth process (run);
t_E = Piece-proportional, non-reducible part of the first time t_1 (corresponding to the above piece-proportional costs);
$1-t_E$ = time portion reducible by quantities;
α = process exponent;
n = number of the processes carried out.

Also here, as in a) above, a distinction is made between a non-reducible time part t_E, the piece-dependent costs $PCe+MtC$ and a reducible part $(1-t_E)$ that corresponds to the one-time set-up costs PCs.

Corresponding to the time periods according to Fig. 7.5-4, that are valid on the average for many other investigations [DeJ56], the results are $t_E=0.315$ and $\alpha=0.322$. The non-reducible part of the first time t1 is therefore 31.5%. Thus Eq. (7.5/2) becomes:

$$t_n = t_1\left(\frac{1-0.315}{n^{0.322}}+0.315\right)\left[\frac{\text{time}}{\text{piece, process}}\right] \qquad (7.5/3)$$

Thus, the needed time t_n is reduced by around 20% for each doubling of the quantity (scatter range: 15–25%).

c) Cost reduction through optimized design

Products are designed very differently, depending on quantity to be produced. On one hand longer design times and thus higher design and development costs are reasonable for larger quantities, since they are spread over the larger quantity (see paragraph a) above. On the other hand, the improvements thus achieved are worthwhile, even if they are small on a per-piece basis (Sect. 7.5.2d). As an example here we need only point out a size comparison of industrial and automobile transmissions. Industrial transmissions are build much larger, since they do not have to be, or can be, so highly optimized in the detail (life span, weight, space, connections, etc.) as automobile transmissions.

d) Cost reduction through more effective production processes

For every production process, there is a range for the produced quantity, within which, relative to other processes, it costs the least. This is shown in Fig. 7.11-5. Production processes that are more efficient (i. e., which show low per-unit costs, *MC*, for larger lot sizes) usually have higher one-time costs. These can be capital expenditures that are then divided by the quantity of the products made during the service life, or added to the unit costs in the form of depreciation. These one-time costs per piece are, in general, lower for processes that are more efficient. Ordinarily, the production times, which govern the per-piece production costs, are considerably shorter. It is therefore a matter of choosing the optimal production processes for the given quantity, and to design the product for those processes.

Figure 7.5-5 gives an example of quantity degression in the manufacture of passenger car engines [Bro66a, Bro96, Der71]. The per-piece production costs

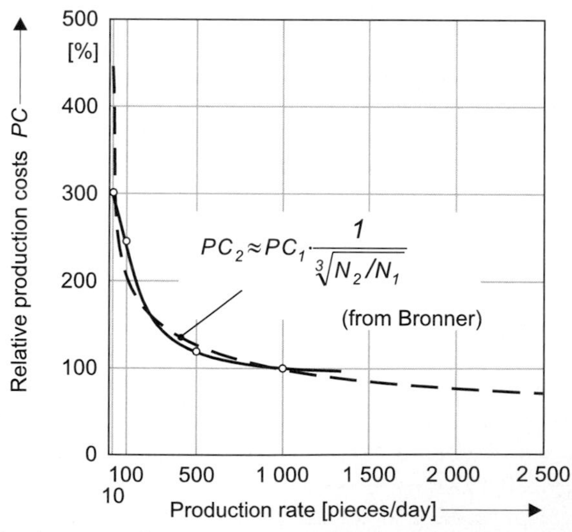

Fig. 7.5-5. Production costs of a passenger car engine depend on the production rate [Der71]

drop down to approximately 25% if, instead of 10 engines per day, 1 000 are produced. The equation shown by Bronner for the relative production costs resembles the equation for training effect (7.5/3)

$$PC_2 = PC_1 \left(\frac{N_2}{N_1}\right)^{-0.322} \approx PC_1 \cdot \frac{1}{\sqrt[3]{N_2 / N_1}} \qquad (7.5/4)$$

and shows a **reduction of around 20% for each doubling in quantity**. In a first approximation, one can use the reciprocal of the ratio of the third root of the quotient of the two quantities. This cost reduction is largely independent of the product type. Although it considers only the production labor and overhead costs (not the material costs), it has nevertheless proved itself in estimating the **manufacturing costs** (see below) in the first approximation. This is because the material costs have a weaker quantity degression, particularly for standardized parts, since they are already produced in large quantities.

e) Cost reduction due to quantity discount

Quantity discount is common when large amounts of raw material or parts are purchased. In investigating gears, Ehr82a, Ehr82b] determined that, depending on the order quantity and negotiation talent of the procuring company, the price ratios for common, totally similar steels could range from 1:1.5 to 1:2 (Fig. 7.13-2). In addition, differences in the annual delivered quantity are important. Fig. 7.9-11 shows the price drop for various stainless steels, depending on the order quantity [Eck77]. Without additional negotiations, there can be a 35% price difference between order quantities of 5 000 kg as opposed to 100 kg.

Also, an investigation by Boston Consulting revealed that for the electric motors and turbo-chargers produced by the Brown-Boveri Co., **a doubling of the output resulted** an average **cost (price) reduction of approximately 20–30%** [Tre78]. This is also true for their semi-finished products such as laminated sheets and semiconductors, and is inflation-adjusted (Boston Consulting empirical data). Since these products and materials are charged as material costs in finished products, this corresponds to a long-term quantity discount over an observation period of several years.

The rationalization measures carried out in the course of time probably also make a difference of a few percent per year. It is common knowledge in the automobile industry that manufactures' suppliers have to reduce their prices by a few percent per year.

7.6 Influence of size and dimensions

The size of a product has as strong an effect on the costs of the product as do the concept and quantity. It is an old engineering rule that size reduction (i. e., **small design**) will lower the costs as long it not carried to extremes. For example, in airplanes, racing cars, missiles, or satellites the development, testing, prototype costs, and costs for special production processes might cancel the effect of the

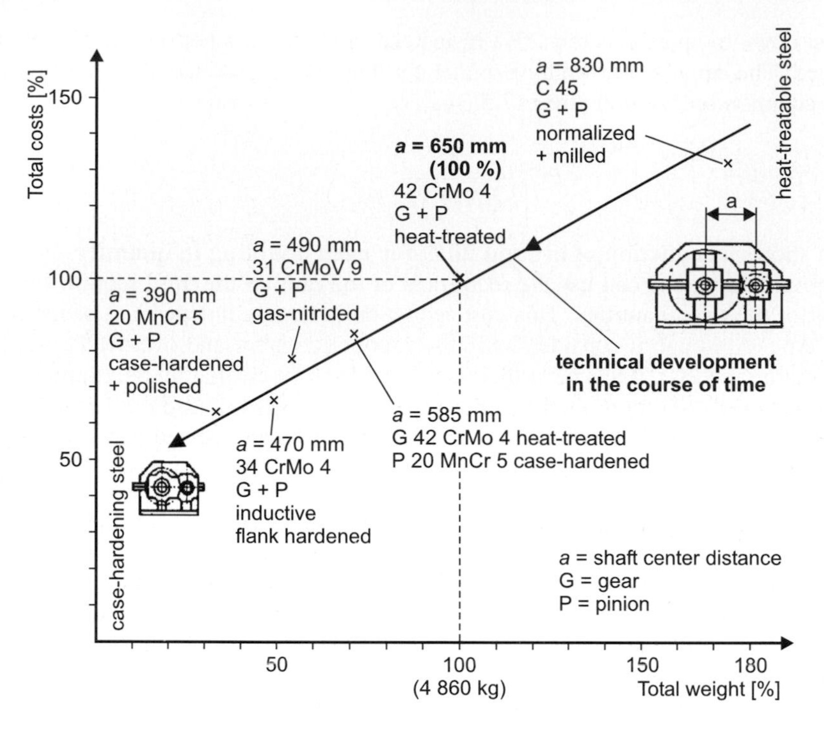

Fig. 7.6-1. Cost, weight and size reduction for a gear train (for $T_{1\mathrm{nom}} = 21\,400\,\mathrm{Nm}$; pinion speed $n_1 = 500\,\mathrm{rpm}$; speed ratio $i = 3$; welded housing, single-unit production; data from [Nie83])

reduction in material costs by lightweight design. In such cases, more expensive special materials are often necessary, so that only the weight is reduced, not the costs. **Figure 7.6-1** shows such correlations for gears. Slightly more expensive materials that are case-hardened and ground, reduce the weight and the manufacturing or the total costs (see Sect. 7.8).

7.6.1 Overall growth laws for costs

a) It is obvious that a machine's material costs (MtC) increase as a first approximation, proportional to the material volume, therefore with the third power of the linear dimension or size ratio $\varphi_L = L_1/L_0$. Here L_0 is the typical linear dimension of the basic (initial) embodiment and L_1 that of the succeeding embodiment; see Sect. 7.12.5.3. This is valid for geometrically similar components, and it is understood that the material costs per unit volume are constant. However, they are actually not so. On one hand, the production costs for processing the material (e. g., semi-finished material) grow at a rate less than $\varphi_L^{\,3}$ (for example, proportional to the surface area, thus $\varphi_L^{\,2}$). Also, for material purchased in large quantities, there is a quantity discount. On the other hand, the raw materials must be produced differently for forming very large parts (e. g., shafts use free-form forged instead of drawn material). Thc material cost per piece (MtC1) is assessed, therefore, based on that for size 0:

$$MtC_1 = MtC_0 \cdot \varphi_L^{2.4...3} \tag{7.6/1}$$

In a cost analysis of gears (Sect. 7.13.3) [Ehr82a, Fis83], made of 16 MnCr 5, it was found that the weight-dependent costs grew proportional to $\varphi_L{}^{2.4}$ (Ø 50–200 mm), or to $\varphi_L{}^{3.0}$ (Ø 600–1 500 mm; Fig. 7.6-3).

b) The **production costs from total time units** (*PCe*) for machining increase proportionally to the surface area (i. e., with $\varphi_L{}^2$) for finish machining and grinding and proportionally to the volume cut (with $\varphi_L{}^3$) for rough machining. According to Bronner [Bro66a, Bro96, VDI87] we can write:

$$PCe_1 = PCe_0 \cdot \varphi_L^{1.8...2.2} \tag{7.6/2}$$

According to Bronner, the exponent 1.8 in the above equation is valid for mass production, and exponent 2.2 is valid for single-unit production. According to Lindemann [Lin80], an exponent of 1.8 works for single-unit production on a bar or engine lathe for some rough machining, and goes up to 2.0 when a large amount of rough machining is involved. An FVA gear investigation described in Sect. 7.13.3 found an exponent of 0.8 to be valid for small gears of 50–200 mm diameter in single-unit production, and 1.9 for large gears of 200–1 000 mm diameter. Apparently for small gears the idle times, which grow nearly not with the size, are dominant. It is assumed in these more thorough investigations that all portions of the total time units t_e (production time t_m, idle time t_i, recovery time t_{re} and extra time t_x) change similarly (classification of the manufacturing times is shown in **Fig. 7.6-2**). Since in industry one seldom identifies production and idle times separately, rather the single time t_e is considered. Thus, a practical and detailed investigation is hardly possible. As shown in Fig. 9.3-7, different manufacturing processes have different exponents [Pah79, Rie82].

c) The **production costs from setup times** (setup costs) *PCs* of the production are not usually established like the above production costs from the manufacturing process by using the physically necessary time required, since they vary more widely [Ehr82a, Käs74, Lan72]. It is clear that the setup costs increase with the size of the work piece because larger machines and fixtures require more time for work preparation. As an example, large work pieces can only be moved with the help of a crane. Thus, they increase, not continuously but in steps, depending on the production equipment used.

Our own investigations [Ehr82a] on machined parts (gears) of a few kg to approximately 1 500 kg showed that as a first formulation the mean value is:

$$PCs_1 = PCs_0 \cdot \varphi_L^{0.5} \tag{7.6/3}$$

For gears, the exponent increases with size [Ehr82a, Fis83]. For 50–200 mm diameter, it is 0.14, for 200–1 000 mm 0.56, for 1 000–1 500 mm as high as 1.8. Other investigations [Rie82] show it varying between 0 and 0.5.

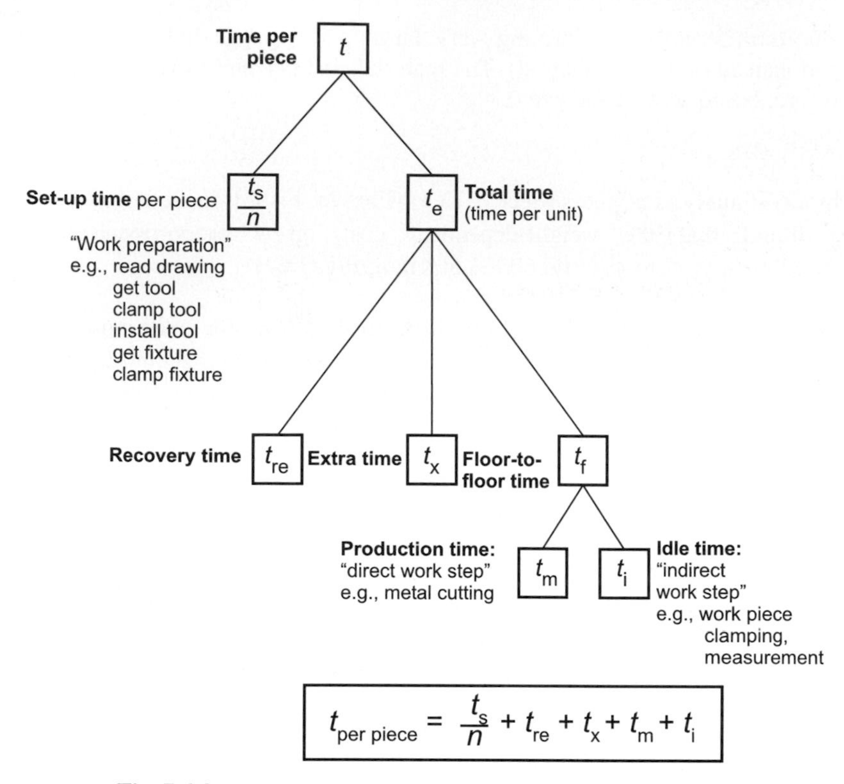

Fig. 7.6-2. Classification of standard times, according to REFA

d) If we summarize the findings from the cost analysis of gears concerning the **influence of size**, we get the mean value of all company calculations, as shown graphically in **Fig. 7.6-3**. Strictly speaking, the following statements are valid only for single-unit production of gears. It is claimed that they are applicable also to other similar critical parts, assemblies and products (for example, pump rotor disks, turbine and engine parts). This is true, provided a standard production method is followed and that similar production technology is used. For important parts it is advisable to check the company's internal cost growth laws and cost structures, similar to those in Fig. 7.6-3 and Fig. 7.7-1.

Figure 7.6-3 shows that for small parts manufacturing costs increase at first slowly, proportional to the linear ratio $\varphi_L = L_1/L_0$. Later they increase more steeply, with φ_L^2 and finally for very large parts, with φ_L^3 proportional to the volume. The basis for that is the similar relationship of the three cost categories: With small parts, the material costs may be neglected, however they dominate for large parts, growing according to φ_L^3 (i. e., proportional to volume). On the other hand, with small parts the setup costs predominate, increasing with approximately $\varphi_L^{0.5}$. We see that they form a band on top of the material costs, *MtC*, and production costs based on total times, *PCe*: For small gears they constitute 80–90% of the manufacturing costs, with large ones only a few percent. The influence of quantity is accordingly different, as

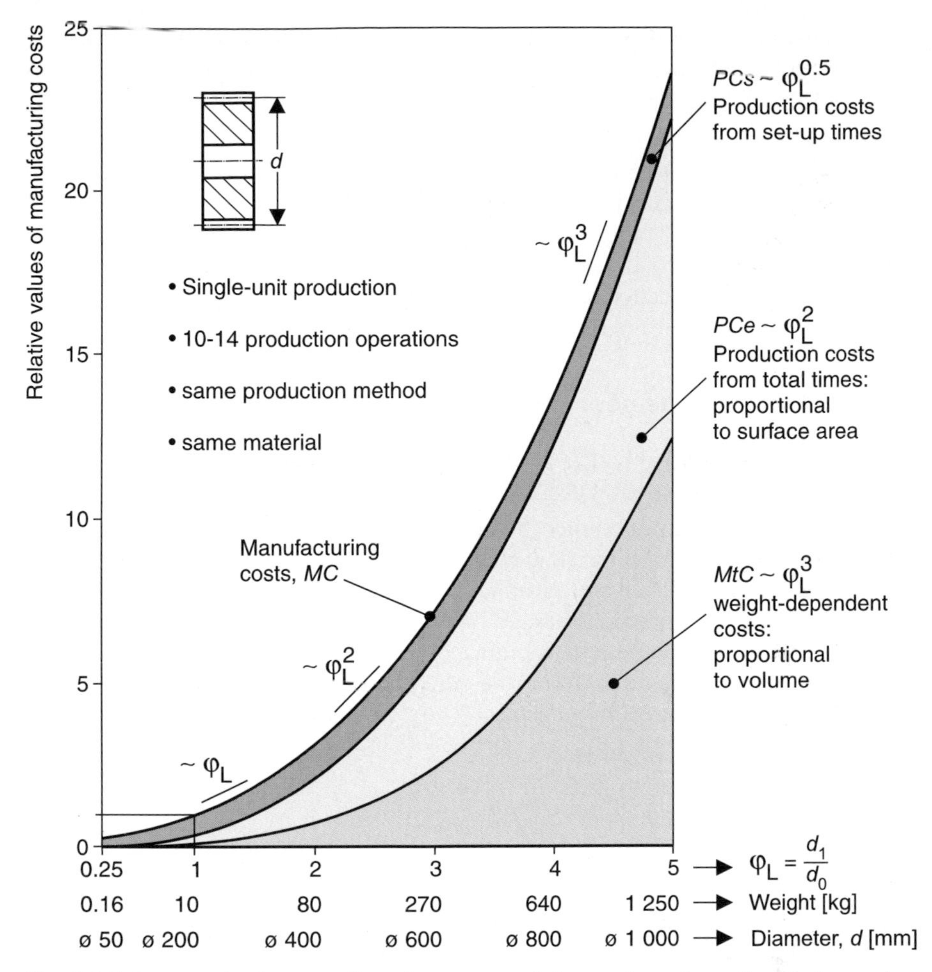

Fig. 7.6-3. Increase in the manufacturing costs and the portions thereof, with increasing size, for single-unit production of case-hardened and ground gears [Fis83]

explained in Sect. 7.7. From the curve, the following findings can be derived, which are well known to some extent. These are formulated like a rule:

> ➔ **Small design** decreases the manufacturing costs, particularly in the case of large parts in single-unit production. This is also valid for small **and** large parts in series production. In series production, small design therefore always leads to lower costs.

The last statement regarding series production is based on the high setup costs of small parts that decrease when spread over a larger quantity. So what are left are the strongly size-dependent material costs *MtC* and production costs from total times *PCe* (Fig. 7.7-1).

The cost structure of small parts in series production is similar to that of much larger parts in single-unit production. Accordingly, one can then also estimate starting from the weight.

➔ The **weight cost calculation** [$/kg = f(weight)] is especially suitable for estimating the manufacturing costs of large parts (products) in single-unit production, or for all parts (products) made in series production. The necessary condition is that only the size is changed, but the design layout and production remain virtually identical (Sect. 9.3.2.1).

7.6.2 Influence of the geometrical relationships of active surfaces

In the case of cylindrical bodies in particular, whose active surface is at the periphery, the design decision always arises as to how the ratio of width b to diameter d is chosen. Examples of such bodies are gears, friction wheels, radial bearings, drum filters, or tangential flow machines such as tangential blowers, through flow water turbines, paper and printing machines, etc. Should one rather make narrow, disk-shaped or long, "sausage-like" shapes? It is clear that for the latter there are technical limits due to bending or twisting effects. As we can realize from following, there is no uniform rule other than it depends on the growth laws as to which shape is more advantageous.

a) Spur gears

In a cost benchmarking study, as described in Sect. 7.13.3, the manufacturing costs of hardened and ground spur gears made of 16 MnCr 5 from 12 gear companies were analyzed. The question was, which gear sets (speed ratio = 3.55) based on the torque capacity ($/Nm) are the best: narrow (b/d of the pinion = 0.3) or wide (b/d = 1.2). It turned out that it makes almost no difference in the cost per unit torque whether the gear sets are narrow or wide. This is due to the fact, at least for very large gear sets, that the torque based on the Hertzian pressure increases in first approximation with the diameter squared and linearly with the width. Therefore:

$$\varphi_T = \varphi_d^2 \cdot \varphi_b = \varphi_L^3$$

The manufacturing costs increase roughly proportional to volume, since the material and heat treatment costs (Sect. 7.6.1) are dominant. Thus, we again get:

$$\varphi_{MC} \approx \varphi_d^2 \cdot \varphi_b = \varphi_L^3$$

In the first approximation, therefore, the quotient of costs and torque, $/Nm, remains constant.

For small sizes (i. e., small torques) the parts of the manufacturing costs containing φ_L^0, φ_L^1 and φ_L^2 predominate (Fig. 7.6-3). Accordingly the manufacturing costs change only little with measurements. That is valid both for the width and for the diameter. This behavior is to be expected.

Experience from the practice shows that gears are more economical when the greater the ratio *b*/*d* is technically feasible. The insignificance of *b*/*d* described here, on the manufacturing costs of gear trains based on torque, is therefore opposite to the experience of gear manufacturers. Nevertheless, this is confirmed by the investigation, as shown in Fig. 7.13-14. According to that data, nearly cube-shaped housings are around 13% cheaper than long, narrow ones containing a gear set transmitting the same torque. The tendency to "design for width" is caused therefore by a decrease of the housing cost and not by the gears. If the function (for gears, the torque; for filters, the filter performance) increases proportionally, not to volume but to the surface area, as shown in the following, other ratios come about.

b) Drum filters

Vacuum and pressure filters have rotating drums for solid/liquid separation processes. The decisive active surface for the function, which carries the filtering material, is the cylindrical surface of the drum. The vacuum or pressure moves the solid/liquid medium through this material. The cake being left on the material is removed at the periphery. The filter performance is proportional to this active surface area πdb. Here also we have to make the design decision whether to build narrow disks or wide rolls for the same filter surface.

In **Fig. 7.6-4**, the complete manufacturing costs of 30 filter designs are shown plotted against the filter surface area. We see that the manufacturing costs drop up to approximately 65% per m^2 filter surface when the ratio of drum width *b* to drum diameter *d* of $b/d = 1.2$ instead of $b/d = 0.6$ is used. Wide drums therefore manifest lower-cost filters than narrow disk-shaped drums for the same filter surface.

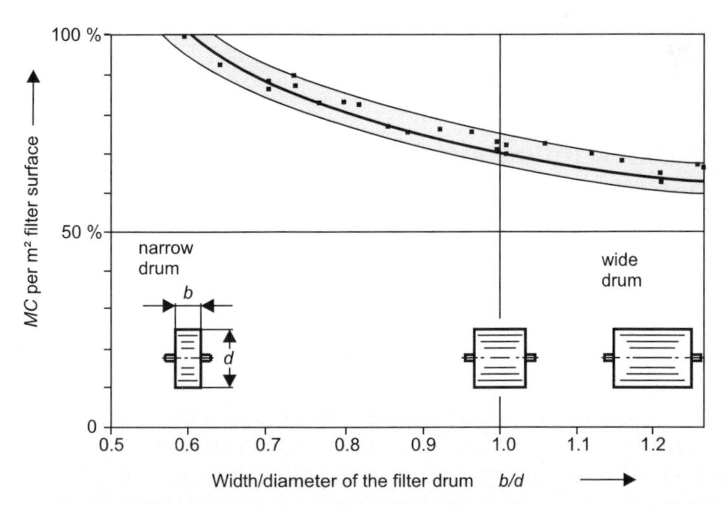

Fig. 7.6-4. Influence of diameter *d* and width *b* of a filter drum on the manufacturing costs per m^2 filter surface (BHS Pressure and Vacuum Filters)

The probable basis for these relationships between measurements and manufacturing costs is the following: In the first approximation, the manufacturing costs are proportional to the surface of the can-like filter drums of welded stainless steel sheets:

$$MC \sim \text{Surface of (end faces + cylinder)} = \frac{\pi d^2}{4} \cdot 2 + \pi\, db$$

The filter performance is $P_F \sim \pi db$. That results in

$$\frac{\text{Manufacturing cost } MC}{\text{Filter performance } P_F} \sim \frac{\pi \dfrac{d^2}{2} + \pi db}{\pi db} \sim \frac{1}{2\dfrac{b}{d}} + 1$$

Therefore, the manufacturing costs based on filter performance decrease as b/d increases. There is a similar result if the manufacturing costs are set proportional to the drum volume. If the manufacturing costs as well as the filter performance were to increase proportional to the filter surface πdb, then there would be no influence of b/d. The ratios would act similar as shown above in the case of gears. Now the end faces add a part increasing with d^2. This is to be held small relative to the cylindrical surface. Small end faces, however, mean long slim filters.

c) Through-flow water turbines, paper and printing machines

Similar investigations on drum-shaped water turbines of the Ossberger system (through-flow turbines) showed the same trends. For a given performance increase (i. e., an increase in blade active surface) it is more economical to bring about the blade surface increase by widening the drum (an option that might have technical limitations) instead of an increase in diameter. Experiences with cylindrical paper machines and printing presses are similar.

In [VDI97], by applying dimensional analysis, a variety of examples of the influence of measurement ratios on costs are worked out.

7.7 Combined influence of size and the production quantity

As shown in the previous section, size influences each of the three cost components, production costs from setup times *PCs*, production costs from total time units *PCe* and material costs *MtC*, to a different extent. The quantity *N* (lot size) for its part has an especially strong effect on the setup costs per part, since these are simply divided by *N*. There is thus a close connection between the two parameters in the case of single-unit and limited-lot production. The observations are valid for machine tools, with and without NC operation. If the programming cost is included in the setup cost, it adds up to significant costs in the NC production. Idle time costs (handling, measurement times) are reduced with NC programs and shifted to the one-time costs (Sect. 7.5.1b).

7.7.1 Formal relationships

We can now combine Eq. (7.5/1) with Eqs. (7.6/1) to (7.6/3) and get the manufacturing costs per piece of a size *i* for a lot size *N* in the case of single-unit and limited-lot production:

$$MC_{iN} = \frac{PCs_{01}}{N} \cdot \varphi_L^{0.5} + PCe_0 \cdot \varphi_L^2 + MtC_0 \cdot \varphi_L^3 \left[\frac{\$}{piece}\right] \tag{7.7/1}$$

Here it is assumed that the production process does not change (conventional machining processes) and that the one-time costs C_{one} consist only of the set-up costs *PCs* (Sect. 7.5.1b).

In reality the quantity effect will be even greater, since the production costs from total time units also decrease with the quantity *N* (training effect, other production processes) and since the material costs also decrease with the quantity (Sect. 7.9). A general approach (see Fig. 7.7-6) must be established for the given production process:

$$MC_{iN} = \frac{PCs_{01}}{N^\alpha} \cdot \varphi_L^\delta + \frac{PCe_{01}}{N^\beta} \cdot \varphi_L^\varepsilon + \frac{MtC_{01}}{N^\gamma} \cdot \varphi_L^\varsigma \left[\frac{\$}{piece}\right] \tag{7.7/2}$$

We factor out the manufacturing costs MC_{01} for single-unit production for size 0, so that only the cost elements remain in the brackets (lower-case alphabet) from (7.7/1). These represent the cost structure:

$$MC_{iN} = MC_{01} \left(\underbrace{\frac{pcs_{01}}{N} \cdot \varphi_L^{0.5} + pce_0 \cdot \varphi_L^2 + mtc_0 \cdot \varphi_L^3}_{W} \right) \left[\frac{\$}{piece}\right] \tag{7.7/3}$$

Accordingly, the starting point for the calculation are the cost structure for single-unit production ($N=1$) of the size 0 (basic embodiment)

$$1 = pcs_{01} + pce_0 + mtc_0 \tag{7.7/4}$$

and the absolute value of the manufacturing costs for this size MC_{01} in single-unit production. From this, the new manufacturing costs MC_{iN} may be calculated in a specific field of this size 0 and not too large a lot size deviation (no new production processes may be used). Since the cost structure is a notable means for recognizing focal points for reducing costs, the new cost structure is especially interesting. We label the expression in parentheses in Eq. (7.7/3) as *W*. The new cost structure becomes:

$$\frac{MC_{iN}}{MC_{01} \cdot W} = 1$$

$$= \frac{1}{W} \left(\frac{pcs_{01}}{N} \cdot \varphi_L^{0.5} + pce_0 \cdot \varphi_L^2 + mtc_0 \cdot \varphi_L^3 \right) = pcs_{iN} + pce_i + mtc_i \tag{7.7/5}$$

The individual terms on the right side of Eq. (7.7/5) thus represent the parts of the new cost structure.

Note: In this book **capital letters (*MC, MtC, PCs, PCe*)** are used for absolute quantities and **lower-case letters (*mc, mtc, pcs, pce*)** for quantities as fractions or percentages (usually referred to *MC*).

7.7.2 Calculation example

Task:
Following data are given for the component of a double gear coupling shown in **Fig. 7.7-1**:

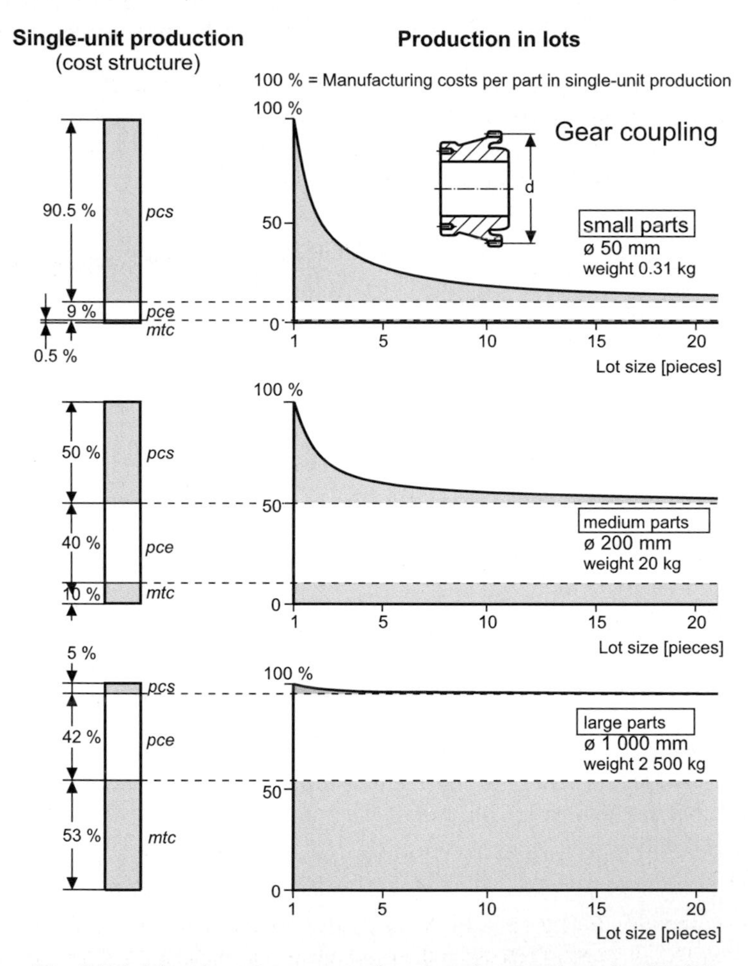

Fig. 7.7-1. Manufacturing costs as function of part and lot size [Fis83]

- **Technical data:**
  - Pitch diameter $d_0 = 200$ mm,
  - Weight $m_0 = 20$ kg,
  - Material: 16 MnCr 5,
  - case-hardened and ground,
  - 13 production operations;
- **Cost structure (single-unit production):**
  - 50% of the production costs from set-up times pcs_{01},
  - 40% of the production costs from total time units pce_0,
  - 10% of the material and heat treatment costs mtc_0 (Fig. 7.7-1).
- **Costs:**
  - Manufacturing costs for single-unit production $MC_{01} = 500$ \$,
  - Cost structure $pcs_{01} + pce_0 + mtc_0 = 1 = 0.5 + 0.4 + 0.1$

Sought are the weights, manufacturing costs, the new cost structure and the variation of the manufacturing costs for quantities ranging from 1 to 20, and for pitch diameter of 50 to 1 000 mm. The purpose of this is to get rough values for quick project planning, as depicted in Fig. 7.7-1.

Result:
The example calculation given here is for a pitch diameter of $d_1 = 1\,000$ mm and quantity $N = 1$:

The linear size ratio becomes:

$$\varphi_L = \frac{d_1}{d_0} = \frac{1\,000 \text{ mm}}{200 \text{ mm}} = 5$$

Therefore, the new weight is:

$$m_1 = m_0 \cdot \varphi_L^3 = 20 \text{ kg} \cdot 125 = 2\,500 \text{ kg}$$

The expression W in Eq. (7.7/3) for lot size $N = 1$ is:

$$W = \frac{0.5}{1} \cdot 5^{0.5} + 0.4 \cdot 5^2 + 0.1 \cdot 5^3 = 1.12 + 10 + 12.5 = 23.62$$

The new manufacturing costs, according to Eq. (7.7/3), become:

$$MC_{11} = MC_{01} \cdot W = 500\,\$ \cdot 23.62 = 11800\,\$$$

and the new cost structure according to Eq. (7.7/5) is:

$$1 = \frac{1}{23.62}(1.12 + 10 + 12.5) = 0.05 + 0.42 + 0.53 = fkr_{11} + fke_1 + mk_1$$

Thus the manufacturing costs of the coupling component consist of the following three parts: 5% production costs from set-up times, 42% production costs from total time units and 53% material costs and weight-proportional heat treatment costs (e. g., hardening). These and other results of the calculation are displayed in Fig. 7.7-1 (see also Sects. 7.12.5.3 and 9.3.5).

We recognize the following important points from this example:

- **Large parts** (Fig. 7.7-1, bottom) have high portions of material costs and weight -proportional heat treatment costs. This comes about (also shown in Fig. 7.6-3) due to their growth proportional to φ_L^3. The production costs from total time units are the second important part. On the other hand, the production costs from setup times are almost negligible. Thus, there is hardly any reduction in the manufacturing costs with quantity.
- **Small parts** (Fig. 7.7-1, top), in contrast to large parts, have a negligible portion of materials cost (less than 1%). Also the production costs from total time units are proportionately still very small. On the other hand, the production costs from set-up times are dominant. The quantity degression is corresponding high. With the set-up costs at 90% of the total, the per-piece manufacturing costs of two identical parts are 45% lower.

The design rules resulting from these findings are formulated at the end of Sect. 7.7.3.

7.7.3 Example of spur gears, other parts, and rules

The numerical values used in the equations in Sect. 7.7.1 are based on cost analysis of gears from 12 companies making transmissions, as cited in [Ehr82a] (Sect. 7.13.3). The influence of quantity and size on manufacturing costs is depicted in **Fig. 7.7-2**. Costs for large gears were not calculated for large lot sizes in the investigation, since that is unrealistic. The modification of the cost structure is shown **Fig. 7.7-3**. We see here also how the production costs from setup times decrease and the weight-dependent costs increase. **Figure 7.7-4** shows how the cost elements of the production processes change with size. Specific production processes such as "mill teeth" and "grind teeth" dominate for small gears. From such cost structures the focal points for cost reduction can be recognized.

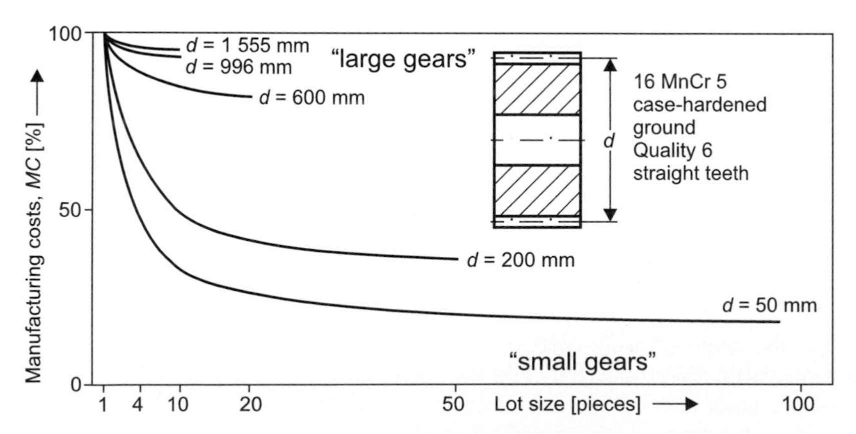

Fig. 7.7-2. Influence of lot size and pitch diameter on the manufacturing costs of a spur gear [Fis83]

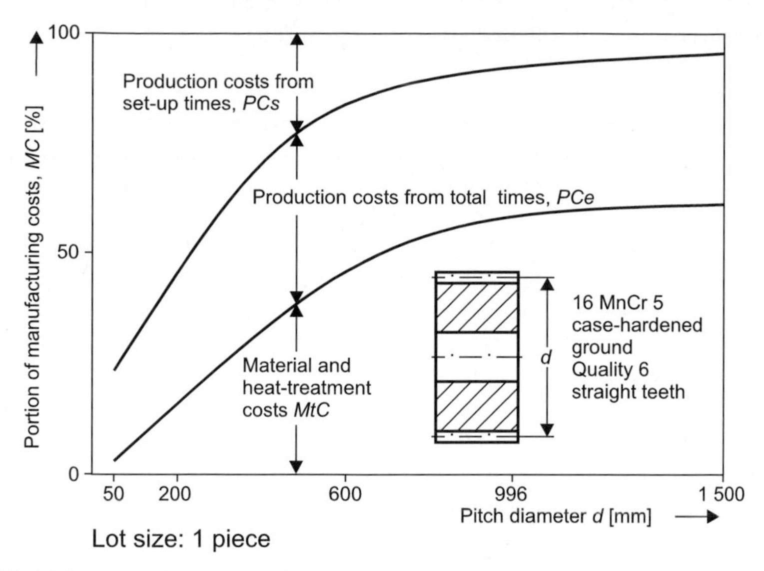

Fig. 7.7-3. Manufacturing costs of spur gears as function of pitch diameter (*lot size: 1 piece*) [Fis83]

The influence of both the size and the quantity on the manufacturing costs is represented in **Fig. 7.7-5** ("cost cube"). At the bottom are the material or weight-dependent costs, *mtc*. We see how from the lower left corner (single-unit production, small parts) they increase both with the lot size and with the part size. They constitute up to 50–60% of the manufacturing costs (the upper boundary of the

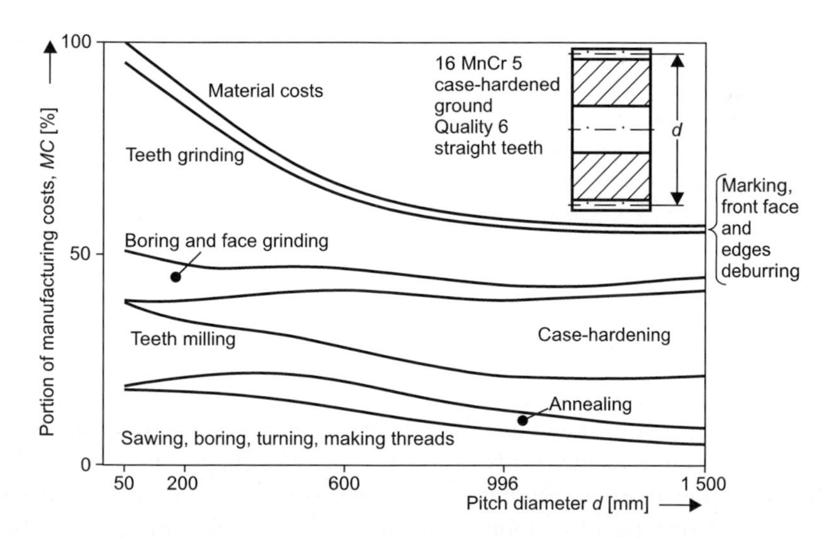

Fig. 7.7-4. Contributions of the production processes in the manufacturing costs of spur gears (*lot size: 1 piece*) [Fis83]

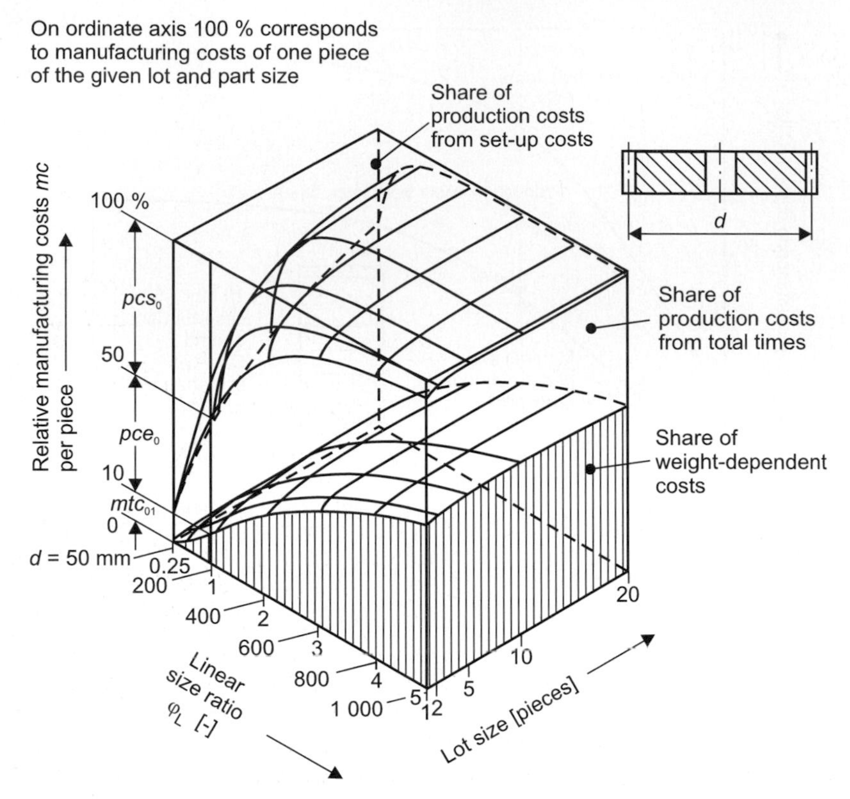

Fig. 7.7-5. Relative manufacturing costs for spur gears according to lot and part size

cube corresponds to 100% manufacturing costs per piece). In the space above *mtc* we see how the production costs from total time units *pce*, increase with size. Correspondingly, the share of setup costs *pcs,* which are at the top, decrease steeply in both directions.

The axes are chosen according to the conditions for single-unit and limited-lot production. If the lot size were 100 or 1 000 pieces, of course other more efficient production processes would be used, which reduce the setup and total time units (**Fig. 7.7-6**). Thereby the cost structures of small series-produced parts also conform to those of large parts in single-unit production, so that to a certain extent, the same rules for the cost reduction apply to both.

It is worth noting that the cost cube represents nothing other than the cost structure W in Eq. (7.7/3) and was obtained from the initial cost structure of a gear with 200 mm pitch diameter ($pcs_{01} = 0.5$, $pce_0 = 0.4$, $mtc_0 = 0.1$). So, it suffices in practice to use the cost structure of a representative part (or assembly) in order to determine the costs of a geometrically similar part in different quantities.

Obviously, the absolute manufacturing costs of this one part are necessary for this purpose. Using Eq. (7.7/1) the calculations for gears from the 12 companies cited in [Ehr82a] were checked. For large sizes there was specially good agree-

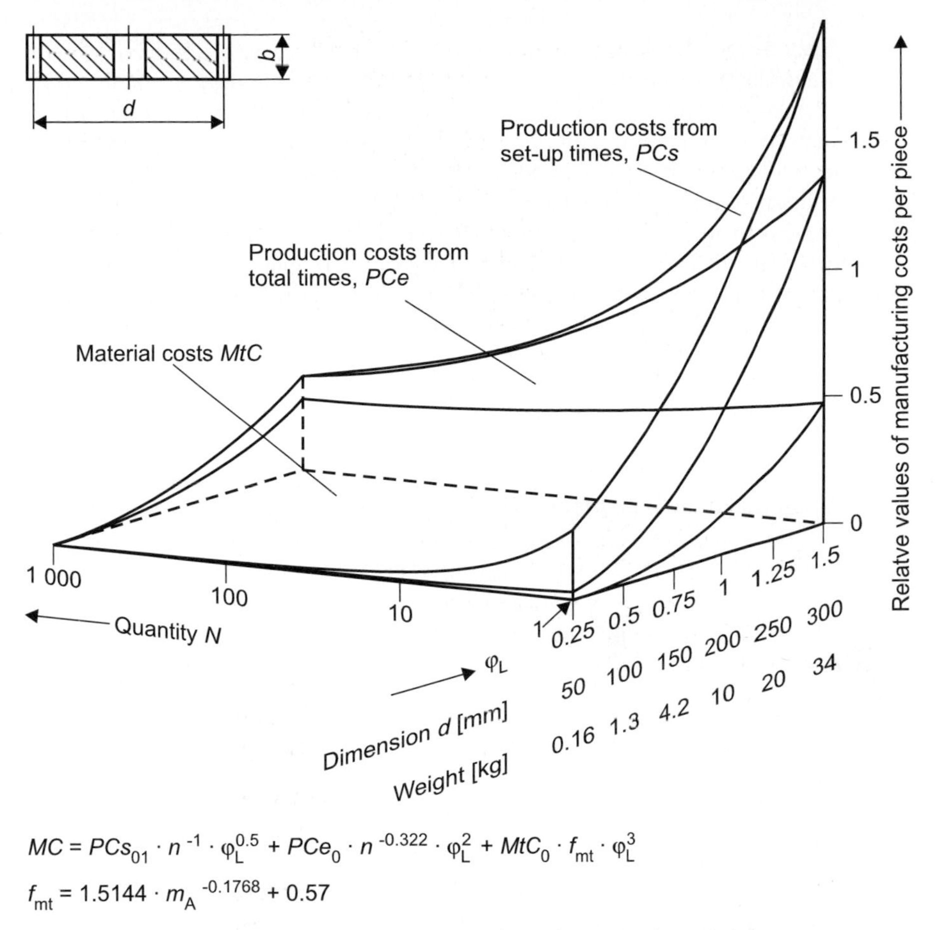

Fig. 7.7-6. Manufacturing costs depend on quantity and size

ment. This agreement is better if we stay close to the reference part (e. g., for a size ratio φ_L range of 0.5 to 1.5) and not, as in Sect. 7.7.2, use a greatly different ratio $\varphi_L = 0.25$ or 5. Figure 7.7-6 shows the information for a quantity of up to 1 000 units (large parts are seldom made in such large quantities). The effects of the quantity degression over such large numbers were taken into account in the material costs by the formula for factor f_{mt}. The equation implements the small quantity surcharges or large quantity discounts, depending on the acquired quantity m_A, as shown in Fig. 7.9-11. For the production costs from setup times *PCs*, the quantity degression according to Fig. 7.7-5 was considered, with ($N^{-0.322}$). For rolling bearings, bolts and screws, similar trends were found as for gears (see also Sect. 7.12.5.3 and 7.12.5.5).

For low-cost design, the following rules, which are valid for similar parts and similar type of production, can be derived from the findings presented above:

➔ **Very large parts and products** (weight 1 000 kg and up) have predominantly material and heat treatment costs. It becomes a matter of saving material or using low-cost material (Sect. 7.9). The production costs from total time units that are in the foreground must be reduced [e. g., by reducing the number of production processes and using more efficient production methods (Sect. 7.11)]. In addition, a more easy-to-process material should be tried. There is hardly any influence of quantities with such parts, if the production process remains the same and no new introductory costs *IC* (Sect. 7.5.1) arise. The parts can be designed and produced individually. Part families, size ranges and modular designs play a role, if in that way the effort in design and production planning can be rationalized. In addition, if production processes with high introductory costs can be used (e. g., model costs for casting, or tool costs in sheet metal forming) (Sects. of 6.3.2 and 7.12.3 to 7.12.6).

➔ **Small parts and products** weighing a few kilograms, in **single-unit production** of items in size ranges and modular systems, generally have production costs due to setup times that can be greatly reduced by larger lot sizes. Designs using same or similar parts, part families, size ranges and component system design are the most important methods for cost reduction (Sect. 7.12). A further means in case of machining is the decrease of the number of the production steps [e. g., through integral design (Sect. 7.12.4.3)]. Then the setup costs also decrease. Saving in materials cost is in general not worthwhile for single-unit production of small parts.

➔ The cost structures **of small and medium-sized parts and products** in the case of **large-scale production** resemble those of large parts in single-unit production (see above). The material and production costs from total time units are dominant. The production costs per piece from setup times become small. They are, however, not negligible over the total quantity!

7.8 Influence of the loading

Loading refers to the nominal loading of active surfaces of the machine, as it is carried out during embodiment. It is primarily a matter of:

- Mechanical loading (tension or compression in energy transfer with solid bodies). The limits of loading are breakage, deformation, wear, or galling.
- Thermal loading. The limits are burning, scaling, cracking, or warping.
- Corrosive loading. The limits are the many types of corrosive damage [Pah97].
- Flow loading. The limits are erosion, cavitation, and at times, abrasive wear.

A machine becomes smaller when more suitable loading or higher performance materials are used. With smaller size the manufacturing costs drop, especially for

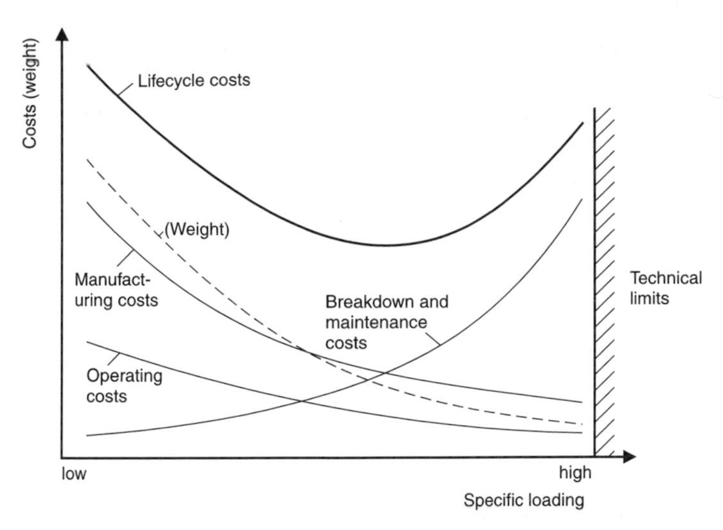

Fig. 7.8-1. Lifecycle costs depend on the specific loading

large machines (see Sect. 7.6, Figs. 7.6-1 and 7.6-3). In general the operating costs of a machine will also be lower, due to smaller losses, thus the lubricant used, the maintenance and repair costs. As **Fig. 7.8-1** shows schematically, the breakdown and maintenance costs will progressively increase due to lower reliability that is to be expected for higher loading. Accordingly, the **lifecycle costs** hit a **minimum**. To find this point is particularly difficult in the capital goods field, since experiments to determine the relationship between occurrence of loss, the repair and maintenance costs arising from that, as well as the loading are virtually impossible.

In general, designers pay the highest attention to reliability, since damages strongly lower customer satisfaction and the design gets the most blame for that. As long as the product development department is made responsible for costs only to a small extent, they will happily design something more expensive, rather than less safe.

It is known that the manufacturing costs of machines drop after the prototype is made (e. g., by introducing more rationalized production processes). On the other hand, after damages occur and complaints come in, expensive modifications are often introduced which stay on in all subsequent orders. **Oversized parts** are not detected from the start, since they never lead to failures. Accordingly, they always remain too expensive. It is known that automobile manufacturers find out during salvage which parts are sold most frequently in scrap yards. These were thus apparently oversized and too expensive.

The influence of loading on size and total costs is obvious from the transmission example in Fig. 7.6-1. For the same torque, the shaft center distance of 83 mm decreases to 390 mm, hence by about 53%. The gear loading (tooth force per mm width) increases by a factor of 4.5. The total weight reduces to 19% and the total costs to 48%. This large increase in loading can only be borne safely when stronger material (e. g., case-hardened steel) is used instead of the normal machine steel. The case-hardened gears are no less reliable than the large gears made of ordinary steel, so that by increasing the loading there are genuine decreases in manufacturing and lifecycle costs, as suggested in Fig. 7.8-1. Only if the size decreased while keeping the same material would statistically more failure rates be expected and a corresponding increase in lifecycle costs.

➔ Through an **increase of loading,** the size of the product and thus likewise its **manufacturing costs** can generally be **lowered** This is often possible by using stronger materials or appropriate surface treatments. It should be kept in mind that the lifecycle costs do not increase as a result of lower reliability.

7.9 Influence of material

7.9.1 Significance of material costs

In material costs, we include purchase parts and semi-finished products (i. e., the raw materials) as entered in company accounting (Fig. 8.3-2). The emphasis in this chapter (in particular Sect. 7.9.2) is on the raw material costs.

Material costs including the costs for purchase parts, make up around 43% (15–60% [VDM95]) of the total costs in the machine industry (**Fig. 7.9-1** and 8.4-2), thus a large part of the entire costs. For large machines in single-unit production the material costs (i. e., raw material) can make up to 50–7% of the manufacturing costs (Sects. 7.6 and 7.7, Fig. 7.6-3). The tendency toward outsourcing (see Sect. 7.10) (i. e., concentrating on core competencies, parts and processes) leads to rising material costs whereby purchased parts become important. A second fact that makes material costs significant is they are about the only actually variable costs over which the designer has control. There is nothing more direct and more cost effective than, for example, to not buy a part that is not necessary, to provide thinner sheet steel, or to buy a part of more favorable quality. Due to social legislation, labor costs can not be reduced in many countries in the short run, for the

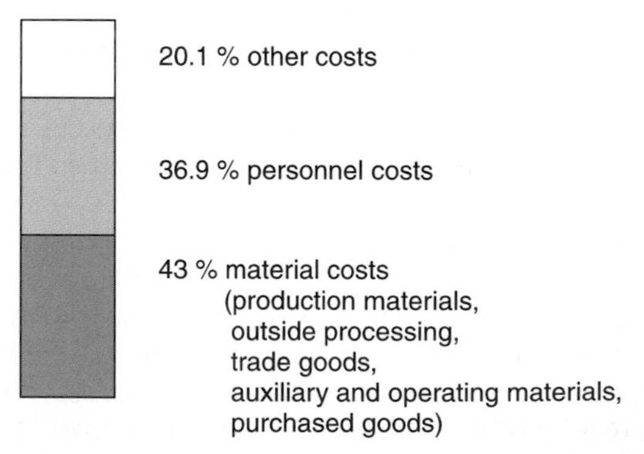

Fig. 7.9-1. Overall costs in the machine industry, according to cost categories [VDM95]

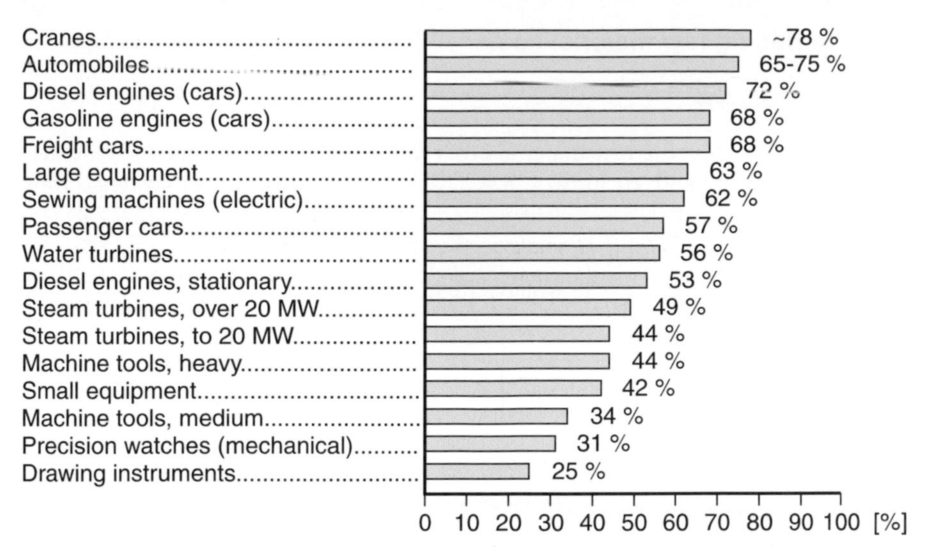

Fig. 7.9-2. Cost of materials as percentage of manufacturing costs [VDI77]

company as a whole. They can only be shifted from one product to another (maybe up to now procured from outside).

With the importance of the **material costs** which in the machine industry, according to Fig. 8.4-2, are **four times as high as the labor costs**, the typically intense endeavors toward production rationalization appear in a different light. Accordingly, a much closer cooperation of purchase and logistics, production planning and design should be striven for (Sect. 7.12.3.1b). The importance of purchasing in material costs is considerable, as shown by Fig. 7.13-2, (f). Even for standard case-hardened steel (16 MnCr 5), among 12 gear manufacturers, the prices vary by a ratio of almost 1:2. With other not so common materials or with purchased parts, the differences are still greater. **"The profit lies in purchasing"** still holds true.

The share of materials cost in series production is especially high. As **Fig. 7.9-2** shows, the materials cost share for automobiles is usually about 70% (including purchased parts, and referenced to manufacturing costs). If better production processes reduce the production time, the material costs remain at first a parameter that can be little influenced. This holds above all for raw material (Fig. 7.7-6). Purchased parts often contain high production costs. However, larger quantities make a cost decrease possible. For example, industrial consumers get 80–90% discounts on rolling bearings. The less production costs the material contains, the less the price can react to the order quantity.

Material costs are high for large, simple machines (i. e., machines that require few production steps). In Fig. 7.9-2, approximately 78% of the manufacturing costs of cranes are for material; for railway freight cars, materials comprise approximately 68%. At the other end of the scale, drawing instruments have approximately 25% material costs. For components, the focal points for influencing costs derive from the cost structure (material/production costs from total time units or setup times, as shown in Fig. 7.7-3 and Fig. 7.7-4).

7.9.2 Reducing raw material costs

7.9.2.1 Overview

Possibilities for reducing raw material costs follow from the approach to calculating material costs as given in Sect. 8.4.2 (differentiating overhead costing). Raw

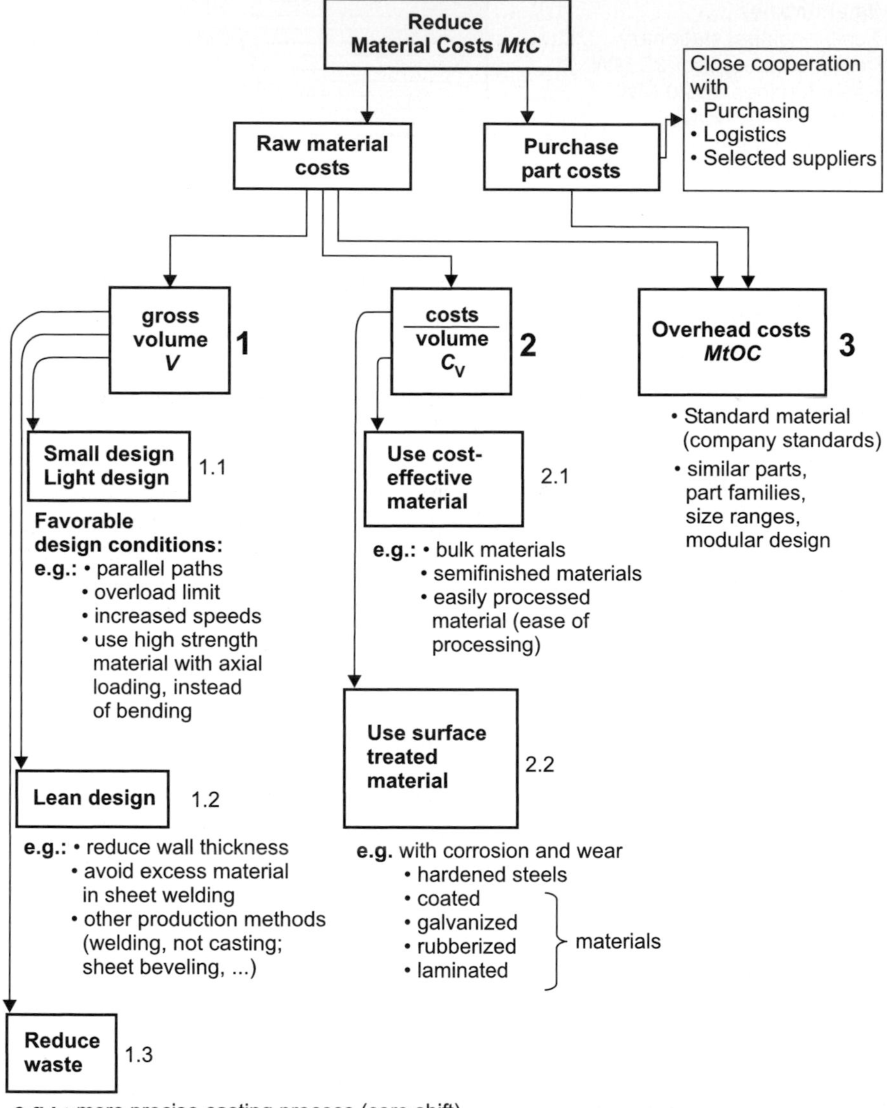

Fig. 7.9-3. Design focused on materials cost, and pertinent rules

material costs are calculated primarily from material volume (**gross volume**) and the specific material cost C_V (**cost per unit volume**) (see **Fig. 7.9-3**). Of course, we can also base the material costs on weight. The calculation method that design follows works essentially with volume.

It is a matter not just of using a material of lowest per volume cost (i. e., minimizing the value C_V). An expensive, high-strength material can so reduce the necessary volume that in spite of the higher material price the overall material costs $V{\cdot}C_V$ are reduced.

The material costs per unit strength are usually the deciding factor. That holds in particular for pure tensile stress. If we find the value

$$\frac{C_V}{S_u}\left[\frac{\$}{m^3}\cdot\frac{m^2}{N}\right],$$

for different steels, then we see that customary high-strength steels are somewhat less expensive than low-strength steels (**Fig. 7.9-4**). This is also true in case of screw fastenings, where high-strength screws in general are more economical than low-strength screws (Sect. 7.11.5.5).

This is valid only to a limited extent, since both the strength values and costs show large variations. If there are further demands such as sufficient toughness (usually higher for low-strength steels), added wear resistance, good workability or weldability, or if the type of loading changes, then it is impossible to make a general statement.

If there is only one place at the component boundaries where strength matters, as in case of bending and torsion loading, then the material costs C_V referred to volume become more important. This holds even more if the component has little force loading but is used only as a separator (paneling, facing, lining), or if it is a matter of keeping the deformation small so that a low loading is chosen. In the

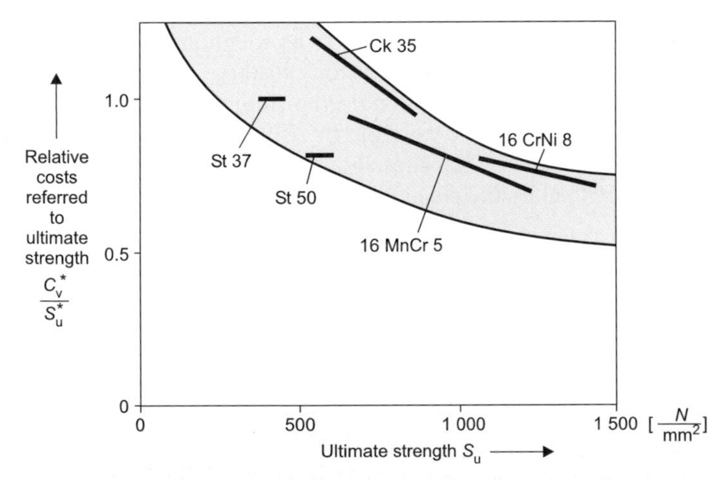

Fig. 7.9-4. Effect of strength on relative costs: high-strength steels are somewhat more economical than low-strength steels (relative costs C_V^* [VDI77], relative tensile strength S_u^*; star means relative to round stock USt 37-2)

following, the procedures for design focused on materials costs are reviewed, using the numbering in Fig. 7.9-3.

7.9.2.2 Reduction of material volume

Procedure 1.1: **Small and lightweight design** (see Fig. 7.9-3)

Small design refers to procedures that reduce the size of products, in particular by effecting a change in the product concept, as for example:

- **A parallel arrangement of active surfaces** results in a division of power into several paths and thus leads to a size decrease. **Figure 7.9-5** shows how increasing the number of power paths in a transmission, results in a decrease in size, since for each power path only the corresponding fraction of the power is transmitted (influence of concept). Size, weight and thus material costs drop. This is also clear from Fig. 7.3-2. However, the part quantity increases and with that the logistics and the assembly costs. Other examples are multi-disk clutches with a large number of active surfaces, rolling bearings and free wheels with as many rolling/gripping bodies as possible.
- **Overload limits** ensure that the whole device need only be designed for a defined throughput and not for the rarely occurring peak value. Examples of this are devices such as slip clutches, hydraulic couplings, shear couplings, isolation switches, fuses, bypass controls or over-pressure valves.
- **Rotation speed increase** for rotating energy conversion machines is a measure constantly used for decreasing size and weight. Examples are engines (Fig. 5.2-3), turbines and electrical machines. At a specified power $P = T \cdot \omega$, the torque T can be reduced by increasing the rotation speed ω. Forces or lever arms become smaller, thus decreasing size and weight. In addition, the decrease of the moment of inertia is desirable in machines where rotation speed frequently changes or that contain reciprocating masses.
- **High-strength material** is of lower cost, as mentioned above in Fig. 7.9-4, particularly for axial (tension/compression) loading, since the uniform stress makes use of the whole cross-sectional area. Indirectly, there are further savings as shown in **Fig. 7.9-6**, for example by using high-strength fasteners, since the whole design space then becomes smaller. Figure 7.6-1 shows how strongly gear design is affected by the change from heat-treatable steel to casehardening steel.

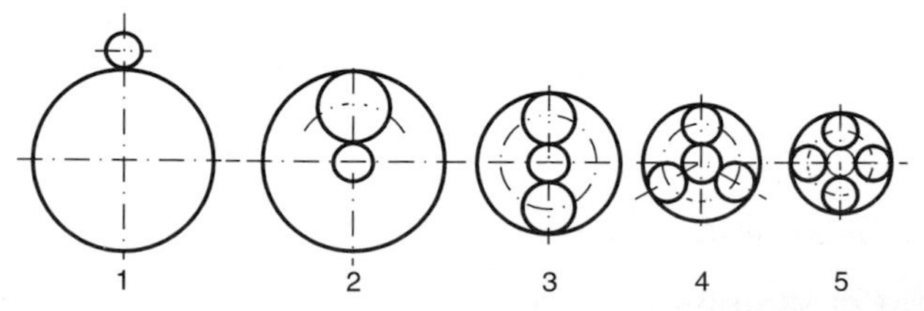

Fig. 7.9-5. Size comparison between a standard and an epicyclic gear train for the same transmitted moment and same gear ratio

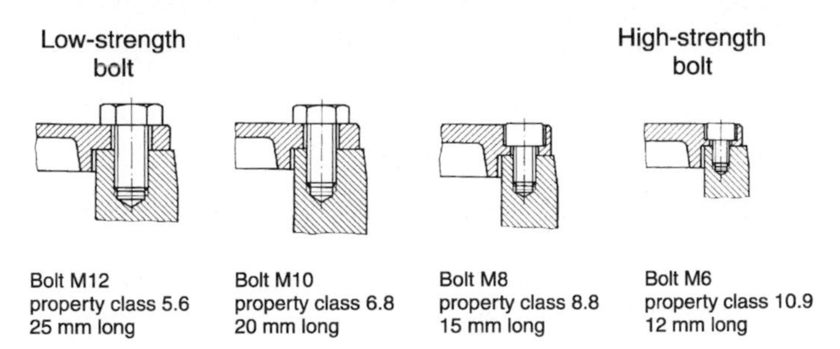

Fig. 7.9-6. Influence of the bolt strength on weight and material costs: Higher bolt strength reduces the weight of the component and the material costs. Socket head bolts (DIN 912) are especially advantageous

- Strive for **axial loading** instead of bending or torsion so the material is uniformly stressed throughout the volume. In the same sense, strive for force flow lines for the shortest possible path between the loading and support points and aim for symmetrical shapes.

The procedures for **lightweight design** (1.1 in Fig. 7.9-3) are also complex. Lightweight design is achieved not only by simple choice of a lighter material (e. g., light metal/lightweight design, therefore "material-lightweight" design), but also by the form-efficient use of high-strength materials ("form-lightweight" design). Weight decrease is not always to be equated with cost reduction. For products that include mass that is not directly used, a materials cost savings is feasible through simple weight reduction (lean design, Fig. 7.9-7). If actual lightweight design is carried out, however, the costs of development, computation and testing, as well as the higher production investment can exceed the savings in materials cost so that the component costs again become very high. Then lightweight design is no longer beneficial through savings in manufacturing costs, but only through a decrease in lifecycle costs (Chap. 5, Fig. 7.8-1). For example, the weight decrease of an airplane results in increased fuel economy or produces higher revenues by carrying more paying freight. The weight saving of 1 kg in an airplane is balanced by 1 million $ of design, development and test-costs (Airbus).

The formulation of a target is essential in design – for costs it is the cost target, for weight the weight target, etc. For this purpose, a very radical procedure has proved itself: **Set the target value for a product by averaging the most favorable individual values that can be found from your competitors**. If, for example, it is a question of designing lightest combustion engine, the weights of the lightest components from the competition are added up, regardless of whether they match or not. Thus, a weight target is formulated. Car manufacturers follow the same procedure for costs in automobiles. The costs of parts and assemblies from rival firms are obtained, followed by in-house calculations.

Procedure 1.2: **Lean design** (see Fig. 7.9-3)

Lean design means, as suggested above, a simple material volume saving (**"lose" weight**) without changing the constructive embodiment or the material type. Examples are decreasing sheet metal thickness, wall thickness, or the sheet metal overlaps

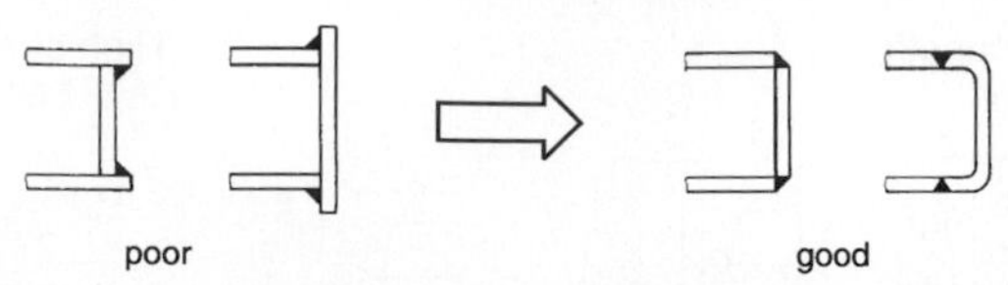

Fig. 7.9-7. Avoiding sheet steel projections in welded designs (from K. Tuffentsammer)

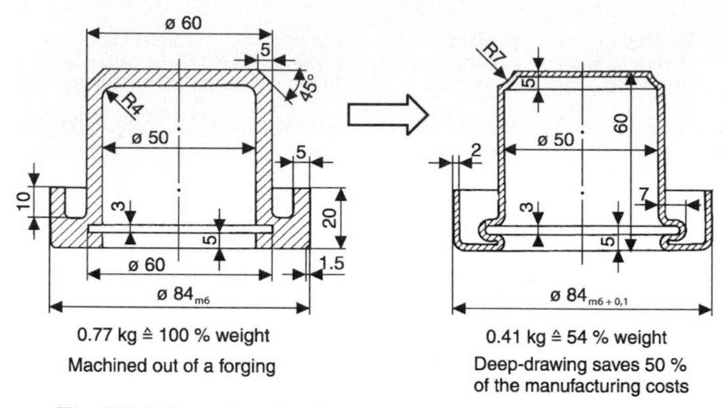

Fig. 7.9-8. Deep drawing instead of drop forging (from RKW)

and projections in welding (**Fig. 7.9-7**). That can be done with lightly loaded products (e. g., housing) after simple testing. Lean design is particularly important with large machines or large quantities, and when the price per unit volume of the material is high (e. g., for heat- or corrosion-resistant steels or copper alloys).

A further step in decreasing the material volume is a **change in the production process**, for example from casting to welding. For large vessels and housings, weight decreases of up to one-half are common. With sheet metal, thinner walls are possible than with casting. As shown in **Fig. 7.9-8**, the processing for a container was changed from machining a forged piece, to deep drawing. This resulted in saving around one-half on the weight as well as on the manufacturing costs. The wall thickness was reduced from 5 mm to 2 mm.

Procedure 1.3: **Reduce weight used and waste** (see Fig. 7.9-3)

Reducing the weight of material used and the waste during production are often neglected in design.

It is a matter of producing near-net shapes (i. e., reducing the starting volume of the raw material as compared to the finished volume). In addition, some expensive machining process must usually be applied to remove the excess material. **Examples of such steps** are:

- **Choose near-net shape production processes** (e. g., casting, forging, sintering, and sheet-metal design). See Fig. 7.11-24, along with Fig. 7.12-15.
- **Castings** should be made more exact, such as by avoiding core shifting during casting by providing better core support. This saves machining costs later, often also prevents blowholes, and thus rejects.

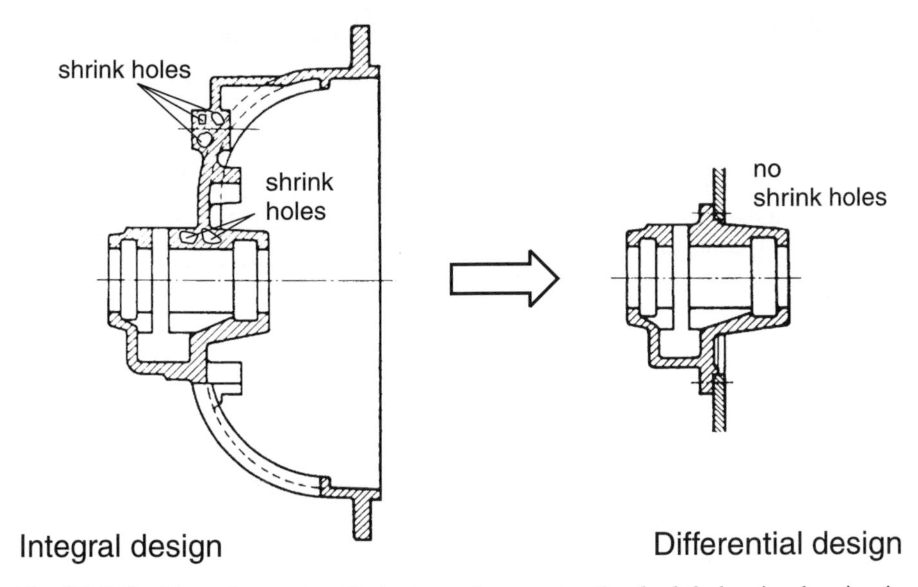

Fig. 7.9-9. Partition of a casting likely to produce waste. *On the left:* bearing housing in one-piece (integral design), manufacture and machining also expensive due to possibility of rejects. *On the right:* bearings separated (differential design), easier and lower cost production, particularly with series production on special machines (Sect. 7.11.2.5b)

- **Do not machine parts out of a solid piece**, but use forged or cast semi-finished parts. Restrict the worked mass (raw weight, finished weight) by proper production planning and purchasing, and avoid unnecessary machining. This is strongly influenced by quantity (rules in Sect. 7.7.3).
- With **stamped parts**, pay attention to use of the sheet (fewer rejects). Figure 7.11-34 (bottom) shows how waste can be minimized and material saved by minor changes in shape. CAD programs are available for this.
- The **cost of rejects** in the machine industry constitute 0.5–1.5% of the total costs (or 1.5–4% of the total material costs) [VDM78b]. Rejects are rejects. In such cases, it can be cheaper to split the integral design into two or more parts (differential design), as shown in **Fig. 7.9-9**. In spite of additional processing and interfaces and the assembly expense, such a part can become more economical. That, however, does not become evident until the first scrap is produced (see also Sects. 7.12.4.3 and 7.11.8).

7.9.2.3 Reduction in material costs per unit volume

If the material volume V cannot be reduced, a material should be chosen that is more economical per volume (smaller value of C_V in $/cm^3) so that the material costs $V \cdot C_V$ go down. According to Fig. 7.9-3, the following procedures should be considered:

Procedure 2.1: **Using lower cost material with a lower C_V value** (see Fig. 7.9-3)

Mass-produced materials are more economical than are less common materials. According to some machinery manufacturers, this is the case for ferrous materials, for example:

- General structural steels, St 37, St 42, St 50, St 60;
- Heat-treatable steels, C 10, (Ck 10), C 15, C 35, (Ck 35), C 45, (Ck 45), 25 CrMo 4, 42 CrMo 4, 37 MnSi 5, 34 Cr 4, 41 Cr 4;
- Case-hardening steels, 16 MnCr 5, 20 MnCr 5, 15 CrNi 6, 18 CrNi 8;
- Nitriding steel, 31 CrMoV 10;
- Stainless steels, X 5 CrNi 18 9, X 10 CrNiMoTi 18 20;
- Gray cast irons, CI 20, CI 25;
- Nodular graphite cast irons, NCI 40, NCI 45, NCI 50;

Cast steels, CS 45, CS 52, CS 60, CS 62, CS-20 Mn 5, CS-22 Mo 4, CS-C25.

To avoid high procurement and warehouse costs, material types should be limited to those that are common in the factory (company standards). For the selection, relative cost values prepared in-house are valuable since material prices differ by around 20% for inquiries from purchasing (in special cases, even more, Fig. 7.13-2). Relative cost tables must be produced internally. They must be updated from year to year (Sect. 4.6.3). It is not necessary to have a precise numerical value for C_V^* in \$/cm^3, but only rough ratios between the customary materials or families of materials.

Examples from VDI 2225 [VDI77] of **relative material costs** C_V^* for families of materials are given in **Fig. 7.9-10**. The different material types within each group give rise to the scatter in the data. As a base value, ($C_V = 1.0$), USt 37-2 (DIN 1013) steel rod in the diameter range of 35–100 mm, in lots of 1 000 kg, has been used (Sect. 4.6.3). We recognize that case-hardening steel and heat-treatable steel are a little more expensive. Stainless and high temperature steels cost many times more. In particular, the costs of copper-base alloys lie an order of magnitude higher. For material choice such relationships are essential to know. The following ratios for cast ferrous metals have long been known:

Gray cast iron :	Nodular cast iron :	Malleable cast iron :	Cast steel
1 :	1.2–1.5 :	1.7 :	2.0–2.5

Strength, general technical data and relative costs can be shown in combination in the **company standards**. A graphical representation is very helpful.

The importance of the **quantities acquired** for a given purchase price is shown in **Fig. 7.9-11**, particularly the high surcharges of up to 30% when purchases are in small quantities. These result from the one-time costs for the delivery. For different procured quantities and probably different negotiation skill of purchasing [Ehr82a, Fis83] were found to be the probable causes of scatter in case of common materials (e. g., Fig. 7.13-2).

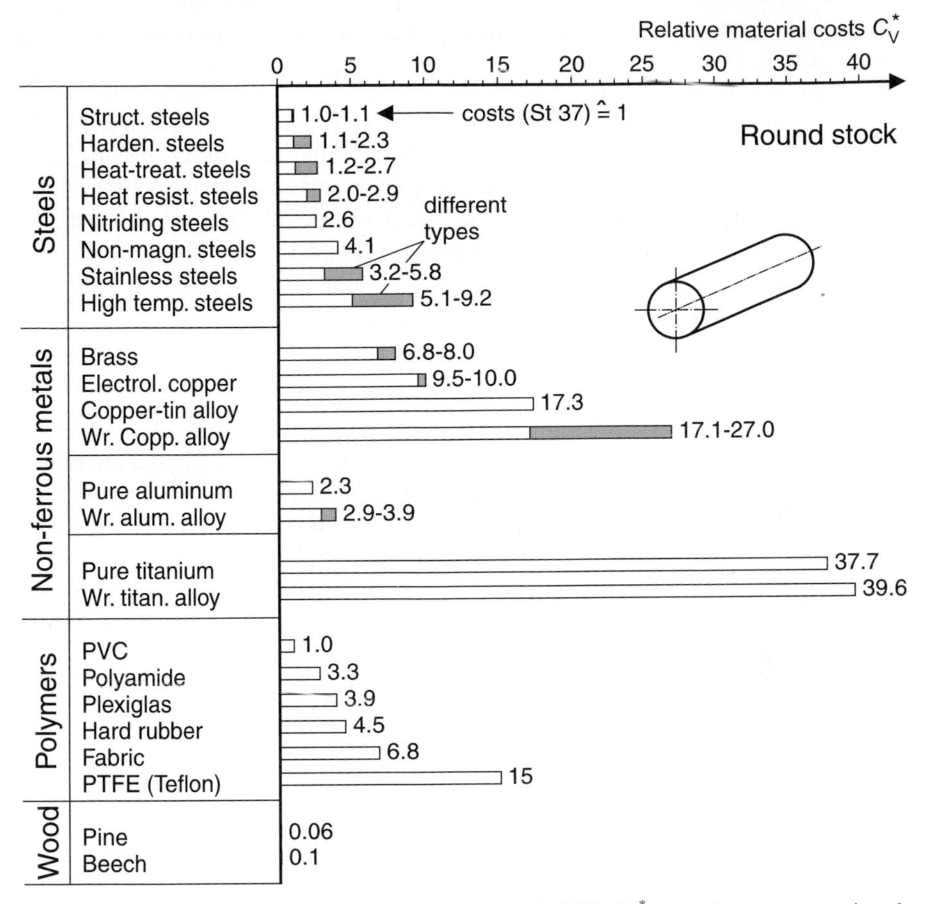

Fig. 7.9-10. Examples of relative costs of material [VDI77] (C_V^* = relative cost per unit volume, referred to USt 32 round stock)

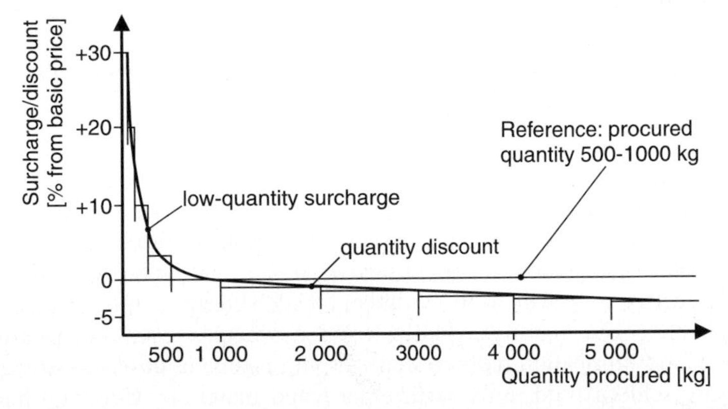

Fig. 7.9-11. Low-quantity surcharge and quantity discount for stainless steels [Eck77]

For semi-finished material and standard parts there are certain **sizes** sold at the **lowest prices** in supplier price lists, resulting from size surcharges. For every section and profile (e. g., round stock, flat steel, U-shape) a price minimum exists, depending on the material, probably for the most popular size. Small sizes are expensive because of tight tolerances to be kept. In the case of large sizes, it is often a question of items made to order, or higher storage costs are charged because of the infrequent production. Thus, steel rod of 20–30 mm diameter and square sections of 20–30 mm per side are usually the most economical.

➔ For semi-finished material, there is a price minimum at specific sizes, which is close to the most common sizes.

Regarding production costs, the **processability of** the **materials** is fundamental. Therefore, not only are low costs per unit volume to be considered in the material choice, but also its machinability, weldability and formability. This is especially important in the case of large components and the manufacture of large quantities where, besides the material costs, the costs based on machining time play a substantial role. We can formulate the following rules for machinability:

➔ Materials that have lower strength (hardness) are in general easier to machine. Gray cast iron behaves very nearly like a medium-strength steel.

➔ Copper-base alloys, polymers and in particular light metals, are in general easier to machine than steel. High-strength, austenitic steels or special cast steels (stainless and/or heat-resistant) are difficult to machine.

Procedure 2.2: **Use material that can be surface-treated**

Machines become non-functional mostly from wear and corrosion and not due to breakage. Wear-resistant or corrosion-resistant materials are often expensive. Stainless steels cost three to six times as much as common structural steels (Fig. 7.9-10). It is then more economical, by a separation of functions, to use low-cost carbon steels for strength and to let only the surfaces handle the wear and corrosion protection (Fig. 7.11-4).

- **Wear protection** against regular wear, galling, surface fatigue, pitting, erosion and cavitation:

  Total wear protection against abrasive or jet wear is not possible even by using special steels such as Manganese steel. The surface can be protected through metal-deposition, flame-hardening, plating with hard steels or covering with protective materials such as rubber. With sliding or rolling surfaces (for example in gears, bearings, guides, friction wheels, chains) higher material hardness and lubrication reduces wear. From reasons of cost and strength hard steels are seldom used, only surface-hardened materials. Common hardening procedures are protective layers, Boron-hardening, case-hardening, gas nitriding, liquid nitriding, flame and induction hardening.

- **Corrosion protection** against steady corrosion, split, contact, boundary surface, vibration crack and stress crack corrosion:

  Complete corrosion reduction is possible by using of suitable materials (for example through use of stainless steels, aluminum alloys or polymers). For procedures for surface protection, see Fig. 7.11-4. Pahl and Beitz [Pah97] show design measures to counteract corrosion.

7.10 Decision between in-house development and production, versus outsourcing

7.10.1 Overview

The tendency of product manufacturers to strengthen their respective **core competencies** and to assign peripheral activities to suppliers (Sect. 6.2.3) [Wil96] is a result of increasing and worldwide competitive pressures, short-term order fluctuations and increasing product complexity (corresponding to the knowledge and variants explosion [Ehr97]). An essential basis for outsourcing was always to use the advantages of having a specialized supplier, with regard to costs, technology and schedules. Thus, the decision to be made is between **in-house production** and **purchasing** (make or buy, Sect. 7.10.3).

In the machine industry, the in-house production fraction was about 60% in 1993; in the automotive industry, approximately 35–50% [Bir93]. Today the decision is also seen in another light: Whole products are assigned to suppliers, including product development, production, assembly, and testing (**system suppliers**). It is therefore a matter of more than just external production; it concerns the degree of performance capability in the company. **Performance capability** (strength) can be defined in the technical field as the value-added portion in the production process that is provided by the company itself (**production capability**) and the value-added portion of the product development performance (**development capability**)[9].

7.10.2 Advantages and disadvantages of outsourcing

The farming out the work gains advantages, but must also expect disadvantages by allocating the effort to other companies. These must always be weighed against each other [Bir93].

Advantages:

- More flexibility at varying utilization, due to lower fixed costs.
- Low costs through low prime costs.

---

[9] Added-value describes the total output/performance (sales revenues and changes in assets) minus the advance payments (materials purchase, external services, interest on capital) [Hein95]

- Faster delivery and reaction capability.
- Special know-how of the supplier in a given specialty area.
- More capacity for the company's own core competence.
- Presence in the foreign market (with prescribed delivery range).
- Less capital investment in own production plants, therefore lower fixed and higher variable costs.
- No dependence on own equipment.

Disadvantages:

- Loss of know-how to the suppliers and at times even to the competition.
- Possibly unsatisfactory quality.
- Design-for-manufacturability more difficult, since suppliers not known.
- Supplier loss potential due to:
  - bankruptcy/sale/strike, or
  - defects in the logistics chain of the suppliers.
- Expense for coordination with the suppliers, perhaps for their qualification (higher process costs).
- Relationship of trust with the suppliers must be created first.

We must therefore always consider where the core competencies should lie. Higher in-house performance capability is not necessarily better. After the work is moved to the outside, the company should concentrate on a few suppliers and build up long-term partnerships. It is not enough to out-source only the production. Product development and production should then be analyzed, together with the supplier. This basis forms a close (at times contractual) cooperation between product development and purchasing departments of the company (e. g., through mutual transfer of employees for a limited time, as practiced by system suppliers in the automobile field; Sect. 3.3.2).

7.10.3 Decision between in-house production and purchasing (make or buy decision)

The "make or buy" decision depends, first, on the basic strategy of the company. There are successful, even competing, companies in the market which have no in-house production; they only assemble and test. There are others that produce everything themselves. Every decision introduces its advantages and disadvantages that must be planned for accordingly. The following notes are meant to stimulate discussion about **whether to perform the task yourself:**

- the more strategically important it is (core competence),
- the more innovative it is,
- the more frequently it ensues (quantity effect; however, then calculate the costs: see below),
- the smaller the possibility to standardize is, and
- the more uncertain its planning may be.

Since these strategies cannot be applied to decide on all cases, the decision is made during the design phase and is often wrong, because of an ignorance of the **cost correlations** [Bro66b, And86, Män90, VDM78a, Mel92]. It is generally not correct to compare the cost prices with one's own manufacturing or even total costs (full costing), and then to go for external purchase, if one's own manufacturing costs are higher than the cost prices (purchase price + procurement). Manufacturing costs include significant fixed costs which the company still incurs even if the product is purchased. It must then bear its own fixed costs as well as pay the fixed costs and profits of the supplier. Therefore, for cost comparison variable and fixed costs must definitely be categorized (direct costing, limit costing, Sect. 8.5.1). Furthermore, the modifications of process costs must be estimated, if processes charged as overhead costs are affected (e. g., materials management, logistics, product development). Overall, it depends on recognizing and comparing the pertinent decision-relevant costs [Sei90].

Finally, the decision is not simple enough, as suggested above, based only on lowering the costs [VDM78a]. Rather, a range of **company policy perspectives** play a role, such as liquidity problems, risk considerations (dependence on supplier and logistics), targeted outsourcing etc. According to their significance, the decision must therefore be made at times by top management. A point evaluation of the criteria that are not directly cost-related can be used in preparing for the decision.

Independent of cost accounting, the **following cases** can be distinguished:

- The decision should be against **external procurement** (i. e., **production must be in-house**)
  - if the company places value on absolute secrecy or on the security and expansion of know-how;
  - if it turns out that no supplier is available, who might be a possibility on the basis of technology, quality or schedule;
  - if great risks are involved in transportation.
- The decision should be against **in-house production** (i. e., **external procurement must** take place)
  - if the product can only be procured from suppliers because they have legal property rights or are the only ones to have the necessary know-how and/or the necessary quality standard,
  - if one's own company cannot supply the needed product of the specified quality or quantity, or within the required time.
- The decision about **in-house production or external purchase** requires a **cost comparison**
  - if own company has the necessary capacity, but external purchase appears to be of a lower cost due to the bid price;
  - if the company with certain investments, could take up the production and/or needed additional (more qualified) workforce;
  - if the production could occur in subsidiaries of the company, that practice separate accounting;
  - if the quantity to be produced increases, but not enough to call for a new production process;

- if the products are only seldom needed, or if one's own production equipment is not used any more and their place is needed for new production equipment;
- when two products, A and B, could both be produced in-house but capacity is available only for one;
- and in many other situations (complexity in the own production, wildly varying demand, etc.).

The following **rules** can be formulated for the "make or buy" **decision:**

➔ **Production in own facilities**, if the real (variable) per piece costs in the company are less than the cost prices. Further process costs are still to be assigned to the cost price, if for example in design and development or in quality assurance, additional measures to maintain the quality of the supplier become necessary. In mass production companies there are "suppliers' product development programs".

➔ With product parts, which are not representative of the product expertise, look at whether **purchase** is more **economical** than in-house production. The quality, the delivery time and the dependence on suppliers should be considered. In the cost comparison for in-house production, only the costs that really arise in the company are to be used (how much money goes out in the alternative cases?).

7.10.4 Cost-driven design for cases of uncertain manufacturing facilities and inadequate cost transparency

Due to specialization and globalization, product development processes are increasingly separated from the production processes (wholly, or in part). With that there is a danger that the product may become less than optimal and unnecessarily expensive.

Problem:

It is obvious that low-cost designing becomes difficult if the production of assemblies or whole products occurs at a supplier (the details of which are not known) but the design and development takes place in-house.

Neither the material procurement, the production facilities, nor the cost accounting of the supplier are known. The intellectual wall shown in Fig. 3.2-2 between design and production becomes high and seemingly insurmountable. This becomes especially difficult if the manufacturer is sought after in an foreign country whose quality and cost conditions are hard to assess. This happens particularly with plants for which customer-specific parts, specified in that country, are required.

The seemingly much simpler case is the usual one, in which purchasing inquires with shop drawings at potential suppliers and decides on the most favorable cost price. However, even here the supplier's more favorable manufacturing costs and prices can be obtained if the product designers know the production advantages of the supplier.

Problem solution:

Apparently, it is thus a matter of lowering the above-mentioned wall of ignorance between design (the client) and production (supplier) or even better, eliminating it. The procedure is therefore the same as in in-house manufacture, as described in Chaps. 3 and 4. Product development, part production, and assembly must come together mentally. For the supplier, it is very beneficial to be located nearby.

That sounds unusual at first and seems perhaps to be impossible as long as the strategy of the cooperation between the customer and the supplier is not changed. Automobile manufacturers are doing this already.

The following must be achieved:

- Close **cooperation** between the manufacturer's product development, materials management (purchasing), and marketing departments. Marketing and sales are to sell, if possible, to in-house standards and to accept few special wishes from the customer. In this case, some deliveries are less problematic.
- Do not send inquiries off to arbitrary, unknown suppliers. Instead, search from a few generally dependable **preferred suppliers.** This can come about through test offers, concept competition (with system suppliers), visits and negotiations, or as part of supplier days. The selected suppliers should be supplying over a relatively long term to create a continuity in the relationship and thus a (at times, contractual) partnership. How long they remain favored must be tested repeatedly by comparing their offers with those of other suppliers. Companies in worldwide commerce with large product lines practice that within the framework of the global sourcing.
- With these preferred suppliers (often contract-bound), design is carried out in **teamwork**, toward a mutually established cost target. They supply their own production, assembly, and cost accounting knowledge (see Sect. 4.3.2). The cost accounting is disclosed [Stu94, Romm93].
- Alternatively, after the agreement on price the customer determines only the rough shape of the product. The function-critical quantities, tolerances, and material properties are noted by the customer and agreed upon. **Detail design and production documentation** are then defined by the **preferred supplier** according to production facilities and raw material, and provided to the customer [Lin93a, Deb98].
- A further step is that the customer's **production and cost consultant** gets to know the especially favorable production conditions of the preferred supplier as well as the restrictions, and passes on this knowledge to the customer's design group. It is advisable in this case that the customer maintains some production planning, perhaps within the framework of cost calculation.

- **Offers** must, at the start and repeatedly later, be analyzed and **checked** with regard to their contents (quantity and value schedule). For that, very detailed offers must be requested from several suppliers. It is also useful to make statistical comparisons over a longer period, as well as cross-comparisons with different providers.

It is clear that the following **basic rules of production, assembly, and low-cost design** form the basis of the target cost-driven work. These rules arose from the experience in low-cost design **without** a **dependable cost analysis** being possible. They are already available elsewhere in the book, but are deliberately repeated here in a summary form. They are applicable both for in-house as well as for external production.

They should be particularly helpful:

- If no cost accounting is possible.
- If no cost analysis of a comparable product exists, or is possible.
- If it is not known who makes the product.

A: Basic rules for design focused on manufacturing costs (Fig. 4.5-7):

➔ **I Clarifying the task and procedure**

- Set up a **team:**
  Design, production planning, purchasing, service, and master operators from production and assembly who have thorough knowledge of the real problems and time expenditures in production, assembly and logistics. Designate a team leader as the party responsible for cost reduction. Plan the procedure.
- Determine **total target costs** and divide these into components, if possible (estimates). Which costs would have to be lowered?
- Analyze the product with regard to the **probable dominant cost elements**; where are the **potentials for cost reduction**? Which assemblies/parts? Production/material or purchase costs? Which processes/production steps especially need a lot of time? Which ones present the most problems? Which ones are perhaps exaggerated? Where are the elements with savings potential? Which qualities are valued by the customer, which are not? Note down everything clearly! How does the competition do it (Benchmarking, see Sect. 7.13)?

➔ **II Search for solutions**

- Function: Fewer or more functions? Function integration?
- Principle: Other principle (concept)?
- Form design: Size decrease? Fewer parts (Integral design)? Factory-internal standardization: Similar parts, repeating parts, part families, size ranges, modular design?
- Material: Less material? Less waste? Lower cost material? Standardized material, purchase parts?

- Production: Other, fewer production steps? Other fixtures, equipment? Less accuracy? Assembling variants? In-house or external production?
- How does the competition do it? How is it done in similar fields (analogies)? How does nature do it (bionics)?
- Application of other rules for cost reduction (from this book)

➔ **III Select solution**
- If no comparative cost calculation of the favorable solution alternatives is possible, the only choice is to **estimate** the probable production and assembly expenses for the alternatives that have been found (work station, labor, or machine hourly rates should be known. Sect. 9.2).
- Much can be **estimated roughly** from provisional values (e. g., costs per m², costs per kg, costs per classified part).
- **Inquire** about externally obtainable sizes even if they are later produced in-house. Usually it is possible to get the prices and thus costs of suppliers more rapidly than from inside the company.
- Store the bid data and evaluate if required.

B: Rules and measures for manufacturing cost reduction, if the cost information is insufficient

B1: Universal measures for product cost reduction

- **Consider boundary conditions:**
  - single-unit/series production
  - small/large parts
  - low-cost, standard materials/expensive, special materials
- At bid release/order negotiation, if possible, agree upon few or **few "sharp" demands**, functions, tolerance limitations, warrantees, acceptance conditions, rules to be kept or standards.
- Limiting to **core functions/core parts.** Farm out all peripheral items:
  - Externally produced, or also assembled outside.
  - Externally designed, and produced outside.

  (Grant the freedom for cost reduction to the supplier.)
- Search for **other suppliers** or develop one, if there is no competition at the time of the first search.
- **Design for the proper quantity:**
  - **Prototypes** (lot size = 1) to be made of standard materials in conventional mechanical production. (Important for this is a short realization time, high reliability and a high degree of compatibility with later production.)

- For **series production**, in particular **limited-lot production**, **redesign** previous **prototypes** for lower cost primary productionprocesses and deformation processes (e. g., casting, injection molding, sintering, sheet metal forming, etc.)

- **Reduce part count:**
In most cases, this results in a cost reduction. Every new part brings about organizational launch costs, logistics costs, setup costs; it requires interfacing, connection elements (processes) and corresponding assembly costs. Therefore, integral design is generally preferable.
The opposite (i. e., to separate into two or more parts – differential design) is rational only when a part is too large for processing, assembly, and transportation, or because of its complexity there is danger of rejects. This is also the case if expensive material is not needed at all active surfaces of the part, or if only certain active surfaces need to be replaced as parts wear.
Reduce part count by:
  - Designing with similar and repeated parts.
  - Designing in size ranges, modular design.
  - Application of primary production processes (casting, injection molding, sintering).
  - Application of deformation processes (sheet metal forming, forging, extrusion, rolling).
- For small quantities expensive tools, forms and models are not economical.
- For **simple** and **robust** (mistake-proof) production, design:
  - parts with few production steps;
  - parts that can be processed without retooling, but easy to machine;
  - parts should have few machined faces, all at the same height (to mill through), not oblique to each other and similar (for example, having the same boring diameters);
  - dimensions that are measurable, and if possible, with coarse tolerances;
  - as far as possible, parts suited to existing production machines (e. g., for the available machining center).
- Simple design for **assembly:**
  - Few parts, few variants.
  - Pre-assembled, separately testable assemblies.
  - Simple connections (without additional parts), easily accessible.
  - Assembly in one direction (sandwich design).

B2: Specific rules for single-unit and series production

- **Mechanical production of small parts in single-unit production** (for example up to 5 kg weight) of common materials (structural steel, etc.) Here, the set-up costs are dominant (e. g., 60–90% of the *MC*). The material costs are proportionately very small (a few percent).

For overhead costs, likewise, "one-time costs" dominate, such as drawings preparation, launch costs (in the machine industry each part or item number can amount to a few thousand dollars, in vehicle manufacture, 10 000 $ or more).

- During assembly, the handling, logistics, material supply costs are dominant.

➔ **Rules:**

  - **Part count decrease** through similar parts, repeated parts, integral design, part families, size ranges; modular design (thereby,indirectly, also an increase in number, with a reduction in set-up and one-time costs.)
  - Few, and only **simple connections**.

- **Large (heavy) parts in single-unit production** (e. g., a few hundred kg and up) of common materials, **likewise, parts for serial and mass production** (also small parts.)

  Here, portions of the material costs and production costs of components form the dominant share of manufacturing costs. Set-up costs are small.

➔ **Rules:**

  - **Reduce material costs** through small design (avoid over-sizing by using FEM analysis, for example), increase of speeds; use of high strength materials (usually only slightly more expensive); use of low-cost common materials, when there are no high demands. In the case of **series production,** choose material saving, near-net-shape production processes, such as casting, forging and deep drawing.
    Strive for small material thicknesses. Pay attention to direct force lines between loading and support points (strive for axial loading, avoid bending and torsional loading; see Fig. 7.9-3).
  - Reduce production costs from total time units. In case of **series production** use near-net-shape production processes (casting, injection molding, sintering, forging, sheet metal forming, etc.) Avoid more expensive production operations (loose tolerances). Little machining. Easily processed material.

7.11 Influence of the production process

7.11.1 Overview

The production costs form a large portion of costs, comprising approximately 28% of the total costs in the machine industry, next only to the material costs (about 38%) [VDM95]. Accordingly, the choice of the production processes for parts and/or their sequence, together with assembly, has considerable influence on production costs. In most cases, however, the production and assembly processes and the types of materials are prescribed. This is because experience has proved certain processes to be optimal, and the pertinent know-how and investments exist in the company.

However, if production and assembly are to be redesigned, the **inter-linked features** shown in **Fig. 7.11-1** must be taken into account because "everything is connected to everything else". For example, the choice of the production process depends on what material the parts are made of and whether they are large or small, simple or complex, super-exact or ordinary, and whether the surface is to be shiny, smooth or rough. Should the parts be assembled at all, or can the product be made at a lower cost as **one** part, using integral design? Otherwise, must the fastening methods allow for disassembly, perhaps because spare parts may be needed, or it is required for recycling [Mat57]?

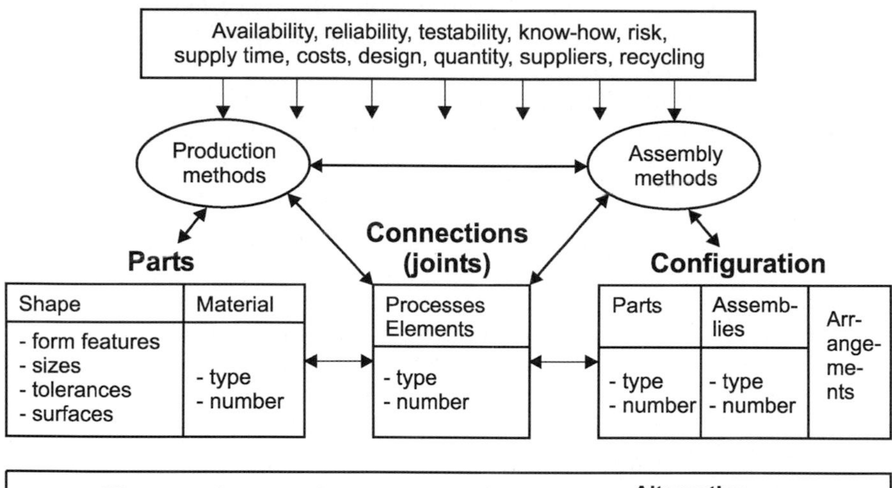

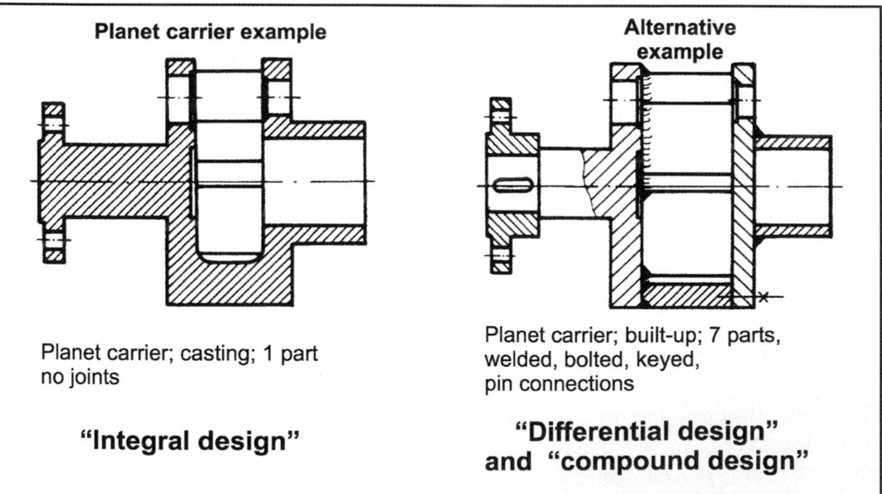

Fig. 7.11-1. Influences on the choice of the production process

The **example** of a **planet carrier** for a planetary gear drive, shown in Fig. 7.11-1, makes the complexity more apparent. The planet carrier (at the right) is of differential design, assembled from many semi-finished parts (seven preprocessed parts) and/or welded and permanently joined. What are the costs of preprocessing, the welding, the reworking, and the assembly of the mounted coupling? What is the delivery time? Is it not better to cast it all in one piece, as shown on the left? How much does the pattern for the mold cost, and the casting and the subsequent processing? For how many of the same planet carriers must the costs be found? We see that there are a number of difficult questions to be answered. The distinct technical properties such as deformation behavior, strength, and testability have not been addressed yet. See also Fig. 10.1-9.

From this, it is clear that a complex situation cannot be optimally solved with a simple, final decision. A team of people from product development, production, assembly, quality assurance, purchasing and cost control must collaborate, in several stages, from the initial to the detailed decisions (Sect. 4.3.2). This is shown in **Fig. 7.11-2**. Only after the "detailed decision" in which all details for material, production, assembly, fastenings are set, can a **design oriented towards production, assembly, and costs** be carried out. Often that must also come about iteratively: During designing it might be found that, for example, assembly would present problems, or that for a certain production process there is no flexibility. Therefore, something in the production or assembly must be changed.

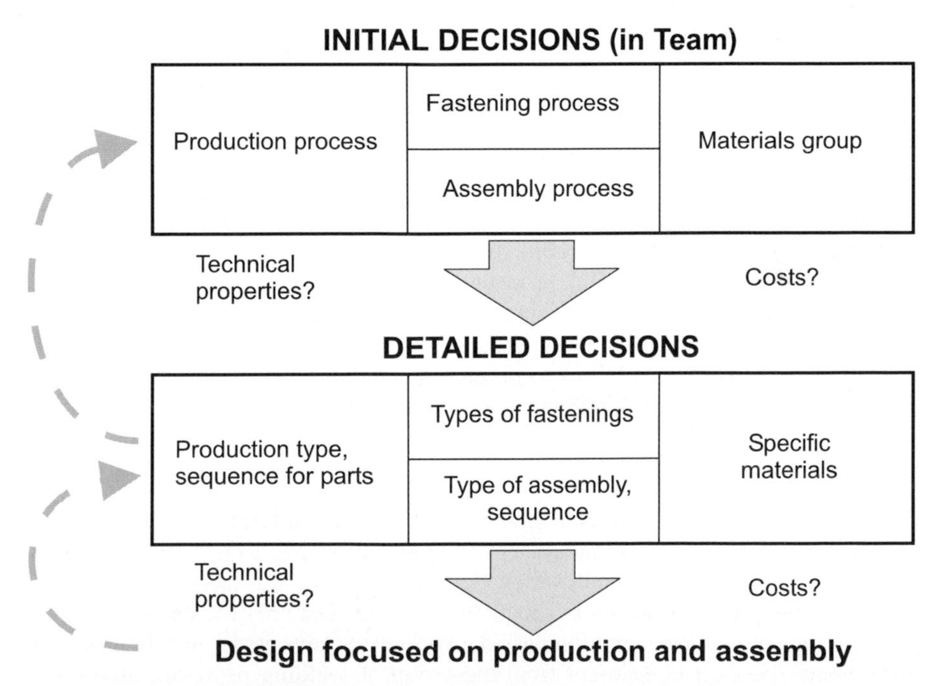

Fig. 7.11-2. Initial and detailed decisions in form design, focused on production and assembly. Without the knowledge of production, assembly, fastening methods and materials, a low-cost design is not feasible

Information possibilities for production

1 Personal information

- Designer gathers information in the company on production and assembly processes and requirements.
- Designer is advised by the **production specialist**: In the design department for routine meetings at specific times, or in continual contact with design.
- **Interdisciplinary team** (e.g., product development, production, assembly, purchasing, quality assurance, cost control) supplies the necessary information.
- **Supplier** or **specialist** discusses meaningful changes with the designer.
- **Seminars, videos** on new production technologies; lectures.

2 Written ("paper") information

Guidelines, good-bad examples, shop standards, machine data sheets, books, journals, design rules, old drawings and documents.

3 Computer-based information

- **Computer queries** on shop standards, repeated, standard and purchase parts, machine data, design rules, old designs, production plans.
- **CAD macros** for standard parts, design zones; repeated parts are imported directly into the drawings, rules incorporated into design.
- **Production and assembly simulation,** to avoid collisions, achieve time-optimal movements.
- **Design-for-manufacturing CAD System** [Mee98].

Fig. 7.11-3. Information possibilities for production-focused design

That means that **without sufficient knowledge about production, assembly, joining processes and materials, low-cost design cannot be carried out.**

Figure. 7.11-3 describes which possibilities exist regarding direct personal information and teamwork to procure the **necessary information,** including cost comparisons. It is the task of the product developer to understand the effects (parameters and cost drivers) of the company's own production processes, as well as the supplier's alternative production processes. That may take years.

Information is provided below to prepare for and support the decision for production-focused design, according to Fig. 7.11-2.

a) Choice of the production process

It is important to know the **typical characteristics of production processes.** From the large number of possible processes (overview shown in **Fig. 7.11-4**), some of the important processes, about which cost information was accessible, are reviewed in the following sections. The picture is to be used only for a rough orientation, to **suggest looking at alternative production processes**, and to discover more about those. It is apparent from the group of welding processes alone that specialists are to be called upon for advice, from within the company or at the supplier. In Fig. 7.11-4, only eight out of about 250 different processes are mentioned.

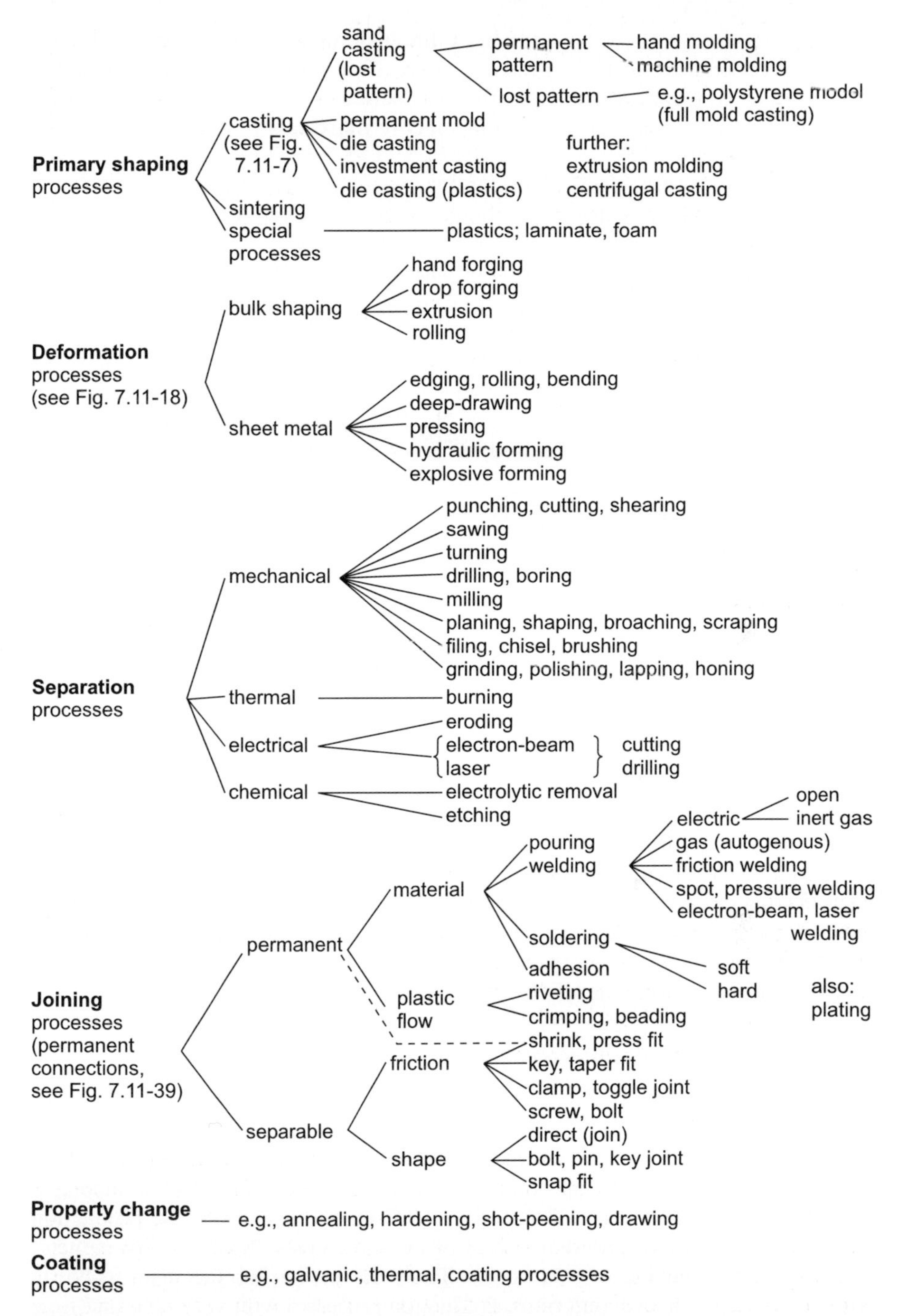

Fig. 7.11-4. An overview of the customary production processes (classification according to practice requirements, only in part from DIN 8580, and Bauer)

Furthermore, there are processes **for high dimensional accuracy** (e. g., machining processes, investment (lost-wax) casting, or precision forging); processes **for large quantities** (e. g., permanent mold casting, pressure and die casting, extrusion, deep drawing, broaching); and processes **for specific materials** (e. g., sintering, injection molding, etc.).

In the case of **series production,** costly production lines (e. g., transfer lines, special machines) largely determine the design details. Changes for cost reduction then are nearly ineffective.

In this case a constructive revision of parts must be reviewed with the supplier **before the acquisition of a new production line:** Which changes lead to a reduction of the cycle time for the part? Which lead to a reduction in the capital expenditure? It is therefore advisable to gather solution ideas for lowering costs long before the purchase of the new production line.

For choosing a production process in the company from the viewpoint of the costs, the **variable costs** must be used for the **comparison** (Sect. 8.5.1).

b) Structure of the production costs

The production costs PC are the sum of production labor costs PLC, overhead costs POC and special direct costs of production $SPDC$ for part production and assembly (Fig. 8.4-2).

$$PC = PC_{parts} + PC_{assembly}$$

$$= (PLC + POC + SPDC)_{parts} + (PLC + POC + SPDC)_{assembly}$$

The ratio of labor to overhead costs for large investments (e. g., machining centers, and gear-cutting machines) shifts more toward overhead costs. For simple workstations (e. g., assembly in single-unit production) it is about equal.

According to part and lot size and the type of part, the essential points of production costs can be distributed differently (Sect. 7.7) in total or setup times. Then the cost structures of production sequences for typical parts are necessary to be able to recognize the affecting key points and to choose more suitable production processes or to do further design work.

c) Crossover quantities for economic production

Crossover values of quantities produced are essential in selecting production processes with regard to the expected costs. These are the quantities for which one type of manufacturing becomes more economical than another. **Figure 7.11-5** shows how the manufacturing costs per piece decrease as the number increases (for its basis, see Sect. 7.5 or 7.7, and Fig. 7.7-1). In Fig. 7.11-5, the production processes A to C have different values of one-time costs (Sect. 7.5: investment, setup, pattern, template, or tool costs). Thus, the curves for the manufacturing costs per piece fall at different rates. Production processes with very high one-time costs (e. g., permanent mold casting versus manual sand casting) often have smaller production and idle times, so that the production costs from total time

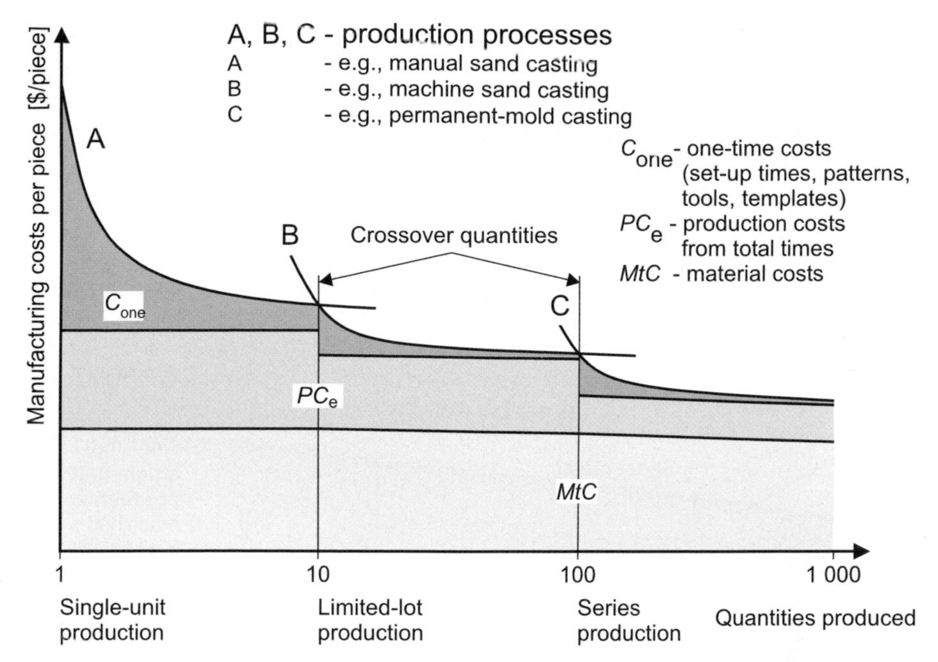

Fig. 7.11-5. Reduction in the manufacturing costs with increasing quantities, by use of production processes A to C that lead to lower costs (see Fig. 8.5-2)

units, *PCe,* drop sharply. The material costs, *MtC*, were assumed in the figure to be almost constant, which is not the case in practice. There are discounts for large quantities (Fig. 7.9-11). More efficient production processes (precision casting or drop forging instead of machining from the whole) often generate less material waste. Thus, the material costs per piece decrease. We recognize from Fig. 7.11-5 that there are limiting values (i. e., crossover points of the cost curves A to C) for different production processes that indicate which production process becomes more economical as the quantity (lot size) changes (Fig. 10.3-5). From the figure we see also how the material costs become more prominent with increasing quantities (Fig. 7.7-6). The example in Sect. 10.3 shows the calculation of the crossover quantities using concrete data.

Figure 7.13-13 shows with an example of casting/welding, how the crossover quantities can be dependent on one of the many parameters.

d) Production-based shape design

Form (shape) design is strongly dependent on the type of the chosen production process, since for every production process (and range of quantities produced) the lowest cost production-based design is different (Fig. 7.11-2). It would therefore be a mistake if we start with an initial welded design, round off the sheet-metal corners, and call that the ideal cast design (**Fig. 7.11-6**). On the other hand, gray cast iron enables certain designs (even with many ribs and of almost arbitrary form) and allows for optimum force-flow lines, in large quantities and without

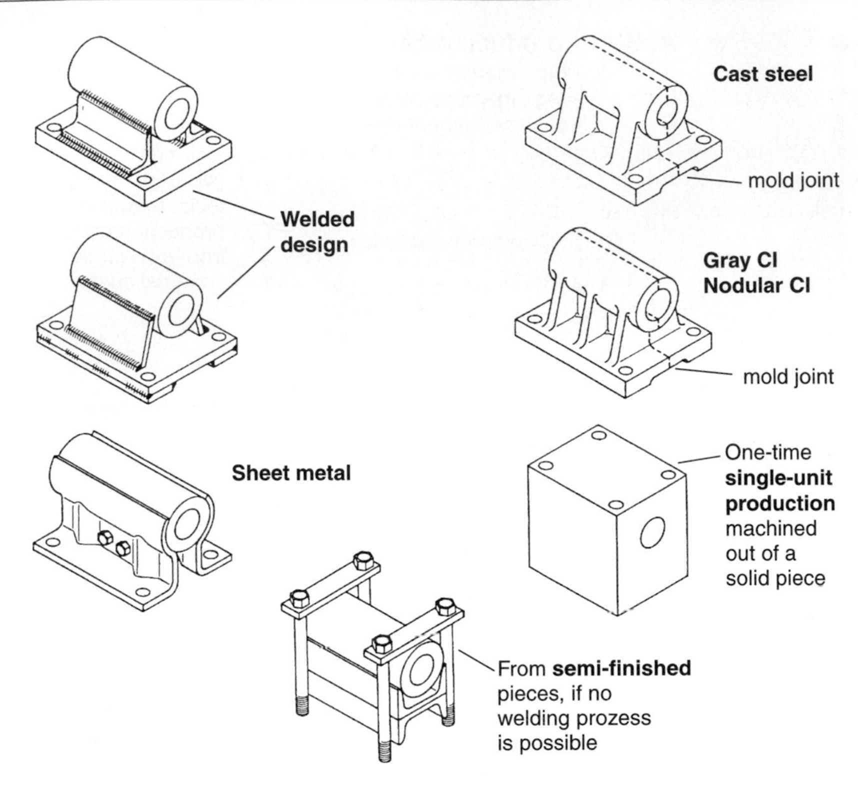

Fig. 7.11-6. Typical "production shape design" for the bearing pedestal example

additional cost. In contrast to this, every rib in a welding adds extra costs (to cut sheets, to prepare the joints, fitting, welding, and cleaning). Furthermore, there are big differences between the shapes of cast steel and gray cast iron parts. Due to easy flow of the liquid iron during casting, the walls can be thinner and more corners, recesses, and openings are possible. Cast steel, with its strong tendency to shrink and form blowholes and cracks, must have thick walls, with wall thickness increasing toward the runner. If we make an analogy to known architectural forms, a cast steel housing corresponds to the bulk and ponderousness of an early Roman church, a gray cast iron one to a Gothic cathedral, and a welded housing is like a simple, smooth, functional concrete building. Similarly, there are typical designs for all production processes (i. e., typical die-castings, die-forgings, or deep-drawn shapes that designers use as mental models during form designing).

The sequence during designing is therefore that shown in Fig. 7.11-2: First the choice of the part production and assembly processes, then the production and assembly-focused form design.

7.11.2 Primary production processes

7.11.2.1 Most important casting processes

The important casting processes are listed in **Fig. 7.11-7**, along with the data for their respective ranges of application. Since these are subject to continuing development, they are to be used only as rough guidelines.

- **Sand casting**: single-unit, small-lot production;
  (**manual casting**) weight range: 100 g to 100 000 kg;
  accuracy: medium to rough in mm range
  (e.g., ±1 mm for GCI);
  smallest wall thickness: 5 ± 0.8 mm GCI, NCI; 6 ± 1 mm CS.
- **Sand casting**: small to large series;
  (**Machine casting**) weight range: to 5 000 kg;
  accuracy: medium.
- **Permanent-mold casting and/or low-pressure casting**:
  large series, e.g., 1 000 pieces and up; weight range: to 70 kg,
  accuracy: fine to medium (0.2 mm for GCI);
  smallest wall thickness: 3 mm GCI, NCI.
- **Die casting**: very large series, specially for light metals
  e.g., 3 000 pieces and up);
  weight range: to 50 kg;
  accuracy: fine (e.g., 0.03-0.1 mm for Aluminum alloys);
  smallest wall thickness: 0.8-3 mm for Aluminum alloys.
- **Shell-mold casting**: (molds of sand and polymer binder are removed
  from the pattern, rejoined and filled with sand)
  medium to large series (a few 100 pieces and up);
  weight range: 1 g to 150 kg;
  accuracy: fine.
- **Investment casting**: wax/plastic melted out; series production (e.g., 50 pieces
  and up); weight range: 1 g to 10 kg;
  accuracy: very fine, in 1/10 mm range;
  smallest wall thickness: 1-2 mm;
  little rework necessary; also high-strength materials.
- **Full-mold sand casting**: with polymer foam patterns (melted out, or
  reusable a few times after rework), for single-unit production;
  weight range: 100 g to 100 000 kg;
  accuracy: medium to high (in mm range, to 1/10 mm).
- also: **centrifugal, continuous and composite casting**.

Fig. 7.11-7. Overview of the important **casting processes**

7.11.2.2 Factors affecting the costs of cast parts

Fundamentally, the following three cost elements have to be taken into account for a ready-to-use cast part (see Fig. 7.13-15):

- Pattern costs
- Costs of the unfinished cast part
- Processing costs (mechanical working)

The **pattern costs** are introductory costs, *IC*, which are divided by the total number cast, N_{tot}, to determine the piece costs, as described in Sect. 7.5. With small numbers, the pattern costs are dominant; the cast part should then be designed for a low-cost pattern. In practice, N_{tot} is very uncertain – the customer usually overestimates. The **costs of the unfinished cast part** are dominant for large cast parts due to the growth law of the material costs portions (see Sect. 7.6, Fig. 7.6-3) and with large numbers, N_{tot} (low pattern cost per piece). The design should then be to keep material costs low. Higher pattern costs (for example, many ribs in the case of thin walls) can be tolerated. For large quantities, N_{tot}, the **costs of mechanical work** are important in the case of small parts. Then the pattern costs per piece are small and the material costs, because of the small size, are still not too high.

The above three cost elements are influenced particularly by following parameters:

- Total quantity N_{tot}
- Size, with material costs per unit volume
- Quality requirements

These and some other parameters are now dealt with in more detail:

a) Effect of quantity

Ranges of the quantities produced (Sect. 7.11.1c, Fig. 7.11-5) for the specific casting processes show roughly (N_{tot} = entire number to be cast) the following:

- Sand casting, manual (wood pattern) from $N_{tot} \geq 20$ pieces for small parts, from $N_{tot} \geq 2$ pieces for large parts (see Fig. 7.13-13). For fewer pieces, consider welded design
- Sand casting (manual) with polymer foam pattern from $N_{tot} \geq 1–2$ pieces
- Sand casting (machine) from $N_{tot} \geq$ of 50 pieces
- Permanent mold casting from $N_{tot} \geq 200–1\ 000$ pieces
- Die casting from $N_{tot} \geq 500–3\ 000$ pieces

This is dependent on weight in all cases! (See Fig. 7.13-13).

These are obviously only provisional values. The limit numbers depend on the size (weight), the complexity, the quality requirements, and the kind of material. In many foundries, the influence of quantity in the quotation is traditionally small, since the pattern (usually belonging to the customer!) is not included. For larger lot sizes, statistical investigations by Pacyna [Pac80] show considerable cost differences. For example, for machine-cast gray iron parts (not including the pattern costs) the cost per piece decreases as follows: Manufacturing costs which are 100% for a lot size of 10 become 85% with lot size 100, 70% for lot size 1 000, and 60% with lot size of 10 000 pieces.

A qualitative cost comparison between casting and welding processes is shown in **Fig. 7.11-8**. Both the welded part and the unfinished cast part display a falling cost curve, since setup costs and training-effect, which have a cost-reducing effect, are included. Compared to the welded part, the limiting quantities turn out to be

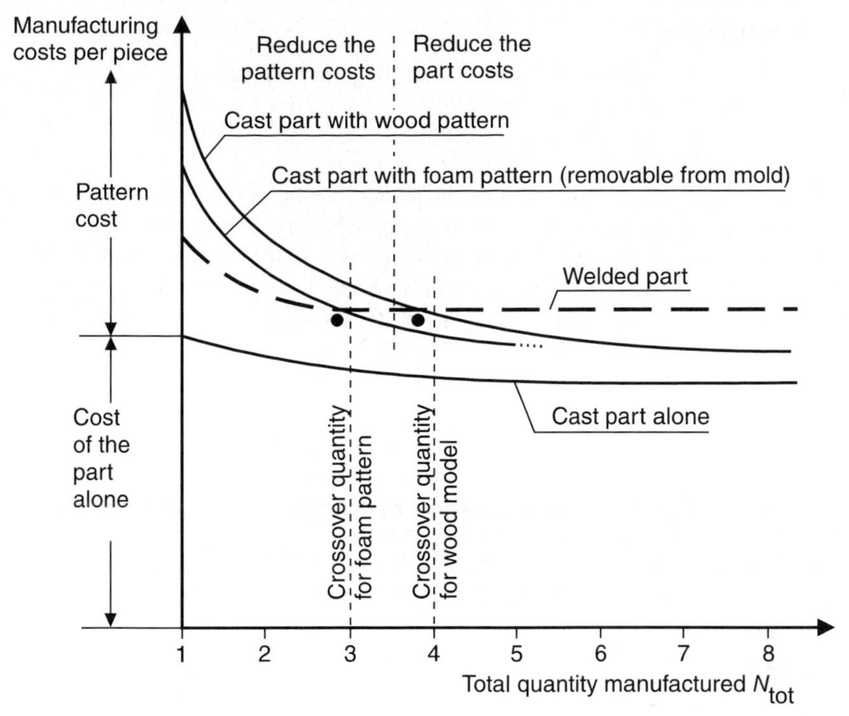

Fig. 7.11-8. Qualitative manufacturing cost comparison of welded and cast parts [Haf87]

three pieces for the removable polymer foam pattern and four for the wood pattern. We see that because of the cost components, for small numbers N_{tot}, it is a matter of lowering the **pattern costs**, whereas for large quantities it is the unfinished part costs that should be reduced (see also Fig. 7.13-15). This depends, again, on the size of the part to be cast. As Fig. 7.13-12 shows with the example of large cast gears, the pattern cost as a percentage decreases greatly with size. The reason is that the manufacturing costs of wood patterns generally increase far less rapidly with size than those of machined parts (see Fig. 7.13-11). Since there are no data from investigations on pattern costs, they must be estimated or inquired about. A formula for estimating cost of wood patterns is shown by Pacyna [Pac82a]. Rules for low-cost pattern design are shown in Fig. 7.11-11a. The gist of it is to aim for simple patterns, without or with few cores and with few projections and ribs. Similar rules are valid for welding. Each rib must be specially produced.

In contrast, for a cast piece that is made in large quantities, it is a matter of making the **unfinished housing** i. e. the cast piece itself of low cost (in many cases, light). In view of higher pattern costs, it is better to plan for more ribs if that should lead to reduced wall thickness and weight. Properly placed ribs exhibit hardly any higher production expense in form design and casting (except for, perhaps, in fettling). The material flows easily into the mold. This is very different in the case of welding (Sect. 7.11.5.3). Casting therefore allows for very unrestricted shapes. That makes itself felt only when basic technical rules are broken; such as "a cast piece should be simple to take out from the mold".

b) Influence of size

The size (weight) is the fundamental parameter that affects cost. In earlier times the **quotation prices** of foundries were based only on the weight: A heavy part was expensive, a light part cheap. Accordingly, the emphasis was on designing the part light, regardless of how complicated it became.

Today, weight still has the greatest influence on the price quoted, but from experience with similar parts cast previously, complexity is also taken into account [Pac80, Pac82b]. Complexity increases with, for example, a great deal of core work, or large sizes, or parts with small wall thicknesses, or special quality requirements. To a certain degree, the pertinent lot size is also considered. Material enters into the picture as follows: Between gray cast irons (GCI 10 and GCI 25) there is hardly any difference, while nodular cast iron (e. g., NCI 40) costs on the average 1.2 to 1.5 times as much, and cast steel (e. g., CS 52) costs 2 to 2.5 times as much. Within cast steels alone, there are differences in price of the order of 1:20 (e. g., corrosion-resistant, high temperature cast steel is expensive) [VDI77]. **In general, to keep costs low on external orders, a casting should be designed to be light, with perhaps a high degree of complexity.**

Data, such as given in Fig. 9.3-1, can be prepared for machine parts and assemblies of interest, for the purposes of cost calculations or quotation prices. Then, **weight-cost calculations** can be carried out, as described in Sect. 9.3.2.1.

c) Influence of quality requirements

With cast steel parts, in particular, quality requirements play a dominant role because of the costs of testing and the necessary finishing and reworking. The relative costs of parts with the same complexity and demands are indicated in **Fig. 7.11-9**.

Unnecessarily acute quality requirements should be avoided.

Requirements	Relative costs
• Without special requirements (as-cast)	1
• Normal requirements (no quality grade)	2
• Enhanced requirements (quality grade II-III)	3
• High requirements (quality grade I, special requirements)	4

Fig. 7.11-9. Relative costs for castings of different quality classes (DIN 17 245)

d) Additional influences

Determine the company-specific influences on the costs of cast parts by consulting with a competent foundry specialist. This concerns, for example, the costs of cores, the influence of molding box filling, and the subsequent heat treatment. The latter might be annealing for stress relief, particularly with plate-like parts made from GCI, if there is danger of warping or cracks. This annealing makes for additional costs of 10–20% (more information in [Ehr83, Ehr85]).

7.11.2.3 Cost reduction by using full-mold casting processes

a) The technology

For single-piece production (or, at most, a few pieces) with manual sand casting, the use of polystyrene foam patterns has proved useful. A true copy of the cast piece is made from this material by milling or cutting with an electrically heated wire, and/or by adhesion. Shrinkage is taken into account. With the full-mold casting process, the pattern is left in the molding box and evaporates (lost foam) while the cavity fills with fluid metal. Since unwanted residue sometimes remains on the casting, with the hollow-form casting process the polymer foam pattern is burned off before pouring. If it is (rarely) made for reuse, some more castings can be made after repairs [VDI78].

b) The cost reduction

The more complicated and larger the piece is, the greater are the cost advantages of a polymer foam pattern as opposed to a wood pattern. Only the necessary wooden core boxes for making the cores are expensive. Furthermore, the maintenance and storage of the wood patterns must be considered. Depending on the handling during the removal of the pattern from the mold or the storage conditions (warping from humidity), this can cause up to 30% of the pattern costs.

According to [VDI78], the costs of polymer foam patterns are only 30% of the costs of a wood pattern; other sources claim it to be only 10–20%.

c) Form design guidelines for lost patterns

To be practical, avoid any subdivision of the pattern and thus avoid lifting-out tapers and undercuts. Likewise, the cores and their supports are eliminated. There is great design freedom in this approach because only the molding must be thought of (filling with sand). In addition, steel pipes, bolts, cans, and wear strips can be molded in. The casting weight can be 100 g to 100 tons and higher. The moldable materials are arbitrary.

7.11.2.4 Rules for low-cost form design of castings

The rules depend on the production operations, which in turn determine the cost central points in a concrete case. They extend from the pattern and template preparation, to casting, up to the fettling and the mechanical work. Which production operations drive the costs can be recognized from cost structures of similar components (Sect. 4.6.2). For this, of course, the size and number must also be similar. At minimum, it must be clear whether the cost focus is in the pattern costs, the costs of unfinished casting (perhaps in the material costs), or in the costs for the mechanical refinishing (see Sect. 7.11.2.2 and Fig. 7.13-15).

Such a cost structure for cast steel parts is shown in **Fig. 7.11-10**. For **small cast steel parts** of high quality, it is not advisable to invest much effort to try to lower the material costs. It is better to design the cast part for simpler subsequent processing (welding, fettling), since more than 50% of the manufacturing costs lie

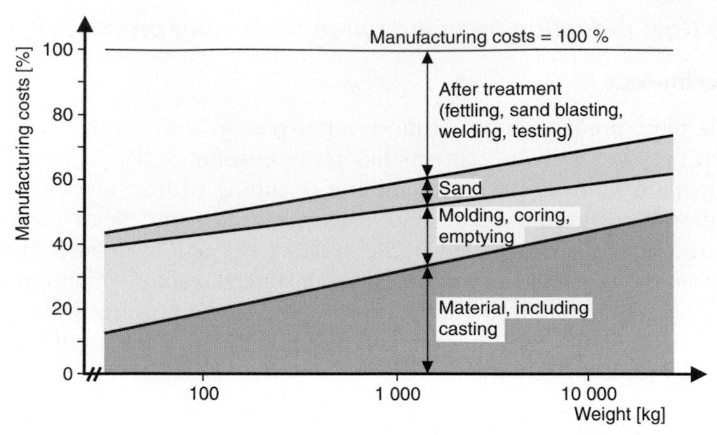

Fig. 7.11-10. Cost structure of the manufacturing costs of cast steel parts

there. For large cast steel pieces, it is different because welding is necessary to fix the defects in the casting.

For **large, heavy cast steel pieces**, on the other hand, attention must be paid to reducing the material costs. With FEM analyses, for example, the technical limits of material loading should be pushed to reduce the wall thicknesses (Procedure 1.1 in Fig. 7.9-3). Likewise, as far possible, choose low-cost materials.

As long as there are no special changes in technology, such cost structures are valid for a relatively long term. However we must be aware of the large spread which was implicit in the Fig., and which characterizes the "individuality" of the given casting. It is understandable how fast the cost structure shifts if a casting material 8-times more expensive than that in Fig. 7.11-10 is used. Then the material costs for small components also become important. Thus an individual cost analysis should be required before starting on cost reduction [Ehr85].

Summarizing, from [Ehr83], the following statements can be made from the investigation of cast steel parts (from 30 kg to 20 000 kg).

The share of **production costs** (sand, molding, emptying, fettling, testing, and other handling) in **casting manufacturing costs** increases with

- increasing quality requirements
- increasing part complexity

The share of material costs in casting manufacturing costs increases with

- increasing material price, e. g., specialty steel; with very expensive materials this part can constitute about 95%
- increasing part weight

The variety of **form design rules** for low-cost castings are gathered together in **Figs. 7.11-11a–c**.

Design rules	poor	better
1. Low pattern costs (for few casts and small parts)		
Simple patterns and cores made from planes, cubes and cylinders (for small parts and low quantities N_{tot} = 1 to 3)	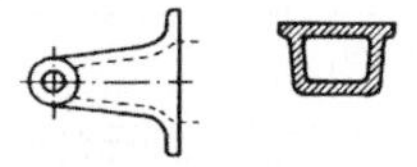	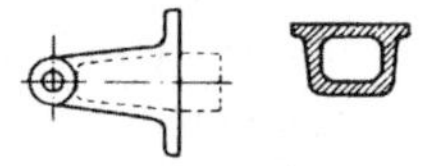
One-piece patterns, as far as possible without cores ("rib, rather than hollow casting"), few mold boxes	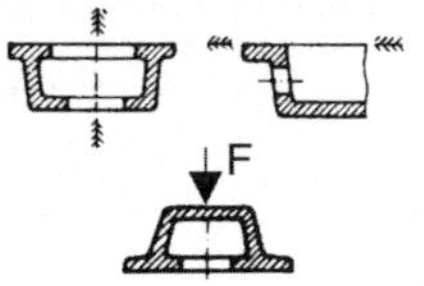 F	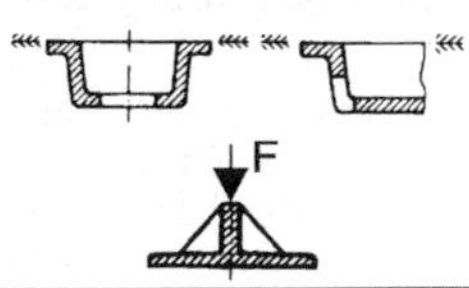 F
Taper (1:10 to 1:50) at mold joints provided in the pattern (DIN 1511) for easy removal of patterns and cores	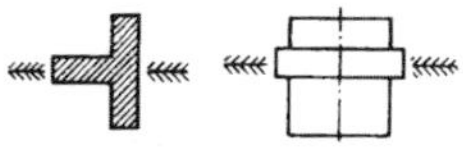	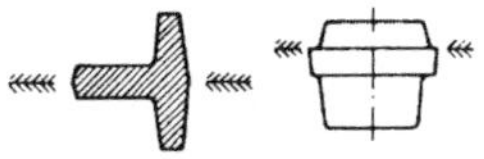
Avoid **reentrant angles**, thus inserts	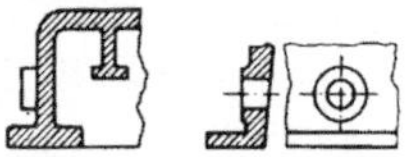	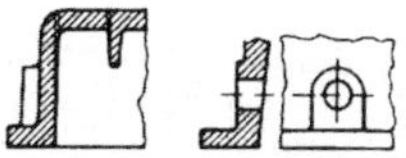
Make parts **symmetrical** instead of "right" and "left" handed and save mounted parts	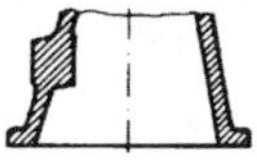	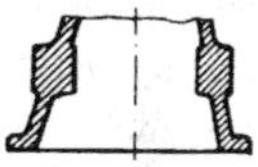
Without marking (automatically) processed **castings, position and machine near important surfaces** (staggering patterns and cores does not lead to thin walls)		
2. Low wastage (high quality casting)		
Securely placed cores, wide core marks (sand floats on iron)	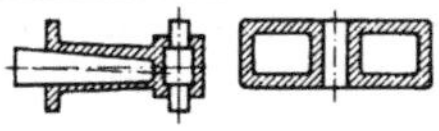	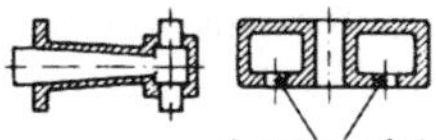 4 core prints
Uniform wall thicknesses no masses of material, no junctions (blowholes, cracks), uniform corner radii	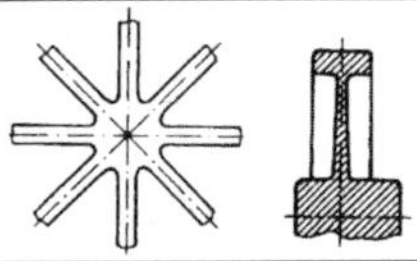	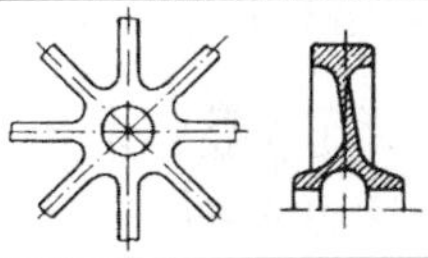
Inclined, not horizontal **surfaces** in the casting (gas removal)	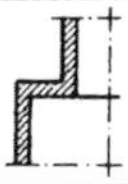	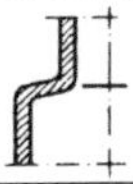
No constrictions in section between runner and riser (good flow, back flow during shrinkage, impurities to float up to riser), more important for high shrinkage: 2 % for CS, MCI; 1.5 % for NCI; 1 % for GCI	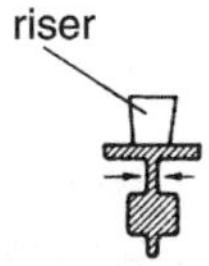 riser	

Fig. 7.11-11a. Form design rules for castings (from [Pah97], K. Tuffentsammer, W. Riege)

Design rules	poor	better
Gradual transitions, no sharp corners		
Partition of large, bulky pieces		
Avoid **thin** walls that project too far (breakage danger in transport)		
Stiffening ribs thinner than the stiffened walls (should solidify earlier)		
Inclined and curved wheel **spokes and ribs,** to avoid shrinkage stress and cracks; bending allows for more deformation	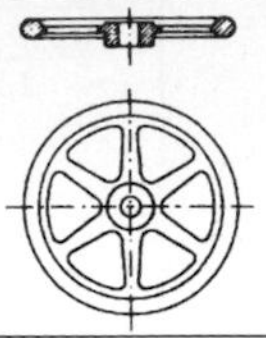	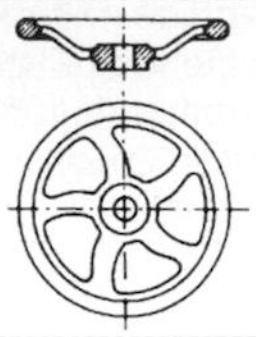
Casting stresses can be reduced by supplemental annealing of the parts		
3. Lower cleaning costs (up to 30 % of the manufacturing costs)		
Avoid **hard-to-reach corners;** provide **large openings in hollow parts**	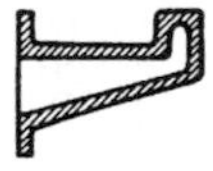	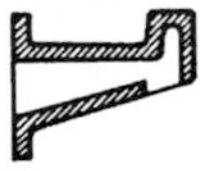
Allow space for runners and risers on the casting, or tolerate risers/runners (dirt will float to top)	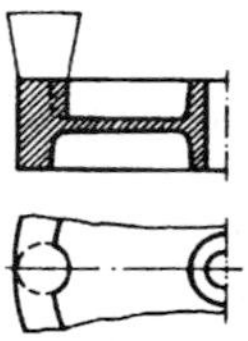	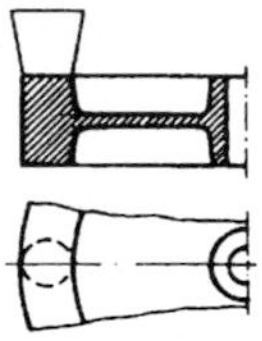
Place runners/risers on level surfaces; they are easier to remove from there than from uneven surfaces		
Leave mold joint flashes, or provide for special ribs to carry the fins	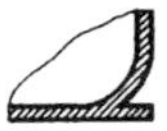	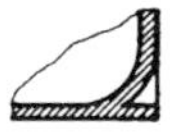
Sharp and right-angled flash areas are hard to grind; obtuse angles are better	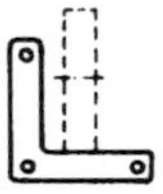	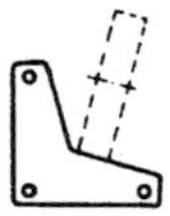
4. Low finishing costs (for small parts, large quantities: 60-80 % *MC*)		
Place mold joints such that **flash** does not stick out on unfinished surfaces and can be easily removed	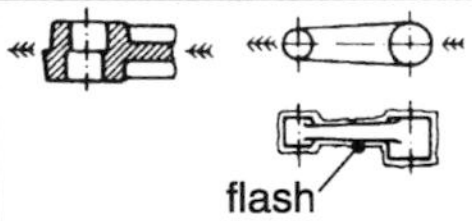 flash	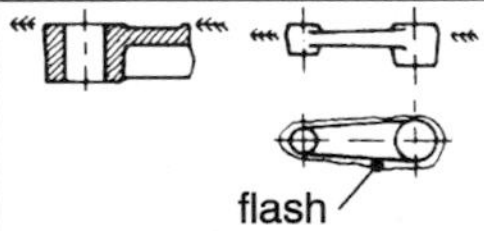 flash
Keep in mind the **path** of finishing tools and cutters	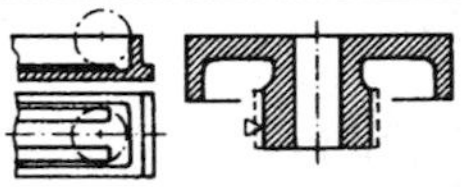	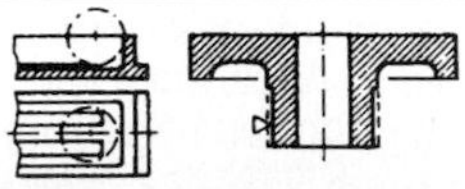

Fig. 7.11-11b. Form design rules for castings (from [Pah97], K. Tuffentsammer, W. Riege)

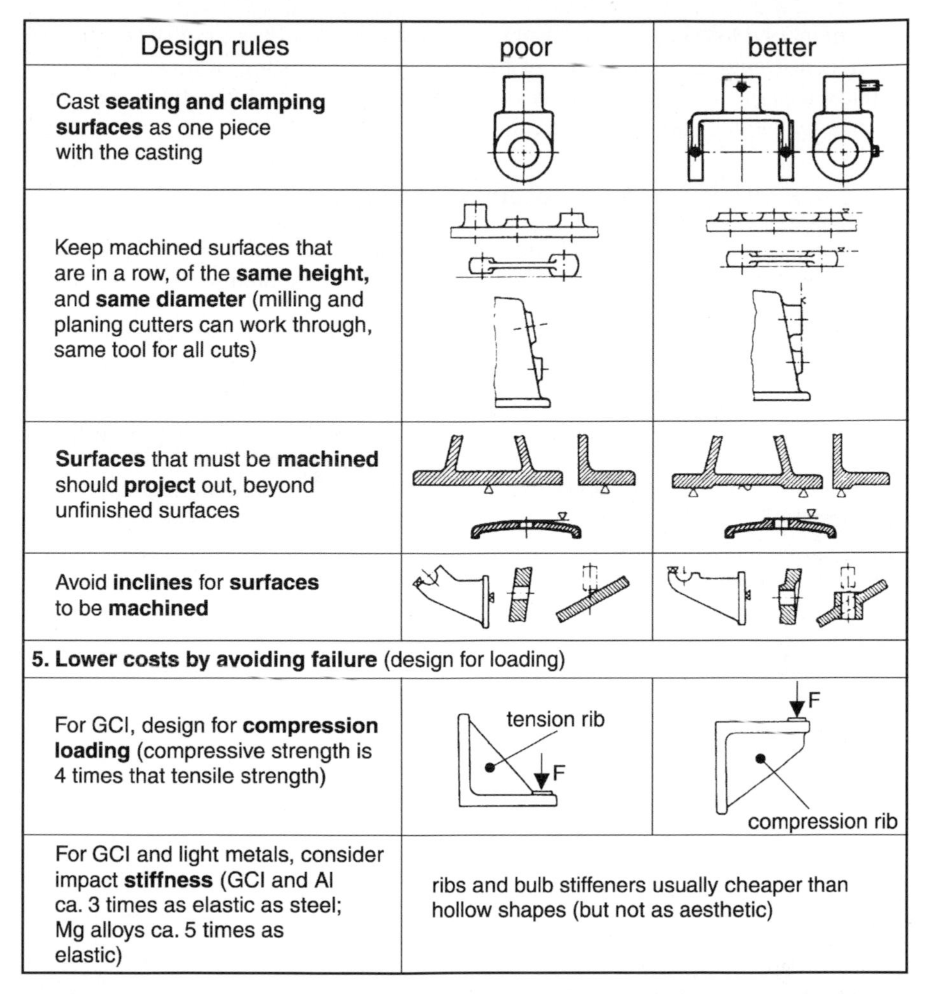

Design rules	poor	better
Cast **seating and clamping surfaces** as one piece with the casting		
Keep machined surfaces that are in a row, of the **same height,** and **same diameter** (milling and planing cutters can work through, same tool for all cuts)		
Surfaces that must be **machined** should **project** out, beyond unfinished surfaces		
Avoid **inclines** for **surfaces** to be **machined**		
5. Lower costs by avoiding failure (design for loading)		
For GCI, design for **compression loading** (compressive strength is 4 times that tensile strength)	tension rib F	F compression rib
For GCI and light metals, consider impact **stiffness** (GCI and Al ca. 3 times as elastic as steel; Mg alloys ca. 5 times as elastic)	ribs and bulb stiffeners usually cheaper than hollow shapes (but not as aesthetic)	

Fig. 7.11-11c. Form design rules for castings (from [Pah97], K. Tuffentsammer, W. Riege)

With **castings that require a reusable pattern** (e. g., with sand casting) **or a permanent mold**, it is a matter of shaping the part so that, in spite of the great design freedom, it complies as much as possible with the casting technology. The mold joint must be settled on first. Then attention must be paid to the possibility of removing the pattern and the casting from the mold. The smallest shape detail (for example, a projection or a rib placed parallel to the joint face) can cause considerable extra costs.

In the case of full-mold casting, these points of view are largely disregarded. Often a "ribbed casting" can replace the cored, expensive "hollow casting".

For shape design, the choice of the **casting material** is important because there can be notable differences: GCI and NCI flow easily while cast steel (CS) freezes more rapidly and has high shrinkage. CS has 2% shrinkage whereas with GCI it is about 1%.

As mentioned earlier, it is advisable to consult a foundry specialist or pattern maker concerning the fine points of form design of a casting, following the preliminary design. Minimizing the costs should also be considered for the pattern, the casting, and the subsequent machining.

7.11.2.5 Examples of form design of castings

The following will show how redesign lowers costs in casting design, (example a), and how reversing the process can have a positive effect (example b).

a) Figure 7.11-12 shows an example of the **redesign** of a machine part (a bending tool for a packaging machine) assembled **from 11 separate parts** into **one** part made by investment casting, weighing 300 g. Investment casting of parts is especially useful for complicated forms, since due to the high dimensional accuracy the subsequent processing costs are generally saved. The processing time for this part was reduced from 7.75 hours to 3 hours (i. e., by 61%) and the manufacturing costs dropped by approximately 72%! In many cases, a prototype is made in a right- and a left-sided design. If the machine in then made in high volume, however, it is sometimes forgotten to redesign it for the larger quantities produced (Sect. 7.12.4.3).

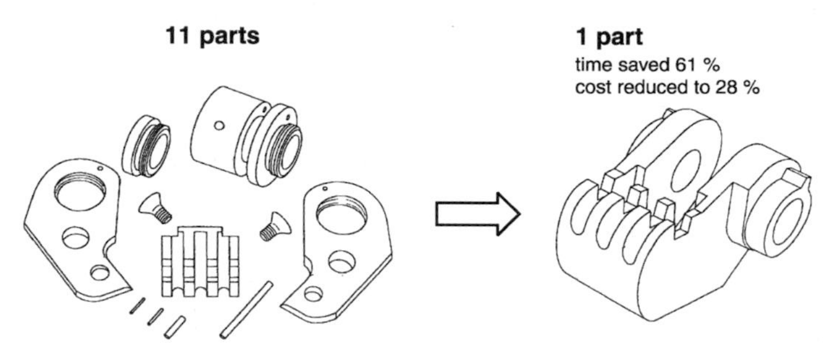

Fig. 7.11-12. Integral design through investment casting: Bending tool for packaging machine, from 11 parts *(on the left)* to only one part *(on the right)* (from RKW)

b) The **redesign** of a **casting** (integral design) into a combined **cast/sheet metal form** (differential design) for a bearing housing is shown in Fig. 7.9-9. The incidence of rejects for the complex casting due to blowholes was so high that the reduction in the cast part had a positive effect on the actual bearing housing. The shield (i. e., the connection to the machine) was implemented as a bolted steel plate, and thereby both costs and delivery time were saved.

7.11.2.6 Low-cost form design of polymer parts

a) Shape, material, and the production process ("the triad")

Polymers are used increasingly in technical applications and are replacing metals, since they can be made with **engineered material properties.** They can be light;

they can range from being elastic to rather stiff; they can be electrically and thermally insulating (more often than not); and are often wear- and corrosion-resistant. In contrast to metals, however, their strength, thermal, and long-term stability and hardness are generally still problematic. The costs per part are low if parts from mass-produced polymers can be made in high volume by injection molding, blow molding, or foam processes.

Engineered material properties were mentioned because in this fast-developing field the collaboration of the **"triad"** of suitable **shape**, suitable **material**, and the **production process** must be optimized [Zol96]. Polymer parts with oriented fibers offer the best example of such collaboration.

Figure. 7.11-13 shows the influence of quantities on weight, the production process, and the costs of a truck side panel of size 1 000×500 mm, of fiber-reinforced polymer. The figure is meant to stimulate the thinking and collaborating with specialists about the alternatives and their properties. In the actual case,

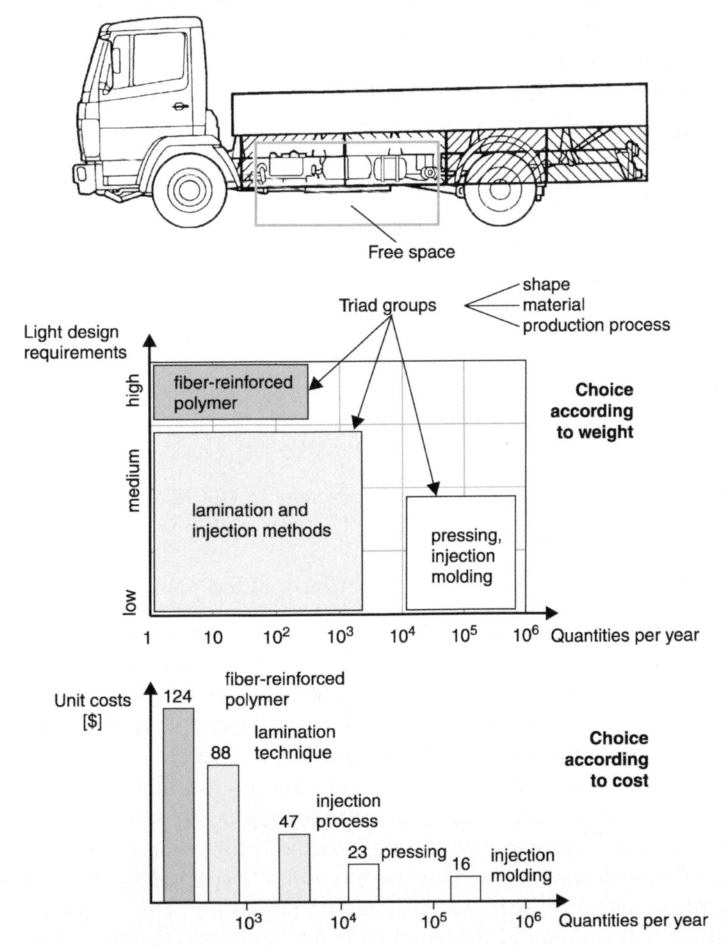

Fig. 7.11-13. Example of development and selection of polymer products [Zol96]

a ribbed plate made of polypropylene PP GF 30 with non-oriented glass fibers, with 2.5 mm wall thickness and 40 mm rib height, was selected. It is produced with compact injection molding.

b) Polymer injection molding technology

In what follows, **only thermoplastics** are considered for **injection molding,** and only as regards the cost information for design. In injection molding the polymer is melted by application of heat (by heating and/or screw-feed) and injected under pressure into the mold cavity of the tool. Upon freezing, the part is removed from the mold.

Advantages of polymer injection molding:

- Very complicated parts that were previously made from a large number of individual parts can be formed as integral parts made in one work cycle and without any subsequent processing. The assembly costs are lowered drastically (Fig. 7.12-14). The polymers' high elasticity relative to the strength enables forming snap joints and mold-on springs that can replace hinges (Fig. 7.11-48).
- In some cases, materials especially adapted to requirements for elasticity, gliding properties, insulation, and corrosion resistance can be used.

Disadvantages:

- Due to the often expensive die casting tools (tool costs = introductory costs *IC*; Sect. 7.5.1), the part costs come down significantly only in mass production. Small lot sizes are realizable at a low cost only with very simple parts.
- Polymers often have low mechanical strength, do not resist high temperatures, creep under load, and change their properties upon absorbing moisture. The properties are strongly dependent on the process conditions (warping). Thermal expansion is very much higher than for steel (for polyethylene, 20 times as high).

Technical form design rules are given in [VDI79]. The rules stated in Figs. 7.11-11a–c are also valid in part.

c) Rules for low-cost form design of injection-molded parts

Manufacturing costs of injection-molded parts consist of material and production costs. As shown by the cost structure **in Fig. 7.11-14**, the larger (heavier) the parts are, the more expensive is the material per unit volume (weight) and the more the material costs predominate. The production costs depend essentially on the **cycle time.** This is time that elapses in the production of a part (a cavity can produce many parts) on the most expensive injection molding machine. The cycle time is dependent on the material, the tool, and the injection molding machine itself. It physically depends very much on the time required for cooling down the parts. If the parts are removed too soon, they are too soft and lose shape. Since a polymer is a poor heat conductor, **thin walls** should be aimed for. This shortens the cycle time twofold: The amount of heat applied is less since the volume is less, and the

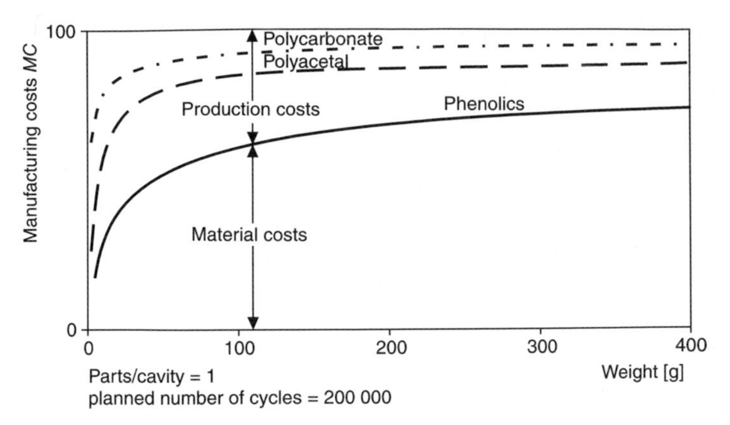

Fig. 7.11-14. Share of production and material costs in the manufacturing costs for polymer injection-molded parts [Kie79]

conduction paths are shorter. Parts prone to warp (plates, gears) and parts of high accuracy, in particular, must be left in the mold longer so that they can cool sufficiently. In the case of plates, this can be overcome by adding ribs.

The **material costs,** as mentioned earlier, depend on the volume of the part or on the weight. From this viewpoint, parts of small volume should be striven for. In the case of otherwise specified measurements, this is realized through **small wall thickness**. Furthermore, a polymer with low material price should be sought; this can be found, for example, from a table of relative costs. It should be kept in mind that, as with ferrous materials (see Sect. 7.9.2.2, Fig. 7.9-4), sometimes a material of higher strength and only slightly higher cost exhibits a large volume reduction and becomes more economical in the end.

As Fig. 7.11-14 shows, the material costs portion of the manufacturing costs increases with weight and becomes so dominant that with larger injection-molded parts the need to pay attention to low material costs becomes the foremost requisite.

A typical **manufacturing cost structure** of an injection-molded part is presented in **Fig. 7.11-15**. The material costs (polypropylene PP) comprise the highest portion of the costs, followed by the machine costs.

To the right are indicated the **steps** (rules) **with which the costs can be lowered**. Making thin-walled and light parts, as mentioned, has twice the effect. There are less material costs and shorter cycle times in the injection molding machine, and thus proportionately lower machine costs.

The **tool costs** per piece are low when the part is simple, does not have any re-entrant angles (lateral slides), and there are many parts per tool (number of cavities in the mold). Unfortunately, the designer does not have any influence on the latter. Furthermore, the tool becomes cheaper if the accuracy demands are kept low. The number produced should be high (e. g., a few hundred thousand pieces).

For large quantities, the tool costs are nearly negligible, or can be regarded as fixed.

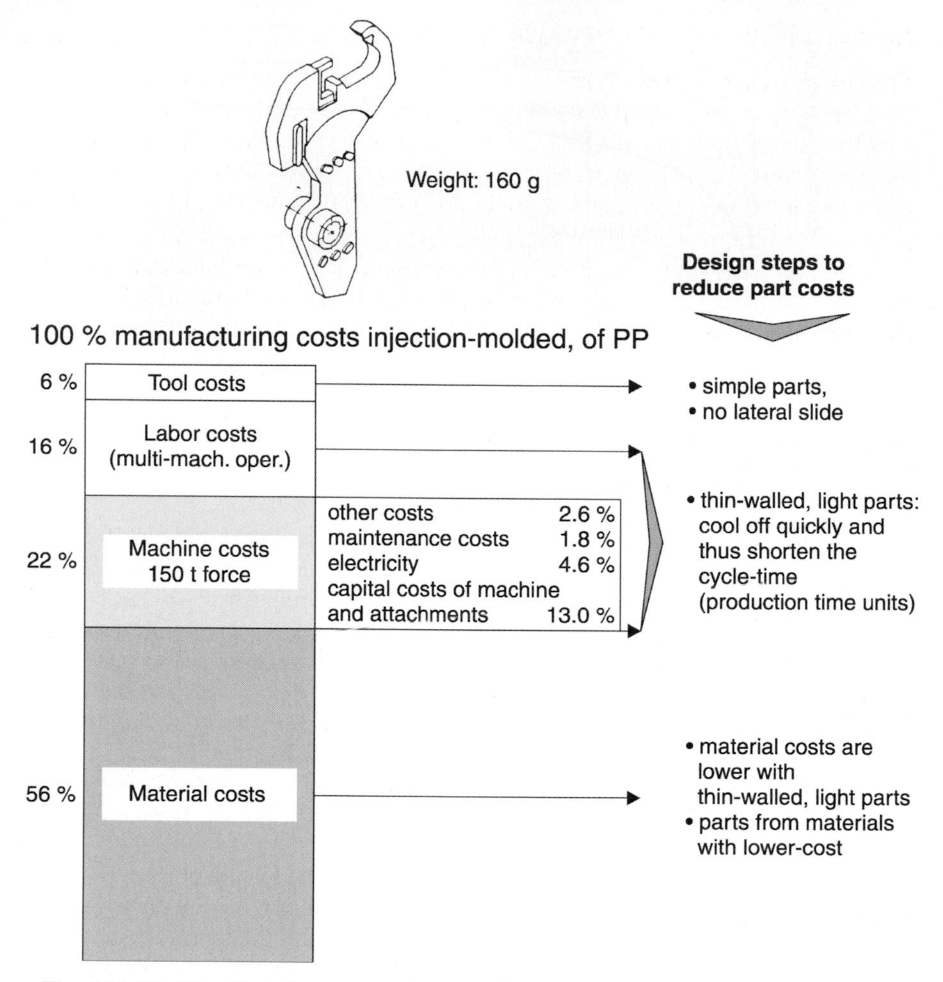

Fig. 7.11-15. Manufacturing costs and cost-cutting measures with a typical polymer part

7.11.2.7 Low-cost design of sintered parts

a) The sintering process

In sintering, iron powder (with possible addition of Cu, Ni, Cr, Mn, and C) is pressed in a usually prismatic mold by a press die, to a semi-finished form. Then it is sintered below the melting point of the main component. This may be followed by an additional press operation (calibration). The process results in very precise, small to medium-sized parts, with little rework and refinishing necessary, mostly in mass production [Der71]. Typical parts made with sintering are pump gears, bushings, friction linings, or porous filter cartridges.

b) Form design rules (Fig. 7.11-16)

The best shapes for sintered parts are not too long and are not stepped, but they may have an arbitrary and complicated profile. Machining of surfaces can generally be avoided, and that makes sintered parts in large quantities more economical than machined parts. For small parts of sizes in the range of a few centimeters, the material costs of ready-for-use parts are only 10–20% of the manufacturing costs. The press and sintering costs (30–50%) and the control costs are significant. It is important, therefore, not to choose too fine tolerances and to save the calibration process. In particular, the dimensions in the press direction should not have close tolerances (for example, IT 13). The parts can also be case-hardened or gas-nitrided.

Design rules	poor	better
1. **Low amount of waste** (good sinter quality)		
Prefer **prismatic bodies** with **uniform dimension** throughout; cross-section may be arbitrary and complicated		
Bodies' **height not much greater** than width (diameter) h/d < 2.5 wall thickness *s* > 2 mm hole size *d* > 2 mm (uniform thickness)	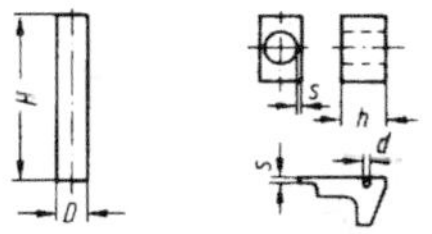 H, D, s, h, d	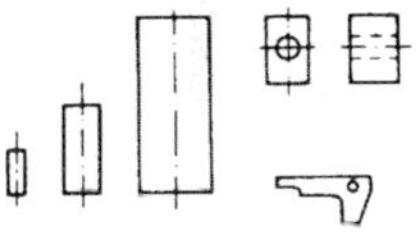
No **reentrant angles,** negative slopes, sharp angles, uniform changes in thickness	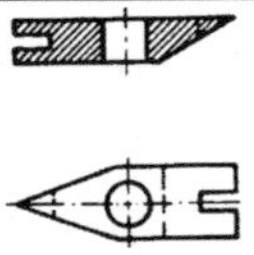	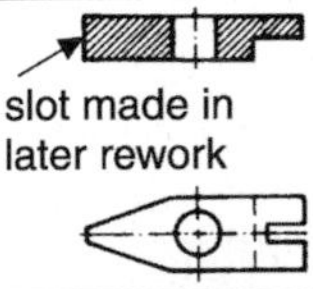 slot made in later rework
Prefer **rough tolerances**	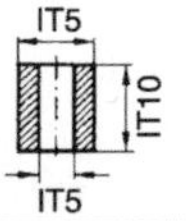 IT5, IT10, IT5	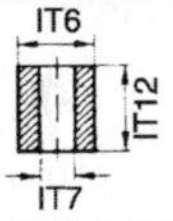 IT6, IT12, IT7
Avoid **fine teeth** and profiles, module m > 0.5 mm (otherwise non-uniform powder filling)	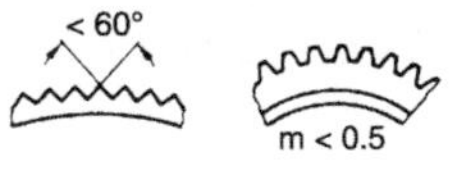 < 60°, m < 0.5	 < 60°
2. **Low tooling costs**		
Through holes to be made **circular** to minimize material use or weight		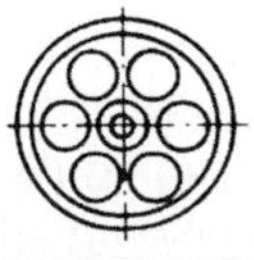
Avoid **step changes in height;** press requires costly tooling with inserts		

Fig. 7.11-16. Form design rules for sintered parts (from [Pah97], H. O. Derninger, G. Hoffmann)

c) Example

- In the automobile and motorcycle industry, sintered **connecting rods** of **powder-forged steel** are used instead of die-forged connecting rods. The mold joint face does not need machining. Instead, the sintered part is carefully cracked into the connecting rod and cap and aligned **without fitted bolts,** mold-precise and in a stable position during assembly (**Fig. 7.11-17**), all at a lower cost. The high fitting accuracy ensures quiet running.

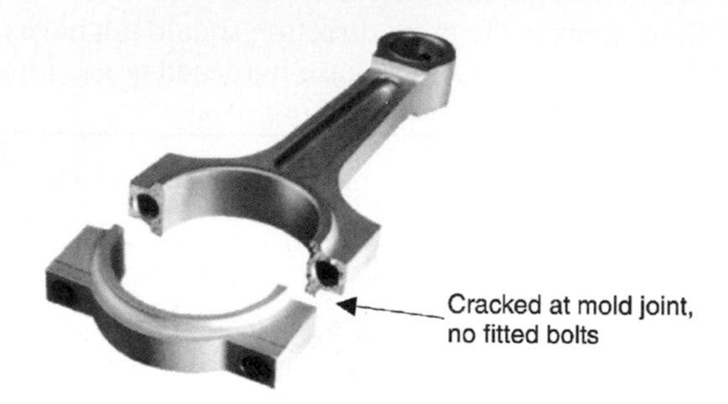

Fig. 7.11-17. Cracked, powder-forged connecting rod (BMW)

7.11.3 Deformation processes

7.11.3.1 Most important deformation processes

The most important deformation processes are reproduced in **Fig. 7.11-18**. **Figure 7.11-19** shows some typical shapes in sheet metal work.

Notable developments in the past several years in sheet-metal processing include rolling (i. e., cold working of sheet-metal strips to custom-built sections), deep drawing, bending combined with nibbling, and laser cutting on complete processing machines (**Fig. 7.11-20**) [TRU96].

Forging is used for achieving an increase in strength, particularly in the direction of grain flow, which should be aligned with the direction of the lines of force. As opposed to parts made by machining, considerable savings are possible through drop forging for larger quantities (examples: steering knuckles, crank shafts, connecting rods, coupling parts, or levers). As with casting (for example, of higher strength nodular graphite cast iron or cast steel), cost investigations of the alternatives are necessary. For plane faced (straight and spiral bevel) gears, for 2 000–5 000 pieces/year and up, gears made by precise forging show cost savings compared to machined gears, and Quality 7 (DIN 3962) is achieved. Straight and twisted turbine blades are also finish-forged.

With **open-die forging** for large pieces and small quantities where material costs make up a high portion, the roughing costs can be reduced by simple preforming. During form design, pay attention to simple shapes (no spherical surfaces), large radii, and not too large cross-sectional changes.

Process	Properties
Open die forging:	Small quantities, accuracy: low.
Closed die forging:	Series production, accuracy: medium.
Cold extrusion:	Cold forming under high pressure, series production, accuracy: high to medium.
Hot extrusion:	For profiles.
Rolling (hot and cold):	For profiles, tubes and sheet.
Sheet rolling:	Cold forming sheet metal rolls into a variety of custom-profiles.
Drawing:	For wire and profiles.
Flanging, rolling, seaming, beading, rounding of sheet metal:	Single-unit and series production; accuracy: medium.
Shape bending, 'V' bending:	accuracy high;
Deep-drawing:	with or without hold down, often in many draws, with annealing of the sheet between drawing, for medium or large series production, accuracy: high to medium.
Spinning:	Sheet forming on spinning mill, up to 50 % increase in strength (1 g to 35 kg).
(Punching, nibbling and cutting are treated with separation processes, sec. 7.11.4)	

Fig. 7.11-18. Most important deformation processes

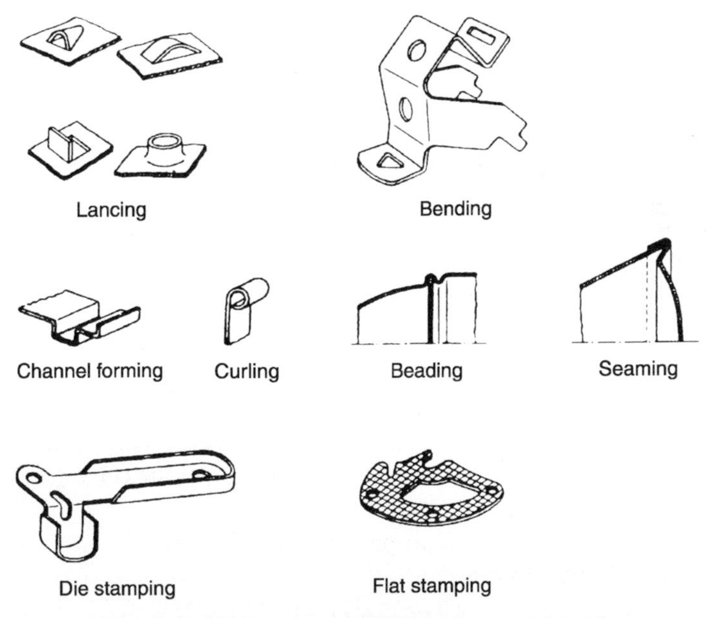

Fig. 7.11-19. Some possibilities in sheet metal work

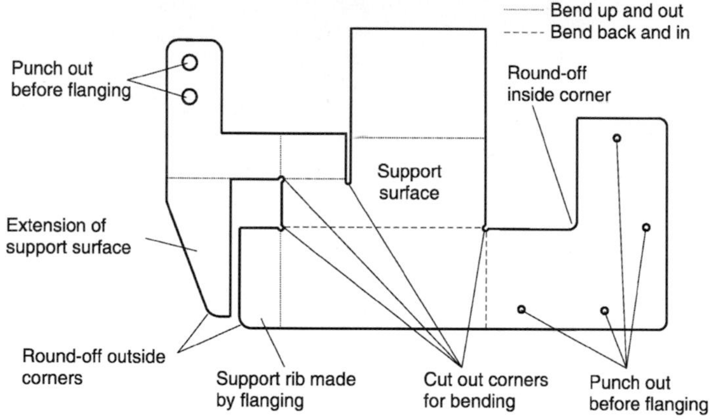

Fig. 7.11-20. Design of a sheet metal part (stamping and bending) [TRU96]

Extrusion can be very economical for simple axially symmetric bodies (shells, and tapered and stepped tubes) and large production quantities, with good accuracy. As with all cold working, this requires strain hardening, which leads to a decrease in toughness.

Sheet metal and plate working, combined with welding, also offers weight and cost savings possibilities for large machine parts (e. g., equipment frames, paneling, textile machine or machine tool stands), when compared to casting. The weight saving is understandable, particularly for big parts, since wall-thicknesses in castings must be larger for reasons of metal flow, than the corresponding sheet metal thicknesses.

7.11.3.2 Form design rules

Form design rules are shown in **Figs. 7.11-21a and b** for **die-forged parts**, in **Fig. 7.11-22** for **cold-extruded parts**, and in **Fig. 7.11-23** for **sheet/plate bent parts.**

Design rules	poor	better
1. Low amount of waste (good quality)		
Provide **generous fillets,** gradual transitions (DIN 7523); avoid too thin ribs, channels and grooves and too small holes (doughy material)	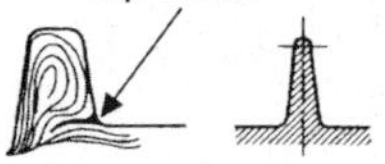 lap/crack	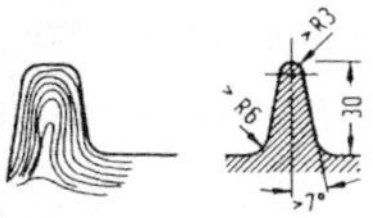 R3 R6 30 >7°
Avoid **sharp changes in cross sections** and cavities reaching too deep into the dies	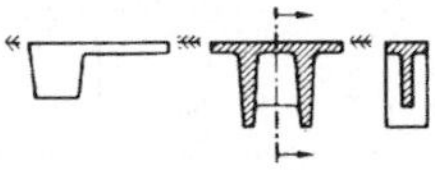	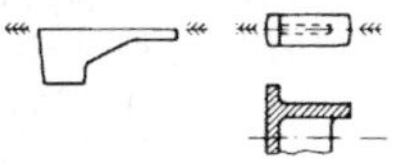
Offset **die lines** for bowl-shaped parts of great depth	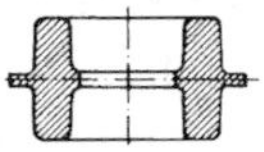	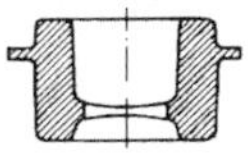
Arrange **die lines** such that the offset is easy to see and the flash/burr is easy to remove	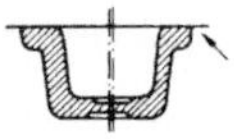	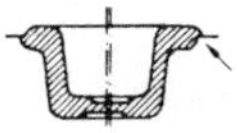
Avoid too thin panels and large, flat forgings	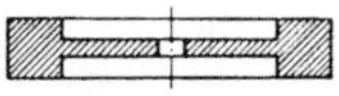	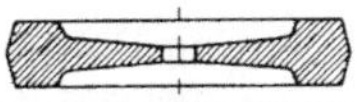
2. Low tooling and production costs		
Avoid **lips** and **projections**	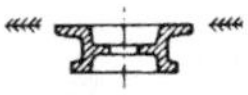	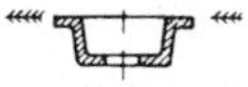
Provide **tapered surfaces** (DIN 7523, sh. 3); e.g., inner surfaces 1:5; outer 1:10	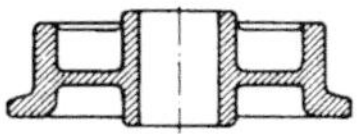	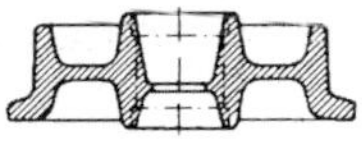
Avoid **buckled die lines** (flash/burr)	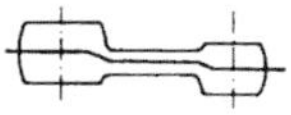	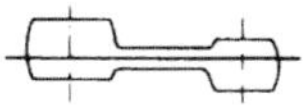
Strive to place **die lines** at about the **mid-plane**, perpendicular to smallest dimension		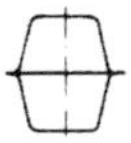
Strive for simple, as far as possible, **axially symmetric parts**; avoid markedly protruding parts (dies made by turning, instead of milling)		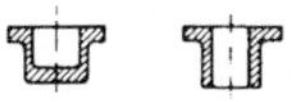
Strive for shapes as are formed in **open-die upsetting**; strive for the final shape in case of large quantities	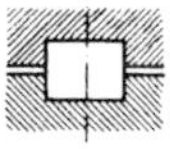	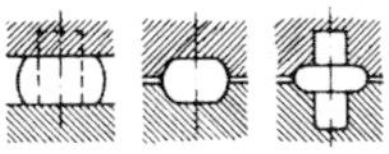

Fig. 7.11-21a. Form design rules for **die-forged parts** (from [Pah97], K. Vieregge)

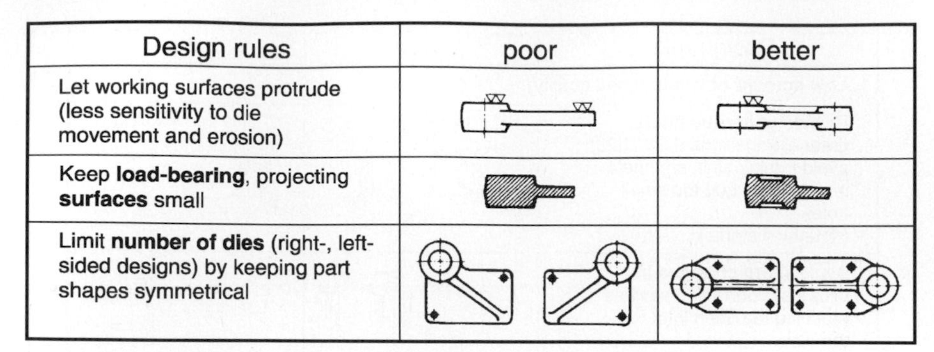

Design rules	poor	better
Let working surfaces protrude (less sensitivity to die movement and erosion)		
Keep **load-bearing**, projecting **surfaces** small		
Limit **number of dies** (right-, left-sided designs) by keeping part shapes symmetrical		

Fig. 7.11-21b. Form design rules for **die-forged parts** (from [Pah97], K. Vieregge)

Design rules	poor	better
Avoid **lips** and **projections**		
Avoid **sloping side walls** and **small changes in diameter**		
Aim for **axially symmetric bodies** without lumps of material; rather make separate pieces and join		
Avoid sharp **changes** in **cross section**, sharp angles and grooves		
Avoid small, long or lateral **holes** and threads		
Aim to achieve **cavity depth** in **one stroke** $h \leq 0.65\ d$ for St, Al, Cu **for multiple strokes**:		h d
Long holes with $d < 10$ mm, better to drill		

h_{max}		s_{min} wall thickn.
10 d	Aluminum	0.08 mm
8 d	Zinc	0.5
6 d	Al Mg Si	1
1 200 mm	Steel	0.1

Fig. 7.11-22. Form design rules for **cold-extruded parts** (from [Pah97], H. D. Feldmann)

Design rules	poor	better
Avoid **complex bent parts** (minimize scrap) make separate pieces and join; compare costs	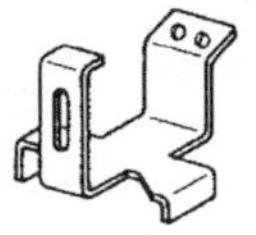	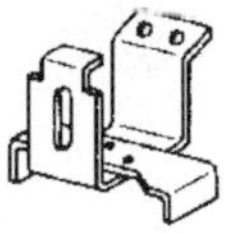
Observe minimum allowable values of **bend radii** (bulges on the inside, stretched on outside), leg height and tolerances	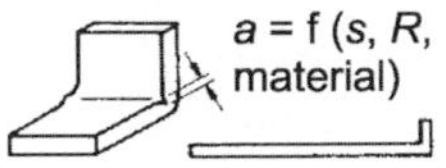 $a = f(s, R,$ material)	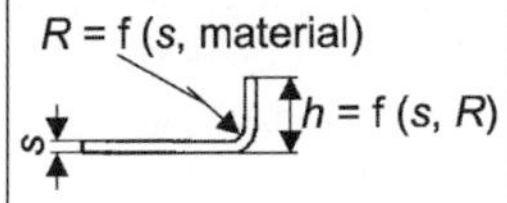 $R = f(s,$ material) $h = f(s, R)$ s
Observe **minimum allowable distance from bend edge** of holes existing before bending	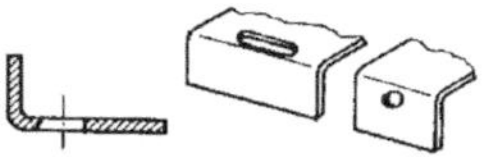	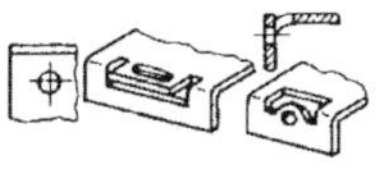
Aim to have **openings overlap** the bend if keeping the minimum distance is not possible		
Avoid **inclined outer edges** and tapers near the bend corner	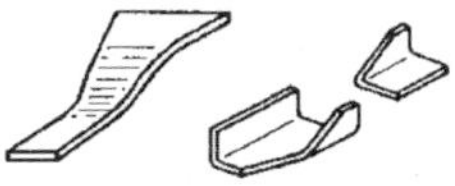	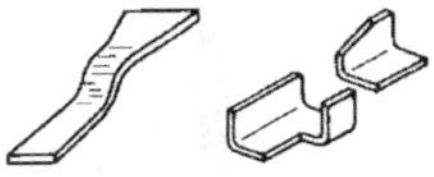
Provide **stress-relief holes** at **bend corners** where bent legs are on all sides	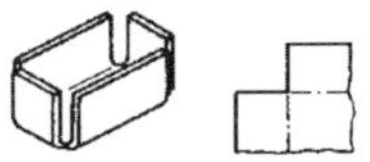	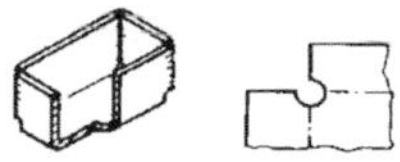
Stiffen **thin plates** (e.g., < 1.5 mm) through beads, seams and folds; possible material savings	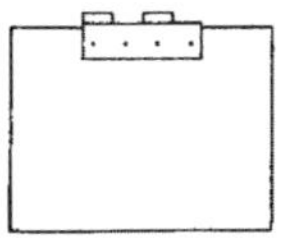	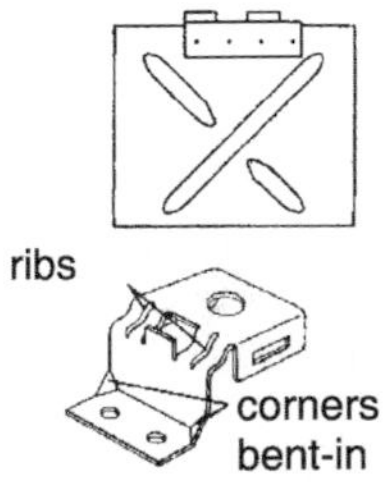 ribs corners bent-in
Bend corner not parallel to rolled direction (less strength)	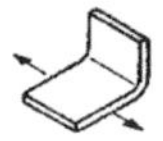	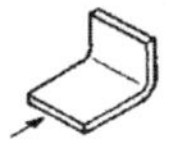

Fig. 7.11-23. Form design rules for **bent sheet metal parts** (from [Pah97])

Forged flange example

With **die-forged parts**, as with castings, an optimum must also be sought between the costs of producing the semi-finished part and the subsequent processing (machining). **Figure 7.11-24** gives an example of a flange. Dies that increasingly approach the near-net-shape are more expensive, but for large quantities produced N_{tot}, they lead to overall lower costs (Sect. 7.5.1). The processing costs that recur

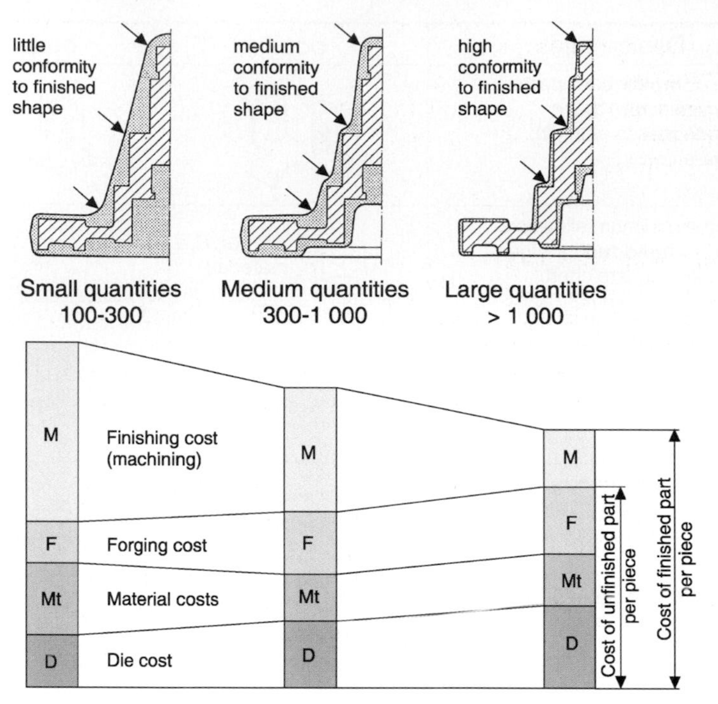

Fig. 7.11-24. Advantage of expensive near-net-shape dies for large quantities (from Voith)

for every part decrease, as do the material costs. In Sect. 10.2, a plate-welded centrifuge stand is shown as an additional example of optimizing between the cost of an unfinished housing and its mechanical processing.

7.11.4 Separation processes

7.11.4.1 The most important separation processes

An overview of the most important separation processes is provided by **Fig. 7.11-25**. Machining processes are generally characterized by high accuracy and are used at least for the finishing of parts. According to [Spu82], machining processes form a very large part of the entire production volume. Approximately two-thirds of the parts manufactured by machining are rotary parts.

It is uneconomical to make **a part** in its **rough form** by using machining processes for large parts and/or in large quantities. As described in Sects. 7.6 and 7.7, the material costs are so prohibitive that machining a part from a whole block should be avoided. An approximate part form is achieved by primary and shaping processes and by weldments. On the other hand, for small parts (particularly in single-unit production), template or pattern costs and setup costs dominate, so that at times machining processes are the only option and greater material losses are accepted. The material costs are then small in any case (for jigs and fixtures, one-time test setups (see Fig. 7.11-12, on the left.)

- Machining processes:
  - turning,
  - planing, shaping, broaching, chipping,
  - drilling/boring, reaming, countersinking,
  - milling, filing,
  - grinding, lapping, honing, polishing,
  - sawing;
- Punching, shearing, cutting, nibbling;
- Gas cutting, water-jet cutting;
- Electrical discharge, electrochemical milling, electron beam-/ laser cutting or drilling.

Fig. 7.11-25. Most important **separation processes**

In general, the starting point is the rough form (block, round stock, forging, etc.) and the finished part is approached in several stages (rough, fine, and final machining) depending on the final quality required (**Fig. 7.11-26**). So that the material is not deformed by the stresses caused during machining, the part must be annealed for stress relief. Furthermore, heat treatment (hardening, quenching, and tempering) can occur before the final machining (Sect. 7.13.5).

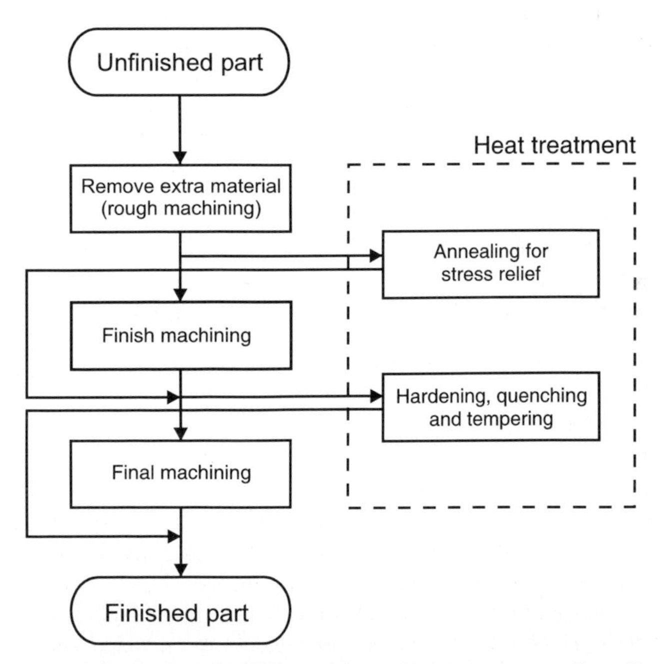

Fig. 7.11-26. Work flow during production with machining processes (from Spur)

7.11.4.2 Parameters that affect costs in machining processes

Important parameters that can be altered for cost reduction are the following:

- **Machining processes and their capabilities:**
  With regard to material removal, there are alternative processes (e. g., milling instead of planing) that have different effects on costs. The same is also true for achievable accuracy and roughness. An example for gear finishing is given in **Fig. 7.11-27**. The five processes each have their technological advantages and disadvantages, with different costs that must be weighed. Following that, superfinishing and carbide milling, with almost the same quality, are more economical than grinding. **Figure 7.11-28** shows another example. The boring operation with the same tolerance is much more expensive on a boring mill than on a drilling mill (boring and reaming). For the influence of dimensional tolerances and roughness, see Sect. 7.11.6.
- **Type of tool:**
  The tool type affects the operating time and thus the production costs. Designers generally have no influence on what type of tool is used (e. g., high-speed steel, carbide, or ceramic lathe tools). However, for a milled part, for example, they can encourage the use of a cylindrical mill instead of an end mill.
- **Size, quantity:**
  Since in machining the production costs basically increase proportionally to the machined surface (Sect. 7.6.1), these should be kept small, particularly for large parts in single-unit production, and for parts of every size in series production. For example, machining could be restricted to edges. For larger quantities (by standardization), the setup costs, which are often high for small parts, can be reduced (see also Sects. 7.6 and 7.7).

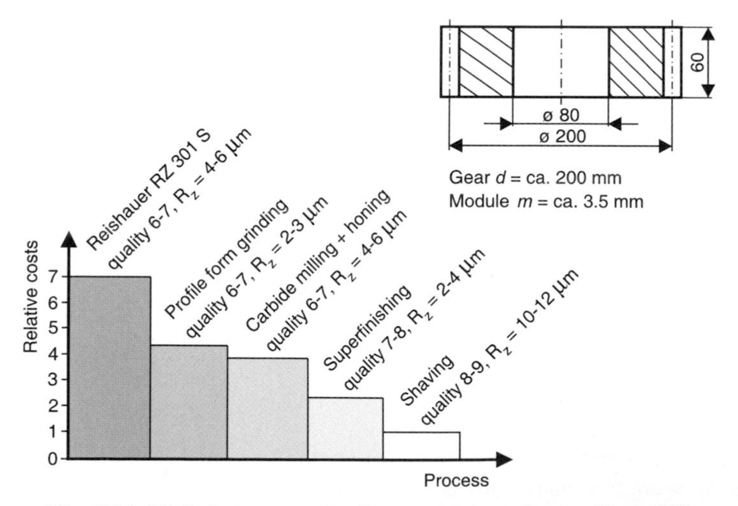

Fig. 7.11-27. Relative costs for fine machining of gears (from ZF)

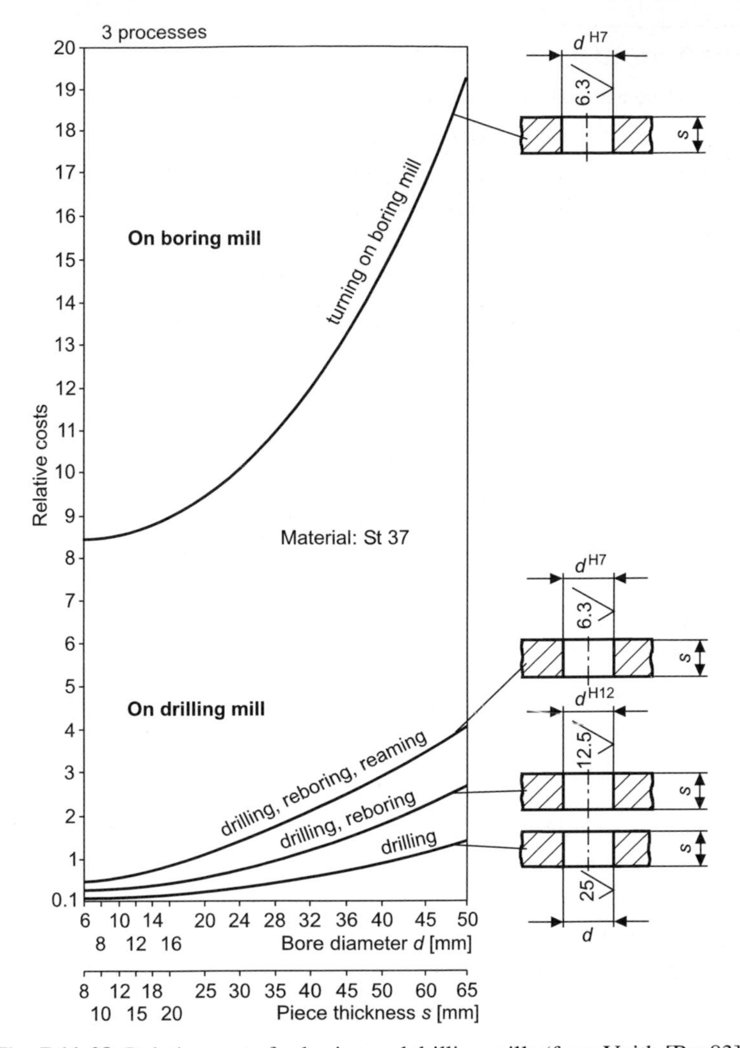

Fig. 7.11-28. Relative costs for boring and drilling mills (from Voith [Bus83])

- **Material:**
  There are considerable differences in the machinability of materials and, thus, the ensuing production costs. For steel, the rule of thumb is that the higher the ultimate strength, the greater is the machining time. Low-carbon steels and case-hardening steels are easier to machine compared to stainless steels, heat treatable, and nitriding steels.

7.11.4.3 Form design rules for machining processes

The rules are subdivided into rules for decreasing waste and rules for decreasing the tool and production costs (from setup, machining, and idle times).

The rules shown in **Fig. 7.11-29** are **generally** valid for machining processes.

General form design rules for machining

- Remove as little material as possible (work surfaces should protrude).
- As little fine machining as possible (allow for rough surfaces).
- Use as rough tolerances as possible (as good as needed, but as low-cost as possible).
- Work surfaces should not be steeply inclined to one another (requires tilting table for machining).
- On each part, use similar geometric shapes (same hole diameter, threads, fillet radii, same tapers), same tools and (standardized) gauges.
  Think in terms of production families (Sec. 7.12.4.2).
- Parts more likely to end up as scrap are often cheaper to process if they are divided into smaller parts (differential design), produced separately, then assembled into the complete part (Figure 7.9-9).
- Do all the machining in one set-up. This is cheaper and more accurate than to retool or remount (Figure 7.11-35).
- Provide for use of cutting tools of large radii, which enables higher cutting speeds, e.g., in milling.
- Provide for ability to clamp the workpiece rigidly, to resist high cutting forces.
- Mark dimensions from one coordinate origin (show angles measured from a coordinate parallel to the axis).

Fig. 7.11-29. General form design rules for machining processes

The form design rules for conventional machining are depicted for:

- turning in **Fig. 7.11-30**
- boring/drilling in **Fig. 7.11-31**
- milling in **Fig. 7.11-32**
- grinding in **Fig. 7.11-33**
- stamping, cutting in **Fig. 7.11-34**

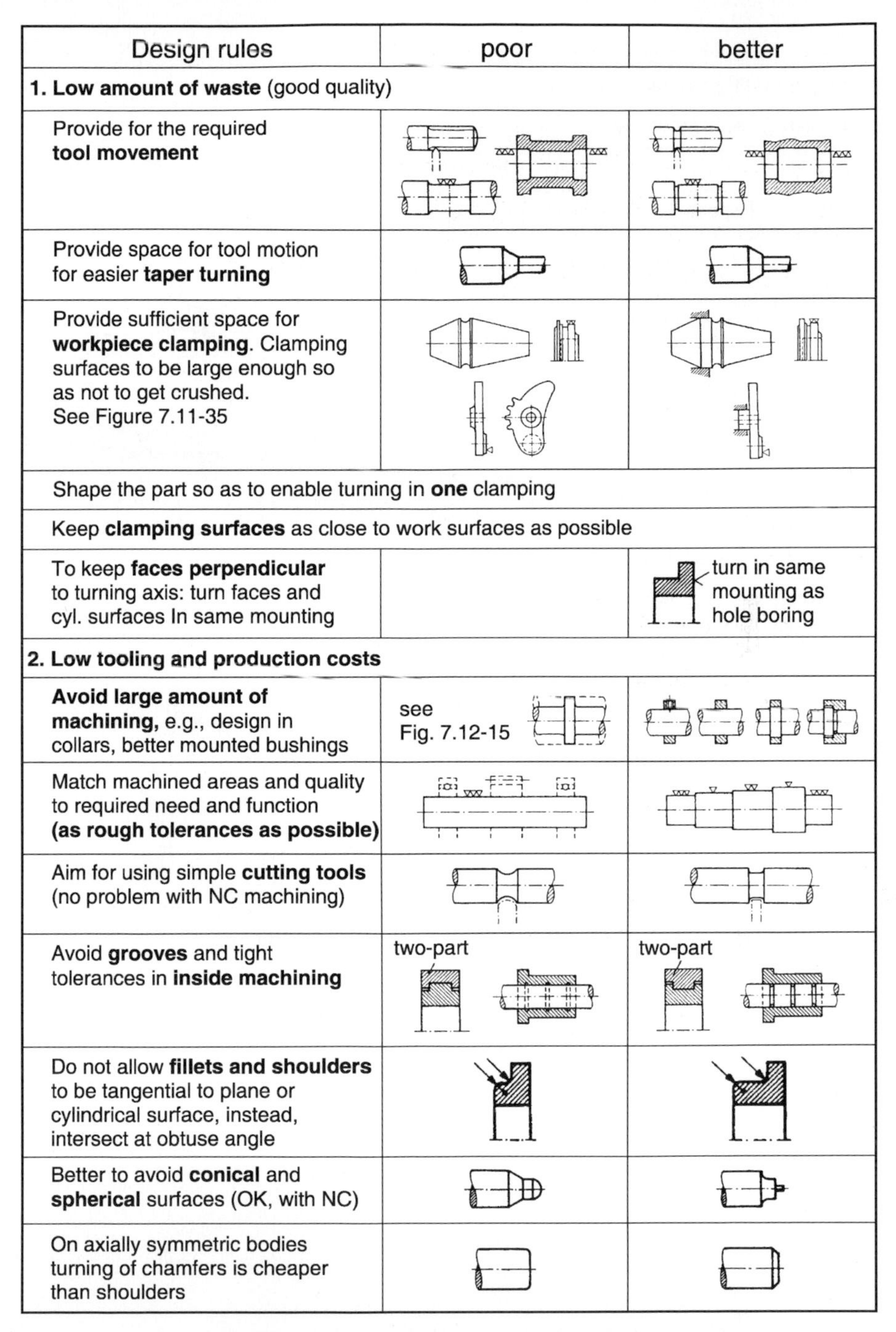

Design rules	poor	better
1. Low amount of waste (good quality)		
Provide for the required **tool movement**		
Provide space for tool motion for easier **taper turning**		
Provide sufficient space for **workpiece clamping**. Clamping surfaces to be large enough so as not to get crushed. See Figure 7.11-35		
Shape the part so as to enable turning in **one** clamping		
Keep **clamping surfaces** as close to work surfaces as possible		
To keep **faces perpendicular** to turning axis: turn faces and cyl. surfaces In same mounting		turn in same mounting as hole boring
2. Low tooling and production costs		
Avoid large amount of machining, e.g., design in collars, better mounted bushings	see Fig. 7.12-15	
Match machined areas and quality to required need and function **(as rough tolerances as possible)**		
Aim for using simple **cutting tools** (no problem with NC machining)		
Avoid **grooves** and tight tolerances in **inside machining**	two-part	two-part
Do not allow **fillets and shoulders** to be tangential to plane or cylindrical surface, instead, intersect at obtuse angle		
Better to avoid **conical** and **spherical** surfaces (OK, with NC)		
On axially symmetric bodies turning of chamfers is cheaper than shoulders		

Fig. 7.11-30. Form design rules for **turning** [Pah97]

Design rules	poor	better
1. Low amount of waste (good quality)		
Provide level **entry** and **exit surfaces** for inclined holes		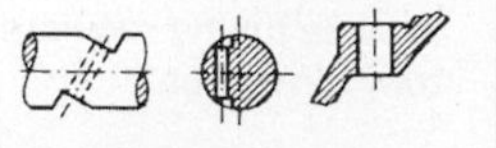
Provide for **space** for **movement** of drill chuck and tool		
Materials of **equal hardness** in drilling hole at interface of joined parts (else, hole distorted)	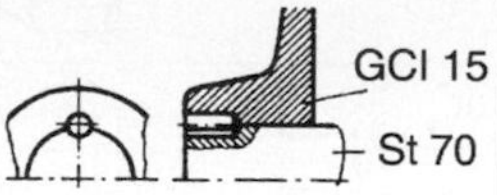 GCl 15 St 70	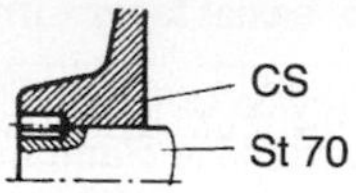 CS St 70
2. Low tooling and production costs		
Aim for **through-drilled holes**; avoid blind holes, if necessary, provide special cover	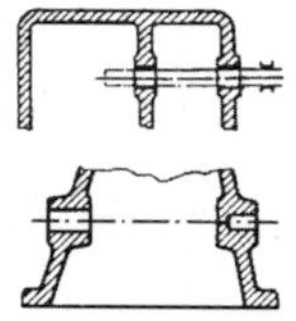	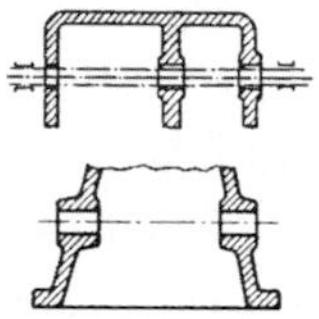
Blind holes, as far as possible, end with drill-bit tip; through-drilled holes are cheaper	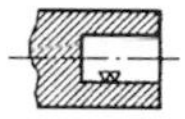	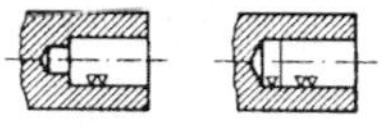
Avoid **stepped holes**		 locating ring
Avoid **inclined holes**	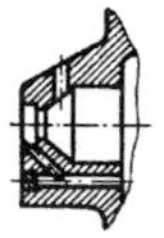	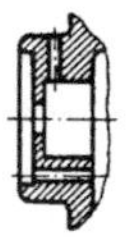

Fig. 7.11-31. Form design rules for **boring/drilling** [Pah97]

Design rules	poor	better
1. Low amount of waste (good quality)		
Clamping surfaces as **close** to **work surfaces** as possible (more accuracy and lower cost due to possibly faster material removal)	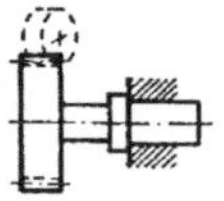	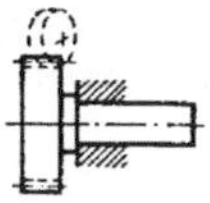
Think of **clamping options**		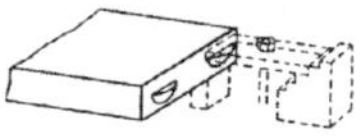
2. Low tooling and production costs		
Arrange **surfaces** to be at **same height** and parallel to clamping surface	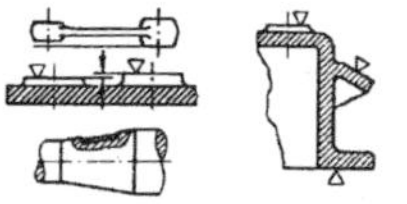	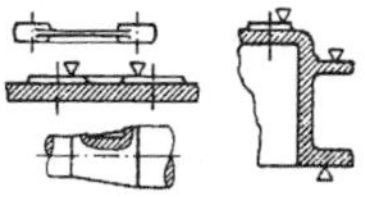
Strive for **flat milled surfaces**, form cutters expensive; choose dimensions so that gang cutter may be used	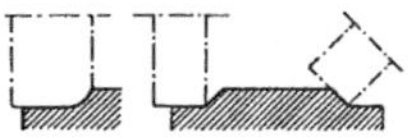	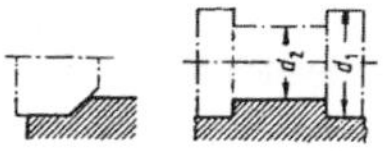 d_1 d_1
Milling through with face mill is cheaper than sectional cutting with end mill (separate into 2 parts, cross wheel)	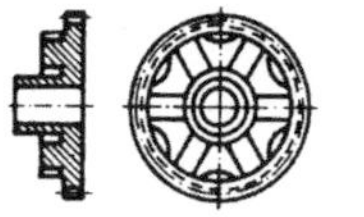	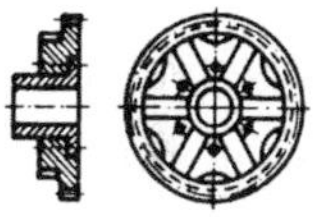
Provide for **through-cut grooves** in face milling; face milling is cheaper than end milling	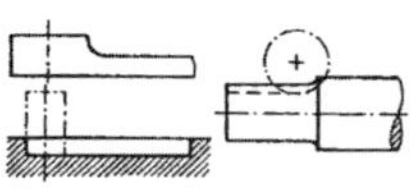	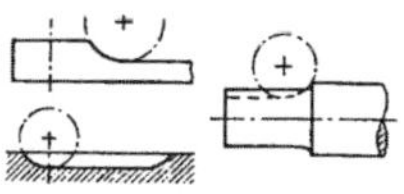
Match the cutter diameter to **tool movement**. Think of **clamped length** of workpiece	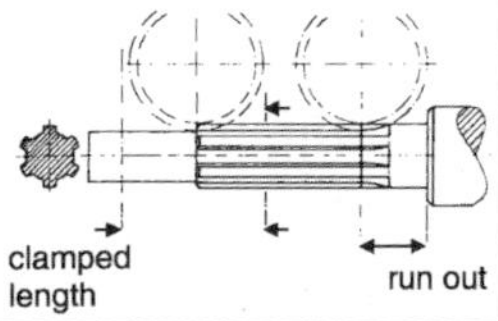 clamped length run out	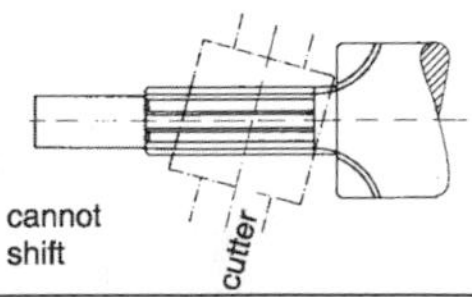 cannot shift cutter
Similar parts to be shaped so, they may be clamped together and **machined simultaneously**	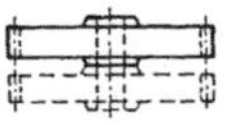	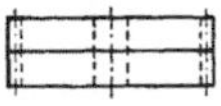
Mill with **inserted-tooth** instead of cylindrical cutter		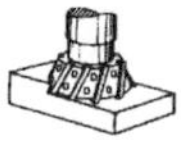
Aim for symmetric shapes for **broaching,** else the broaching tool runs off-course	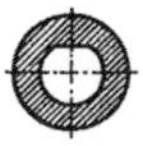	
Entry and exit faces to be perpendicular to broaching axis	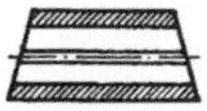	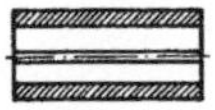

Fig. 7.11-32. Form design rules for **milling** [Pah97]

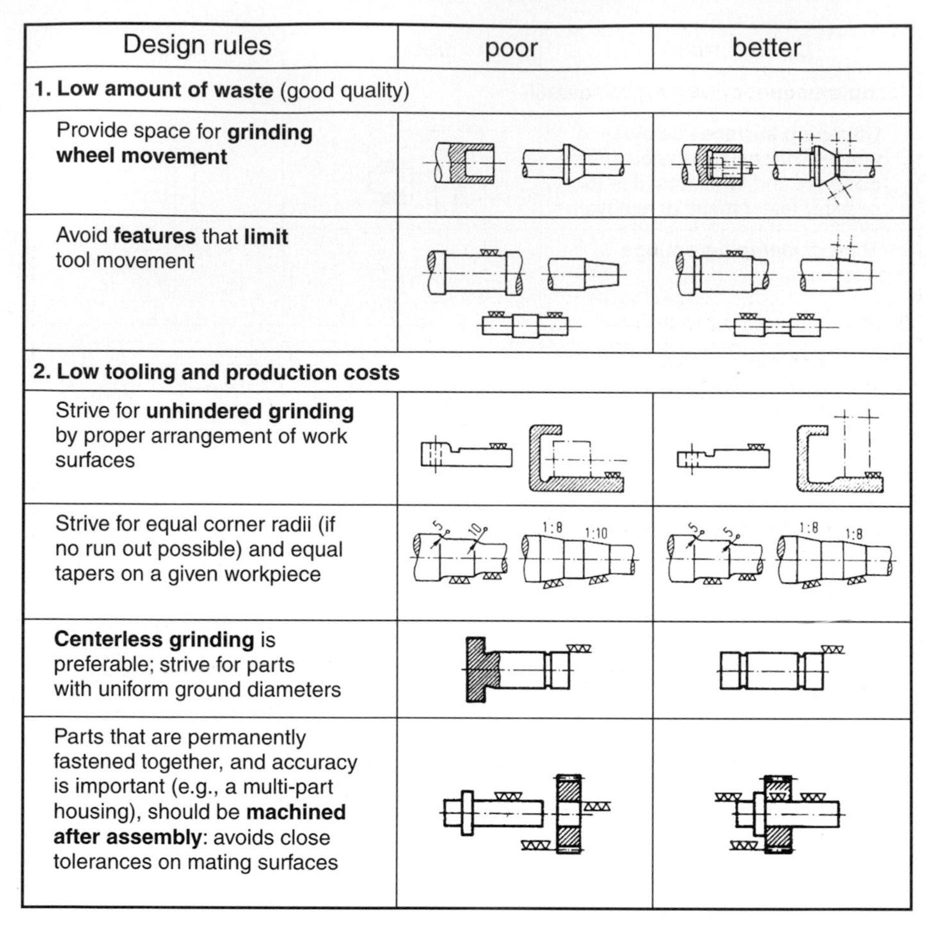

Design rules	poor	better
1. Low amount of waste (good quality)		
Provide space for **grinding wheel movement**		
Avoid **features** that **limit** tool movement		
2. Low tooling and production costs		
Strive for **unhindered grinding** by proper arrangement of work surfaces		
Strive for equal corner radii (if no run out possible) and equal tapers on a given workpiece	5 10 1:8 1:10	5 5 1:8 1:8
Centerless grinding is preferable; strive for parts with uniform ground diameters		
Parts that are permanently fastened together, and accuracy is important (e.g., a multi-part housing), should be **machined after assembly**: avoids close tolerances on mating surfaces		

Fig. 7.11-33. Form design rules for **grinding** [Pah97]

Groove nut example

The nut shown in **Fig. 7.11-35** was previously made in two setups. First, the inside fit Ø 80^{H8} was machined. Then the nut was clamped on the inside for cutting the outer thread. The design group, together with production planning, decided to increase the length, enabling the screw thread and end face to be machined in one setup after it had already been turned with a rough tolerance. The inside diameter was machined to a tolerance of ± 1 mm. Result: a considerable decrease in cost.

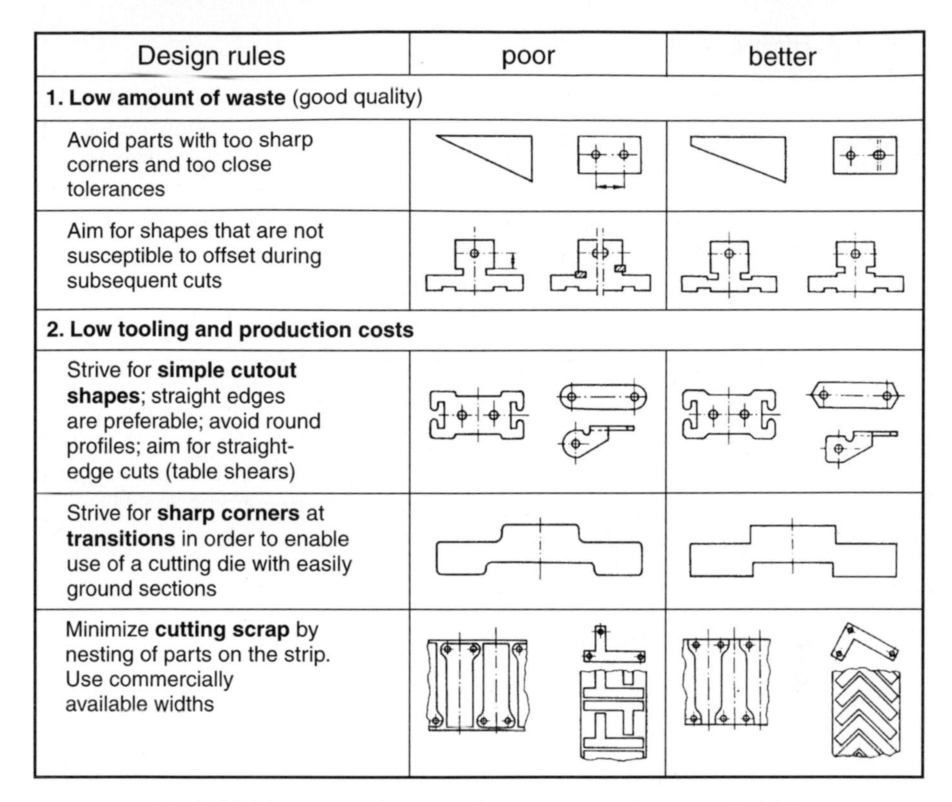

Design rules	poor	better
1. Low amount of waste (good quality)		
Avoid parts with too sharp corners and too close tolerances		
Aim for shapes that are not susceptible to offset during subsequent cuts		
2. Low tooling and production costs		
Strive for **simple cutout shapes**; straight edges are preferable; avoid round profiles; aim for straight-edge cuts (table shears)		
Strive for **sharp corners** at **transitions** in order to enable use of a cutting die with easily ground sections		
Minimize **cutting scrap** by nesting of parts on the strip. Use commercially available widths		

Fig. 7.11-34. Form design rules for **stamping and cutting** [Pah97]

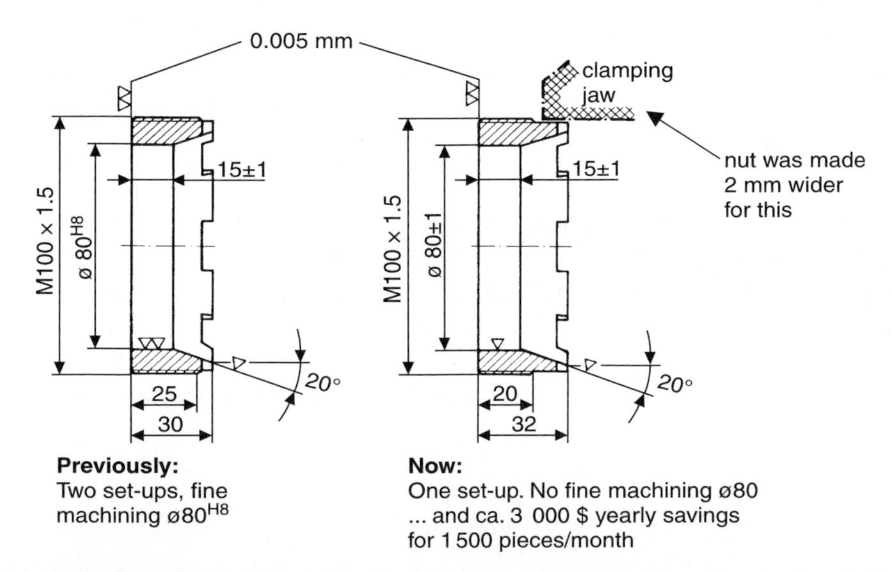

Fig. 7.11-35. Groove nut: through a redesign of **one** clamping surface, only one clamping is required (from RKW)

7.11.4.4 High-speed milling and grinding

In recent years, machining processes using extremely high cutting speeds (HS) have been introduced in the practice. The cutting speeds in HS milling are usually 5–10 times higher than in conventional milling [Schu96]; in HS grinding the surface speeds of the grinding wheel reach 60–200 m/s, with a machining output 10–20 times that of conventional turning, milling, and surface broaching [Fer92]. That reduces the production time, which is of particular interest from the cost viewpoint for large parts with substantial volumes to be machined.

Alternatively, in **HS milling,** for the same processing time the feed marks are closer, so that a better shape of the required contour is achieved and the cost of manual refinishing is reduced [Schu96].

With **HS grinding,** the roughing and final finishing can be carried out on **one** machine. A superior finished quality is achieved without the high costs of manual refinishing. For a cost comparison with conventional grinding, it is important to note that the capital expenditure is higher than with conventional machines, so there must be sufficient number of parts for processing, with large volumes to be machined [Ver94, Fer92].

7.11.4.5 Stamping and nibbling

Modern stamping and nibbling machines can be put to a variety of uses, having high processing speeds and a commendable cost/performance ratio [TRU96]. Furthermore, these machines are designed as combination machines with bending fixtures, so that parts such as those shown in Fig. 7.11-20 can be produced directly from the sheet on one machine.

For CNC machines there are CAD programs that make it possible to design a part complete with NC programming, and also to make it immediately in one production sequence.

The dominant share of machine costs with this technique are the fixed costs (depreciation, interest, space costs) that amount to approximately 90% with one-shift operation and approximately 80% with a two-shift operation. The variable costs (tools, maintenance, and energy) are therefore minor. For calculation of manufacturing costs, the personnel costs must also be considered.

Accordingly, stamping and nibbling, in conjunction with NC and welding technology (for example, in the textile and machine tool industry) has replaced the customary cast design of housings to a certain extent, for cost reasons (see examples in **Fig. 7.11-36**).

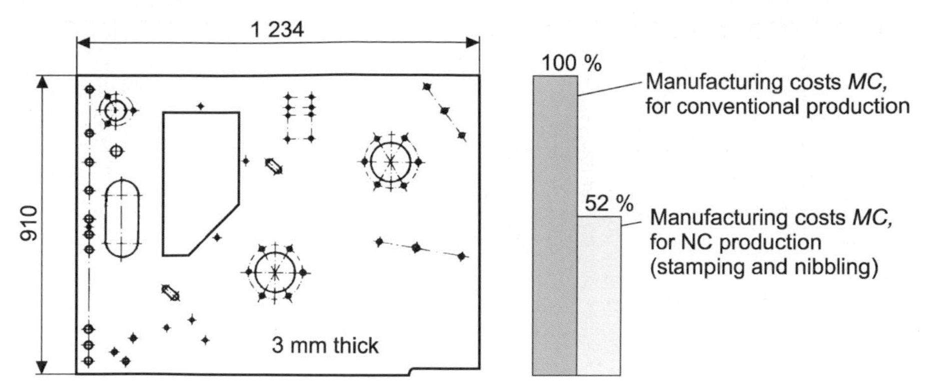

Drum side wall made from plate is significantly cheaper with NC production for 1 400 pieces per year, as opposed to hydraulic stamping and hand drilling (Fahr).

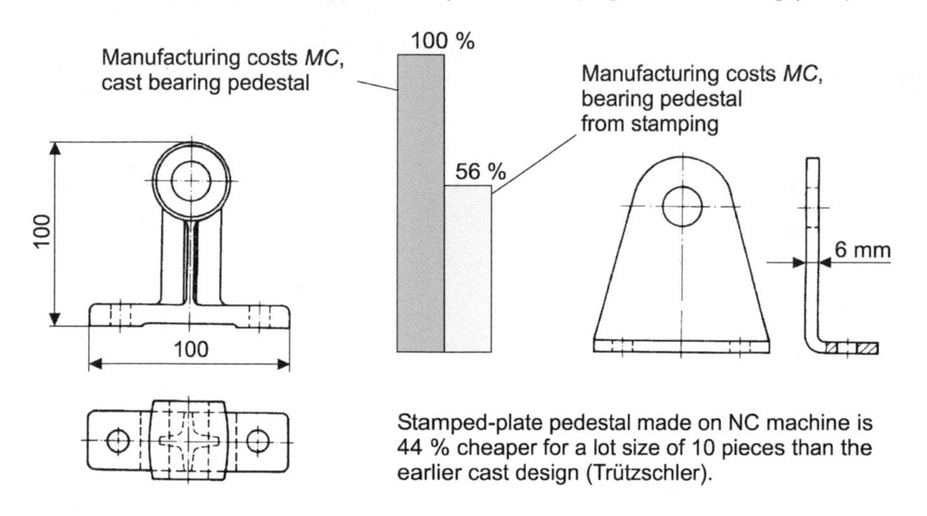

Stamped-plate pedestal made on NC machine is 44 % cheaper for a lot size of 10 pieces than the earlier cast design (Trützschler).

Fig. 7.11-36. Examples of low-cost sheet-metal designs (from M. Geiger)

7.11.4.6 Gas, laser, plasma and water-jet cutting

a) Comparison of the processes

All four cutting processes are characterized by the fact that, unlike the mechanical separation processes, virtually no forces are applied on the part. With NC control, almost any shape can be cut. These processes compete with each other, depending on the materials to be cut and according to requirements on the cut edges.

A comparison of their **properties** (also relating to stamping and nibbling) is shown in **Fig. 7.11-37**.

Oxyacetylene cutting is the traditional cutting method for carbon and low-alloy steels up to 1 m thickness. However, the cutting speeds are relatively low.

Process	Materials and thicknesses	Average cutting speed	Cut edges	Comments
Gas cutting (e.g., with acetylene gas and oxygen)	Carbon and low-alloy steels 3 mm to 1 m (not thin plates under 2 to 10 mm for high-alloy steels and Titanium)	4 m/min	Rough	• High heat input • Refinishing necessary in very exacting situations
Laser jet cutting	Structural steel to ca. 20 mm high-grade steel to ca. 10 mm Aluminum to ca. 6 mm polymers, laminates, glass to some extent	20 m/min and higher	Nearly vertical, generally without burrs, cut width 0.1-0.5 mm roughness < 100 mm	• Usually no refinishing necessary • Very narrow ridges and sharp points also possible • Practically no thermal warping
Plasma jet cutting (DC electric arc, with cutting gas: Argon, Nitrogen, ...)	Primarily high-alloy steels, light metals, non-ferrous metals 3-100 mm thick (thinner than 1-3 mm difficult)	4-6 m/min for thin, low-alloy steels	Tapered edge	Refinishing necessary in very exacting situations
Water jet cutting, abrasive (a few thousand bars pressure, with Corundum, or Quartz powder)	Structural steel and high-grade steel to ca. 100 mm Aluminum, polymers, laminates, glass, masonry, textiles, corrugated board	A few tenths of a m/min	Nearly no indent radius, no thermal effects, no burrs; roughness and angularity depend on material thickness and cutting type	Usually no refinishing needed; materials remain in original form (no buckling as with sawing, shearing), almost no temperature increase
Stamping/ Nibbling	Structural steel and high grade steel to ca. 10 mm Aluminum polymers, laminates to some extent		No indent radius with low roughness, certain angularity of edge, little burring	With stamping, hardly any refinishing needed; cost-competitive with laser cutting; large produced quantities due to stamping tool costs

Fig. 7.11-37. Comparison of **cutting processes** (in part from [TRU96])

As a separation process, flame cutting is competitive with **sawing.** Sawing is generally more economical with round and rolled sections, while flame cutting is better for sheet metal, particularly when more than just straight cuts are involved. The costs for flame cutting are nearly proportional to the cut length and increase less than proportionally to plate thickness. It is necessary to hold the cut length to a minimum by, for example, designing parts that have a common cutting edge on a side, or by having the waste result in a usable part.

Virtually all materials can be cut with abrasive **water-jet cutting,** but at lower cutting speeds. **Laser cutting** provides unequalled cut edge quality [Eng93, Gie92].

b) Costs of cutting

The **costs** of **cutting** must be established by a (bid) cost comparison. Because of the varying conditions, no general statement can be made. In addition, laser and water jet cutting are currently undergoing particularly rapid technological development.

Figure 7.11-38 shows that in laser cutting of high-grade steel the machine costs are approximately double those for structural steel, primarily due to the high consumption of the cutting gas.

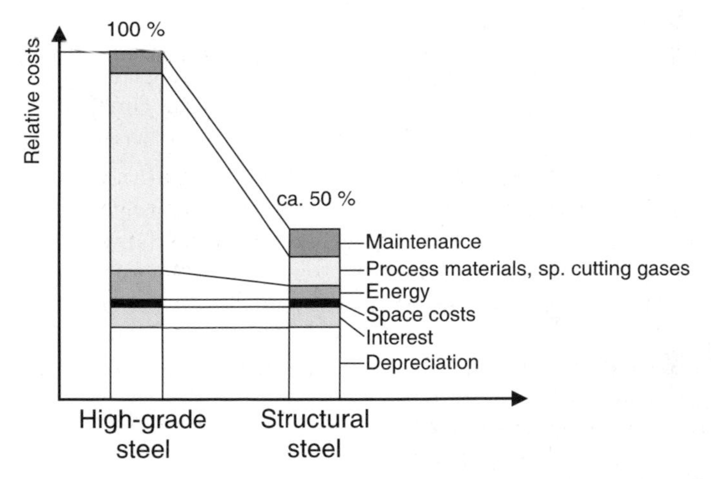

Fig. 7.11-38. Laser cutting of high-grade steel is twice as expensive as that of structural steel [TRU96]

7.11.5 Connections (joints)

Machines consist of machine parts that relate to each other by way of connections (joints) or connection elements. Every part needs at least one fixed connection to every adjoining part, if it does not move in a bearing. Thus, a machine is considered as a system that consists of elements and their relationships. The relationships are the connections. The importance of the connections is clear from Fig. 7.11-1. From the class of fixed (not moving, force-transmitting) connections, we are concerned here not only about material-based, inseparable (permanent) connections.

These comprise welding, soldering, adhesion, but also about separable joints that are friction-based and shape-based, as well as fasteners such as screws, clamps and couplings.

According to industry use [Bau80], we do not speak here of "joining" (DIN 8580). The classification of connections is practical. For a systematic treatment, refer to Roth [Rot96, VDI82]. Movable connections (for example, bearings, guides) and those for which there are no useable cost assertions are not treated here.

7.11.5.1 Most important fixed connections

The most important fixed connections are shown in **Fig. 7.11-39**. Fixed connections are understood to be those that do not allow any movement between the joined parts (see Fig. 7.11-4).
Because of lacking data, **function costs** that are meaningful relative to connections cannot always be compared. For a connection, all the important costs for the function (e. g., "connect = transmit × N force (or moment) in y-direction") must be considered (purchase parts, production, material, assembly costs, and assembly/disassembly costs for the customer regarding maintenance). As Fig. 7.11-44 shows with the example of a thrust take-up for a rolling bearing, the function costs (i. e., the entire manufacturing costs) of a connection can be several times the purchase part (standard part) costs. (We might say: "Let us put a retaining ring there. That costs only 40 cents". However, the manufacturing costs could be 4 $.)

Furthermore, particular connections have **cost-reducing conditions** due to the physical properties of the system environment. They are offered as the "customary" low-cost connections. Examples of such joints are:

- **Beaded fastening** for plastically deformable sheet metal that is not too thick. This is not possible for brittle spring steel sheet or for polymers. Also pertinent to this are lock, notch, crimped and tooth joints (Fig. 7.11-4).
- **Snap fits** for polymers due to their low modulus of elasticity relative to the strength.
- **Press fits** for shaft-hub connections due to the encircling of material and its elastic behavior.
- **Spot welding** for thin sheet metal (Fig. 7.11-40).

a) Permanent connections (usually material-based)

- **Soldering**:
  - soldering (with temperature < 450°C),
  - brazing (with temperature > 450°C, high strength, for most large series (continuous ovens));
- **Adhesion:** cold/hot adhesion;
- **Welding** (see also Sec. 7.13.4)**:**
  - **oxyfuel (gas) welding**: e.g., for thin sheets and tubes, thickness (0.5…1 to 5…15 mm);
  - **open electric-arc welding**: universal welding process for sheet thicknesses of ca. 1 to 2 mm, usually with shielded electrodes;
  - **gas metal-arc welding** suppresses slag and oxide formation, universal welding process; MIG, MAG, WIG processes;
  - **forge** or **press welding**: heating by electric current or gas, followed by pressing together, for joining round stock, profiles, bolts on sheets etc., usually for series production;
  - **friction welding**: similar to press welding, but heat is generated by friction: for joining round stock, profiles, bolts on sheets, usually for series production;
  - **spot welding, roll welding**: joining, specially of thin sheets, bolts on sheets; sheet thickness usually < 6 mm for steel, < 3 mm for light metals, single-unit and series production;
  - **electron beam welding:** for special materials, very localized heating, heat reaches great depths;
  - **laser welding**;
- **Plastic shape-based connections**:
  - riveting;
  - beading;
  - clamping;
- **Shrink/press connections** (see Sec. 7.13.6)**:**
  - e.g., shaft-hub connections; usually permanent friction joint, as far as not statically expandable shrink joints (oil-pressure joints) are used.

b) Separable connections (usually shape and friction-based)

- **Friction-based joints**:
  - screws/bolts (are also shape-based);
  - key/pin connection;
  - clamp/compression joint
- **Shape-based joints**:
  - snap joint (partly also friction-based);
  - bolt, pin, feather-key connections.

c) Important applications in machine industry

- **shaft-hub connections** (see Sec. 7.13.6);
- fixed **shaft couplings.**

Fig. 7.11-39. Most important **fixed connections** (see also Fig. 7.11-4)

Relative costs for sheet-metal connections

In **Fig. 7.11-40** different sheet-metal connections are compared with regard to their relative costs. Cost of the lowest cost connection, sheet metal spot welding was set at 1.0.

Connection	Relative cost	
• Spot welding	1	
• Adhesion (Araldit)	1.7	
• Riveting	2.6 - 3.5	(3 mm steel or aluminum sheets, lot size 200, production costs exclusive of material costs)
• Welding (electric arc)	2.9 - 4.4	
• Screw/bolted joint	3.6 - 4.4	
• Brazing	3.7 - 6.9	

Fig. 7.11-40. Relative cost of sheet-metal joints (from G. C. Schulze, Siemens)

The relationships change if the costs are based on mechanical properties (function costs).

7.11.5.2 Low-cost design of welded assemblies (conventional arc welding)

a) General

With welded assemblies such as housings and stands, manufacturing costs are found in two elements: the **unfinished part** and its **machining.** With small units of a few hundred kg weight, both costs are important. For large units of a few thousand kg weight, the cost of the unfinished part becomes more dominant because the material portion increases approximately with the third power of the length (Fig. 7.13-15, left). Therefore, with large welded assemblies it is desirable to save on material (thinner plates, more ribs and braces) even if thereby the production costs for the unfinished part increase. Our own and other investigations [Bei82a, Pah82] point to the following important **items** that **affect** manufacturing costs of the unfinished part:

- Welding process
- Type of seam
- Weld volume (= joint sectional area x joint length)
- Number of welded sheet metal and semi-finished parts made
- Size (weight) of the welded assembly

Material that is easy to weld is a basic requirement. As always, it is advisable to think through the production sequence before the design is finalized and to examine to what extent the proposed design hampers or supports the individual production steps (**Fig. 7.11-41**). We must also keep in mind that according to Sect. 10.2 (Fig. 10.2-2) the actual welding constitutes only a small portion of the manufacturing costs of a welded assembly (e. g., a housing).

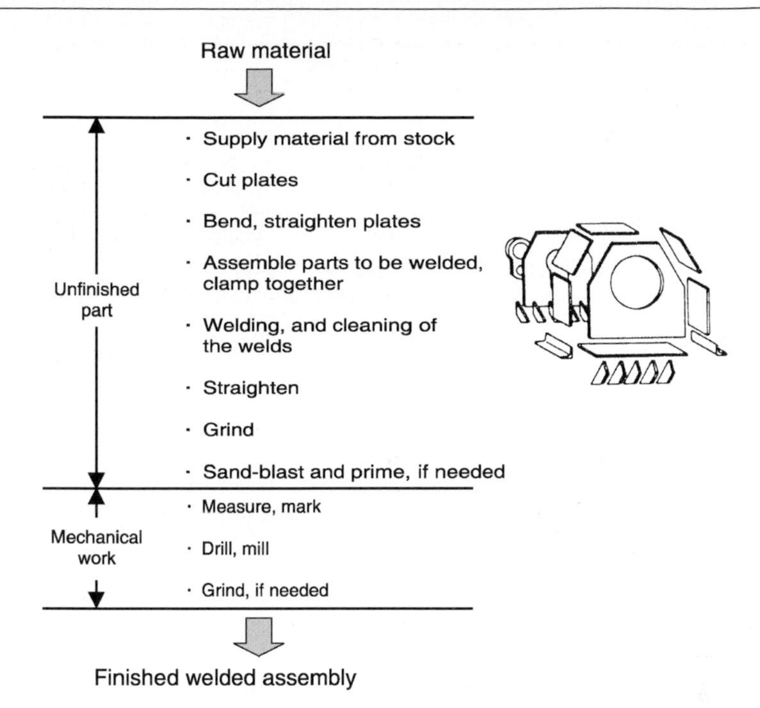

Fig. 7.11-41. Production operations in the manufacture of a welded assembly

b) Welding costs

Inert gas welding (MIG, MAG) brings about important time savings over electric arc welding, so that it is generally more economical in spite of higher workstation cost rates. Only for large weld thickness and length is submerged arc welding more economical than gas-metal arc welding. Furthermore, joints with small weld cross-section *A* require shorter manufacturing times: The double-vee butt is better than the fillet weld. This information refers only to production welding process. From Fig. 10.2-2 it is seen, however, that welding constitutes only 15–20% of the costs for the unfinished housing and only 7–9% of the manufacturing costs of the complete housing. The individual steps of welding are considerably more cost intensive – cutting out sheets, preparing the seam, assembling and fixing the plates to be welded, straightening and grinding. The shape of the assembly must be designed accordingly.

c) Examples

- A **redesign** from a cast to a welded design is shown in **Fig. 7.11-42**. We recognize in the upper part of the figure, as in the next example, the importance of the reduction in part numbers for cost reduction ("bending and edging instead of welding"). This measure has a positive effect on many production operations.

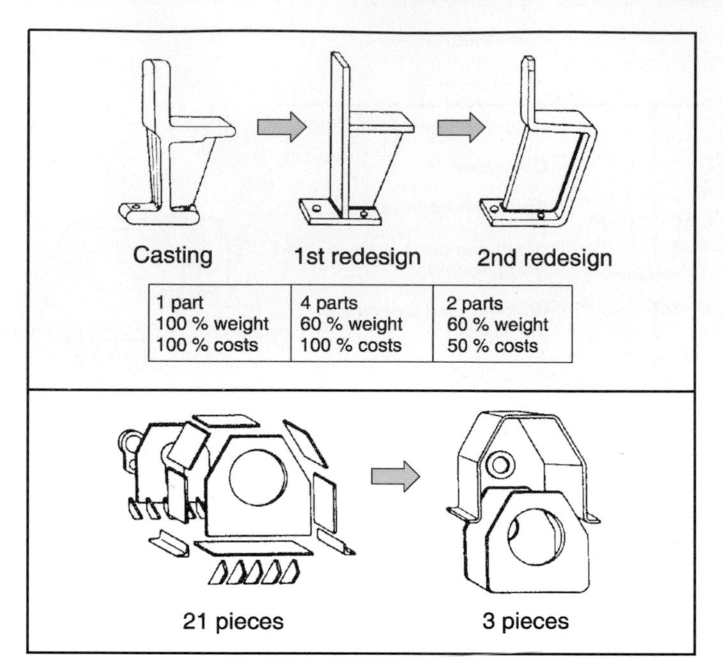

Fig. 7.11-42. Use of fewer bent parts in welding, instead of many parts

- The cost reduction on a **welded housing** for a centrifuge is discussed in Sect. 10.2. Here the teamwork was especially fruitful between the specialists (design, production planning, cost calculation, welding). In spite of the same size and weight – primarily through edging and bending – the manufacturing costs can be reduced by about 50%.

d) Form design rules

Form design rules for welded assemblies are compiled in **Figs. 7.11-43a and b**. Since attention must be paid to the cost calculation of the suppliers, which in the first place is a weight-based calculation (Sect. 9.3.2.1), the following rules apply:

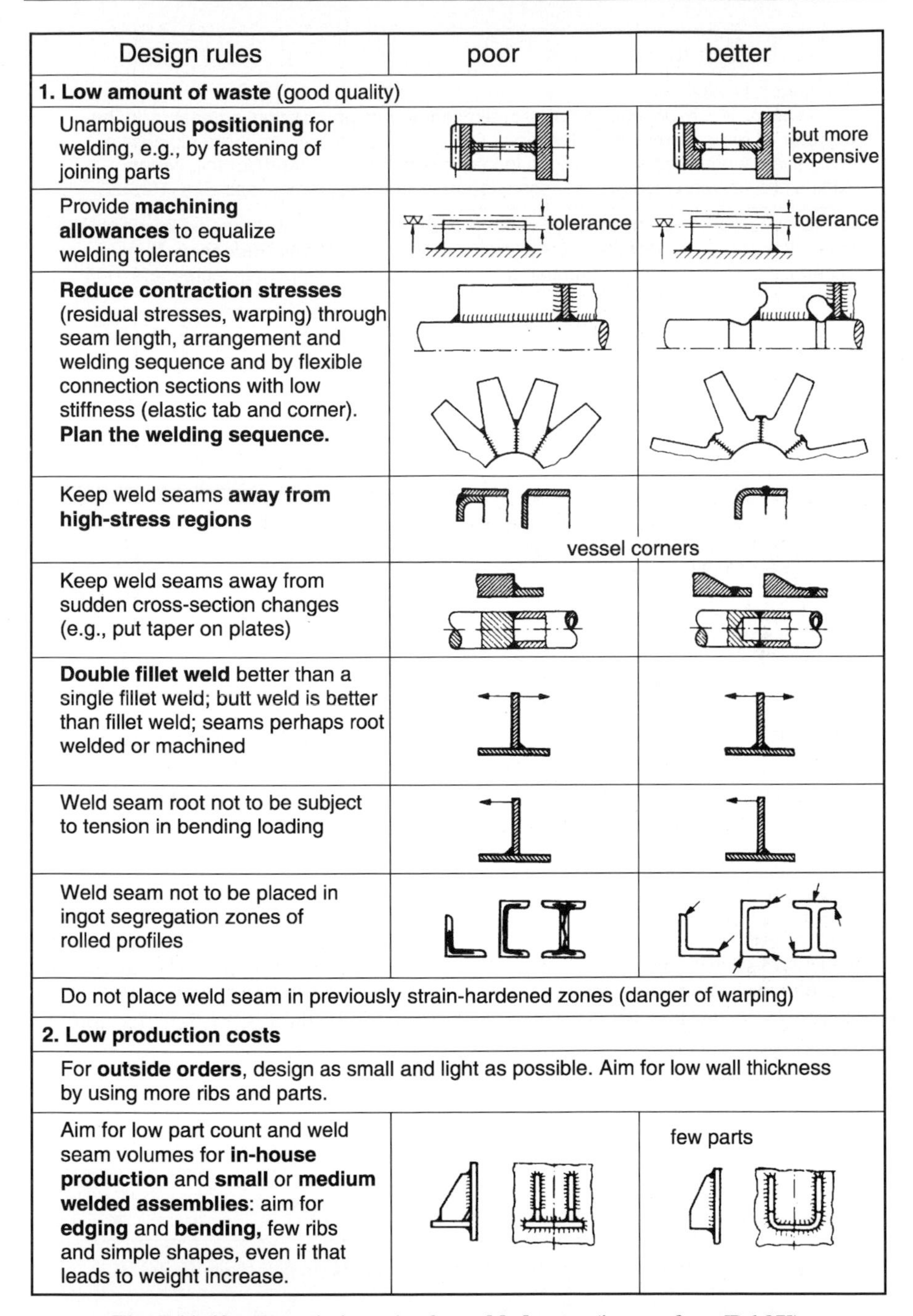

Design rules	poor	better
1. Low amount of waste (good quality)		
Unambiguous **positioning** for welding, e.g., by fastening of joining parts		but more expensive
Provide **machining allowances** to equalize welding tolerances	tolerance	tolerance
Reduce contraction stresses (residual stresses, warping) through seam length, arrangement and welding sequence and by flexible connection sections with low stiffness (elastic tab and corner). **Plan the welding sequence.**		
Keep weld seams **away from high-stress regions**	vessel corners	
Keep weld seams away from sudden cross-section changes (e.g., put taper on plates)		
Double fillet weld better than a single fillet weld; butt weld is better than fillet weld; seams perhaps root welded or machined		
Weld seam root not to be subject to tension in bending loading		
Weld seam not to be placed in ingot segregation zones of rolled profiles		
Do not place weld seam in previously strain-hardened zones (danger of warping)		
2. Low production costs		
For **outside orders**, design as small and light as possible. Aim for low wall thickness by using more ribs and parts.		
Aim for low part count and weld seam volumes for **in-house production** and **small** or **medium welded assemblies**: aim for **edging** and **bending,** few ribs and simple shapes, even if that leads to weight increase.		few parts

Fig. 7.11-43a. Form design rules for **welded parts** (in part, from [Pah97])

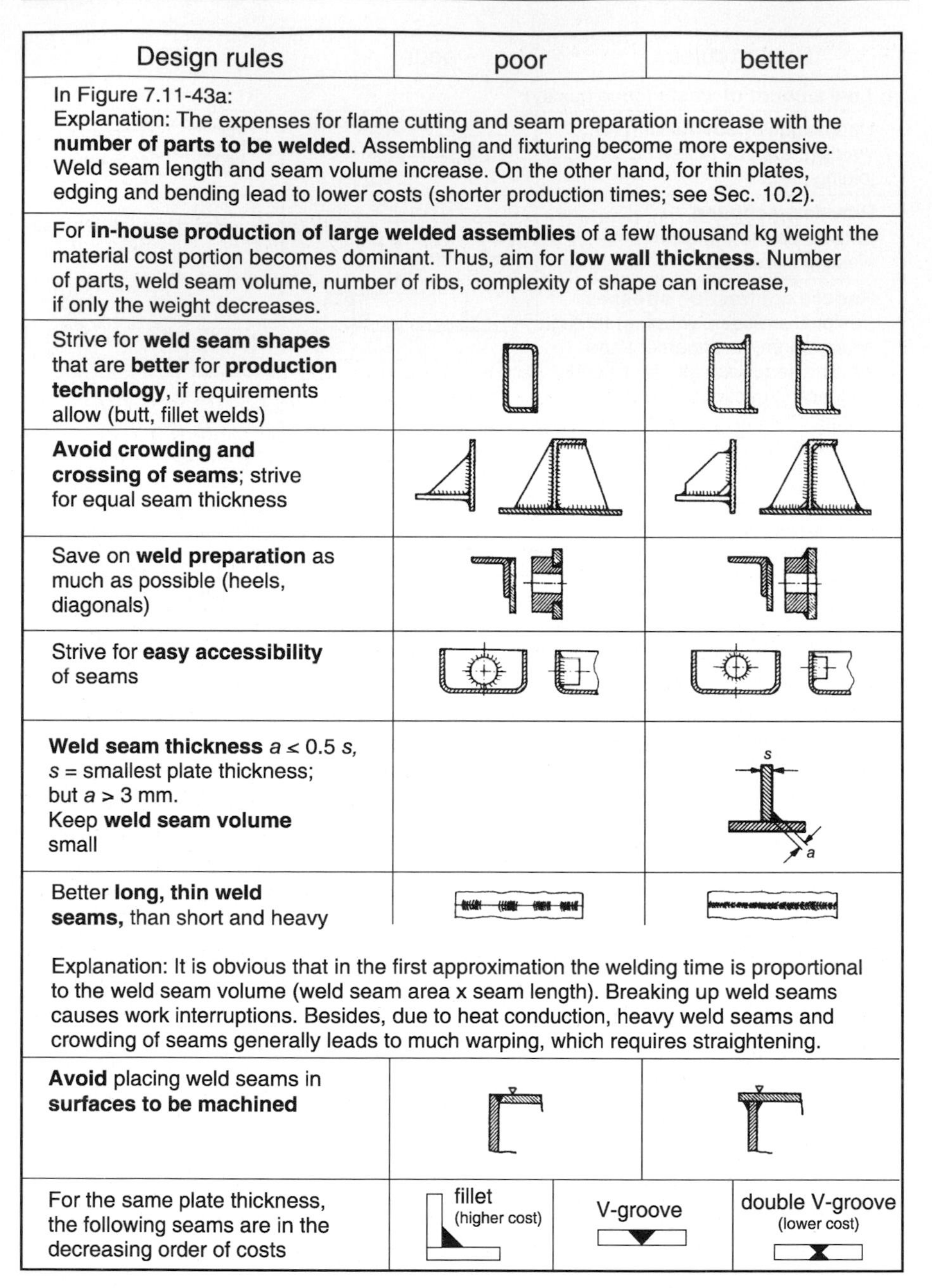

Design rules	poor	better
In Figure 7.11-43a: Explanation: The expenses for flame cutting and seam preparation increase with the **number of parts to be welded**. Assembling and fixturing become more expensive. Weld seam length and seam volume increase. On the other hand, for thin plates, edging and bending lead to lower costs (shorter production times; see Sec. 10.2).		
For **in-house production of large welded assemblies** of a few thousand kg weight the material cost portion becomes dominant. Thus, aim for **low wall thickness**. Number of parts, weld seam volume, number of ribs, complexity of shape can increase, if only the weight decreases.		
Strive for **weld seam shapes** that are **better** for **production technology**, if requirements allow (butt, fillet welds)		
Avoid crowding and crossing of seams; strive for equal seam thickness		
Save on **weld preparation** as much as possible (heels, diagonals)		
Strive for **easy accessibility** of seams		
Weld seam thickness $a \le 0.5\ s$, s = smallest plate thickness; but $a > 3$ mm. Keep **weld seam volume** small		s a
Better **long, thin weld seams,** than short and heavy		
Explanation: It is obvious that in the first approximation the welding time is proportional to the weld seam volume (weld seam area x seam length). Breaking up weld seams causes work interruptions. Besides, due to heat conduction, heavy weld seams and crowding of seams generally leads to much warping, which requires straightening.		
Avoid placing weld seams in **surfaces to be machined**		
For the same plate thickness, the following seams are in the decreasing order of costs	fillet (higher cost)	V-groove double V-groove (lower cost)

Fig. 7.11-43b. Form design rules for **welded parts** (in part, from [Pah97])

Rules for **low-cost welded assemblies:**

➔ Design as small and light as possible on **orders from suppliers**. Aim for low wall thickness, through more ribs and parts.

➔ With **in-house production and small to medium sized welded assemblies,** pay attention to keeping number of parts and weld volumes small: favor edging and bending, few ribs and simple forms even if that leads to more weight.

➔ With **in-house production of large welded assemblies** of a few thousand kg weight, material costs become dominant. Therefore, pay attention to small wall-thickness. Part numbers, weld volume, rib number and complexity of form can increase only if the weight decreases.

7.11.5.3 Laser and electron beam welding

a) The processes

Both processes work without addition and auxiliary materials and make very narrow, deep seams (during laser welding a protective gas is desirable).

The processes are easy to automate. The heat applied is low and thus there is little warping of the parts. Subsequent treatments are unnecessary. In particular, by **electron beam welding,** a large number of material combinations of almost arbitrary workpiece geometry can be welded together [Schl89].

An essential disadvantage of electron beam welding, vis-a-vis laser welding, is that it is generally necessary to put the parts in a vacuum chamber. Therefore, the procedure has proved itself particularly for small parts in precision engineering field for high-quality welding.

Since this is not a restriction in **laser welding** and powerful welding machines are available (up to 40 kW for CO_2 laser) laser welding is increasingly used due to quality and cost reasons [Ben93]. However, the capital expenditure is high and considerable personnel expertise is needed. Welding job shops offer economical advantages for less-frequent users [Hei94, Kin94a].

b) Design information [TRU96]

Steels with < 0.25% C are **easy to weld,** otherwise preheating is advisable. Cr-Ni steels and titanium (with protective gas) are very well suited. Nonferrous metals are less easy to weld with a CO_2 laser; it is better to use a Nd-YAG laser (see also Fig. 7.11-37).

The **seam preparation** must be precise, since the laser beam has a diameter only 0.3–0.6 mm and therefore the butting edges must be parallel and flush. Coatings of carbon, lacquer, rust, etc., must be thoroughly removed.

7.11.5.4 Adhesion

Adhesion can be used as a low-cost fastening, when following requirements and conditions are given for the parts to be joined:

- Joining of very different materials (for example metals with organic material) when large forces are not involved
- Existence of large joint surfaces (e. g., for interior trim in automobiles)
- Impermeability to liquids or gases

Cleaning and degreasing of the surfaces, the long hardening and loosening time of the adhesive, the quality inspection, and the disposal of the pickling solution and adhesive residue, are all cost-intensive.

Automated adhesion (also, the more rapid hot adhesion) leads to better adhesive quality and savings on the adhesive through appropriate measuring and mixing devices [The89, Hab97].

7.11.5.5 Screws, bolts and other connection elements

Screws and bolts are the most commonly used joining method, followed in distant second place by rivets.

a) Function costs (see Sect. 7.11.5.)

As purchase part costs, the cost of the most standardized connection elements (screws, nuts, rivets, snap rings, etc.) constitute only a few percent of the costs of the entire product. That means it may be neglected as a "C-part" cost in accordance with Pareto analysis (Sect. 4.6.2). However, as **Fig. 7.11-44** shows with the example of an axial locator for a rolling bearing, they are important because the overall costs of the standardized part generally amount to many times the purchase part costs.

The **function costs** of **Alternative A** (thrust protection of a ball bearing with a groove nut) in the figure are set to 100%. This is the sum of the purchase price for groove nut and lock washer as well as the machining and assembly costs – the entire manufacturing costs of the connection. The result is that the purchase part costs (groove nut and lock washer) constitute only one-third of the function costs of the connection.

Alternative B costs only 40% of solution A, complete with the retaining ring. Here the purchase part costs (retaining ring) constitute only about 10% of the function costs.

Bauer [Bau91] estimates that the cost prices for purchased connection parts are usually under 10% of the function costs.

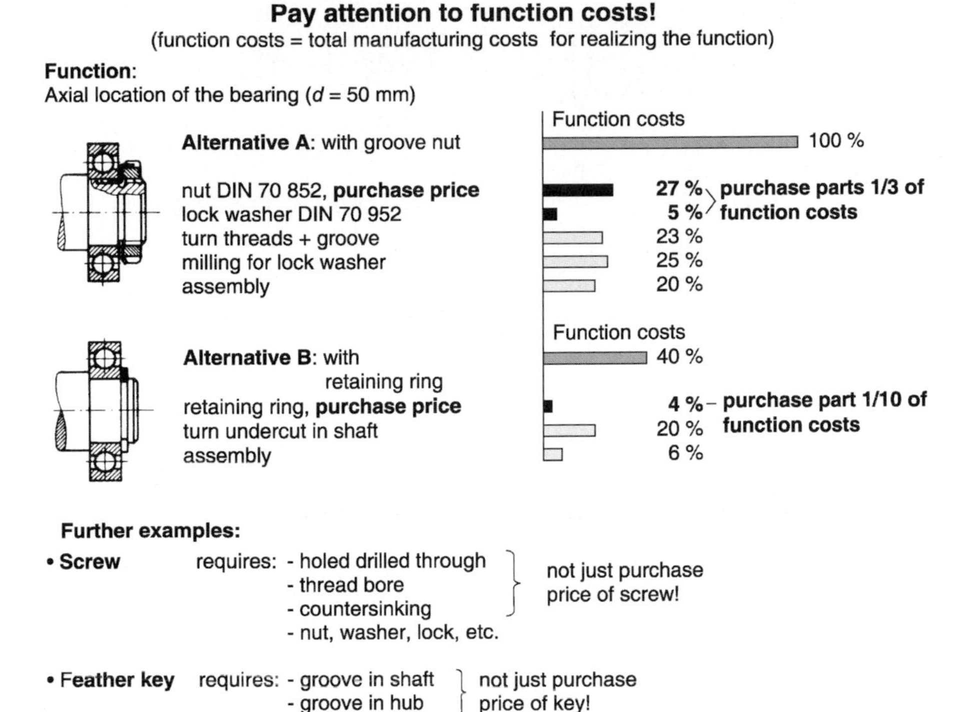

Fig. 7.11-44. Function costs for thrust take-up of a ball bearing (here, function costs are the total manufacturing costs of the connection). They amount to 3 or 10 times the cost of purchase parts (see Sect. 7.11.5.1, beginning)

b) Relative costs and rules

As mentioned in Sect. 4.6.3, relative costs are to be developed inside the company, considering all affected costs and with the information on all meaningful binding conditions. Their timely updating [Bau91] is important. The relative cost diagrams in this book are to be understood in this sense only as stimulation for your own investigations. Published under the Voith Company name, they arose primarily in the production of capital goods for small lot sizes and for large parts (paper machines, water turbines). Busch [Bus83] describes their development.

Figure 7.11-45 shows the relative costs of the entire **bolted joint** (production and assembly costs, without making the part stronger, as per Fig. 7.9-6, for different shapes. As a through bolt, the simple hexagonal bolt is the most economical. Cutting threads in in-house production is always more expensive than the mass production of screw threads in nuts. Added outlay is always a cost increase. The following comes from another company's investigation:

> ➔ **Through hexagonal bolts** with nut and **hexagonal bolts with threads cut in the opposite part,** are the lowest cost bolted joints, provided that additional materials usage is not needed (Fig. 7.9-6).

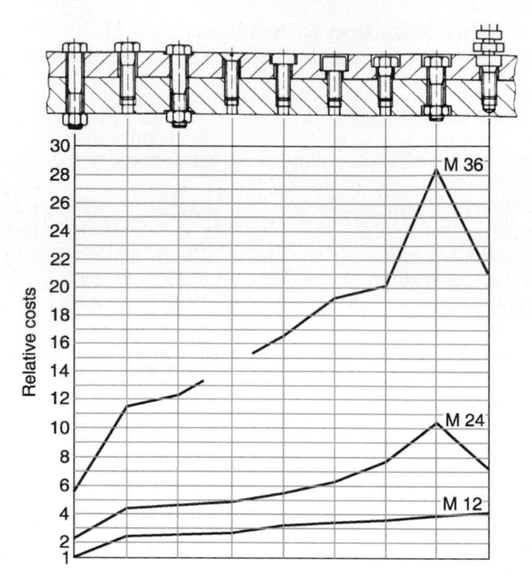

Fig. 7.11-45. Relative cost values of bolted joints for different sizes (from Voith)

Investigations of Voith and [Bau91] showed the following: As regards the benefit-cost ratio (yield strength/purchase price), **bolts of the strength class 8.8 and 10.9** are the most advantageous. Bolts of the class 5.6 are the most expensive solution for all diameters and lengths. This statement becomes even more significant when the additional costs of the larger bolted parts are considered (Fig. 7.9-6).

The following rules apply for the bolt costs only, not for the entire connection:

➔ For a desired pre-tension, the **higher strength bolt** is more **economical** than those of less strength, if all bolts (classes 5.6, 8.8, 10.9) are in stock.

➔ The hexagonal bolt DIN 931 is of lower cost than the pan head hexagonal bolt DIN 912. Considering all the parts bolted together, the **socket head bolt** might be the **lowest-cost connection** in most cases**,** as per Fig. 7.9-6.

➔ If stock and type variety are to be limited by factory standards, the decision to employ **only 8.8 class bolts** is reasonable, as long as the bolts used remain generally under M30.

The most technically appropriate **screw/bolt locking** devices are ratchets and adhesives.

For **pin connections,** clamping tube and straight dowel pin types are most economical (no reaming of the hole.)

> ➔ **Clamping tube** and **straight dowel pin** types are the lowest cost pin connections and are cheaper than straight and tapered pins, which require reamed holes.

In the case of **axial locators for shafts and hubs,** the **retaining ring** (from DIN 471) with a washer is the lowest-cost device. It also resists more force than the split-pin with washer. Most expensive is the groove nut with lock washer, as already shown in Fig. 7.11-44.

7.11.6 Dimensional tolerances and roughness

a) Dimensional tolerances

Designers know that tight tolerances are expensive; the design should therefore follow the rule "only as precise as necessary". Much more that is generally applicable in this regard is not known. Since the published company investigations start probably from other prerequisites (limiting conditions are rarely specified), there are great quantitative differences (**Fig. 7.11-46**). If the production costs for the manufacture of a hole with IT 11 are set to 1, then for IT 4 a cost increase between 2 times and about 15 times can be assumed. A more precise statement than that given above is not possible.

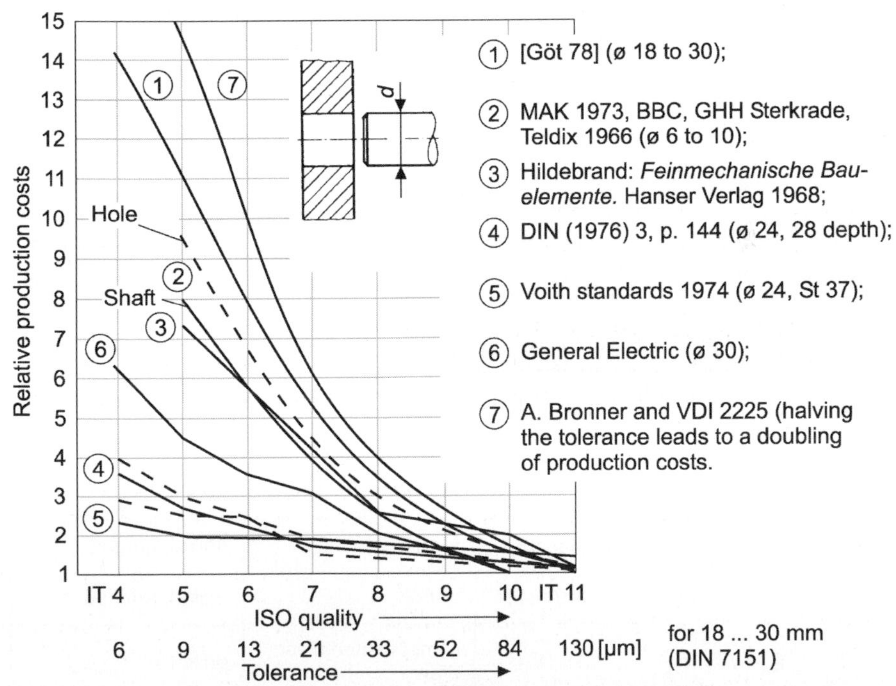

Fig. 7.11-46. Scatter of the relative cost numbers for ISO tolerances for hole diameter 18–30 (*all dimensions in mm*), DIN 7151 (IT 11 was chosen as the basis)

However, as in the example of the FVA cost analysis of **gears**, when 10 companies used identical technical data to calculate the costs of gears of varying quality (4–7 from DIN 3961–3967), a similar picture emerged [Fis83]. If quality 6 is set at 100%, then for a pitch diameter of 200 mm for DIN-quality 4, the manufacturing costs are about 5% to 35% higher. Time differences of 1:4 are the cause for these differences in the milling of the tooth flanks, even with the use of same milling machines (see also Sect. 9.3.7 and Fig. 7.13-2). If the outlying points are left out, a rough rule can be formulated thus:

➔ For case-hardened and ground spur gears of 200 mm diameter, the manufacturing costs of quality 6 (following 5 and 4) increase around 10% on the average; for gears of 1 000 mm diameter, only around 2–3% [Fis83].

It is typical that within the manufacturing costs the costs increase less for tolerances of large parts than for smaller parts. This is because the material costs are higher (Sect. 7.6), the production costs are proportionately less, and thus also the costs for keeping the tolerances smaller. However, if the production costs are chosen as a measure of comparison, the difference becomes greater.

The increase of the costs with lower tolerances is easy to understand if that necessitates additional production processes (**Fig. 7.11-47**). Since all processes have

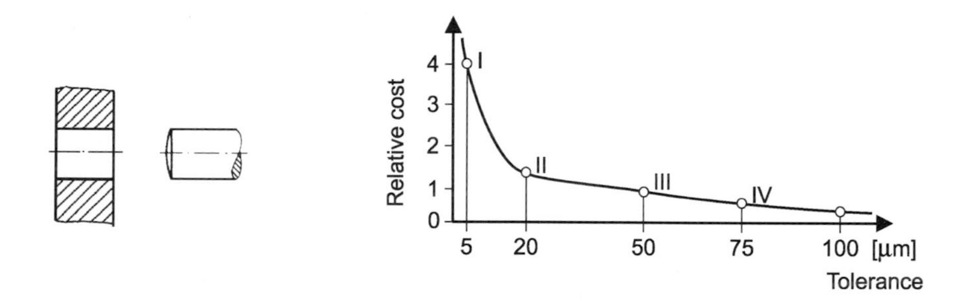

	Required tolerance (medium play)	Fitting possibility	Machining process for: pin	hole
I	$T_m = 5\ \mu m$	H5/h5	turned, ground, lapped	drilled, ground, honed
II	$T_m = 20\ \mu m$	H8/h8	turned, ground	drilled, ground
III	$T_m = 20\ \mu m$	H10/h10	turned	drilled, reamed
IV	$T_m = 75\ \mu m$	H11/h11	semi-finished clean, drawn	drilled

Fig. 7.11-47. Rise in production costs with decreasing tolerance (from S. Hildebrand)

a limit on attainable dimensional tolerances and roughness, the company must ascertain what tolerances the different machine tools and/or production types can achieve. Figure 7.11-28 shows how the tightening of tolerance and the roughness produced in boring require additional operations. Machining on the boring mill is the most expensive.

Examples of tolerance avoidance in form design:

- Figure 7.11-35 shows the redesign of a ring whose internal diameter was changed from Ø 80^{H8} to 80 ± 1. It is apparent how important it is to carefully look at the production sequence during design, and perhaps to consult with production specialists.
- **Figure 7.11-48** shows how the number of parts can be reduced by going to an injection-molded polymer part by means of integral design (fewer joints and fits). Polymer resilience makes close fits unnecessary (compensation of tolerance by elasticity) (See Sect. 7.11.2.6b).

improper **good**

a) rotary switch for stop-motion mechanism

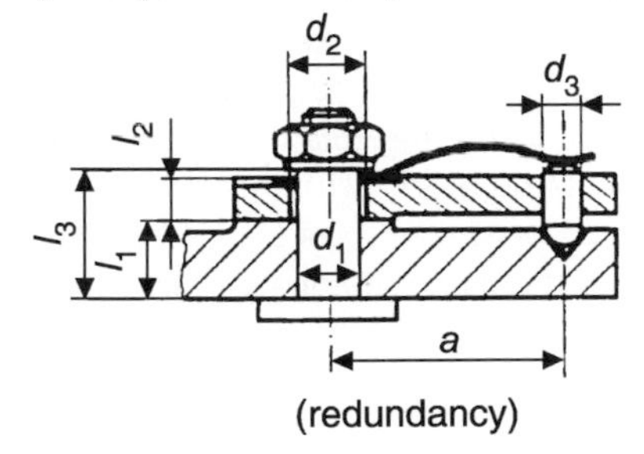

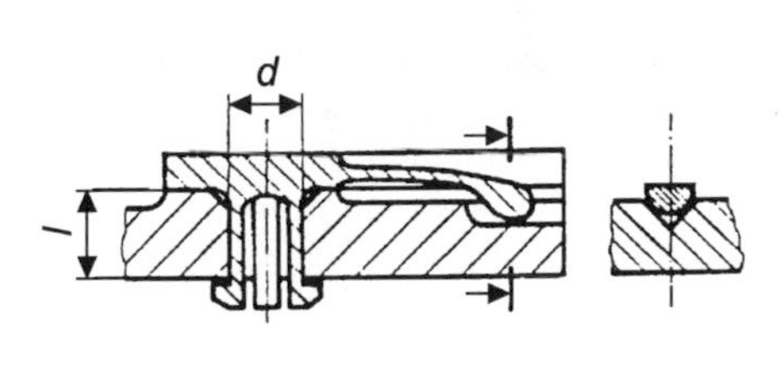

b) stop lever of an indexing plate

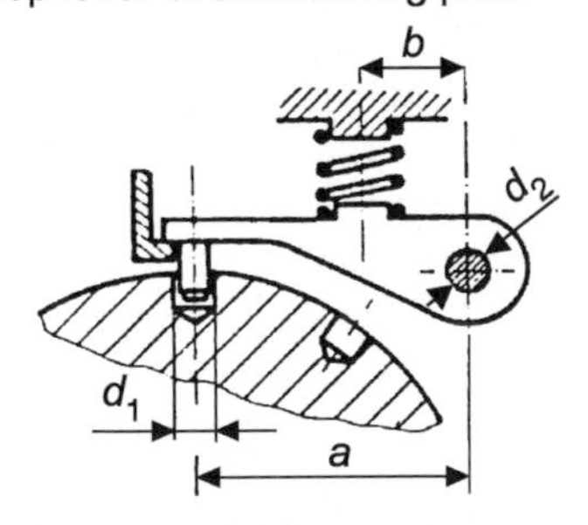

Fig. 7.11-48. Tolerance-averting form design (from R. Koller)

b) Roughness

Regarding the costs for reducing the roughness, essentially the same is true as for dimensional tolerances. The scatter in the information from different companies is large (**Fig. 7.11-49**). According to Lindemann [Lin80], the influence of finish

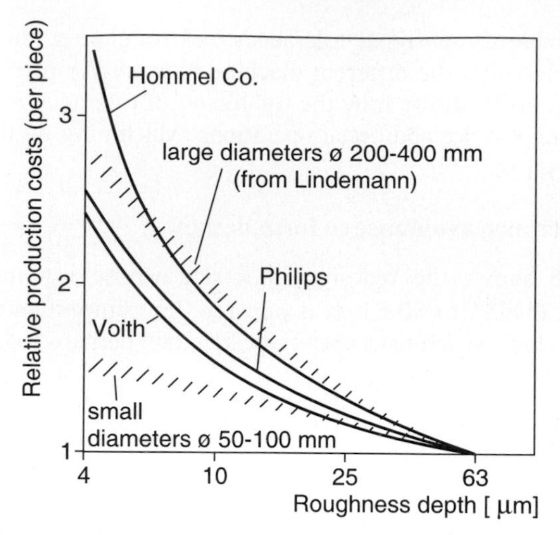

Fig. 7.11-49. Scatter in the relative cost data for the production of varying roughness for turning [Lin80]

machining during turning, and with that the roughness, is dominant if one considers only the production costs (material costs do not change) and if the rough-machining costs are small. On the other hand, in considering manufacturing costs the effect of roughness is not dominant any more, particularly with bigger parts. If, however, the production process must be changed or special processing becomes necessary to attain a required roughness, the costs increase similarly to that shown in Fig. 7.11-47.

7.11.7 Assembly

7.11.7.1 Importance of the design-for-assembly

Assembly costs can comprise up to 50% of the manufacturing costs [Stö75, Gle96], particularly with products with large number of parts (e. g., in precision engineering) or high complexity (e. g., measuring instruments, machine tools). In the machine industry, small quantities with a variety of changing assembly tasks are the rule, hence there is hardly any automation to lower costs. According to [Stö75], the activities during assembly are divided as shown in **Fig. 7.11-50**.

Inadequate part production or unsuitable tolerances achieved thus far are an unsatisfactory state of affairs. A lot of fitting and adjusting needs to be done, referred to here as the subsequent and adaptive processing. This comprises the greatest part of all assembly activities, approximately 43%. To “adapt during assembly” is therefore expensive. The actual assembling constitutes only about a quarter of the entire assembly time and the joining itself only 10%!

The designers already decide upon many assembly processes during concept development, and certainly during embodiment: the number of parts to be assembled, the types of connections, and the adjustment, safety and control possibilities.

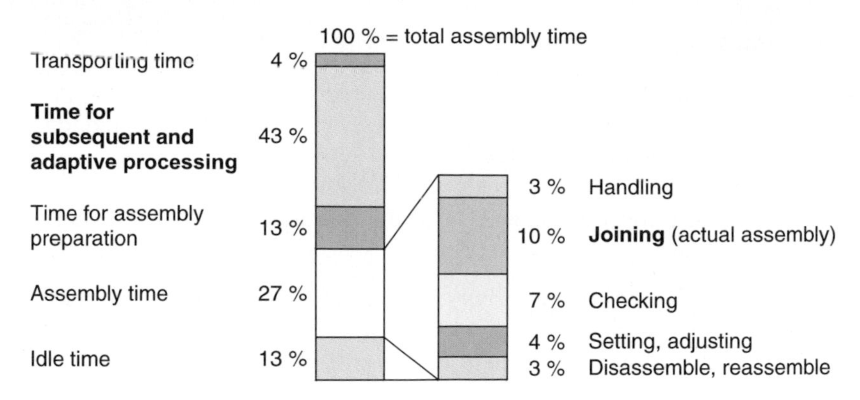

Fig. 7.11-50. A realistic activity structure for assembly in the machine industry (single-unit and limited-lot production) [Stö75]

Due to their technical experience with assembly, the designers must think through the assembly process in all details (this is best done in the team or with an employee from the assembly department) and determine which means (tools, fixtures, machines, measuring devices, etc.) can be employed. According to [And75, Stö75, Gai81, Lot86, Bäß88, Ehr93a, Sto94], here are two principles that the design group can think about:

- **Avoiding the assembly processes altogether:**
  For this, a **decrease in the part count** is particularly effective. This shows how important the decisions preceding the production are, which cannot be compensated for by the greatest efforts in production. When parts that were formerly produced separately are now combined into one part by integral design (e. g., by casting, injection molding, or sintering), there is nothing to assemble (Fig. 7.11-12, Fig. 7.11-54, Sect. 7.12.4.3).
- **Simplification of assembly processes**, basically in three ways:
  - Direct simplification by having more favorable features with regard to assembly, such as geometry and material (e. g., rougher tolerances and suitable connection methods) (see Figs. 7.11-53 and 7.11-53).
  - Ergonomically better shape design, so that the assembly process better corresponds to human abilities.
  - Do shape design such that the use of technical resources becomes possible (e. g., fixtures, motor-driven tools, or automatic machines). This is especially economical when by product standardization (for example, modular design, Sect. 7.12.6) the repetition frequency of the assembly operation increases.

In designing, a compromise is to be found between part production costs and assembly costs, for example, producing parts with close tolerances so that no adjustments are needed during assembly. Additionally between the original assembly costs and assembly (disassembly) costs during maintenance, repair, and disposal (see recycling, Sect. 7.14).

7.11.7.2 Parameters affecting the assembly costs

During the assembly process, parts (part systems) become elements of subassemblies (more complex subsystem) and these, in turn are than joined together as elements of the machine (complex system).

a) Accordingly, the costs are dependent in the first place on the following **variables**:

- Parts count and their joining properties dependent on geometry, surface and material
- Number of subassemblies and their joining properties at the interfaces to other subassemblies or parts
- Connection processes

According to [And75, Pah79], the **flowchart** containing the steps indicated in **Fig. 7.11-51** can be set up for the assembly. The rules shown in Figs. 7.11-52a–d are organized based on this flowchart.

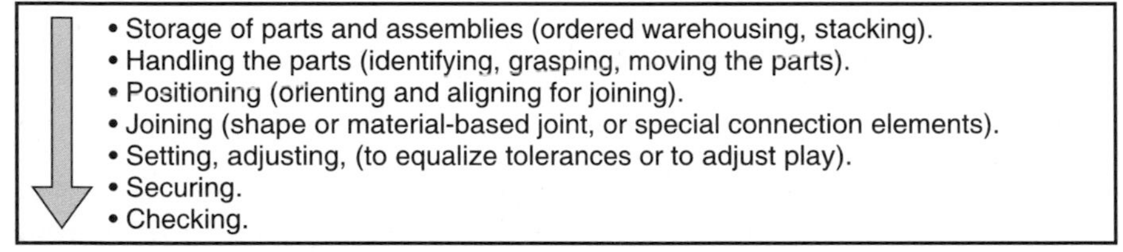

Fig. 7.11-51. Steps in the assembly [And75, Pah79]

b) Hard-to-handle parts cause **technical problems** and thus increase costs; they are characterized by following properties:

- Extreme weight (mass)
- Extreme dimensions and/or size differences
- Coarse tolerances
- Confusing shapes (for example springs, clips, retaining rings, cables)
- High sensitivity
- Extreme physical or chemical properties

Confusing shapes or parts are often avoidable by suitable form design. Figure 7.11-53b provides some suggestions. Confusing parts are generally made of wire or sheet metal (for example, retaining rings, clips, coil springs). Slots and holes should be smaller than the material thickness. Sharp corners and angles should be avoided. The feasible sequence from production should be incorporated in the assembly.

c) **Organizational problems** are considerably more significant as regards costs, as shown in Sect. 7.13.7 by an industry-wide investigation of "the faulty mechanism assembly" [Hub95a]. One of the most time- and cost-intensive disturbances in single-unit and limited-lot production is part unavailability at assembly time (see also Fig. 7.13-24). With a smooth-running serial assembly, this is no longer a problem, but others can come up.

d) In order to **reduce** these technical and organizational **problems,** following **measures** have proved themselves:

- **Assembly-oriented product design and development in a team** with assembly experts.
- Apportioning of the entire assembly process into **pre-assembly** and **final assembly**. Accordingly, sub-assemblies must be designed that can be pre-assembled and tested. With that the complexity of assembly is reduced.
- Establishment of **assembly groups** that are self-responsible, together with whom the materials flow, the time progression and quality control are optimized.

7.11.7.3 Rules for low-cost assembly

Figures 7.11-53a–e shows a collection of a many rules by which assembly costs can be lowered.

These rules are set up for manual assembly, but are also considered prerequisites for mechanized or automated assembly. Since the number of the assembly operations depends directly on the number of parts, elimination and integration of parts are the most important measures (see integral design, Sect. 7.12.4.3).

Furthermore, since the subsequent processing and adaptive work give rise to a large portion of the costs, product standardization is important in order to be able to use, for example, fitting parts in stepped sizes for critical and always repetitive fitting operations. Gages and fixtures for processing otherwise carried out during assembly, and also for the assembly itself, can then be made ready. The form design rules (Fig. 7.11-52) are also usable as a checklist for judging the suitability of designs for assembly.

In **Fig. 7.11-52** the technical and organizational measures for low-cost assembly are clearly summarized.

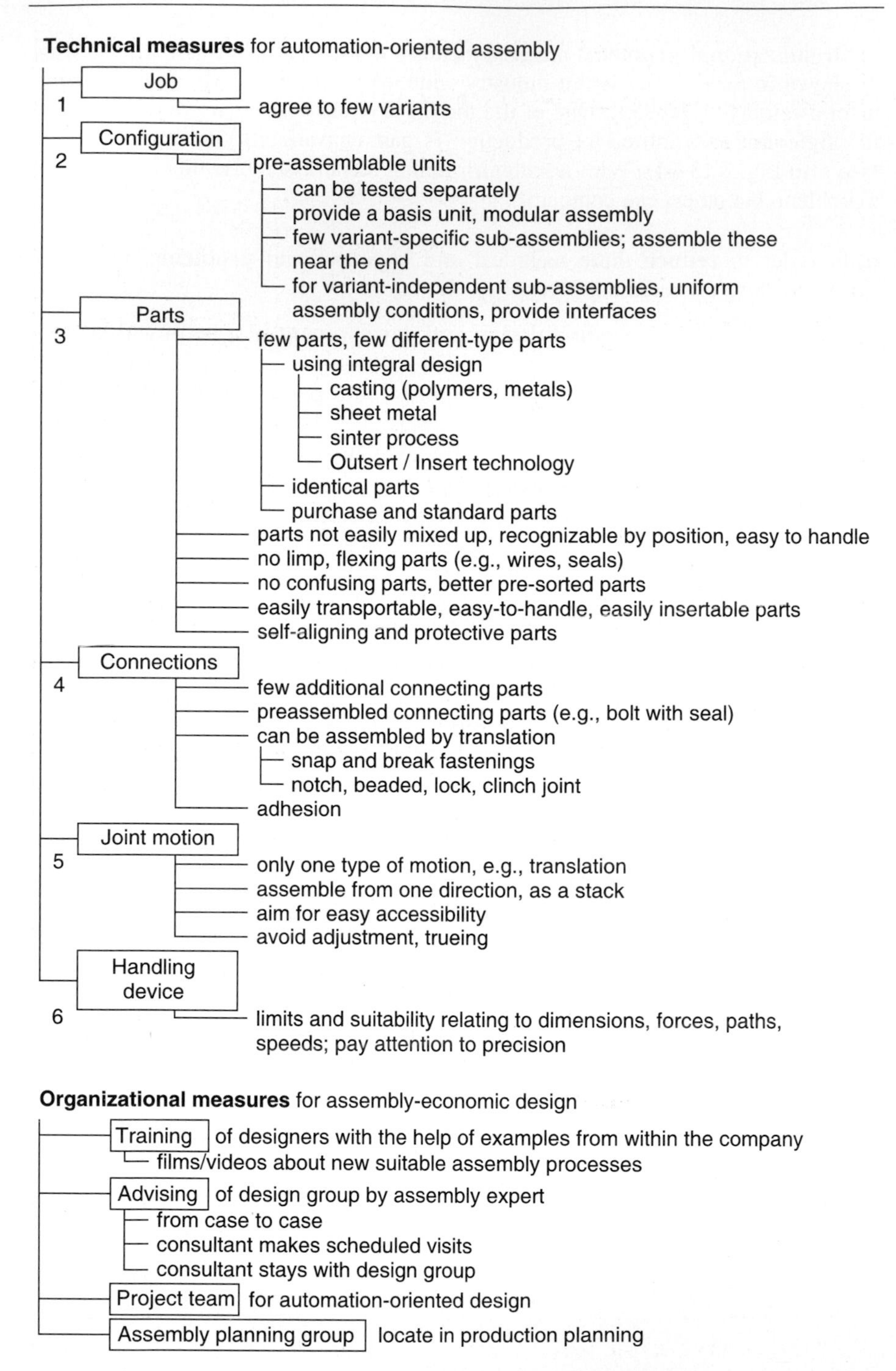

Fig. 7.11-52. Measures for low-cost assembly [Ehr93a]

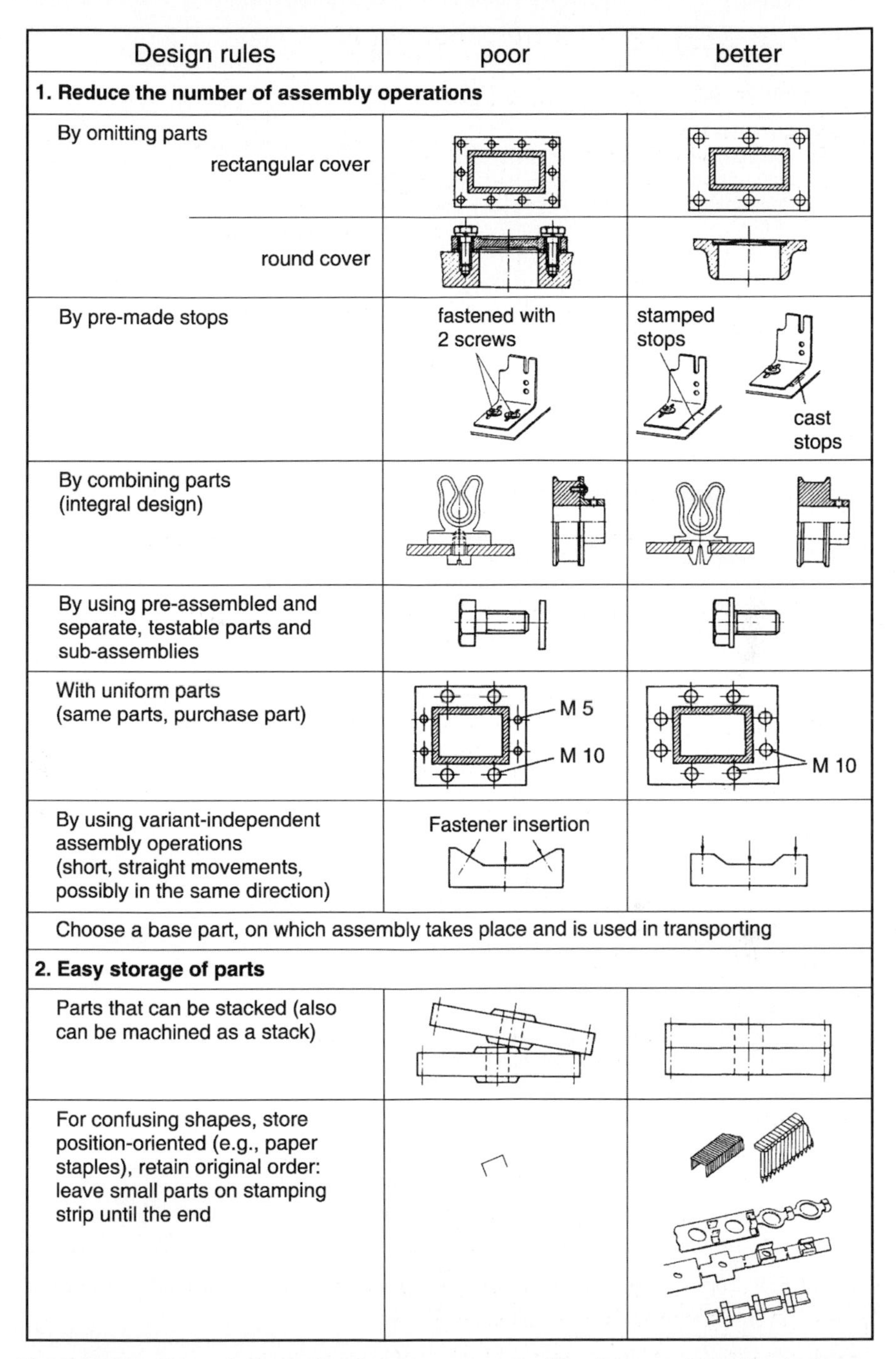

Design rules	poor	better
1. Reduce the number of assembly operations		
By omitting parts rectangular cover		
round cover		
By pre-made stops	fastened with 2 screws	stamped stops cast stops
By combining parts (integral design)		
By using pre-assembled and separate, testable parts and sub-assemblies		
With uniform parts (same parts, purchase part)	M 5 M 10	M 10
By using variant-independent assembly operations (short, straight movements, possibly in the same direction)	Fastener insertion	
Choose a base part, on which assembly takes place and is used in transporting		
2. Easy storage of parts		
Parts that can be stacked (also can be machined as a stack)		
For confusing shapes, store position-oriented (e.g., paper staples), retain original order: leave small parts on stamping strip until the end		

Fig. 7.11-53a. Form design rules for low-cost **assembly** (in part, from U. Andreasen, K. H. Beelich, G. Pahl, Th. Stoeferle, VDI 3237)

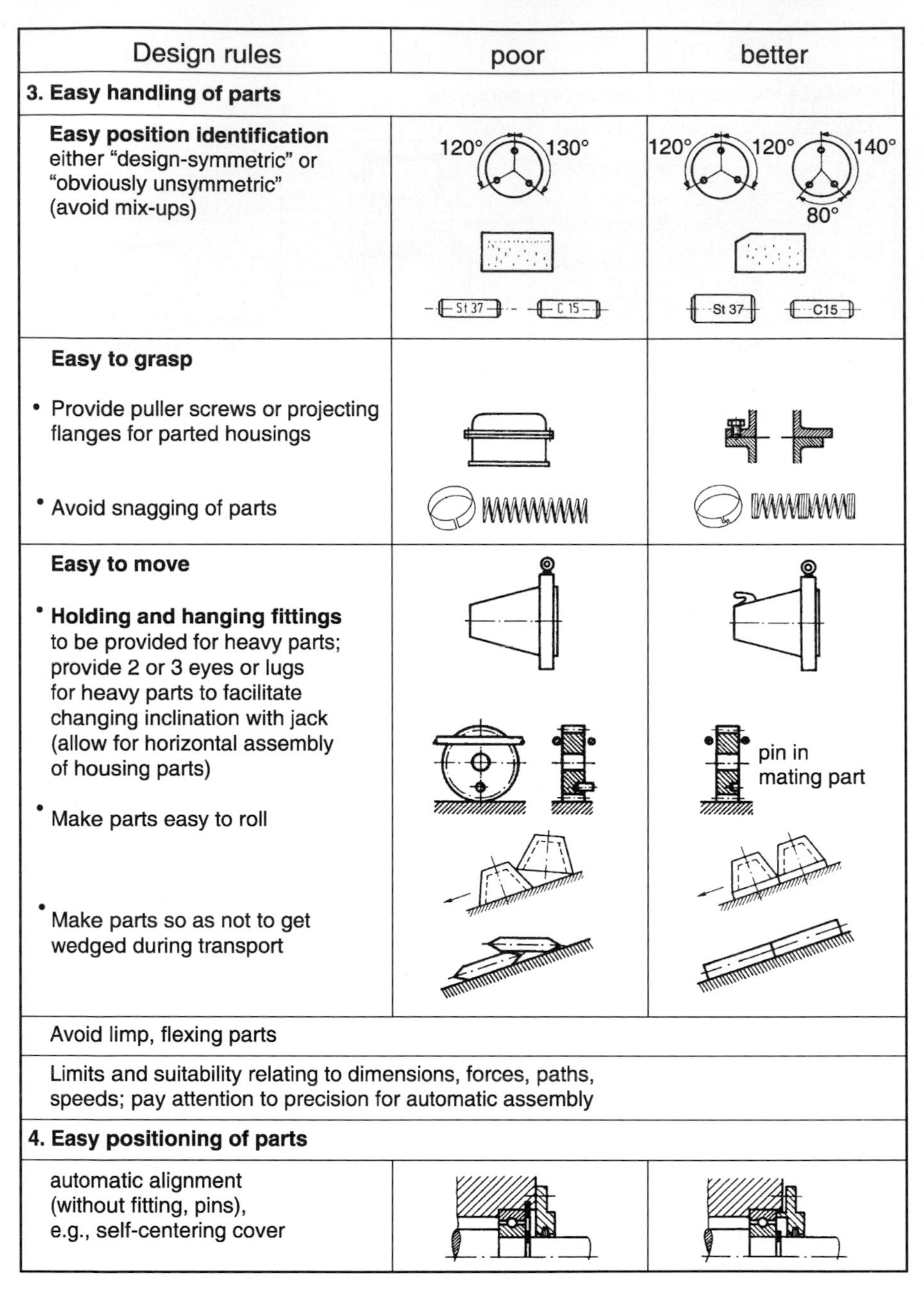

Design rules	poor	better
3. Easy handling of parts		
Easy position identification either "design-symmetric" or "obviously unsymmetric" (avoid mix-ups)	120° 130° St 37 C 15	120° 120° 140° 80° St 37 C15
Easy to grasp • Provide puller screws or projecting flanges for parted housings • Avoid snagging of parts		
Easy to move • **Holding and hanging fittings** to be provided for heavy parts; provide 2 or 3 eyes or lugs for heavy parts to facilitate changing inclination with jack (allow for horizontal assembly of housing parts) • Make parts easy to roll • Make parts so as not to get wedged during transport		pin in mating part
Avoid limp, flexing parts		
Limits and suitability relating to dimensions, forces, paths, speeds; pay attention to precision for automatic assembly		
4. Easy positioning of parts		
automatic alignment (without fitting, pins), e.g., self-centering cover		

Fig. 7.11-53b. Form design rules for low-cost **assembly** (in part, from U. Andreasen, K. H. Beelich, G. Pahl, Th. Stoeferle, VDI 3237)

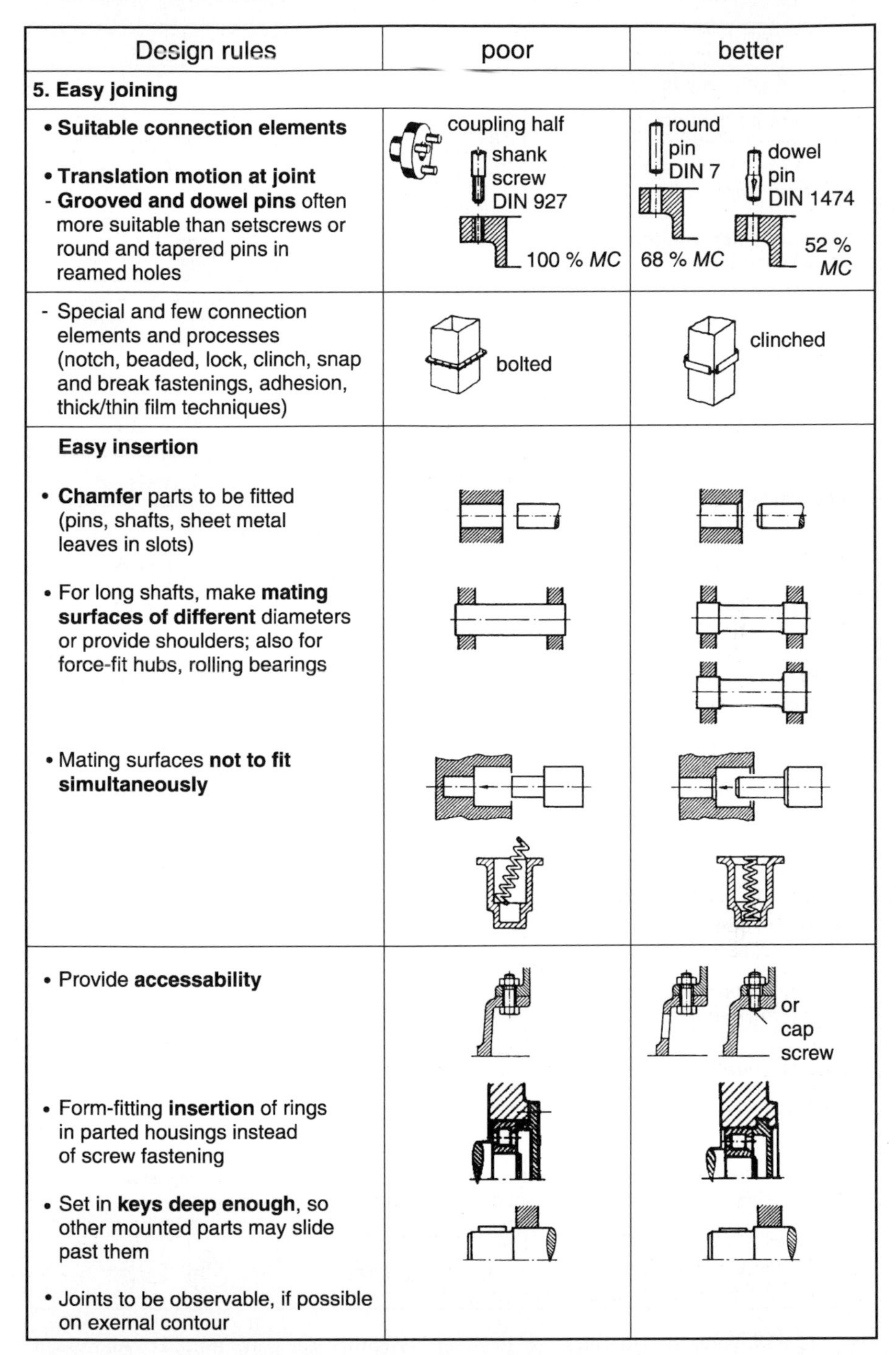

Design rules	poor	better
5. Easy joining		
• **Suitable connection elements** • **Translation motion at joint** - **Grooved and dowel pins** often more suitable than setscrews or round and tapered pins in reamed holes	coupling half shank screw DIN 927 100 % *MC*	round pin DIN 7 68 % *MC* dowel pin DIN 1474 52 % *MC*
- Special and few connection elements and processes (notch, beaded, lock, clinch, snap and break fastenings, adhesion, thick/thin film techniques)	bolted	clinched
Easy insertion • **Chamfer** parts to be fitted (pins, shafts, sheet metal leaves in slots) • For long shafts, make **mating surfaces of different** diameters or provide shoulders; also for force-fit hubs, rolling bearings • Mating surfaces **not to fit simultaneously**		
• Provide **accessability** • Form-fitting **insertion** of rings in parted housings instead of screw fastening • Set in **keys deep enough**, so other mounted parts may slide past them • Joints to be observable, if possible on exernal contour		or cap screw

Fig. 7.11-53c. Form design rules for low-cost **assembly** (in part, from U. Andreasen, K. H. Beelich, G. Pahl, Th. Stoeferle, VDI 3237)

Design rules	poor	better
6. Easy to set, easy to adjust ("produce rough, adjust fine")		
▪ Avoid **redundant fits.** Squeeze. Tapered seats may have neither radial nor axial stops	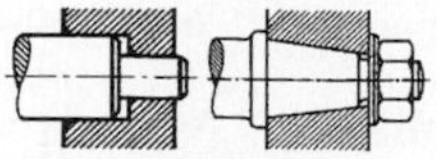	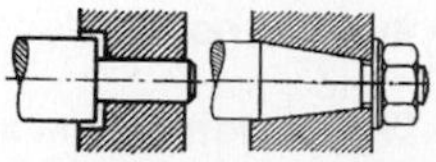
▪ **Do not center with threads.** Avoid special centering fittings	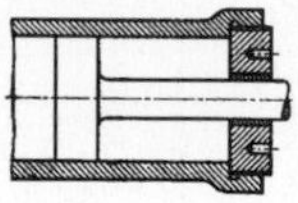	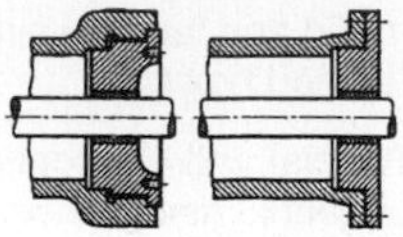
Use stepped **fitting parts**, or smooth **adjustment**	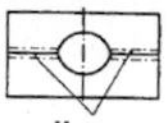 overdimension	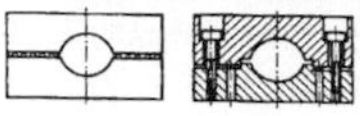
Instead of closer tolerances, use parts **elastically** or **plastically deformable** or **adjustable**	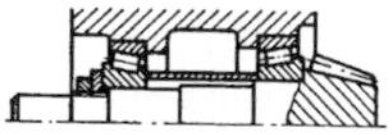	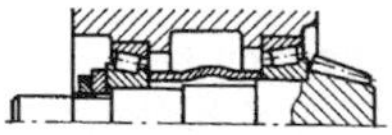
Dowel pin saves use of stepped axle with tight fit; hole can be drilled through	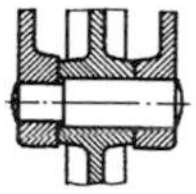	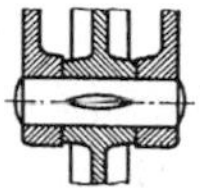
Adhesion of bearing blocks, instead of refinishing for fitting	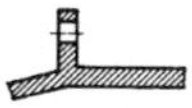	 adhesive
7. Unambiguous, easy protection		
Position locking of housing parts with widely separated dowel pins	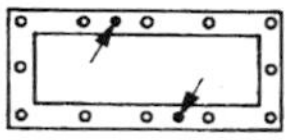	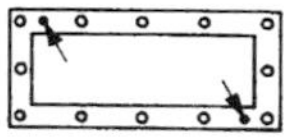
Instead of **special protection** (locking washer), bolt turning prevented by bolt head and housing shape	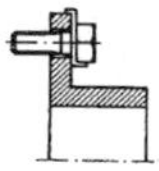	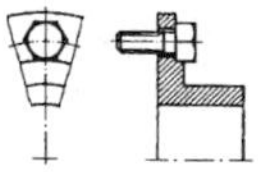
Provide simple safety and **protection elements**	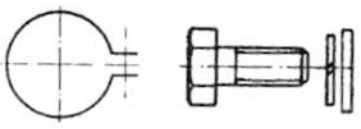	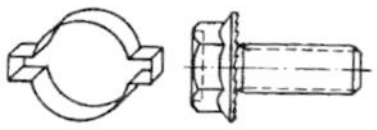
8. Easy to check and control		
"Self-checking" by using different size bolts	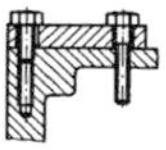	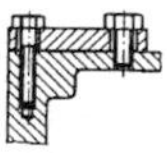
Use standard dimensions and tolerances (tools and gages readily available)		
For unbalanced parts, provide rolling supports and spots for removing unbalance	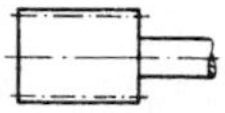	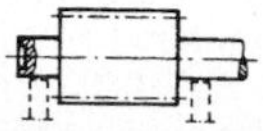

Fig. 7.11-53d. Form design rules for low-cost **assembly** (in part, from U. Andreasen, K. H. Beelich, G. Pahl, Th. Stoeferle, VDI 3237)

Design rules	poor	better
9. Easy disassembly		
Keys can be removed when **worn out**, by a hammer blow at one end		
Provide **holes, grooves** for **disassembly**; keep in mind the accessibility by the tool		groove
Provide puller screws or projecting flanges for parted housings		

Fig. 7.11-53e. Form design rules for low-cost **assembly** (in part, from U. Andreasen, K. H. Beelich, G. Pahl, Th. Stoeferle, VDI3237)

7.11.7.4 Examples of assembly-oriented design

a) The central design idea "no parts, no assembly" is demonstrated in **Fig. 7.11-54** with the example of a **hose clamp.**

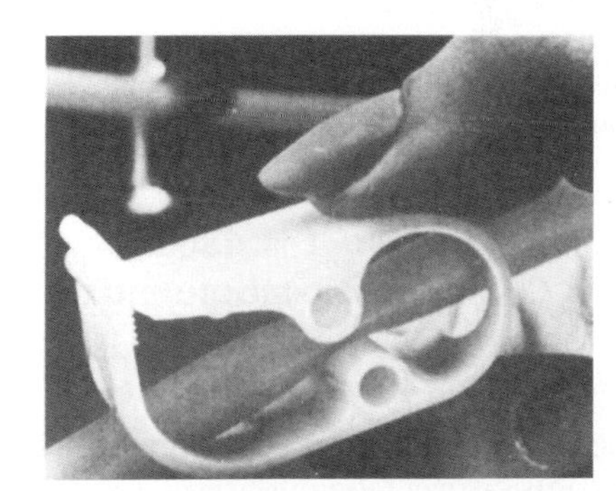

Fig. 7.11-54. Tube clamp in metal from 6 parts or one-piece of polymer (from U. Andreasen)

The former six-piece metal design was replaced by a one-piece integral design of injection-molded polymer, in which the high elasticity of the polymer was exploited to the maximum. The assembly of what is now only a "penny article" is unnecessary, since it is assembled by the user on the hose. We see how very effective are these form design principles.

In the case of the round **cover plate** in Fig. 7.11-53a (top), the initial design with 8–10 parts (depending on the number of bolts) was replaced by **one** standardized part (cover disk DIN 470). The seal against splashing water and clogging is sufficient. With such differences, no further comparison calculations are necessary.

b) As shown in **Fig. 7.11-55**, the **fitting** for **fastening** the automobile brake and fuel lines was realized by applying the rule "reduce the part count" (from three to

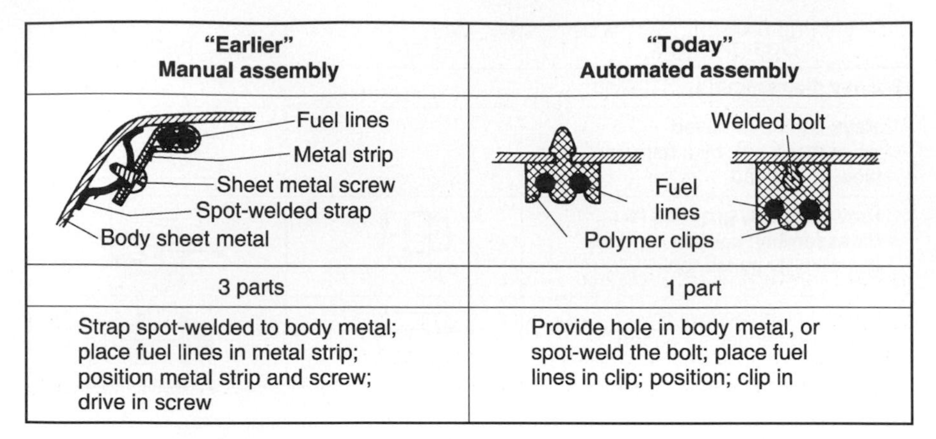

Fig. 7.11-55. Example of the automatic fastening of brake and fuel lines (car, VW Golf)

one, again through integral design, using polymers) as well as the rule "simple rectilinear joint movement" (Fig. 7.11-53c). Thus, the changeover from the expensive manual assembly to the faster automated assembly was made possible.

c) Figure 7.11-17 shows circumvention of the fitting and adjusting steps with the example of a **connecting rod "cracked"** along the die line. This avoided not only the machining of the die line faces, but also the setting and pinning together of the two parts during assembly.

7.11.8 Quality costs, measuring and testing

a) Quality costs

Since the quality of a product represents the agreement of actual and desired properties, this control loop must be closed at the earliest possible point during product realization. That means, the properties must already be established qualitatively and quantitatively during product development and their implementation adhered to by suitable measures (analogous to Fig. 4.4-2). The later the control loop comes into play, the more expensive it becomes (**Rule of Ten**, see Sect. 2.2) [Hal93]. The quality costs associated with that amount to between 5 and 25% of the manufacturing costs in Germany [Sto87]. For some parts and machines (reactors, aircraft industry) the costs of testing exceed all other production costs (see cost of rejects, Sect. 7.9.2.2).

Steering the production processes so that there are no rejects at all is a step forward. This should be done even earlier, during product development, to make the product-based **concept and form design robust** (i. e., error-resistant). This is largely feasible.

A modest example of that is shown in Fig. 7.11-54, a hose clamp. A principle was chosen that is simpler, has fewer parts, fewer production processes, and fewer dimensions to be held. Therefore, the tendency to make mistakes in production must also be less. Further examples are clear: The three-phase alternating current squirrel-cage motor is more robust and also needs fewer spare parts than the direct-current commutator motor with its current transmission brushes. The oscillating armature drive of an electric shaver was more robust, at least in the first few years, than the electric motor drive.

Since there are hardly any universal methods for designing robust and low-cost products, it is left up to the designers' talents.

b) Parts suitable designed for measuring and testing

Regardless of all else, parts and products must be designed for testability to at least partially to reduce the high costs of product testing.

The most important checks and controls in the machine industry are shown in **Fig. 7.11-56**.

To design for testability, the cooperation of employees from quality assurance must be sought. The **joint preparation** of a **measurement and test plan** is a first step that can at times also initiate design changes in the early stage. Essentially, every quantity established by the design group must be tested by the supplier, as the incoming materials, and/or in part production and assembly. Design should therefore consider whether it could actually be measured to the required accuracy [Dut93].

A further possibility of introducing timely measures in the case of important parts is by use of **FMEA** and **QFD** (Sect. 4.9.1).

Some **form design rules** for measurable and testable features in design are shown in **Fig. 7.11-57**. Similar sheets should be created throughout the company.

- Length measurement (distance from surfaces, holes, inner and outer diameters, roundness).
- Angle measurement (surfaces, holes).
- Shape check (e.g., on profile projector, 3D-measurement device).
- Hardness measurement.
- Check for cracks (magnetic, ultra-sound, X-rays, dye penetration).
- Strength testing on co-produced material specimens.
- Tightness testing (water, oil, air, gas).
- Pressure testing (water, oil, air, gas).
- Balancing (static, dynamic).
- Centrifugal test.

Fig. 7.11-56. Important checks and controls in the machine industry

Design rules	poor	better
Aim for **Integral Design**; it gives rise to fewer surfaces to be processed than if there are many parts; thus there are fewer dimensions to be checked; they should not be too big or too complicated for measurement	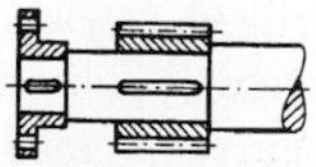	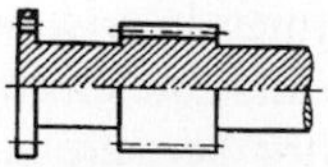
For dynamic **balancing** provide surfaces for rollers of the balancing machine; provide possible place for material removal	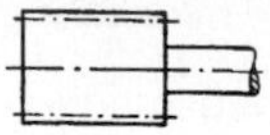	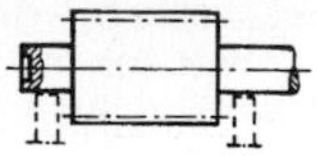
High-speed rotors should be designed axially symmetric (2 keys instead of 1, fastened on inside)	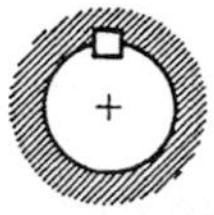	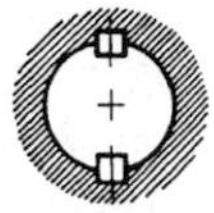
For diameter measurements in gears, bearings, grooves, provide **even number** of measuring **surfaces**	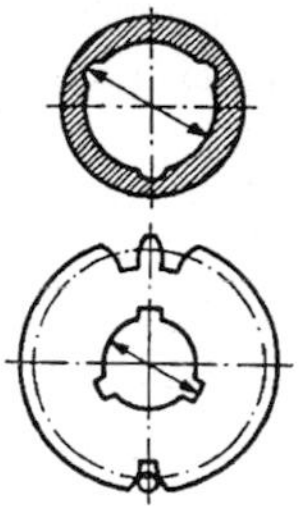	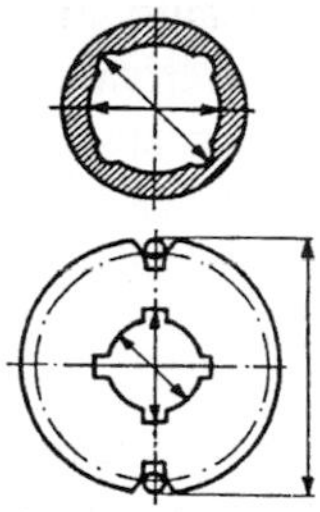
Match **standard dimensions** and **tolerances** to existing gages (snap gages, test prods, clamping devices)	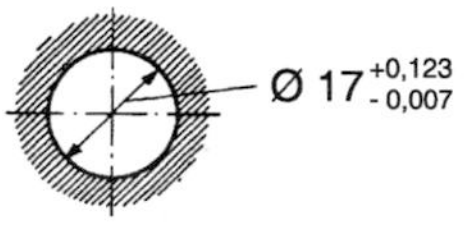 $Ø\ 17^{+0{,}123}_{-0{,}007}$	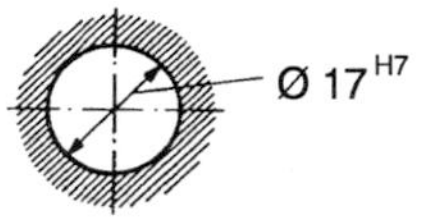 $Ø\ 17^{H7}$
For **aligning,** provide special reference surfaces (e.g., ground surfaces) and show on the drawing	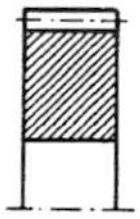	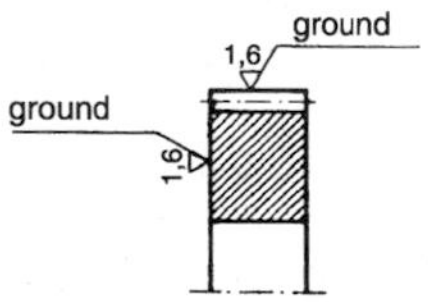 ground 1,6 ground 1,6

Fig. 7.11-57. Some form design rules for measurable and testable parts (partly from G. Reisinger)

7.12 Management of product variants

The change from a vendor market to a customer market has led to increased competition among manufacturers for the benefit of the customer (Sects. 3.1.1 and 4.4.3). Accordingly, less busy market niches are sought and more specific customer wishes are fulfilled. Rapid improvements or design of new products generate new incentives to buy. This market trend increases the variety of product variants while simultaneously leading to smaller number and lot size per **variant**. By a variant we mean an object of similar shape and/or function [DIN77].

An increase in the number of variants and a decrease in the lot size tend to lead to longer throughput and delivery times and problems of quality assurance (Sect. 6.3, Sect. 7.12.2.2), and to higher direct and indirect costs (complexity costs) per variant.

Variant management includes all measures by which the multiplicity of variants (the "variant variety") within an enterprise is deliberately influenced. This is valid, therefore, for products as well as for the affected processes. The aim is the reduction and control of complexity (i. e., minimum internal complexity and/or variant variety while simultaneously offering sufficient number of variants to the outside (i. e., to the customer). Variant management should achieve the following aims:

- Servicing the market **only with the necessary variants** (variation needed in the market [Schh89]).
- Recognition of and **reducing the unnecessary variants.**
- **Decreasing the throughput times** and, in particular, the direct and **indirect costs,** with the necessary variants.

We shall not go into additional aspects of configuration management [Bur93, Say84]. It is a matter of compatibility, versioning and management for change, among others.

The variant variety can refer to:

- **Product variants** that become visible to the customer from the **outside** (e. g., in the variation of performance, size, equipment, materials, or the exterior design).
- **Assembly and part variants** which are **internal,** in different shapes or production and assembly methods.

Next to the variants mentioned above on the product and part level, a large assortment of unrecognized variants often exists in the form of shapes, basic solutions for the same function, and of production processes.

Investigations, particularly in the automotive industry, led to the following conclusions [Eve88]:

- The count of the part numbers increased around 400% from 1975 to 1985
- 50% of the variants are superfluous
- 50% of the investments are complexity-based
- 80% of the activities add to the value only indirectly

In the **machine industry** the product and part variety and thus the complexity of the product creation processes, have also increased enormously. There were corresponding growths in the direct and indirect costs and time expenditures.

To get an idea of how large the possible number of variants can be, we use a simple industrial assembly. **Figure 7.12-1** shows an enameled valve of modular design that is used for filling and emptying a chemical tank. The respective variant count of the components (e. g., size variants) is indicated. For the drive there is

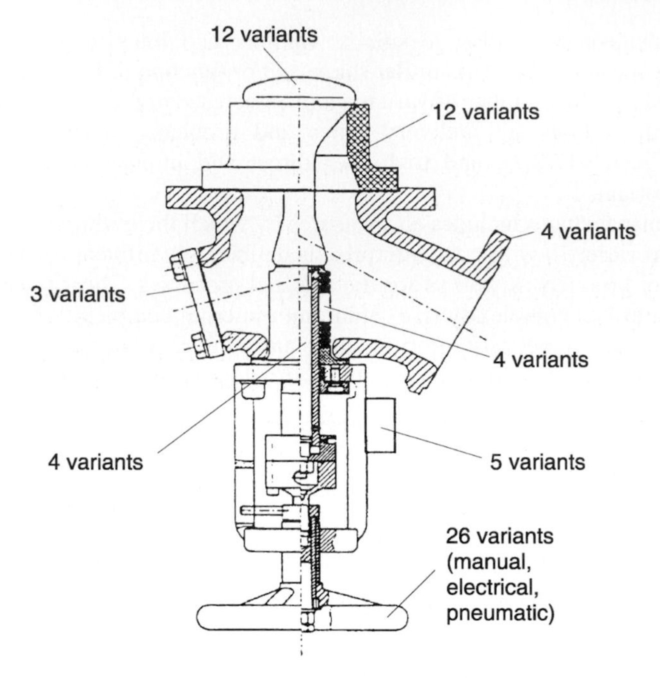

Fig. 7.12-1. Modular design of enameled valves, with 3.5 million useful combinations [Koh96]

a similarly high variant count. From these, 3.5 million useful (compatible) valve configurations can be formed.

If because of the aggressive nature of the chemicals, instead of enamel different materials were used, for which in different countries there are usually different safety guidelines, the variant number would be still higher.

We see that the increase of the part variety, product variety, and the resultant variant variety of the products leads to a systematical thinning of the product line to reduce the number of parts and variants. Additionally, the company processes must be designed such that they are able to deal efficiently with the many part and variant numbers.

Increasing part complexity implies many variant-specific parts and few standardized parts. It is a result of the increased variant variety on product level, a low carryover of existing parts to newer products, and a high proportion of new designs instead of designs that only involve modifications (adaptive designs). A lack of transfer of existing parts results from not having an overview of the product range and the interfaces between the individual components. Possibilities for standardization are often not clear.

The following can all lead to an increase in **production complexity.** A large number of self-designed parts and components; variants arising at an earlier step of adding value; and order- or customer-oriented individualization at an early stage of the production process. The results are small quantities per product as well as a large number and/or diversity of the parts used (part complexity). This results in an increasing number of tasks to be solved in production, smaller lot sizes, and a more frequent readjustment of production, which requires flexible machine concepts that increase process complexity and require greater coordination.

Alongside variant variety is the idea of **complexity**. Design and production are multi-layered procedures. **Complexity costs** are the costs resulting from these multiple layers [Hom97, Ros02, Gem98, VDI98a]. We distinguish between internal and external complexity (for example, complicated legal regulations).

The causes and characteristics of internal complexity are the variant number and/or the extent of the overall program, the structure of the products (part and component number), and the chosen organization of production. To these, we add the customer structure and/or number of customers, the design and production strengths and weaknesses, the number of suppliers, and the number of employees and functions participating in order fulfillment.

An investigation of successful and less successful companies [Romm93] showed that increasing complexity becomes easier to control with the strategy of "simplicity and deciding on the focal points". This affects the five "variety" problem areas represented in **Fig. 7.12-2**.

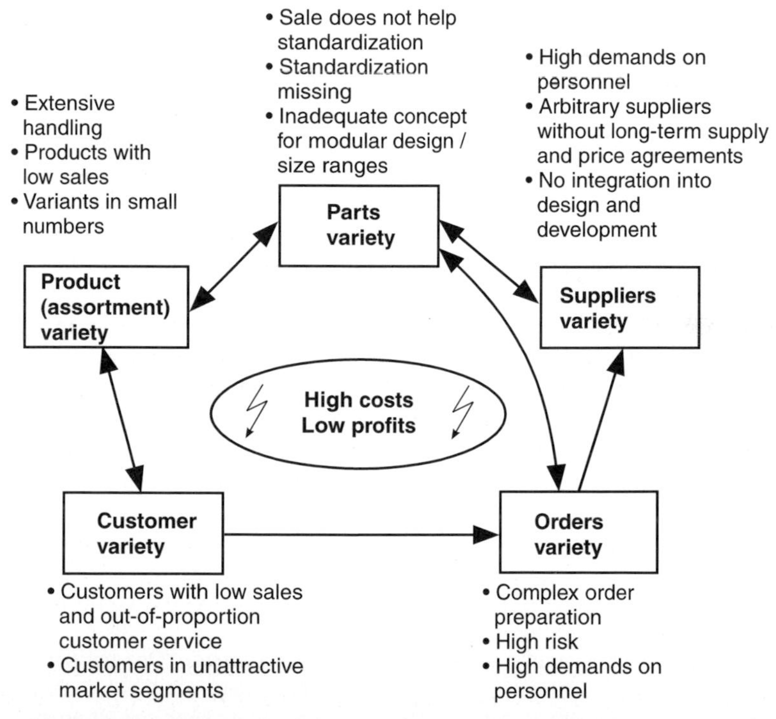

Fig. 7.12-2. Interdependence of the five "variety" problem fields (in part from [Koh99])

Companies that enjoy long-term success are characterized in a statistically significant way by:

- **Less product variety** (concentration on strong products, elimination of products with C-grade sales; Fig. 4.6-4).[10]
- **Less customer variety** (concentration on A and B customers).
- **Fewer suppliers** (integration into the company's design and development, reliance for the long-term). Thus, reductions in capital flow, reduced part variety, and a less extensive inventory are also attained.
- **Segmentation** of production into production and assembly islands **(teamwork)** with responsibility within the group, simplifies planning and logistics [Wil94].
- **Increase of outsourcing** also reduces the number of variants to be planned within the company, since these are shifted to the suppliers (Sects. 7.10 and 6.2.3).

Traditional company aims are high productivity and efficiency, the striving for profit, and the preservation and increase of the company's worth. Nowadays more strategic success factors are added such as a high quality (superior to the competition) or a high-grade service (supplier service, customer service, and additional services).

Furthermore, individual departments in the company have different aims that may conflict with each other. Marketing requires more variants for market and customer-oriented strategic reasons, while production favors a higher degree of standardization and uniformity for efficient production. This complication in the goals requires expensive coordination. A lack of communication, coordination, and cooperation within and between the groups can lead to a dominance of the marketing group relative to design, development, and production. In its own interests, marketing frequently works against attempts at variant reduction. This comes about due to having a purely sales viewpoint, or ignorance of the cost consequences. The customer is promised fulfillment of all individual wishes and the result is an increased number of product variants.

The same holds for the standards group. If they are not effective in checking company-internal standards, the standardization of components occurs too late. This department's potential does not end there, often because the standards department is organizationally in the wrong place, which makes it too weak.

The deficiencies concerning information are closely associated with poor communication. Not having knowledge of the variant variety, inadequate description of product structure, unclear description of variants, and insufficient information flow regarding the recognition and description of the current variant variety all

[10] "Setting limits is the sign of a master": As B. Riebel has said. "One must free oneself from the thought of wanting to take on all achievable orders, all possible customers, to win and supply customers, to work on all the markets, to make or sell all products. Furthermore, one should strive for and support the selected fields of activities, portions of the total market to be handled, the main tasks of marketing, and indeed from the long as also the short viewpoint, in the small and the large" [Rie85].

make it impossible to use the available information. Continuous improvement of future products thus becomes more difficult.

In addition, the lack of design tools and aids that support the development process (e. g., repeat-part search system) becomes apparent.

7.12.1 Causes and consequences of product and part variety

In order to deal adequately with the growth of variants, it is necessary to know in detail the causes and the advantages and disadvantages of having so many variants [Ker99].

Within a company, the causes of variant multiplicity can be classified as external and internal. The external causes result from factors such as market, competition, and technology upon which the company has little influence. Internal causes can be ascribed mainly to organizational and technical deficits that lead to an unnecessary number of variants at the product and part levels.

7.12.1.1 External causes of the growth of variants

A company that wants to succeed through the sale of products stands in a complex field of diverse factors. Each of these external factors has an influence on the large variation at the product level [Hic85]:

- Companies face new challenges from a continuous increase in **market complexity**. The demand for standard products stagnates or decreases. The companies' reaction is frequently to focus on individual products as a market factor. This results in the production of additional variants (more types, special equipment) with the expectation of sales stabilization and above-average margin (complexity trap).
- The **internationalization of the markets** leads in part to the most diverse country-specific demands on products and so also becomes responsible for the variations in the product program of many companies:
  - Internationally, the great diversity of views on the value of design, handling, furnishings, expectation of quality, costs, etc., are fast-changing, particularly in the consumer goods field; this leads to an increase in variety. (See Sect. 7.12.6.10a.)
  - Globalization and deregulation of the markets leads to greater **competition intensity** and dynamics and, thus, to increasing competitive difficulties. Companies see themselves up against a greater number and variety of competing products. So as not to lose favor with customers, the current state of the rival products determines the design and development of their own products. The design and development cycles become shorter. Products are revised and redesigned more frequently so as not to lose contact
  - To remain competitive, many companies see the necessity for **differentiation from the competition**, and **dependable guidance from the market and the customer**. Upon saturation of many markets, the manufacturers respond with

design and development of additional product variants in less-busy market niches in which other rivals may not be engaged (market segmentation). Companies hope for additional market coverage and competitive edge through a broadly distributed range of goods (= high assortment breadth). Thus, there is a diversification of products, particularly if the remaining conditions such as quality, price, delivery time, customer service, etc., do not significantly set them apart from the competition.
  - If adaptive designs are not strategically thought out and matched with each other, a large number of variants arise within a short time. Furthermore, usually it is not possible for a company to completely delete the predecessors of the new variants from the product program. The customer expects the availability of spare parts and performance features over the entire lifetime of the product. Above all, for products with long life spans this results in high costs for the company.
- **Technological progress** also favors the inception of variant-rich products. Not just the changing customer preferences but also the technological progress forces a continuous development of products. These requirements are partly formulated by lawmakers (laws, standards, rules, guidelines, etc.), such as in the environmental field.

  To remain competitive, a company must continually take into account the current state of technology. Since this is generally about adaptive designs, the integration of new technical concepts forces the development of new variants. Many products undergo fast revision or redesign. This goes with product life cycles becoming ever shorter. The aim is to prompt new incentives for the customers. The computer industry is a perfect example: Computers that are bought today are obsolete in few months. Growth of the variant variety results because the available solutions for the predecessor or similar products are frequently not sufficiently taken into account. Often there are product spectra that grow in time, for which no consistent, continual reduction of the part family is carried out. Customers are also often pledged to buy almost all spare parts directly from the manufacturer for years, which demands the long-term maintenance and care of the part family.
- **Different ergonomic demands** on the products due to, for example, anatomical differences in the population.
- **Climatic environmental conditions** that lead to changes in requirements (e. g., cooling, lubrication, sealed packaging).
- **Standards, rules, guidelines, and laws** which differ greatly around the world.

7.12.1.2 Internal causes of the growth of variants

Internal causes of growth in variants result from shortcomings in the company's technical or organizational milieu, which causes unnecessary variant variety at the product and part levels [Schh89, Eve92]:

- Sudden, **impulsive** management **behavior,** with repeated, uncoordinated decisions regarding products, production, etc.
- **Lack of coordination and cooperation** between company divisions

- **Communication flaws** in design and development
- **Difficulties in accessing** relevant information
- **Inadequate description** of product structure
- **Standardization** of the components that is **too late**
- **Lack of** an effective, rapid system for search of **repeat parts and similar designs**
- **Continuous improvement (CI)** processes without control strategies
- Disordered, chaotic **change processes** in all divisions
- Use of **conventional cost calculation procedures** for evaluating costs, instead of the process costing principle (Figs. 8.4-2 and 8.4-10)
- **Experience** available already is **not used**
- **Wrong use of the copying and alteration features** of CAD systems (i. e., fast design but high-cost manufacture)
- **Dominance of marketing** vis á vis design and production. According to practice, "80% of the product variants are thrown into the company by order-hungry marketing from unprofitable customers". Instead of company profit, marketing people often have sales as their target
- Offer of a customer-specific solution as a **"door opener"** for follow-up jobs

The first six points are particularly noteworthy. Due to flawed communication and standardization in the machine industry (even without influence from the market or marketing), new, unnecessary variants are always coming out. Ironically, as far as the orders are concerned, with almost similar products many parts could be identical. This is because designers (let us say A, B and C) who work on almost the same orders do not know much about each other's job, and each lays out details of the design differently. The same thing occurs during work schedule preparation. In addition, the search for same or similar parts that are already available is often very laborious. A way to avoid these variants is described in Sect. 9.4.2.

From the internal causes of the growth of variants, it can also be assumed that the different departments of a company have different requirements and tasks regarding variant management:

- Task of the **marketing department** is to understand customer wishes. Marketing people must reduce the large number of different requirements to a reasonable size.
- **Design and development** play an essential role in the variant situation in a company. Appropriate arrangement of the technical interfaces affects the flexibility of a product. By using the same components in different products, the number of individual adaptive designs can be restricted. Variants should arise as late as possible in the product realization sequence, which calls for an appropriate product concept. This is also called "postponement" (Dav95). Appropriate help should be available to support design and development by supplying the department with information during design and development, or to simplify the design process itself. In working with variants, the type of design of the products being considered plays an important role in this respect. This is because the

measures and approaches for avoiding or controlling the variants are dependent on the processing strength and weakness of the products (new design, adaptive design, or design of variants).

- **Production** should entail the least expenditure. A decrease in mounting and setup times is possible with suitable design of the parts being produced. In a manufacturing company, variant management without collaboration of the production personnel is inconceivable. This department can supply decisive input to design in developing production-oriented part families, for example.
- Similar statements are valid for **assembly**. Timely attention to information on assembly techniques leads to assembly-oriented part families.
- The **customer service department** needs an overview of the variants produced and possible distinct service guidelines for the individual variants.

7.12.2 Advantages and disadvantages of variant variety

7.12.2.1 Advantages of a high variant variety

Customers have different needs that the companies must pay attention to, for reasons of competition (change from sellers' markets to customers' markets). The products must provide benefits to the customer/user, and the **fulfillment** of increasingly **individualized customer preferences** or a customer-specific offering is of great strategic importance. With many customers, there is an aversion to a uniform response to their needs. Next to the use-benefit, there is also a value-benefit to be satisfied by the product, which is made possible by a variation apparent from the outside.

Through appropriate product differentiation, it is possible to offer products in several price ranges and thus address different groups with purchasing power.

In addition, customer loyalty is influenced by the product assortment. A broad and repeatedly updated product spectrum encourages purchase decisions both by new target audiences as well as by follow-up orders from existing customers.

7.12.2.2 Disadvantages of a high variant variety

Typically, the costs of the variant variety have less effect on the manufacturing cost portion of the total costs. They increase the overhead costs in the areas of design and development, marketing, quality assurance, logistics (including materials management), and computation. That is where the activities that used to be necessary only once for a large number of the same products (parts) now arise repeatedly for almost every product sold (see Sect. 6.3.2 and Fig. 6.3-1).

The high (and frequently not transparent) overhead costs are not easy to assign with conventional cost accounting to specific products. An assignment of the costs not based on cause has the danger that costly variants are not identified. The expenses for the realization and administration of the variants are much higher than the subsequent profit.

Since rising product and part counts are also tied to a corresponding increase in the variety of customers, suppliers, and orders, the degree of complexity climbs in all groups and departments, which means an increase in the organizational expense.

In the worst case, the company falls into the so-called **complexity trap**, a vicious cycle stemming from increasing variant variety and competitive disadvantages. By that we mean the following state of affairs:

- The starting point is a relatively simple product program largely consisting of standard products. Stagnating sales set in.
- The reaction of the companies is frequently an enlargement of the product program through niche products and special variants. The desire is an increase of revenues (sales) through increasing variant variety.
- The results are an increase in complexity and product variety and no significant increase of market coverage (cannibalization effect, market saturation, and a market volume that cannot be enlarged at will). Stated even more plainly, there is a disproportionate increase in costs at the same time as revenues are increasing at a much lower rate!
- Thus, the per-piece costs increase for the entire product set without gaining additional revenues; there is a decrease in the company's competitive capacity.
- The problem in this case is that traditional cost accounting systems often miss the overall view of the actual costs (overhead costing); these burden the standard products too much and the "exotic" ones too little.
- A lack of transparency of cause and effect leads to treating the symptoms, not the causes. Many companies attempt to balance this by creating additional variants. The cycle goes on.

Therefore, in many companies the exact opposite situation develops than what was initially desired or anticipated. Consequently, the variant variety presents a big problem.

In the following, we discuss what effects this has on the costs, particularly. The manufacturing costs. The other affected costs are pointed out as well. Reduction of these costs was shown in Sect. 6.3 (Fig. 6.3-4). The variant variety not only adds costs in all fields, but also leads to longer processing times. Therefore, a decrease in the variant variety is important as a starting point for reducing delivery times.

The following processes are accounted for, not in the manufacturing costs, rather in the total costs:

- The **launch costs for new products** and thus for **new parts,** in design and development (see Sect. 7.5.1), production, assembly, sales, marketing, service, and administration. Thus, costs of at least several thousand dollars result from the introduction and administration of a new part and changes in an existing part. Figure 6.3-1 provides information about that. From experience, these costs, with high work division in a company, can be higher by a factor of 5 to 10 [Pok74].
- **Advance payment costs** in design and development, marketing, purchasing, administration, production for the **standardization of products** (for example,

the measures for the variant reduction in Sect. 7.12.4). For the **rationalization of processes** (e. g., organization changes, computer introduction, and recurring costs) as well as costs of **personnel training** and continuing education (e. g., for the introduction of new methods and aids). The costs are increased by these measures in these fields for a short period but are lowered after successful introduction. Guidance from the management that looks only at the short-term and the corresponding cost control therefore prevent such measures. In the same way, the real cost reduction, calculated for variant reduction for products in size ranges and modular designs, for a specific product through the negotiated overhead surcharge over all products is always too low. That is why the quoted price is usually not reduced much. Thus, it reduces the buying incentive in the market, and products in size ranges and modular designs are no more successful than customer-specific special solutions [Jes96]. In fact, companies would then also give up on the standardization efforts and costs. This mechanism is frequently observed in practice.

In the contrast to that, the **manufacturing costs** are **reduced** if the number of variants is limited and thus greater numbers of the same part and assembly are produced (see Sect. 7.5).

This comes about through the reduction in the following:

- **Material direct costs,** *MtDC,* due to quantity discounts (Fig. 7.9-11) which act to lower the purchase prices.
- **Material overhead costs,** *MtOC*, due to reduction in introductory costs and the one-time costs (for example, for purchase order processing, receiving control, ordering, warehouse management, and internal transportation).
- **Production labor costs,** *PLC*, due to reduction in setup costs, more efficient processes, and the training effect.
- **Production overhead costs,** *POC*, due to reduction in introductory costs (for example, for the creation of work schedules, test plans, and NC programming), resulting from reduction in one-time costs (e. g., arrangements for lot production, reduced control processes), and due to more efficient processes. This holds for part production as well as for assembly (the reduction of the overhead costs generally takes effect only in the next accounting period, when the overhead surcharge rates are determined again (Sect. 8.4.2)).

The **potential** for manufacturing costs that may be affected by a reduction in variants is given by Caesar [Cae91] as 15–20%, as shown in **Fig. 7.12-3**. The costs of design and development are included in this figure.

Figure 2.3-3 in Chap. 2 showed with the example of gear trains for agricultural machines, how significantly manufacturing costs can be reduced by reduction in part numbers. In the course of 30 years, the part count in the train decreased in stages by about 30%, through concept changes and appropriate part design. The manufacturing costs fell to one-third, adjusted for inflation. Naturally, there were other factors such as production and assembly rationalization.

To see the influence of the quantities N_{tot} or lot size N on the costs, the reader is referred to Figs. 7.5-3 through 7.5-5, 7.7-1, 7.7-5, and 7.7-6.

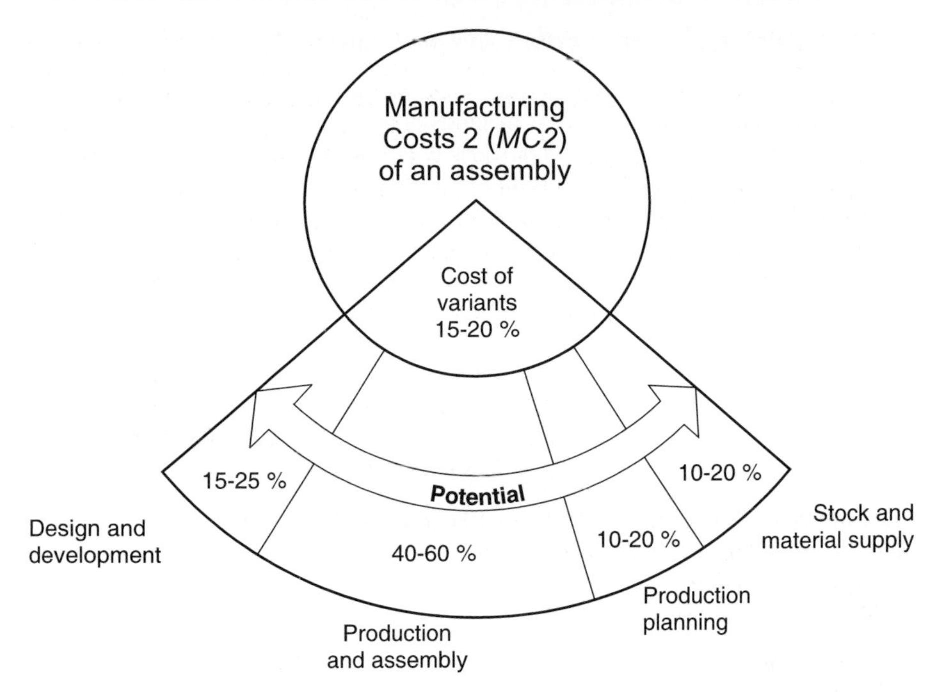

Fig. 7.12-3. Potential for influencing Manufacturing Costs 2 ($MC2 = MC + DDC$; see Sect. 8.4.2) through variant reduction [Cae91]

In the study of a diesel engine manufacturer, it was found that to prepare a new drawing the necessary activities in other departments cost six times more (Fig. 6.3-1). Depending on the degree of the division of labor in a company, a drawing change can cost even more. It is a matter exclusively of administrative expenses. In [Mau01], up to 15 000 $ per part per year are mentioned as “administration costs” in the automobile parts supplier industry. That does not include production and assembly expense.

It is like walking a tight rope: A sufficiently large number of variants are needed to do justice to the different customer wishes. At the same time, the cost increase due to the increasing number of variants (for example, in design and development, part management, etc.) should not be so high that prices become unattractive to the customer. At this point, variant management comes into play.

7.12.3 Steps in analyzing the variant situation

In establishing the correct number of variants, the market requirements and the product range offered must be examined. There are many methods available for providing an overview over the current variant situation, of which some are explained here.

7.12.3.1 Analysis of the product and part variety

Ascertaining the existing situation is the place to start in exploring the **causes for variant and part variety**. Different methods of analysis may be used for this, and depending on the results, remedial steps are derived. These are addressed in greater detail in Sects. 7.12.4 to 7.12.6.

In eliminating the weak points of an (unsatisfactory) initial state, the following is generally logical: The more cause-based, careful and precise the analysis of the existing situation and its conditions is, the more effectively are the measures for improvement recognized. In this respect, many remedial actions of the following sections will be addressed here.

The analysis must relate both to the product program as a whole regarding the products and their part variety. Because of mutual networking and connections, analyzing only in the area of design and development is not enough; most of the company's parts must be included, at least for this purpose.

a) Analysis of the variant variety in the product program (e. g., in the team made up of management, marketing, design and development, cost control, materials, and production)

To get an overview, it is desirable to look into the following for the company as a whole, as well as for the appropriate customer, and as a time history over the last few years:

- **Sales, profit, number of products sold** in specific countries, for specific customers. For this, the customer-specific configurations must be kept in view. It is advisable to take advantage of the remedies mentioned under c), in particular the visualization of "dry" tables and numbers (Sect. 4.6). Here we mention only **Pareto analysis** (a graphical ranking); see Fig. 4.6-4.
- In a similar manner, the company's **own product program** can be compared in a **portfolio diagram** with that of the **competition,** where we usually deal with more reliable estimates (Fig. 4.6-8).

During the analysis of product and part variety, the following questions can be of help:

- Which of the offered variants are the "exotics"?
- Which system components are suitable for use extending beyond one product, for example as part of a modular design?
- How are the individual components produced?
- Can certain process steps be used for several components?
- Can part families be created?
- Which costs are caused by the variants?
- ...

The answers to such questions are important for deciding on the future product range. It is important that an interdisciplinary team provide these answers so that, as far as possible, all the factors are considered, and to uncover possible target conflicts.

Increasingly, so-called product configurators are found on the Internet pages of automobile manufacturers. These configurators offer the possibility of putting together the desired vehicle configuration and to display a likeness of the vehicle to the customer. The customer configurator is a very valuable information source for the car manufacturer. The marketing department gets a survey of the preferred variants without large expense (customer inquiry, market analysis, etc.), and can consider this in the design and development of a new vehicle.

b) Analysis of variant variety for specific types of products

In a team of similar composition as in a), the following can be explored:

- The extent of **part types, part quantities,** and their modifications over the last several years (time history of the growth of item numbers; see Fig. 6.3-2). This can be the motivation to reduce the part count, using integral design or other concepts, as shown in Fig. 7.12-14.
- Furthermore, the **degree of standardization** of a current product should be examined and compared with earlier or similar products. By that we understand, broadly speaking, the scale of the **product standard**. That is, to what extent the product program is standardized regarding types of performance and configurations, or to what extent are, for example, part families, size range of products, and modular design put to use. Also, parts of a product can be designed in this way; the rest is then perhaps customer-specific. **Figure 7.12-4** shows the important advantages, along with some disadvantages of the product standard.

What is a product standard?

Product standard is the binding commitment for a manufacturer, of the product program, of the performance data, the types of configurations and the shape of a product range or the products.
or: a product standard is the factory standard based on the product.

Product standard establishes:

Product program, performance data, types of configurations, part families, size ranges, modular design,
and thus: materials, sizes, tolerances, processing, purchase parts.

Advantages of the product standard

1. **Less project planning, bid and marketing work**, since special solutions are replaced by standard solutions ; faster customer service.
2. **Better technical brochures and price documents.**
3. **Less design work,** since designs already exist for the most part.
4. **Fewer new parts** and thus **lower introductory costs**.
5. **Less work in production planning**, since production documents already exist.
6. **Lower purchase and storage costs.** Ordering procedure exists. Prices are known. Larger lots can be ordered, with more discounts; fewer different parts / groups need to be ordered. The smaller warehouse is cheaper and can be checked more easily; less dead stock remains.
7. **Lower production costs**, since there are larger lots, production and assembly experience; better production control possible with known parts, fewer special machines and fixtures necessary, less wastage.
8. **Lower delivery time**, since lower throughput time in all departments; parts can be produced from stock and fewer mix-ups and mistakes occur.
9. **Lower complaint rates**, since behavior of the product is known, it is realistically planned. Less design and production mistakes occur, and there is better operation and maintenance knowledge (and guidelines). There are clear test conditions for external purchases.
10. **Lower spare part costs and spare parts delivery time.**
11. **Lower manufacturing costs and total costs.**

Disadvantages of the product standard

1. Product does **not** fulfill requirements **optimally.**
2. Manufacturer does **not** react to special customer wishes with **flexibility,** which may encourage **competition** from specialty shops.
3. Customers band together and produce their **own special solutions.**

Fig. 7.12-4. Product standard (see also Fig. 6.3-4)

In a **narrower sense,** the parts of a product are classified thus according to **Fig. 7.12-5**, and their number per class entered (part definitions in Sect. 7.12.4.1).

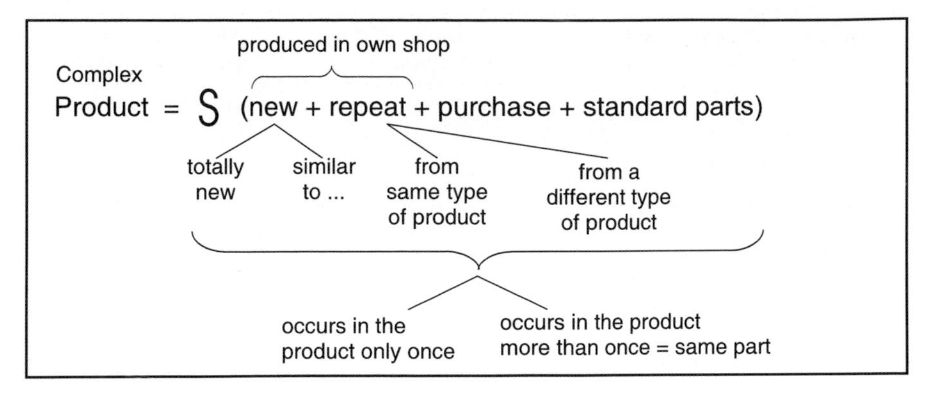

Fig. 7.12-5. Part categories of a complex product (assembly, machine)

- The degree of standardization is calculated according to **Fig. 7.12-6**. The greater the standardization of a product is, the fewer (customer-specific) new parts it contains, which bring about especially high costs (see Sect. 7.12.2.2). For a design revision, it has proved fruitful in practice, to define a **target degree of standardization** based on a forerunner product.

	Previous degree of standardization for known product	Degree of standardization for the new product
Number of different parts (according to category)	71 = 100 %	64 = 100 %
• New parts • Repeat parts from, e.g., similar products • Same parts • Purchase parts • Standard parts (DIN, or factory standard)	29 10 } 2 } 42 5 } 25 }	17 11 } 3 } 47 10 } 23 }
Degree of standardization	42/71 x 100 = 59 %	47/64 x 100 = 73 %

Fig. 7.12-6. Example of defining the target degree of standardization in development of a new product

The **variant tree (Fig. 7.12-7)** gives direct insight into the number of possible and requested variants for a type of product. The figure is self-explanatory; it uses the simple example of an automobile instrument panel. In other cases, the add-on parts are arranged according to their assembly sequence and the pertinent variant variety is presented for every assembly procedure. A basis or carrier

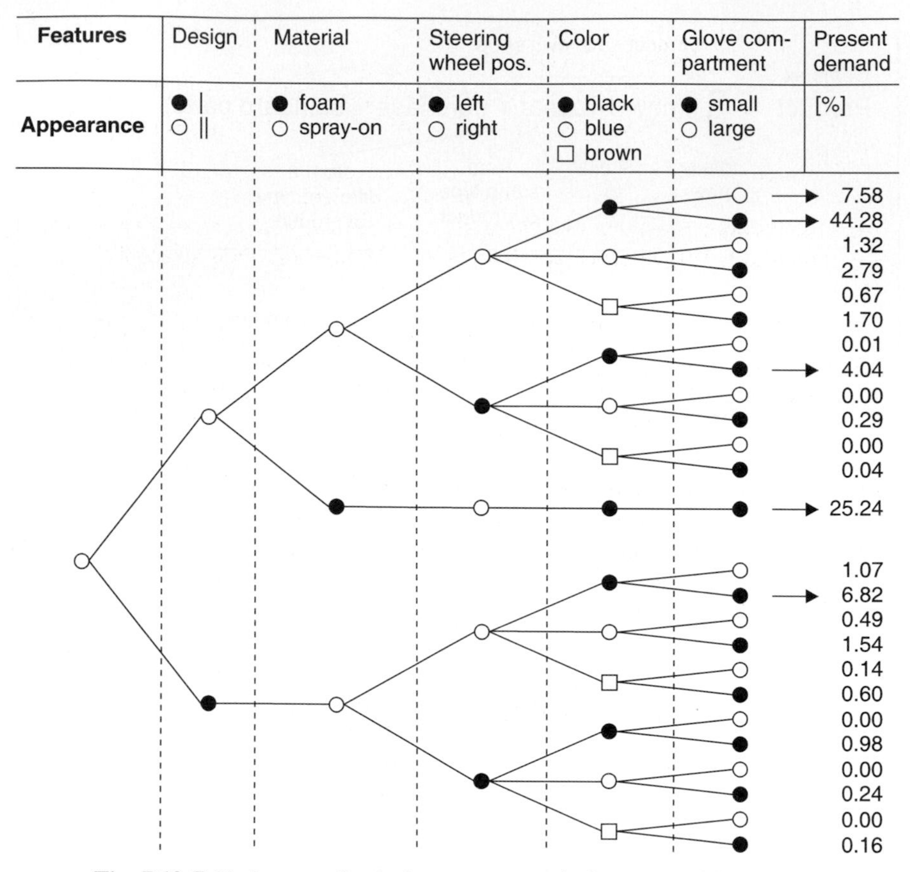

Fig. 7.12-7. Variant tree for the instrument panel of an automobile [Ros96]

component starts this sequence. The add-on parts are mounted on the basis component, and thereby product variants arise. For comparison, it is advisable to set up a variant tree of a competing product.

The variant tree loses clarity with increasing part count or the possible combinations. We can imagine that a depiction of all combinations of an automobile is not that easy to grasp, so a reasonable subdivision of the variant tree offers a better alternative. Along with a Pareto analysis, we can find which products are especially in demand with the customers and which products are "exotics". This information should accordingly be considered in the development of follow-up products. Possible variants can be evaluated with regard to production costs by software systems and, thus, reasonable variant designs may be realized.

For variant reduction, the variants with low sales – provided the customers tolerate it – should be deleted from the sales program [Ros96] or integrated into other components [Schh89, Eve92].

Similarly, one can compare the variant tree of the existing state to that of the desired variant tree. A Pareto analysis is also of interest here:

88% of all designs are realized with 20% of the variants (arrows in the figure). Moreover, 20% of the variants are not sold at all.

Cost reduction measures can be suitably analyzed by means of the variant tree, and considered and introduced jointly by development and assembly departments.

➔ **Variants** should be so designed that they are realized as **late in the production process** as possible, near the end of the final assembly.

A common example is that of the doll shape shown in **Fig. 7.12-8**. With the same die costs for a chocolate Santa Claus and an Easter bunny, the decision is finalized just before the packaging by the appropriate covering foil, about which product it will be. This is true also when a given product is offered under different brand names.

The advantage of late variant generation is that production and assembly are simplified and made less fault-prone. The products not yet split into variants can be produced and assembled in large quantities. The ideal variant tree is very slim in the beginning and divides into many small branches only near the end.

Fig. 7.12-8. The doll figure – an example of late variant generation [Hic86]

- **Material and purchase part variants** can be analyzed in the team with purchasing, materials, logistics, design and development, cost control and production.

  For example, Pareto analysis can be set up on purchases of **types of material** and **finished parts**. The question to ask is: "Why can C parts and C materials not be deleted in favor of A or B parts (e. g., why not use the same screws, bearings, seals, etc., as far as possible)?"

  Furthermore, in the same way, the **suppliers** should be evaluated by Pareto analysis according to sales per supplier, price, quality, and on-time delivery. The aim is to reduce the number of suppliers in favor of preferred suppliers.

An **analysis of the inventory** (handling frequency, value of the stored parts, necessary storage area) has a similar effect. According to [Romm93], it is desirable that expensive parts be stocked the least but ordered more often. Cheap, frequently needed parts are more seldom in stock in larger quantities, or kept as bulk goods without specific orders.

- From production and assembly viewpoints, parts that are similar in form and processing should be gathered into **production families.** These can be parts with entirely different functions, as shown in **Fig. 7.12-9**. Parts are said to be in a production family if they can be produced one after another on the same machine tools without new setups and without new tools and fixtures. With that, an "apparent lot size" is formed which has almost the same effect on the respective production operations as a real lot.

Part drawing	Name	Shape-based key number	Production-based key number
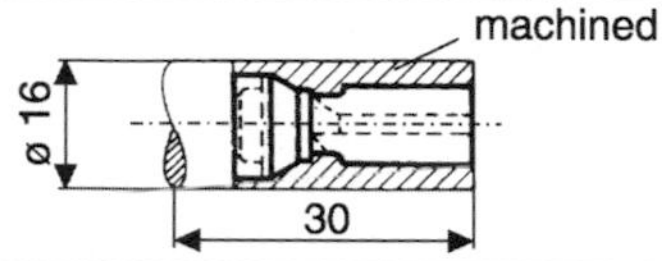 machined; ø 16; 30	Valve	448200508	14104072
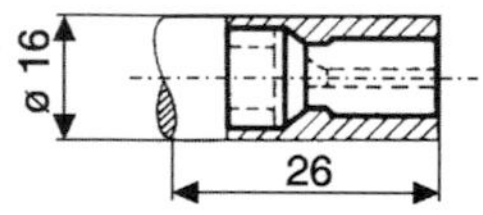 ø 16; 26	Valve	348200508	14104072
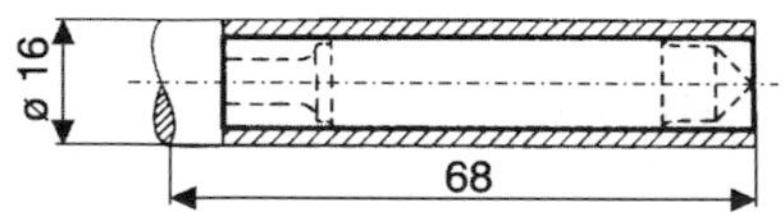 ø 16; 68	Needle	270000501	14102072
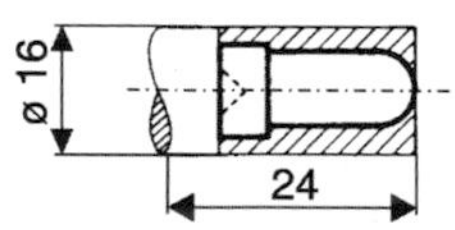 ø 16; 24	Pin	194000501	14104022
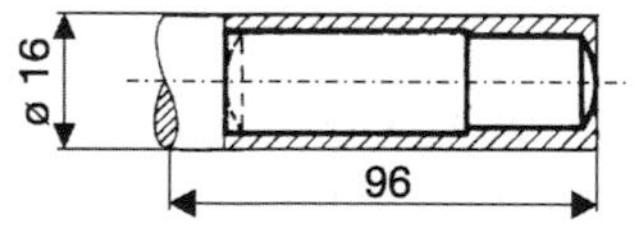 ø 16; 96	Pin	290000301	14104022
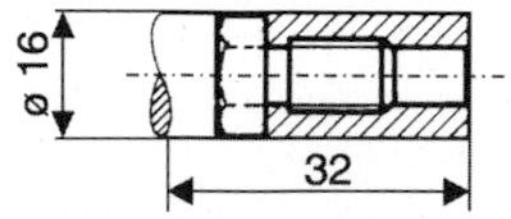 ø 16; 32	Bolt	302220608	14104091

Fig. 7.12-9. Production family for a turret lathe with stock feed; numbers are specific to a company (from R. Wagner, G. Junginger)

- All skilled workers make up production families from self-interest, provided they are free to schedule jobs on the machine to a certain degree. They work on **the** parts, one right after the other, for which reduces the need to retool.

  Furthermore, parts with similar process sequences and the same work schedules can be identified and can therefore be produced according to **group technology**. Group technology means the local and organizational collection of equipment into machine groups on which the parts (a production family) can be processed completely [Hab96].

c) Measures for the analysis

In order to achieve the desired transparency with large quantities and variety of data for group discussion, help is necessary from procedural grounds and for visualization. The data should be available in digital form to sort and present in a simple manner.

Such tools have already been mentioned earlier. Also, depending on the software, they can be directly selected and displayed. The literature describes the well known "seven statistical tools" [Ima93, DIN94] or the "seven new management tools" [Bos91, Kin89], as follows:

Examples of support items (Sect. 4.5):

- Ranking series, Pareto analysis (e. g., Fig. 4.6-4).
- Histograms or bar graphs (e. g., Fig. 7.9-10).
- Correlation diagrams (two or three arguments in 2- or 3-axis representation (e. g., Fig. 9.3-4).
- Portfolio diagram (e. g., Fig. 4.6-8).
- Variant tree (e. g., Fig. 7.12-7).
- Part classification, using keys, numbering systems and feature tables [Bei77, Pfl79].
- Statistical evaluations (e. g., computerized search system, cluster analysis), usual computer tools (e. g., Sect. 9.3.4.1, Fig. 9.3-4).

The following explains the application of the last two items:

- **Parts classification using decimal numbers** [Opi66]:

  Rehm [Reh81] reports on a scheme involving only three digits with which 36 000 parts could be quickly classified. A **repeat parts design catalog (Fig. 7.12-10)** begins with standardized part types, for example a centrifugal pump (VDMA Sheet 24 250). It is further subdivided into shape (form) types, and again split up into shape details (e. g., sizes, materials), similar to a **feature table**.

  The latter are assigned drawing numbers by a computer data list. Existing drawings are used; those not available are created in the system's framework in the course of time. The disadvantage of this hierarchical classification is that future developments should be considered during the design of the system. This can only be inadequately done, and results in a relatively inflexible system. Moreover, a part can be categorized only at one place, which is not desirable sometimes (e. g., a rotating part with milled faces). **Relational databases** [Mül91] are significantly more advantageous.

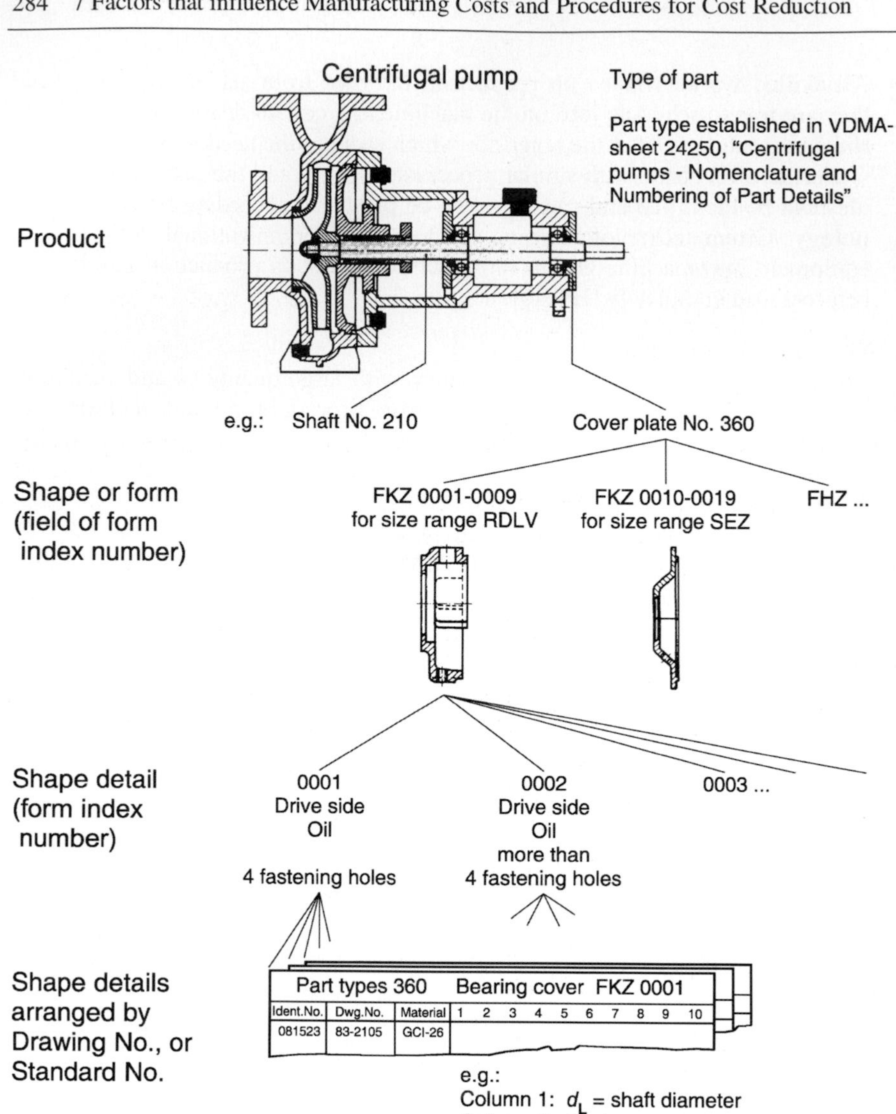

Fig. 7.12-10. Layout of a repeat part catalog for pump parts (from KSB Bremen)

- **Computer-based search system with a catalog of terms (Thesaurus)**: With this flexible system, terms from an open-ended catalog are assigned to the individual parts (e. g., rotating part, covers, 4-holes, GCI). The computer searches for the parts that correspond to a logical combination of search terms [Mül94]. If more terms are combined, for example, with "AND", the resulting quantity is smaller (see also Sect. 7.12.4.1d).

- **Cluster analysis:**
  The computer forms a hierarchy of parts that are most similar, on the basis of quantitative features (e. g., component parameters such as diameter or length [Bei77, Pfl79]) This relatively flexible system is best combined with the one mentioned above. One searches, for example, first with descriptors and forms a similarity hierarchy from the quantity found.
- **Full-text search:**
  Today, full-text search in one's own data resources and on the Internet with the appropriate search engines also shows promise.

7.12.3.2 Interface analysis

An important aspect in variant analysis, and later during the separation of assemblies and modules, are the **interfaces.** They must be defined in a way that is familiar and unambiguous. During interface analysis, we must distinguish between organizational and technical interfaces. In addition, there are other interfaces that can be important, such as legal submission (liability assignment, etc.).

The **organizational interfaces** include the flow of information between the individual subprocesses and the assigned responsibility. The parameters time and quality play an exceptional role in this context. If the quality of information is poor, then later, often with considerable expense, improvements and/or changes must be made. If information is not timely, there are significant failures and breakdowns in the process. These troubles, as well as the changes, lead to delays in schedules, and to cost increases.

Within the setting of an organizational interface analysis, the process sequences should be analyzed. The cooperation between several departments and/or between the company and its suppliers often requires optimization. The organizational interfaces and their arrangement, relevant for the process participants, should be known to them to avoid negative outcomes such as redundant work. Often one also finds starting points for supporting the workflow by information systems during the process analysis.

Technical interfaces deal with the material, energy, and information aspects. Special attention is also paid to the geometrical interfaces and the jointing at the interfaces.

- **Geometrical interfaces** constitute important task contents in the machine and transportation industries. The available part volumes in the product are managed with the help of space management. Because of the generally high interdependencies, this is a very difficult topic with some products, since it is not always obvious which spatial consequences can lead to commitments or changes in specific components or assemblies. Who can foresee that the steering will be affected [Ste98] by modifications of the air-conditioning equipment in a vehicle? What are the effects of movements, deformations or tolerances? DMU (Digital Mock Up – the digital product model) offers increasing support in this regard.
- **Material interfaces** are always important, if material conversion (movement, shape or phase change, etc.) takes place in the system. Moreover, all auxiliary

materials are to be considered. Which material is transferred at which place, with which time behavior and which properties (temperature, pressure, etc.)? Round stock, turned bolts as well as the chips in the lathe are examples of the material passing through a lathe. An example of an important auxiliary material is the coolant fluid.

- **Energy interfaces** involve quantities with mechanical (force, moment ...), thermodynamic (heat flow, temperature, etc.) or electrical (current, voltage, etc.) parameters.
- **Signal interfaces** comprise communication protocols, voltage level of sensors, etc.

The technical interfaces are of absolute importance for design and development. For the interdisciplinary development teams alone, knowledge about the structure of the technical interfaces is especially important. Only thus can it be assured that the individual subsystems can be properly integrated into the total product. This avoids a later extensive revision of the subsystems, which can increase the part count. This is particularly true in conceptualizing a flexible modular product design, where the aim is to generate several variants from a defined number of modules. Interfaces must be ascertained and modified in a suitable manner.

7.12.4 Decreasing the product and part variety

A **decrease in the product variety** is a strategic procedure of management in cooperation with marketing, sales, design and development, production, and cost control. It can be carried out after the analysis of the product program (Sect. 7.12.3.1). In contrast to that, the **decrease in the part variety** is handled subsequently. This involves a decrease of the part count per product and the **number of the different parts**.

The latter is important particularly in single-unit and limited-lot production, since the introductory costs of different parts are important. On the other hand, in series production the absolute number of parts is of greater importancesince introductory costs become small due to the large quantities. However, the costs for joint surfaces, joints, and assembly accrue in full measure for every product made.

To emphasize the importance of the part count, some **representative values** from the practice are mentioned here. An automobile has 10 000 to 20 000 parts; a large diesel engine, 1 500 to 2 000 parts; a paper machine, 120 000 parts; a jumbo airplane as many as 3 million parts.

The most economical way is to avoid the origin of unnecessary variants right at the beginning of the design. If it is desired to reduce variants afterwards, then that makes sense only with frequently asked for (active) variants (parts). With uncommon (passive) variants, the reduction costs can exceed the savings. Process costs (Sect. 8.4.6) should be estimated before and after the reduction. Kohlhase [Koh98] shows the procedure for this in the practice.

We can divide the **measures for decreasing the part variety** into the following, which partly overlap (an outline of the strategies and measures for part count reduction is shown in Fig. A8 in the Appendix, "Guidelines for cost reduction"). They are reviewed subsequently.

- Increasing the degree of part standardization (Sect. 7.12.4.1)
  Use of same parts and repeat parts
- Use of purchase parts per supplier specification (purchase parts$_S$), and per customer specification (purchase parts$_C$)
  Forming design part families (Sect. 7.12.4.2)
  Prefer integral design (Sect. 7.12.4.3)
- Using measures for decreasing the setup costs (Sect. 7.12.4.4)
- Introducing organizational measures (Sect. 7.12.4.5)
- Designing products in size ranges (Sect. 7.12.5), and modular designs (Sect. 7.12.6); these two measures are usually the most effective [Fra87]
- Modularization of products (Sect. 7.12.6.4)
- Use of platforms (Sect. 7.12.6.5)

The use of basic solutions (Sect. 7.12.6.6) and the constructive parameterization (Sect. 7.12.6.7) are also effective, not directly on products, rather in the sense of decreasing variants.

7.12.4.1 Standardization

According to Kienzle, standardization means "the one-time solution of a repetitive technical or organizational procedure, with known optimal means of the state of the technology, at the time of preparation of the standard, through all interested parties. It is always a time-bounded technical and economic optimization".

Standardization exists at different levels:

- Industry-wide national and international standards (DIN, ISO)
- In-house standards (factorystandards)
- Generally usable solution catalogs and other rules, as well as systematical and uniform knowledge representation

This book is not about the general and obvious observation and application of standards, which also has economic advantages. Rather, it deals with the cost savings that are achieved by the use of standard and purchase parts and through the standardization of products.

a) Use of standard and purchase parts

Advantages of using standard and purchase parts:

- **Standard elements are tried and tested**, the interchangeability within the respective standard is assured.
- They correspond (as far as current) to the **state of the art** and are usually economical and **available** from stock **at short notice**.
- **Tools** for use with standardized parts are also partly standardized and need not be expensive to produce (e. g., wrench).

- Within the framework of factory standards, internally developed elements that have proved to be reliable and usable for several products can be additionally committed and standardized.
- Standard elements are developed independent of specific customers, and produced predominantly for the open market.

Disadvantages of using standard and purchase parts:

- The **number** of the standardized design entities, parts, assemblies, or machines **becomes hard to manage**. For many companies, this leads to their own choice of a subset of the standard parts.
- In the course of globalization, the manufacturers are confronted with additional regional and national standards. These require appropriate considerations by the manufacturers and lead to an increase in the number of variants. Thus vehicles, for example, must be adapted to the different markets. This topic affects machine and plant manufacturers, such as in electrical equipment and safety technology.
- **Standards change,** and so lead to the necessity for change in broad areas. This requires considerable expense in order to guarantee that the information being provided to the designers is current. In addition, designs may have to be adapted to correspond to new standards.
- Standards can hinder technical progress, since known solutions are resorted to repeatedly and a systematical solution search seldom occurs.
- On an individual basis, standard parts are often technically suboptimal. Thus, for example, a screw with the nominal size M16 must be used instead of the calculated size M14.3.

Standardization is also possible through, or in cooperation with suppliers of **purchase parts.** Birkhofer [Bir93] describes the increasing online access that includes its direct incorporation into CAD. Purchase parts (and assemblies) (Fig. 7.12-5) can also be standardized parts and are usually more economical, since they are manufactured in large quantities.

Here we differentiate between:

- Purchase parts from supplier specification (purchase parts_S)
- Purchase parts from customer specification (purchase parts_C)

Use of purchase parts according to supplier specifications
(purchase parts_S)
Purchase parts are parts, assemblies, or products that are bought from suppliers. Purchase parts_S are standardized by the supplier and often provide the same advantages as parts that conform to national and international standards. Purchase parts_S are developed largely independently of customers and are produced primarily independently of specific customers (i. e., for the open market). For different orders, there can be also other product variants.

Advantages

- The manufacturer tests the purchase parts$_S$ for specific conditions of use.
- The interchangeability is guaranteed for a certain length of time (according to contract).
- They correspond (as far as current) to the state of the art.
- They are usually economical and available from stock at a short notice.
- Manufacturers change their own standards more easily, according to new findings; at least that is easier than a change in an international standard.
- Purchase parts$_S$ facilitate the job of product design to a great extent, since detailed solutions for subsystems can be bought rather than developed.

Disadvantages

- On an individual basis, purchase parts$_S$ are often technically suboptimal. Thus, for example, a pump from the supplier's catalog with a flow rate of 140 l/min must be used instead of a pump with the desired flow rate of 127 l/min.
- This is true for service and the product users only conditionally. Whoever has searched for a substitute for seemingly a standard part (even only from one manufacturer of, for example, furniture covers) knows the problems.
- If a manufacturers use purchase parts$_S$, they come to depend on the supplier over whom there is no control. The manufacturers may therefore be forced to make design changes. For example, an electronics supplier might inform the manufacturer that a specific part (e. g., a resistor in the previous design) is available only until the end of the current quarter.

Use of purchase parts from customer specifications (purchase parts$_C$)
Purchase parts$_C$, from customer specification, are standardized for the customer or through the customer's influence, and then produced. For different orders there can be also other product variants. Here the particular competence of the supplier becomes important.

The **advantages** with the purchase parts per customer specification are the same as for purchase parts$_S$. However, one disadvantage is no longer there. The customers do not have to buy technically sub-optimal components, since they can define the design.

However, there are the following disadvantages for purchase parts$_C$:

- The core competence for the purchase system components rests with the supplier, thus complicating any change in suppliers.
- Coordinating the technical interfaces beyond company boundaries is more expensive.

- The variant and version control is often difficult due to redundant information. An example is an assembly used in automobiles: A car manufacturer lets the supplier develop the engine control assemblies. This means a continual enhancement of the software package regarding additional functions. Since the purchaser implements even small changes in the software before this is integrated into the assemblies, different software packages exist with the manufacturer and supplier. This requires a large expenditure for coordination in order to describe the functional scope of the respective software packages.

b) Higher degree of part standardization

What is meant by the **standardization degree** was already explained above (in Sect. 7.12.4.1). With a constructive revision or new design, it is a question of holding the number of customer-specific new parts as low as possible and to use, as far as possible, the same parts in the company or obtained from outside. Thus the overhead costs as well as the direct costs (in the sense of Fig. 6.3-4) are kept small. Drawing on Fig. 7.12-5, we provide the following definitions[11]:

➔ Definition: A **new part** is a part newly designed for the given product (internal or external design; "totally new, or similar to ...").

➔ Definition: A **repeat part** is a part that was already used in other products (the same or different product type). Modular design parts are essentially repeat parts, insofar as they are not purchase or standardized parts. A repeat part can also be a part of the company's factory standard.

Repeat parts are defined by part families, products in size ranges or modular designs. According to [Pät77] about 15% of all design parts (without standard parts) were suitable as repeat parts; according to [Wie72], it is between 3 and 6% A procedure for realizing repeat parts is indicated in both publications and in [Bei77]. Changes can cause difficulties if repeat parts occur in products of different departments (Sect. 6.2.2).

➔ Definition: A **same part** is a part that occurs repeatedly in a product. Examples are same housing covers or identical levers or bearings.

In addition, for single-unit production, the number of each part produced can be increased so that the one-time costs (the sum of introductory costs and one-time or setup costs) per part decrease. This is particularly important for small parts.

---

[11] Inside the company, a more precise definition is needed, depending on existing documents; for example, the distinctions between repeat part, factory-standard part, and standardized part from an external standard).

An example for the definition of the **existing** and the **target degree** of **standardization** is depicted in Fig. 7.12-6.

c) A malus[12] handicap

The introductory costs of new parts are not usually accounted for by calculation (Sect. 8.4.6). Therefore, these costs must be taken into account for cost reduction procedures in another way for every new part, for example, by a **malus** [Wip81] (Fig. 6.3-1). In **Fig. 7.12-11** the manufacturing costs of an old part and a new part are plotted against the quantity produced [Mei77]. By the usual cost calculation, the first new part is already produced more economically than the old one ($MC_{new} < MC_{old}$) since the introductory costs are assumed to be distributed (Sect. 7.5.1) over **none** or on **all** product parts. If these are used as a malus, in the example more than three of the new part must be produced (N_{min}) before it becomes economical.[13] The notion of the malus makes sense for all cost reduction activities. In machine manufacturing companies, a malus handicap can amount to around 2 500 $.

A **correction factor** works in the same way for part cost calculation as a malus, which reduces the overhead surcharges with more frequent use of a part: Such parts then become more economical, the others more expensive.

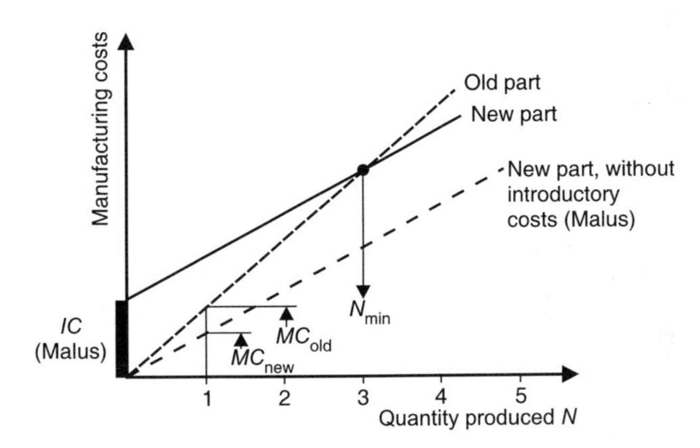

Fig. 7.12-11. Smallest produced quantity N_{min}, for which a new part, including the introductory costs *IC*, becomes more economical than an old part

d) Search systems for repeat and similar parts

A most effective means of achieving the reuse of parts is a simple and fast-acting parts or assembly search system. It must be easier for a designer to search for

[12] Inside the company, a more precise definition is needed, depending on existing documents; for example, the distinctions between repeat part, factory-standard part, and standardized part from an external standard.

[13] For the degree of standardization, (Fig. 7.12-6) the term repeat part is employed in a limited sense: Purchase parts and standard parts are counted separately.

a part and rapidly find a suitable one, rather than designing it anew (see also Sect. 7.12.3c).

Figure 7.12-10 shows a **conventionally** designed system for pump parts that was later computerized. The search system is based on an object features table [Pfl79].

Depending on the CAD system, an effective part administration is provided by **EDM** (Engineering Data Management). Essentially, the parts that are the most similar can be found rapidly, based on the desired part, and these are rendered on the screen as a CAD drawing [Mül91, Koh98].

If work schedules and parts are documented in a **database**, they can be searched according to all available terms; see Sect. 9.4.2, System XKIS, CAD System in Pro-Engineer; it can be searched for features (e. g., bearings, gears) [Rei96a].

An additional very effective measure is the search with **freely entered terms** that are input from the master data field of drawings and/or parts lists. In this case, the software also contains an automatically usable catalog for synonymous terms, as, for example, "roll = bolt = pin = cylinder". The more terms that are entered, the smaller the number of variants that is output. The probability of finding a part similar to the objective part is increased, as is also the probability of not finding any part at all [Mül94]. Not only can parts be found this way, but also assemblies, work schedules, fixtures, and tools [Mül94, Koh98].

7.12.4.2 Forming design part families

a) Design part families, unlike the production families (Sect. 7.12.3 b), are defined as those in which the parts fulfill essentially the same **function**.

b) As an **example,** parts that are geometrically similar, semi-similar, or only roughly similar in shape are depicted in **Fig. 7.12-12**. Parts from a design part family are searched for with a catalog or search system. The search is aimed at:

- existing but not standardized parts, to use these "as-is" (formation of repeat parts)
- similar parts, to be altered
- standardized parts (locating drawings according to types and/or features)
- partially standardized parts (templates and/or macros for entering parameters as needed for each order (e. g., measurements)

Sometimes design part families represent also production-technical part families or production families (cf. Figs. 7.12-12a to c with Fig. 7.12-9).

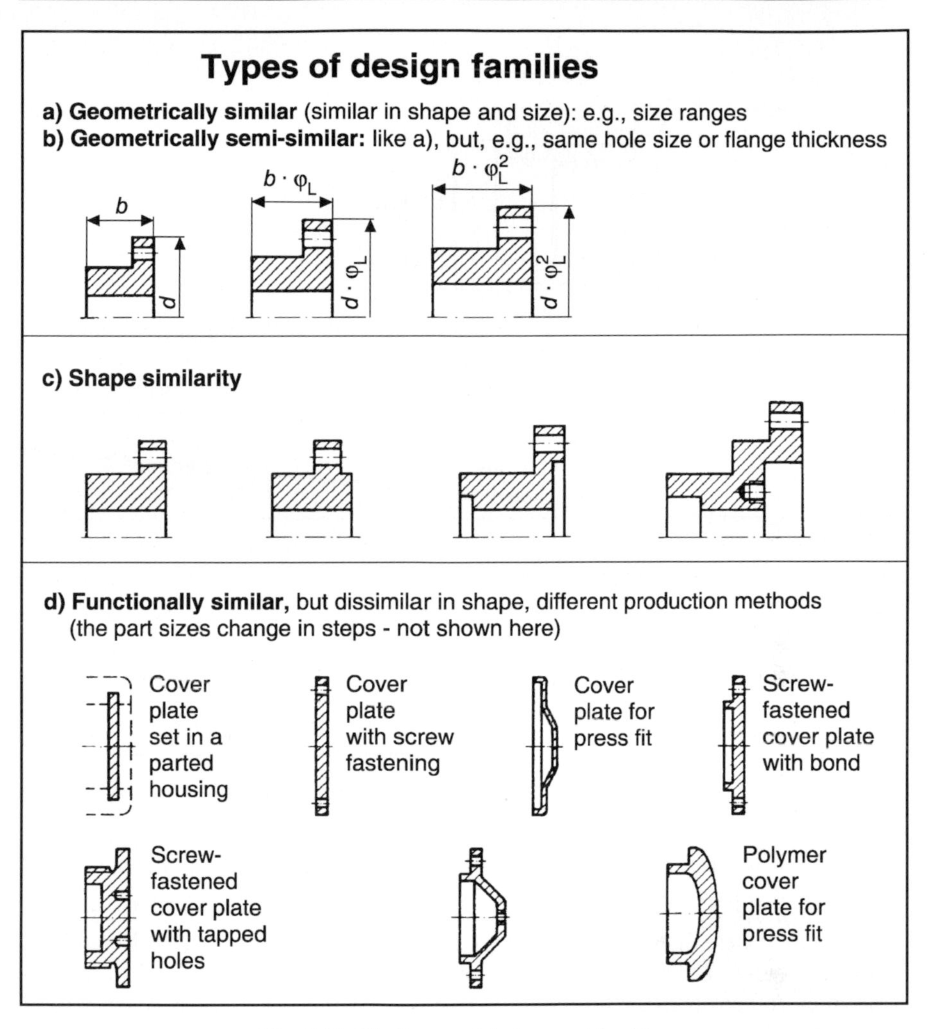

Fig. 7.12-12. Types of design part families

c) The **formation of design part families** comes about **mostly later,** due to a multiplicity of generated designs. It is seldom planned beforehand, which is in fact more desirable. The following example shows the **saving potential** for in-house standardization and part family creation. In **Fig. 7.12-13**, the variant variety of the output coupling flange of an truck transmission was reduced to only two sizes, from hundreds before the rationalization. Before the investigation, there were six main design sizes with 416 dimensions. After the investigation, there were only seven left, when only two sizes of the length proved to be necessary. All other sizes arose due to uncoordinated dimensioning by different designers. Here a communication or an efficient part search system was missing (see Sect.7.12.3 b). The manufacturing cost reduction on as few as the half the products amounted to approximately 1 M$ per year, in today's costs.

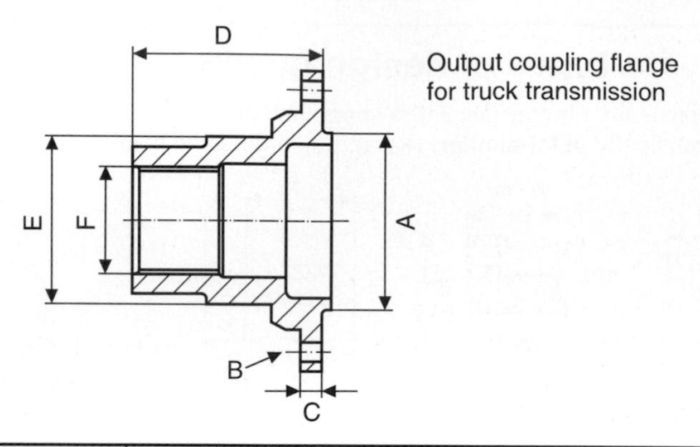

Main design dimension	Number of different dimensions **before** part families established	Number of different dimensions **after** part families established
A Flange centering B Bore diameter C Flange thickness D Flange length E Seal diameter F Inner diameter	176 149 62 21 6 2 } 416 dimensions	1 (d = 90 mm) 1 (8 holes) 1 (9 mm) 2 (60; 69 mm) 1 (d = 60 mm) 1 (d = 42 mm) } 7 dimensions (2 configurations)
Manufacturing costs	100 %	46 %

Fig. 7.12-13. Cost reduction through creation of design part family (from ZF)

Such **"tidying-up"** actions are carried out effectively in four major steps:

1. **Planning the steps**. For this purpose, the number of variants that will remain as necessary is estimated, starting from the existing variant variety. Remember that customers order old spare parts or, on new orders, they stay with the old product design. If it turns out that a majority of the old variants must remain anyhow, it should be considered how many "new variants" would be added through standardization and what their introduction will cost in comparison to the previous state. Practical experience shows that it is more economical to keep the old variant variety (passive variants) and to carry out a variant reduction only with a new design of the product (avoid active variants) [Koh97]. If the decision is in favor of a part family creation, following procedure is called for.
2. The **variants reduction** is then carried out in the following steps:
   - Search for the object to be processed, provided that is not already clear (according to the frequency of occurrence, cost mitigating consequence).
   - Collect the drawings of configuration variants and their manufacturing costs.

- Arrange the designs according to part features [Bei77; Pfl79] (e. g., according to DIN 4000), design types, configuration details, dimensions, and tolerances. It is desirable to reduce the drawings size and to put them in a file or card index [Pät77].
- Review the result with the designer, work process scheduler, vendor, cost controller, and possibly the buyer, with the aim of the limiting the number of variants. Propose geometrical progression (preferred number series from DIN 323) and consider available standards (see Fig. 7.12-19). Estimate the cost effects of the limitation.

3. **Document** the standard either in the factory standards system or in the CAD system as a standard feature. Incorporate it into the repeat-part search system.
4. Publicize the **result** and **establish** the part family.

Figure 7.12-13 shows **cost potentials** that exist, even with a belated part family creation. It is more economical to do the standardization in advance, as in the case for products in size ranges and modular designs. With high potential, it is totally justified if standardization is also carried out for assemblies and parts that experience less demand from the market than planned. It is worth the extra work if they become the future "milk cows" of the company.

7.12.4.3 Favor integral design

a) What are integral and differential designs?

Integral design means the combination of several parts into one part made of the same material; **differential design** is the opposite. Part count decrease becomes possible mostly by a change in the production process. The following primary and deformation process are best suited for integral design:

- Casting, particularly investment (lost-wax) casting and injection molding
- Sintering
- Sheet metal forming
- Forging, deep drawing
- Erosive, electrolytic removal

Moreover, semi-finished products can be the starting points for integral design.
Integral design usually leads to cost savings. As examples show, differential design can also be more economical in special cases (Fig. 7.9-9 and Fig. 7.12-15).

b) When to use integral design, when differential design?

The decision is complex for the following reasons: From **cost viewpoint** the manufacturing costs (including which tool, model and setup costs as well as the assembly and the rejects costs) must be kept in mind. Furthermore, the complexity costs in the sense of the process costs, and ultimately from the lifecycle costs, the costs for transportation and possible spare parts must be considered.

There are other **parameters** to consider, such as size and number of parts as well as production risk and delivery time at the primary and deformation processes that are often used for integral parts. The following designs therefore can only provide rough directions:

- In general, **integral design** is preferable for **small to medium size parts** and for **large quantities.** This is because the per-part tool, model, and setup costs (all introductory or one-time costs) are less important. The reduction in the processing, joining, and assembly costs is greater. Also, if there are no joints, these parts (faces) do not need to be machined or joined. In general, assembly becomes easier, in part because of the falling logistics costs. "Parts that aren't there don't need to be assembled". In the sense of the process costs, the so-called complexity costs (Sect. 6.3.1) come down due to the lower part count. Finally every break-and-join site usually causes quality problems (tolerances, heat and fluid flow, density, deformation, strength, fretting corrosion), which are mitigated due to its absence. There are **limits** if the integral part becomes too complex; if the production technology is not mastered sufficiently (see rejects in Fig. 7.9-9); if the one-time costs do not come down due to too small a quantity (see above); or when the delivery time for the primary and deformation process becomes too long (tool and pattern making takes time). If integral design can be implemented, however, without these pattern and model costs ensuing, it becomes advantageous even for small quantities or single-unit production, as is shown for a small part in Fig. 7.12-15. This is due to the high **share of setup costs,** especially for small parts (Sect. 7.7). If, for example, with integral design three parts reduce to one, the setup costs arise only once in each working cycle. It is obvious how intricately the many parameters are involved in the integral/differential design decision.
- **Differential design** can accordingly lead to lower costs in **single-unit and limited-lot production** for **large parts**. The advantage of the lower production risk was suggested in Fig. 7.9-9. Furthermore, fewer fixed costs arise for models, templates, and specific machines, since the parts can be produced and assembled from easily available and different materials, semi-finished products, and standardized parts. The logistics and assembly costs do thereby increase, so a quality inspection of the individual parts often becomes more advantageous. **Replacing** individual parts, such as due to wear, is often simpler and cheaper since only a few parts have to be replaced. Later changes to the individual elements are easier and cheaper than with integral design since the whole system does not need to be replaced. However, only the parts in question must satisfy the new requirements.

  Based on the effects listed here, by using differential design the schedule risk and the product realization time can be reduced in single-unit and limited-lot production. This is important in making prototypes (see Fig. 7.11-12). The disadvantage of differential design (an increase of one-time costs that are tied to part count) becomes apparent when it deals with **small parts** in **small numbers** without special pattern, template and tool costs. If machining is used for everything, then choose few parts and few production steps, even with more material use (see rule at the end of Sect. 7.7).

c) Examples and rules

- **Integral design**
  - Figure 7.11-12 shows a realization of integral design through casting (investment casting). Due to the resilience of polymers, parts made by polymer injection molding can have flexible hinges and snap fits that otherwise would require additional parts (**Figs 7.12-14**, 7.11-48, 7.11-54 and 7.11-55).

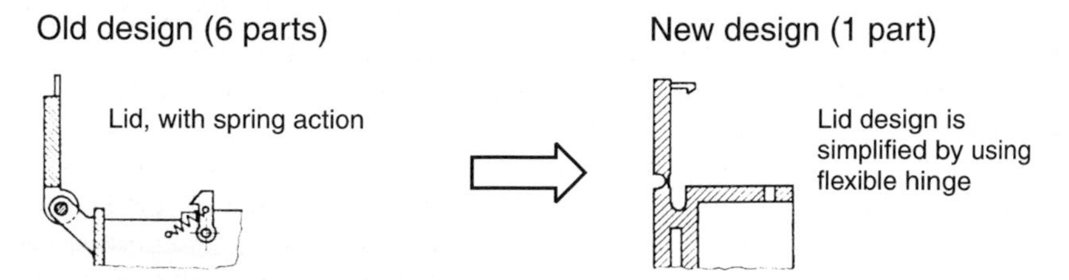

Fig. 7.12-14. Cost reduction through integral design using injection molding

  - Using integral design can eliminate many production operations, as was shown in Fig. 2.3-1 in Chap. 2 with the example of a diesel engine rocker arm. Nine production steps were reduced to three. In both cases the lever is made by forging, but with the lower-cost alternative, not from two parts welded together, but milled out from one part. There is hardly any increase in the materials cost, which amount to only about 5% of the manufacturing costs.
  - With **parts made in large quantities,** the production costs decrease through efficient production processes – lower setup costs and production times (Sect. 7.7, Fig. 7.9-2). The cost of materials then becomes such a large portion of the total cost that one can hardly afford to waste material. Therefore, primary and deformation processes are used to make the parts in near-net shapes. The parts are of integral design, which also saves on production costs by avoiding joints. Hence, the following rule:

> ➔ With **parts made in large quantities,** strive for **integral design,** which approximates the final shape by applying primary and deformation processes. Consequently, there is considerably less material waste.

  - **Figure 7.12-15** shows a **pinion shaft** in differential and integral design [Ehr82a]. A **small shaft** is shown above (Ø 66 mm, 1.6 kg). Compared to the materials costs, the setup costs are dominant in case of single-unit production. Thus, it is not profitable to make the pinion shaft from two parts just to save on the material being removed. The low-cost alternative is the integral design in which the pinion shaft is machined out of one piece (round stock) (Sect. 7.6 and 7.7).

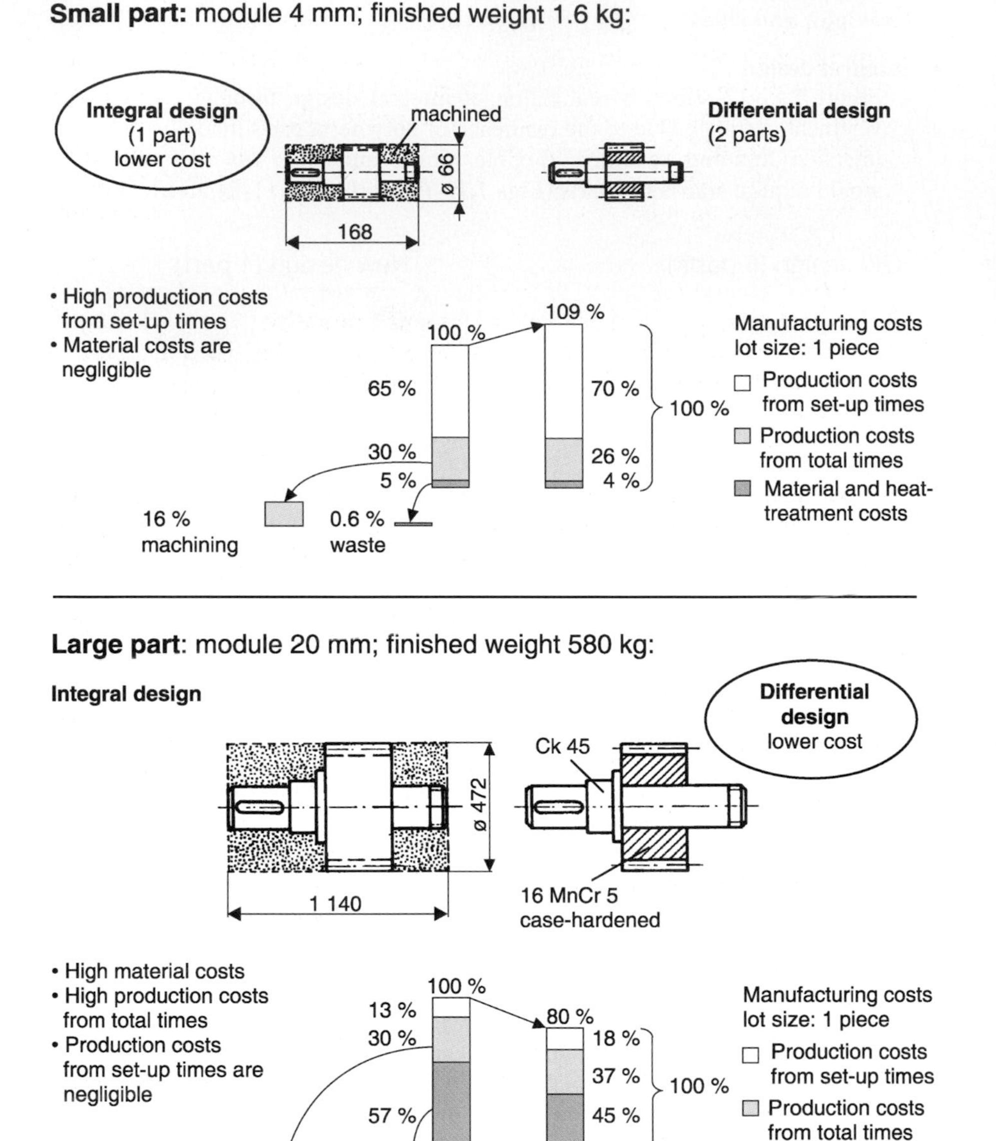

Fig. 7.12-15. Integral or differential design of a pinion shaft with single-unit production from one piece (from company calculations) [Fis83]

Below, in case of the **large shaft** (Ø 472 mm, 580 kg) the circumstances are just the reverse. Here it is worthwhile to make the shaft from two parts (shaft and gear). It costs only 80% of the one-piece, large pinion shaft, for which 28% of the manufacturing costs are spent for the rough machining out of the whole, and the removed material. The size-dependent costs (material, heat treatment, and hardening) are high, almost 60% of the manufacturing costs, so these must be held small. A free-form forged blank as a variant was not examined.

➔ **Integral design** is more economical for **small quantities** and machining from the whole, particularly for **small and medium-size parts,** than the differential design (e. g., fixture design [Bau82]).

- **Differential design**
  The cast **planet carrier** shown in Fig. 7.11-1 is prone to cracks and blow-holes due to differences in material thickness. The rejects costs (including the costs of metal removal by machining) must be borne by the flawless pieces made (similar to Fig. 7.9-9). Differential design can be more economical in this case. For very large parts, differential design is unavoidable for reasons of transportation and the manageable measurements.

  There is a similar situation with the alternatives – **large gears** (more than 1.5 m in diameter) made either as one-piece or in two parts, with a cast wheel body and hardened periphery. In addition, for large diameters it is preferable to plan on differential design with a low-cost cast wheel body (Fig. 7.13-12). The same procedure is described in Sect. 7.12.6.7 for gear transmission housing.

➔ **Differential design** is more economical for **large parts** or with expensive material in single-unit production and for **small quantities.**

d) Strategy of the "one-part device"

We know that it is difficult to radically reduce a familiar *n*-part design, with regard to the part count. In order to facilitate a thought break-through, the strategy was proposed in 1985 of the "one-part device" [Ehr85], which was also dealt with in another way by Paterak [Pat97].

The machine or product (e. g., a transmission in Fig. 7.13-7) are **thought out** and put in the drawing **as a "casting"**. Parts in motion are also cast fixed. All assembly openings of a housing are "cast shut". There are no part assembly joints. Everything is **one** part even if it looks crazy. This is needed because existing shapes have an almost suggestive inertia in our brain; abstraction from what is optically engraved and usual is difficult.

After that a **step-by-step separation** (with a red pencil) into **"minimally necessary part count"** occurs according to the following perspectives:

- Where are different materials necessary?
- Where is a relative movement? (Note: For oscillating parts, elastic joints can sometimes suffice instead of hinges, Fig. 7.12-14).
- How should assembly and disassembly occur?
- Where must spare parts be exchanged?
- Where is separation necessary for transportation?
- Where is the accessibility necessary during later use?

This makes it possible to design, for example, a transmission without a housing parting joint (only with big assembly cover) or without keys at the gear hubs.

e) Assembly interfaces not affected by changes

The interfaces between assemblies or modules of a product should be designed so that in case of job-specific changes to one assembly other neighboring assemblies need not also be changed. Otherwise, additional new parts crop up.

7.12.4.4 Employing measures to reduce the set-up costs

A measure that indeed does not reduce the variant or part count, but makes the variants better controllable, is a decrease in setup costs. In the era of customer-specific production, this measure is becoming generally more important [Pil98]. The setup costs that result from the setup times are an essential segment of the one-time costs (Sect. 7.5.1). **Figure 7.12-16** shows the measures for decreasing the setup times t_s for a product made from several parts. These are reviewed below:

- **Few parts:**
  The importance of the absolute number as well as the number of different parts was mentioned earlier. Integral design (Sect. 7.12.4.3), function integration, and a search for a low part-count concept (Sect. 7.3) are all constructive measures are effective in decreasing the part count. An example of the latter is the change that took mechanical clocks and calculating machines over to the field of microelectronics. Earlier versions had hundreds of parts, while today there are fewer but more complicated parts. Figure 7.3-1 shows how through function integration the part count for a switch was halved. Function integration means that one function carrier (part, physical effect) fulfills several functions. In the calculator example above, a push-button and a spring were eliminated by the flexibility of the polyester film and with the printed connection the contact springs along with their fastening. In addition, by using symmetrical parts to avoid right-hand/left-hand designs, the part variety can be limited.
- **Lot size *N* high:**
  Effective measures for reducing setup costs here are using part families, modular designs, and/or repeat and same parts.

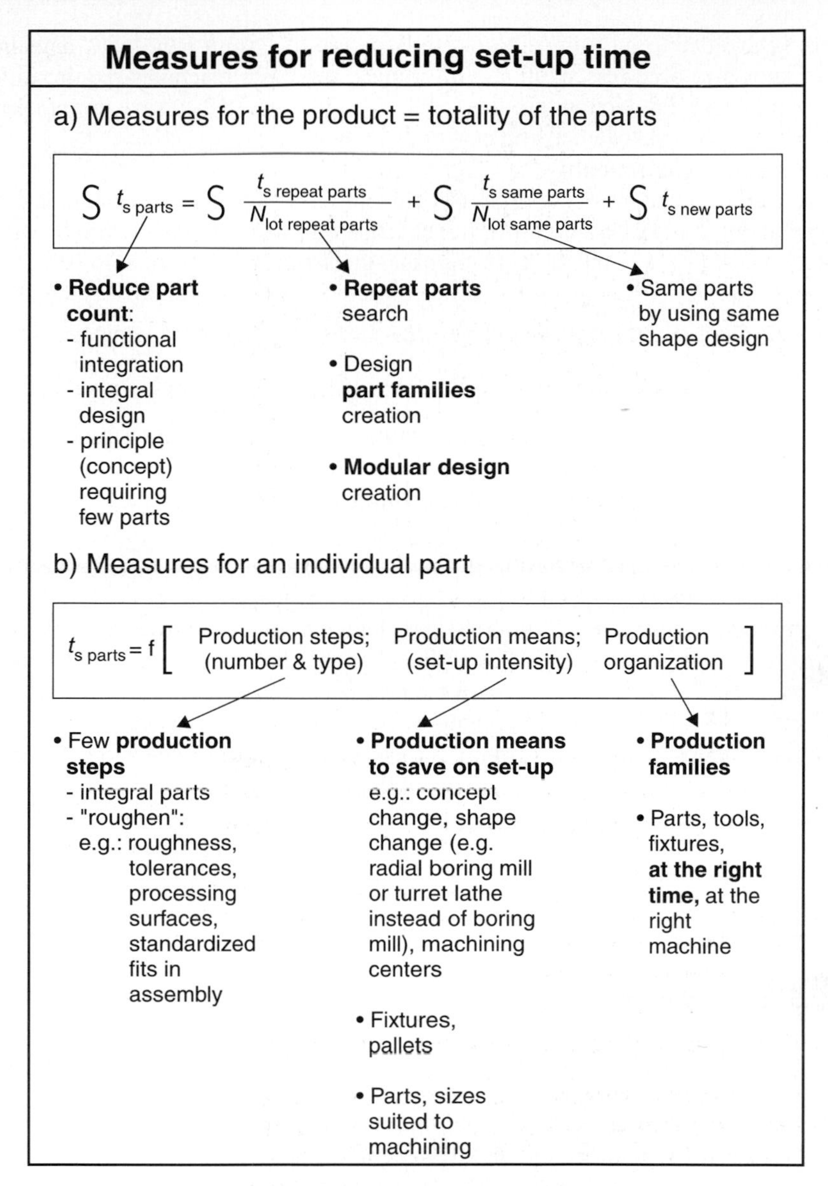

Fig. 7.12-16. Measures for reducing set-up time

- **Few production operations:**
  The set-up costs increase with every production process. Therefore, working with a cost and production consultant, it is advisable to limit the number of the production operations and to avoid operations that need long set-up times.
- **Measures for reducing set-up times:**
  Through appropriate shape design of parts (examples: concrete mixer, Sect. 10.1; welded centrifuge housing, Sect. 10.2), machine tools requiring long setup times

can be avoided (e. g., using a boring mill instead of a turret lathe). NC machines are known to need less setup and idle times, but these machines require initial programming. Production families (Sect. 7.12.3b, Fig. 7.12-9) save on retooling. Design must work together with production planning on these issues.

- **Economical equipment:**
  In the above mentioned examples of concrete mixer and centrifuge housing, the machining was switched to the turret lathe due to its more favorable workstation cost rate. For the same set-up time, the set-up costs become lower.

7.12.4.5 Organizational measures for reducing the part variety

Some of the organizational measures that have already been mentioned are:

- Improvement of the **communication** between designers of similar products and with production, cost accounting and materials management.
- Application of **computer-based information systems**, purchase, and parts made in-house.
- Use of **process cost accounting** for the evaluation of the introduction and change costs (Sect. 8.4.6).
- A **"malus"** handicap (e. g., 1 500 $) per change that must be overcome through cost reduction, which is caused with the change. The malus corresponds, on the average, to the introductory costs for a new part shown in Fig. 7.12-11 (as well as Fig. 6.3-1). It is not valid for changes caused by quality reasons.
- Every product manager who requests a program enlargement should free up a **variant portion** of the same size **for deletion.**
- The dealer (at times, the customer) can also create customer-specific variants later. Only the parts and perhaps fitting instructions are supplied. Thus, for many cars, the dealers mount customer-specific wheels, tires, caps, etc.; for computers, specific components are installed only at the point of sale.

7.12.5 Design in size ranges

Product size ranges are the most effective means to standardize a product over a specific size range and thereby to strictly limit the part variety. Products in size ranges are usually combined with modular design (Sect. 7.12.6). Everyone knows examples from experience: automobile manufacturers cover a certain size range with the different models; also, upon request, different engines and interior furnishings are installed from a modular system.

7.12.5.1 Definition, purpose and effect

a) Definition

A product **size range** consists of functionally identical technical artifacts (machines) which increase in size by steps (Fig. 7.12-12a, **Fig. 7.12-17**).

Fig. 7.12-17. Size range of turbochargers (BBC). Stepping of rotor disk diameter according to series R 40/3, with size ratio $\varphi = 1.18$ [Pah74]

We deal with adaptive design with the following features:

These are identical:

- Function (qualitative)
- Constructive solution
- Most possibly, materials
- Most possibly, production

The following are different:

- Performance data (function, quantitative)
- Measurements, and quantities dependent on those (weight, costs etc.)

b) Purpose and result

The **purpose** is to cover a wide field of application of a product category with minimally different product models in order to achieve following (Fig. 7.12-4):

- **Cost reduction** through a significant decrease in the administrative time (introductory costs) per piece in all departments (increasing the count), as opposed to increasingly different special designs (a ready-made instead of tailor-made suit). This decrease in time is particularly effective in design, production planning, and of course in production. Considerable time is spent initially, working on design, development, and documentation of the product in size ranges (including sales documents). After that, only a little time is necessary for handling of incoming standard orders. For production, for example, the same work schedules can be employed repeatedly. Parts and assemblies can be produced and stocked in greater numbers. The same products are made in larger lots for different customers. Purchasing commands higher discounts with larger quantities of raw material.
- **Delivery time contraction** by a large decrease of the design time (that can constitute about 50% of the entire delivery time for special products). Other factors are the use of available prepared material, fixtures, tools, etc. In addition, the time required in all departments involved in the product is much reduced due to the training effect (see Sect. 7.5.2b).

- **Increase in quality and reliability** because the infant mortality of the product model has been eliminated. There are no more project planning and design flaws; there is better knowledge about testing, starting up and maintenance; and the interchangeability and delivery time are usually beneficial for the spare parts.

These are the **advantages** both for the user as for the manufacturer of the product.

Disadvantages for the product users are that they receive a product often lacking optimal performance data and operating costs for the their needs. In general, these disadvantages are more than made up for by the above advantages. The manufacturer of products in size ranges must therefore consider the following. For the customer of such a product, this is only of interest when the purchase price and delivery time advantages, compared to a special design, are larger than the conceivably higher operating costs.

The **expense** for the **preparation** of all necessary **documents for a product in size ranges** (drawings, parts lists, calculations, sales documents, performance and maintenance rules, etc.) can be considerable (a few man-years). In any case, the less usual types are generally ascertained only as far as it is necessary for the bid preparation ("virtual size range"). Here the methods for the design of products in size ranges can help (i. e., preferred number series and similarity laws). So, starting from the average basic embodiment that was calculated and designed, we can extrapolate above and below, with little extra work (Sect. 7.12.5.5).

The **risk** during the design and development of products in size ranges is that if the products do not make it to the market, there is nothing to show for the investment. Therefore, a thorough **task clarification** and a **market analysis** are necessary before beginning the standardization effort. What are the needs in the potential markets? Where does future design and development go? Which demands are the most important? What are the competitors developing? How far ahead is the competition? What is the influence of this advantage on the lifecycle costs of the customer? By how much can the price be dropped due to lower costs for expected incoming orders, the number that is to be expected per product model? With that, a stimulus is offered to the customer for turning down a custom-made product.

Even technically excellent products made in size ranges can fail to succeed in the market if the risk of lowering the price in the expectation of sizable orders is not taken. It is natural to see the orders first and only then act on the price. Just as the on technical side, an "advance performance" is necessary also on the business side. One must be prepared, right from the beginning, to set the price and delivery time to the expected order size; otherwise the influence spiral of the price (costs), quantity, and delivery time does not get started (Fig. 7.5-1). The cost reduction that was predicted by calculation and what is in fact achieved, usually turns out to be too small due to deficiencies in overhead costing (Sect. 8.4.3). All products that deviate from the previous average are evaluated incorrectly, due to the problems of overhead cost allocation (Fig. 8.4-2; Sects. 8.4.3 and 8.4.4) [Jes96].

c) Example of a product design and development in size ranges: changes over time

An example of the evolution of a product in size ranges is the **gear transmission** with journal bearings, shown in **Fig. 7.12-18**. The project was worked on in the beginning in a preliminary fashion for the first design, and more intensely later, with the increasing success of the product. In Stage 0, the transmission was produced as a special design to specific customer wishes. Delivery time and costs were high but the demand increased. Designers recognized the market requirements. To save work on orders that were always similar for project planning and designing, in Stage 1 the center distance, tooth widths, and the range of gear ratios were prescribed. It was then possible to make standard housings as castings instead of welded housings designed anew for each single-unit order. For specific bearings, sealing rings, and shafts, standard repeat part drawings were created. For customer-specified parts (pinions, gears), standard features were set on the CAD system and were automatically dimensioned after the calculations for the transmission were made. For project planning and sales, the performance and price data could be calculated. The delivery time thus shortened due to faster designing, and

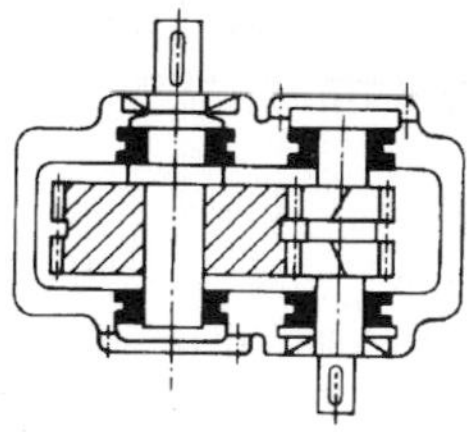

<table>
<tr><td rowspan="5"></td><td colspan="5">Stages</td></tr>
<tr><td>0</td><td>1</td><td>2</td><td>3</td><td>4</td></tr>
<tr><td>Custom design</td><td colspan="4">Design in size ranges</td></tr>
<tr><td colspan="3">Single-unit production</td><td colspan="2">Limited-lot production</td></tr>
<tr><td>Single piece</td><td>Low quantity</td><td>Medium quantity</td><td>Larger quantity</td><td>Continu-ous supply</td></tr>
<tr><td>Housing</td><td rowspan="6">Must be designed anew for each order</td><td>Pattern available</td><td>Produced for stock</td><td rowspan="4">Drive-ratio dependent parts produced for stock</td><td rowspan="6">Produce everything for stock (also possibly assemble)</td></tr>
<tr><td>Bearing</td><td rowspan="3">Standard-part drawings available</td><td rowspan="3">Raw material available</td></tr>
<tr><td>Seal</td></tr>
<tr><td>Shaft</td></tr>
<tr><td>Gear</td><td colspan="2" rowspan="2">Initial drawings available for drive-ratio dependent parts</td><td rowspan="2">Raw material available</td></tr>
<tr><td>Pinion; shaft</td></tr>
</table>

Fig. 7.12-18. Development of a product standard (design in size ranges/modular design) over time

the orders increased. In Stage 2, the decision to produce the housings for fast-selling sizes for stock could be made, and specific material and purchase parts could be ordered. Since the costs dropped, the company could react flexibly in price negotiations, which resulted in more incoming orders. In Stage 3, it was decided to produce speed-ratio-independent parts (shafts, bearings, sealing rings, parts for the oiling system) for stock. For speed-ratio-dependent parts (gears), the raw material (e. g., forged parts) was to be held in stock as inventory. The product was such a great success that in Stage 4 the fast-selling sizes were produced in advance, then assembled and stored in the warehouse. Other sizes could be assembled in response to incoming orders mostly from the stocked parts.

Thus, the desired spiral of interaction between delivery time, costs (price) reduction, and rise in quantities produced developed. The competitors did not stand a chance with single-unit production of such transmissions (Fig. 7.5-1). They had to react with a technically and commercially more attractive product made in size ranges.

In the following, two essential tools are reviewed for the design of products in size ranges: the use of preferred number series, and of similarity laws. The first enables flexible size steps for the product; with the latter, the properties of the products made in size ranges can be assessed quickly without having to design all the units of the series (virtual size range).

7.12.5.2 Preferred number series to facilitate designing in size ranges

a) Concept and purpose

If products are to be designed in size ranges, a fundamental question is, whether the stepping (gradation) will be linear (arithmetic) or non-linear (e. g., geometric). In practice as well as for standardization, both possibilities are used. Here the advantages of the geometrical gradation will be shown by means of the preferred number series.

Preferred number series, according to DIN 323 [DIN74] (**Fig. 7.12-19**), are geometrically stepped series. It is a number series in which, in each decade, every number is a multiple of the previous number by a fixed factor φ (step size). The series are an important aid in choosing size steps in the embodiment of products in size ranges.

Example: If 10 steps per decade are desired, the step size comes out as:

$$\varphi_{10} = \sqrt[10]{10} \approx 1.25$$

With 20 steps:

$$\varphi_{20} = \sqrt[20]{10} \approx 1.12 \approx \sqrt{\varphi_{10}}$$

Unlike an arithmetic series, where the numbers always have a fixed increase (e. g., 1; 1.25; 1.5; 1.75; 2.0; 2.25; …), here they always have an equal percentage

Designation	R 5	R 10	R 20	R 40
Step size	$\varphi_5 = 1.60 = \sqrt[5]{10}$	$\varphi_{10} = 1.25 = \sqrt[10]{10}$	$\varphi_{20} = 1.12 = \sqrt[20]{10}$	$\varphi_{40} = 1.06 = \sqrt[40]{10}$
	1.00	$1.00 = 10^0$	1.00	1.00
				1.06
			1.12	1.12
				1.18
		$1.25 = 10^{0.1}$	1.25	1.25
				1.32
			1.40	1.40
				1.50
	1.60	$1.60 = 10^{0.2}$	1.60	1.60
				1.70
			1.80	1.80
				1.90
		$2.00 = 10^{0.3}$	2.00	2.00
				2.12
			2.24	2.24
				2.36
	2.50	$2.50 = 10^{0.4}$	2.50	2.50
				2.65
			2.80	2.80
				3.00
		$3.15 = 10^{0.5}$	3.15	3.15
				3.35
			3.55	3.55
				3.75
	4.00	$4.00 = 10^{0.6}$	4.00	4.00
				4.25
			4.50	4.50
				4.75
		$5.00 = 10^{0.7}$	5.00	5.00
				5.30
			5.60	5.60
				6.00
	6.30	$6.30 = 10^{0.8}$	6.30	6.30
				6.70
			7.10	7.10
				7.50
		$8.00 = 10^{0.9}$	8.00	8.00
				8.50
			9.00	9.00
				9.50
	10.00	$10.00 = 10^1$	10.00	10.00

Fig. 7.12-19. Preferred number series from DIN 323

increase (for example 1; 1.25; 1.6; 2.0; 2.5; …). In the beginning, the absolute steps are small; later, they are larger. This property of geometrical series fits better with the human feeling than the increase fixed in absolute value, as with the linear step size.

b) Choice of the step size φ

Figure 7.12-20 shows the optimization task involving the manufacturer and the market in setting the step size. The **customers** want a product that has its performance adapted especially to their wishes, with total costs as low as possible and a short delivery time.

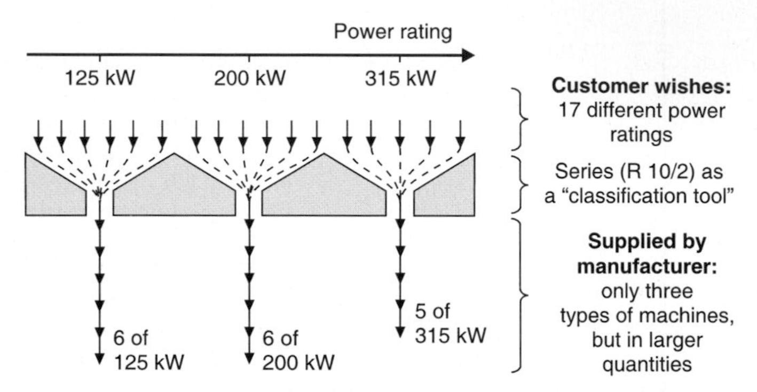

Fig. 7.12-20. Standardization of machine performance: The product standard (size range) is a type of "classification tool" and provides the prerequisite for larger quantities

The **manufacturers want to** produce **as few types as possible, in large numbers,** to attain low manufacturing costs. They therefore make use of size ranges (stepped according to R 10/2, i. e., every second member of R 10), in order to consolidate the diverse customer wishes into only a few channels (channel = size). They make the channel-to-channel step size large, thus resulting in larger quantities of fewer types. However, then it may happen that the customers do not want to buy any more, since the products don't satisfy their wishes well enough with regard to performance [Küh86].

If the manufacturers make the step size small, they offer more sizes so that the customers indeed usually have their performance wishes fulfilled. However, the quantities per category then become so small that the manufacturers cannot offer any great price advantage compared to a customer-specific machine. In both cases, the lifecycle costs can increase (**Fig. 7.12-21**). The optimum to seek is that with the lowest lifecycle costs (Chap. 5).

Step size	One-time costs	Operating costs	Life-cycle costs
Step size φ too large	Large quantities ⇨ manufacturing costs and price low	High, since machine performance and efficiency match poor	High
Optimum	Optimum	Optimum	Minimal
Step size φ too small	High, almost like custom-designs	Low, since machine matches operating conditions well	High

Fig. 7.12-21. Optimization of the step size

Since in general it is difficult to get precise data about lifecycle costs, the step sizes are chosen as follows:

- in the beginning coarse, for example with series R 10,
- late finer, for example with series R 20, and indeed finer,
  - the more precise the specific technical properties must be held, and
  - the more sensitive the market is to large price differences per step.

For capital goods, with large and heavy machines, the step sizes must be finer. There the manufacturing costs (from Fig. 7.6-3; Eq. (7.7/1)) increase with approximately the third power of the linear step size. If this is neglected, then the price jumps between the models are too large. The competition can sell in those gaps (**Fig. 7.12-22**).

Figure 7.12-22 shows how the step size was handled in the development of an epicyclic gear train. At the start of the size range in 1951, five smaller models were stepped according to the preferred number series, R 10; the four larger designs were stepped more finely, using R 20. Five years later, the smaller models were also stepped finely with R 20, since this was justified by the large order quantities; customer wishes could be fulfilled better. Between the gaps of the five

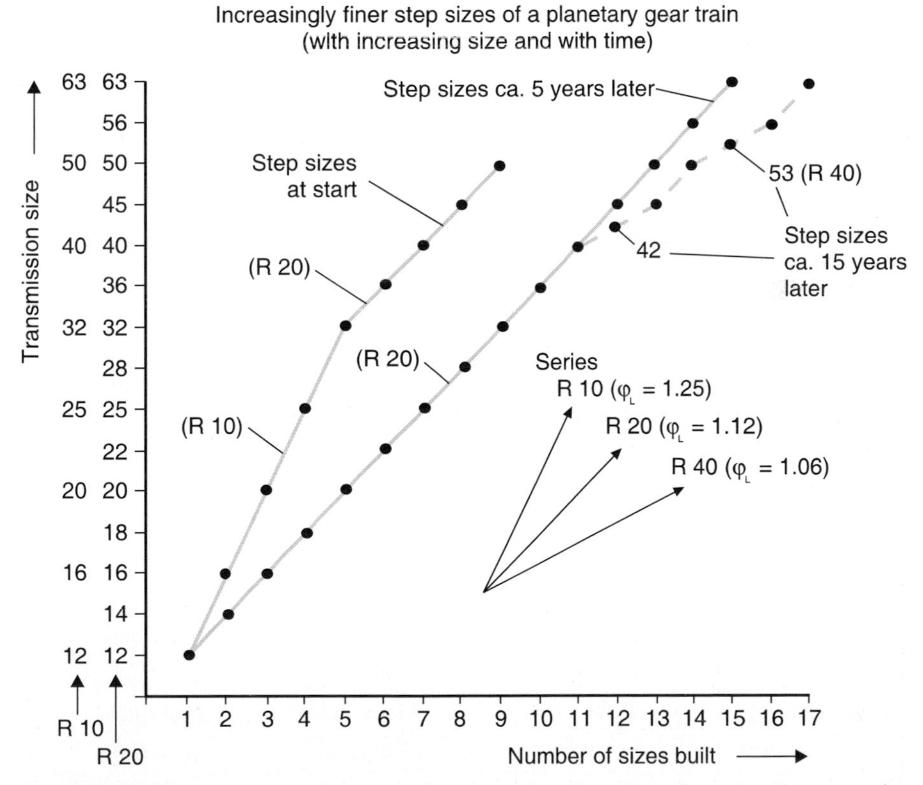

Fig. 7.12-22. Size range gradation according to preferred numbers for epicyclic gear trains (see Fig. 7.12-19)

models of the number series R 10, another four models of number series R 20 were simply moved in – the advantage of flexible refinement! After 15 years, it was noted that the competition got orders for the large, expensive transmissions when they offered transmissions between the R 20 models. Therefore, these transmissions were stepped even finer, indeed carried out using R 40.

7.12.5.3 Similarity laws

In the embodiment of products in size ranges, similarity laws offer product developers the advantage of a quick overview of the technical and economical properties of the individual products. Each product does not have to be designed separately; it needs only its calculations to be checked.

a) Concept and purpose

A **similarity law** shows the physical or economic patterns for products whose properties are similar in well-known design and production aspects, whereby the products differ in size by certain step sizes φ_L (Sects. 7.7 and 9.3.5).

During the embodiment of a product made in size ranges, a model is developed that is at the medial of the performance range, complete in its design (**basic embodiment**). Starting from this all technical and economical data of interest are calculated using similarity laws for all **geometrically similar models** (subsequent embodiment) That saves having to design all the models separately. The purpose of applying similarity laws is the labor saving in design and project planning, as well as having a better overview and control of the technical and economical properties of the size range being planned [Kit90].

There is also a **geometrical semi-similarity**, for which only specific features with certain step sizes change, while other features remain fixed (Sect. 9.3.5).

Only the basic possibilities of the application of similarity laws are shown here [Pah82]. The aim is for the complete geometrical similarity within a size range (pantograph design).

Similarity laws then determine the relationships between the **length step size** (the linear ratio) $\varphi_L = L_1/L_0$ (subscripts: 0 = basic embodiment, 1 = subsequent embodiment) and the remaining parameters of interest in the product. The aim to use possibly the same materials and production types for all members of the given size range, and to maintain a constant stress. This is often only partially achieved.

Corresponding to the basic physical parameters we define the following **basic similarities** for the fixed ratio of a basic parameter (invariant):

Similarity type	Basic parameter	Fixed parameter	Relationship
Geometric similarity	Length L	Linear scale	$\varphi_L = L_1/L_0$
Time similarity	Time t	Time scale	$\varphi_t = t_1/t_0$
Force similarity	Force F	Force scale	$\varphi_F = F_1/F_0$

If the ratios of more than one basic parameter are fixed, special similarities are particularly important here:

b) Static similarity

The step size for the length (φ_L) and for the static force (φ_{Fs}) are fixed, and they are related in such a way that the **stresses** from external static forces F_s (not gravity forces) in all elements are **fixed** ($\varphi_\sigma = 1$).

Stress: $$\sigma = \frac{F_s}{A} \leq \sigma_{allow}$$

Step size for the stress: $$\varphi_\sigma = \frac{\sigma_1}{\sigma_0} = \frac{F_{s1}}{F_{s0}} \cdot \frac{A_0}{A_1} = \varphi_{Fs} \cdot \frac{1}{\varphi_L^2} = 1$$

where, A = area; $$\frac{A_1}{A_0} = \varphi_L^2$$

and thus: $$\varphi_{Fs} = \frac{F_{s1}}{F_{s0}} = \varphi_L^2$$

Therefore, the forces can increase fourfold for a doubling of the product's size ($\varphi_L = 2$) without an increase in the stresses. This makes sense. We already see the limits of deliberations on similarity: Since the material strength declines with size, the limits can not be utilized fully. If additional gravity forces play a role, with the weight W ($W_1/W_0 = \varphi_L^3$), these increase with the third power of the size and can no longer satisfy static similarity, since the forces would increase proportional to φ_L^2 (see above).

c) Dynamic similarity

For machines, dynamic similarity is more important, for which the following basic parameter ratios are fixed:

Step size for the length: $$\varphi_L = \frac{l_1}{l_0}$$

Step size for the time: $$\varphi_T = \frac{t_1}{t_0}$$

Step size for the static forces: $$\varphi_{Fs} = \frac{F_{s1}}{F_{s0}}$$

Step size for the dynamic forces (from mass, inertia laws):

$$\varphi_{Fd} = \frac{F_{d1}}{F_{d0}}$$

The step size for static forces must be the same as that for dynamic forces ($\varphi_{Fs} = \varphi_{Fd}$) so that these forces and their effects on stresses can be handled together.

To find the step size of static forces,

$$\sigma = \frac{F_s}{A} = \varepsilon \cdot E;\ F_s = A \cdot \varepsilon \cdot E \text{, where } \varepsilon = \frac{\Delta l}{l} \text{, the strain.}$$

Since the step size for extension is $\varphi_e = \dfrac{\varepsilon_1}{\varepsilon_0} = \dfrac{\Delta l_1 \cdot l_0}{\Delta l_0 \cdot l_1} = \dfrac{\varphi_L}{\varphi_L} = 1$, then

$$\varphi_{Fs} = \frac{F_{s1}}{F_{s0}} = \frac{A_1 \cdot \varepsilon_1 \cdot E_1}{A_0 \cdot \varepsilon_0 \cdot E_0} = \varphi_L^2 \cdot 1 \cdot \varphi_E \qquad (7.12/1)$$

The step size for the dynamic forces (from the centrifugal force) is

for $F_d = m \cdot r \cdot \omega^2$ with $m = \rho \cdot V$

$$\varphi_{Fd} = \frac{F_{d1}}{F_{d0}} = \frac{m_1 \cdot r_1 \cdot \omega_1^2}{m_0 \cdot r_0 \cdot \omega_0^2} = \varphi_\rho \cdot \varphi_L^3 \cdot \varphi_L \cdot \varphi_\omega^2$$

With the peripheral speed $\upsilon = \omega \cdot r$ becomes $\varphi_\upsilon = \varphi_\omega \cdot \varphi_L$

$$\varphi_{Fd} = \varphi_\rho \cdot \varphi_L^2 \cdot \varphi_\upsilon^2 \qquad (7.12/2)$$

The step sizes of both types of force are then of the same size, when Eq. (7.12/1) and Eq. (7.12/2) are set equal:

$$\frac{\varphi_{Fd}}{\varphi_{Fs}} = \frac{\varphi_\rho \cdot \varphi_L^2 \cdot \varphi_\upsilon^2}{\varphi_E \cdot \varphi_L^2} = 1 \text{ or, if } \frac{\varphi_\rho \cdot \varphi_\upsilon^2}{\varphi_E} = 1$$

$$\text{or, } \frac{\rho_1 \cdot \upsilon_1^2}{E_1} = \frac{\rho_0 \cdot \upsilon_0^2}{E_0} = const = \boldsymbol{Ca} \textbf{ (Cauchy No.)} \qquad (7.12/3)$$

With same material (density ρ, modulus of elasticity E) it then follows that dynamic similarity exists between geometrically similar designs only if at the corresponding points, peripheral velocities v are the same. Since $\varphi_\omega = \varphi_\upsilon / \varphi_L$, then for a doubling of size, the rotation speed may only be half as high. **Figure 7.12-23** describes how the remaining parameters depend on the linear step size φ_L.

The dependencies shown in Fig. 7.12-23 can be indicated on log-log graph and appear as straight lines according to respective powers of step size (preferred numbers marked same distance apart [DIN74] show a logarithmic scale division). One such diagram for a specific product provides a quick overview over the planned size range.

With $Ca = \frac{\rho \times \upsilon^2}{E}$ = const., and for same material, i.e., $\rho = E$ = const., thus υ = const. Under geometrical similarity, with the linear step size φ_L, the following parameters change:	
Speed n, ω Bending and torsional critical speed n_{cr}, ω_{cr}	φ_L^{-1}
Velocity υ; resulting inertia and elastic forces: Strain ε, stress σ, surface pressure p	$\varphi_L^{0} = 1$ (= const.)
Spring rate k, elastic deformation Δl; resulting gravity force: Strain ε, stress σ, surface pressure p	φ_L^{1}
Force F Power P	φ_L^{2}
Weight W, torque T, torsional stiffness k_t, Modulus of resistance EI, GJ	φ_L^{3}
Area moment of inertia I	φ_L^{4}
Mass moment of inertia J	φ_L^{5}
Note: Material utilization and safety are constant only if the parameter influence on material limit values is negligible.	

Fig. 7.12-23. Behavior of physical parameters for dynamic similarity (fixed parameter, peripheral speed υ) [Pah97]

7.12.5.4 Limits for geometrically similar size ranges

As noted above, there are always restrictions on the realization of a strict, geometrical similarity in practice. These arise from the following:

- **Overriding similarity laws,**
  (e. g., through the influence of gravity, thermal processes, etc.)
- **Overriding task statements,**
  (e. g., regarding the human-machine relationship, the human being is more or less fixed in size, while machines can be of very different sizes. Therefore, controls and other interfaces should not increase proportionally with machine size. The lower limits on the size of a pocket calculator are dictated, for example, by size of human fingertips for actuating the keys).
- **Overriding economic demands.**
  Certain parts (suspension eyes, packing boxes, bearing journals, inspection covers) should be in several fixed sizes to be produced in larger quantities.
- **Geometrically similar steps not feasible for parts, tools and production influences.**
  (E. g., semi-finished products, gear milling cutters do not have geometrically similar steps; wall-thicknesses (casting) cannot be made as thin as one likes;

likewise wall thickness of clamped parts (thin rings). Large work pieces in general have a lower strength per unit surface area, since, for example, annealing is more uneven than for small cross sections).

- Demand for **low-cost shape design.**
  An investigation into cast versus welded housings showed that a strictly geometrically similar shape design is not desirable for economic reasons. The portion of materials cost for small housings is low, but high for large housings (Sects. 7.7, 7.13.4, and Fig. 7.13-16). Accordingly, for small welded housings, heavy pieces (slabs) will be welded in the bearing area and seam length and the part count held small (Fig. 7.13-29). In other words, use large quantities of material to reduce production costs. For large housings, it is necessary to save on materials cost (i. e., plan on relatively thin plates and more ribs).

How far the geometrical similarity applies and makes sense has to be tested on an individual basis. Nevertheless, even with the above restrictions, similarity laws can still guide a well-planned development of semi-similar size ranges [Pah84].

7.12.5.5 Example of a size range

a) Task

A customer received a gear train with the technical data indicated in **Fig. 7.12-24** and a price of 55 000 $. Two hours later he called project planning and asked for information to be sent via fax about size, weight and price (one and two pieces) of a transmission of double the capacity (2 500 instead of 1 250 kW). Since there is no documentation on the desired unit, only an estimate can be made by applying similarity laws.

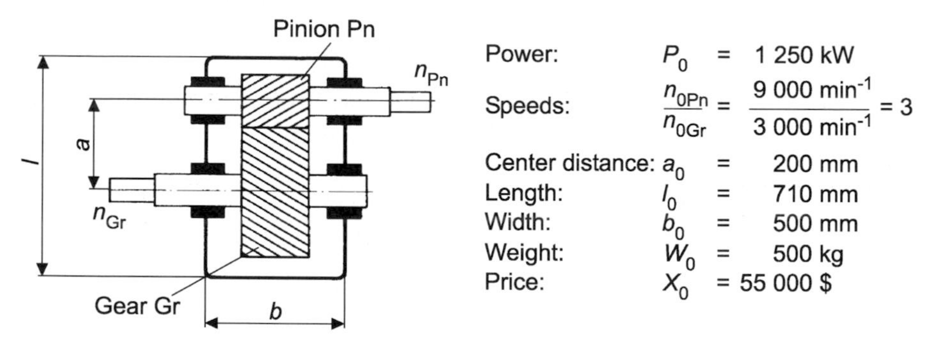

Fig. 7.12-24. Data of the basic embodiment (subscript 0) for a gear train

b) Derivation of similarity laws for size and weight

Only static forces caused by the torques are considered. Dynamic forces (inertia and centrifugal forces) and gravity forces are neglected. Material, constructive solution and production processes will not change. The dimensioning of the gears occurs in first approximation as with friction drives, for the permissible Stribeck pressure K. The module and with that the root stress, independent of the pitch circle diameter, can be varied later.

With normal force F, width of the gears b and equivalent diameter $1/d_E = 1/d_{Pn} + 1/d_{Gr}$ (d_{Pn} = pinion pitch diameter, d_{Gr} = gear pitch diameter) we get

$$K = \frac{F}{b \cdot d_E} \leq K_{allow}, \; F_{allow} \leq K_{allow} \cdot b \cdot d_E$$

Since K_{allow} is fixed, then $\varphi_L = \frac{b_1}{b_0} = \frac{d_{E1}}{d_{E0}}$, $F_{allow} \sim \frac{b_1}{b_0} \cdot \frac{d_{E1}}{d_{E0}} \sim \varphi_L^2$

With that, the transmitted torque becomes:

$$T_{Pn} \sim F_{allow} \cdot d_{Pn0} \sim \varphi_L^3 = \frac{T_{Pn1allow}}{T_{Pn0allow}}$$

and the transmitted power (since ω = const):

$$P_{allow} \sim T_{Pnallow} \cdot \omega \sim \varphi_L^3 = \frac{P_{1allow}}{P_{0allow}}$$

the weight of the transmission: $W \sim V \sim \varphi_L^3 = \frac{W_1}{W_0}$

With that all essential physical relationships have been found from the length step size φ_L (the power P is $\sim \varphi_L^3$, since ω was not set $\sim \varphi_L^{-1}$: Fig. 7.12-23).

The step size for the length results from the required doubling of power.

Since $\frac{P_1}{P_0} = \frac{2\,500 \text{ kW}}{1\,250 \text{ kW}} = 2 = \varphi_L^3$, then $\varphi_L = \sqrt[3]{2} \approx 1.25$, corresponding to the preferred number series R 10.

From that the **data for the transmission sought** (successive embodiment) can be compiled (values in italics, **Fig. 7.12-25**).

	Series	Ascertained step size	Known basic embodiment 0	Sought embodiment 1
Distance *a*	R 10	$\varphi_a = \varphi_L = 1.25$	200 mm	*250 mm*
Width *b*	R 10	$\varphi_b = \varphi_L = 1.25$	500 mm	*630 mm*
Length *l*	R 20/2	$\varphi_l = \varphi_L = 1.25$	710 mm	*900 mm*
Power *P*	R 10/3	$\varphi_P = \varphi_L^3 = 2$	1 250 kW	***2 500 kW***
Weight *W*	R 10/3	$\varphi_W = \varphi_L^3 = 2$	500 kg	*1 000 kg*

The value l = 710 mm is not contained in series R 10, but in R 20 (Figure 7.12-19). Thus every 2nd number from R 20 is used (R 20/2), which provides the same step size: $\varphi_{10} = \varphi_{20}^2 = 1.122 \sim 1.25$

Fig. 7.12-25. Application of similarity laws (preferred number series, see Fig. 7.12-19)

c) Derivation of the similarity laws for manufacturing costs

Next, we answer the question about the calculated price X_1 for the next embodiment 1:

Calculated price from the basic embodiment (subscript 0) was $X_0 = 55\,000$ \$.
At 5% profit, the **total costs** are $TC_0 = X_0 - 0.05 \cdot X_0 = 52\,250$ \$.

Manufacturing costs for $\dfrac{MC}{TC} = 0.70$ (company-internal calculated value, i. e., the manufacturing costs are about 70% of the total costs)
Manufacturing costs of the basic embodiment:
$MC_0 = 0.7 \cdot TC_0 = 36\,575$ \$;

Cost structure of the transmission supplied earlier, according to the information from costing department:

- Portion of production costs made up by set-up costs

  $mcs_0 = 0.2$ of manufacturing costs $= \dfrac{PCs_0}{MC_0}$

- Portion of production costs from production time

  $pce_0 = 0.5$ of manufacturing costs $= \dfrac{PCe_0}{MC_0}$

- Portion of size-dependent (material) costs

  $mtc_0 = 0.3$ of manufacturing costs $= \dfrac{MtC_0}{MC_0}$

Summary of manufacturing costs according to Eq. (7.7/3):

Manufacturing costs of the next embodiment:

$$MC_1 = MC_0 \cdot \left(\frac{mcs_0}{N} \cdot \varphi_L^{0.5} + pce_0 \cdot \varphi_L^2 + mtc_0 \cdot \varphi_L^3 \right)$$

for lot size $N = 1$ piece and $\varphi_L = 1.25$ gives

$$MC_{1(N=1)} = 36\,575\ \$ \cdot \left(\frac{0.2}{1} \cdot 1.25^{0.5} + 0.5 \cdot 1.25^2 + 0.3 \cdot 1.25^3 \right) = 57\,788\ \$$$

Total costs of the next embodiment:

$$TC_{1(N=1)} = \frac{57\,788\ \$}{0.7} = 82\,544\ \$$$

calculated sales price

$$X_{1(\mathrm{N}=1)} = TC_1 + 0.05 \cdot TC_1 = 86\,682\ \$ \cong 87\,000\ \$$$

For lot size $n = 2$ pieces, it is

$MC_{1(N=2)} = 53764\,\$$, therefore, about 7% lower cost,

$X_{1(N=2)} = 80646\,\$ \cong 81000\,\$$

Therefore the customer, after agreeing to the sale, is quoted two prices of 81 000 \$ ($N=2$) and 87 000 \$ ($N=1$).

We see that similarity laws help in the cost estimation of products in size ranges. However, the exponents in Eq. (7.7/3) (Sect. 7.7.1) should be tested for the company's conditions, and used only in the field in which there is experience.

7.12.6 Modular design

Modular designs are developed for limiting the multiplicity of special designs and the associated variant variety that come about in ordinary practice, without going deeper into the methodology. The cost advantages from the consistent use of modular designs are mostly in overhead costs, and there are few cost systems to calculate process costs. The cost savings compared to customer-specific special designs can be found only with difficulty by overhead costing (Sect. 6.3). They are usually also not passed on to the customer in the price. Ultimately, it leads to a mix of standard and individualized special solutions with a great number of unknown expensive variants. Customers do not have the price advantage of the standard solutions and so force the acceptance of their special wishes without extra charge. The following actions should help in correcting this situation.

The narrow modular design view with exactly defined elements is described in Sect. 7.12.6.1. In recent times the practice has been extended through terms such as modularization, platforms, and design principles. These deal with the basic idea of multiple use of known elements in modular design-like form. Therefore, they are also handled in this chapter.

7.12.6.1 Definition, purpose and effect

a) Definition

Modular design refers to a combination system of **building blocks** (elements) of different or same function and shape [Bor61, Bie71].

- By using **elements** with **different functions,** we get an overall system of different overall function (example: a handyman's multi-purpose tool set). Herein lies the difference to a size range: For a size range, the function is always identical; only the size varies.
- By using **elements** with the **same function,** we get a size change, but with same overall function (example: factory halls, bridges). Here there is a common attribute with products made in size ranges. However, products in size ranges do not represent any combination system, but rather a firmly defined system that is built only in different size steps.

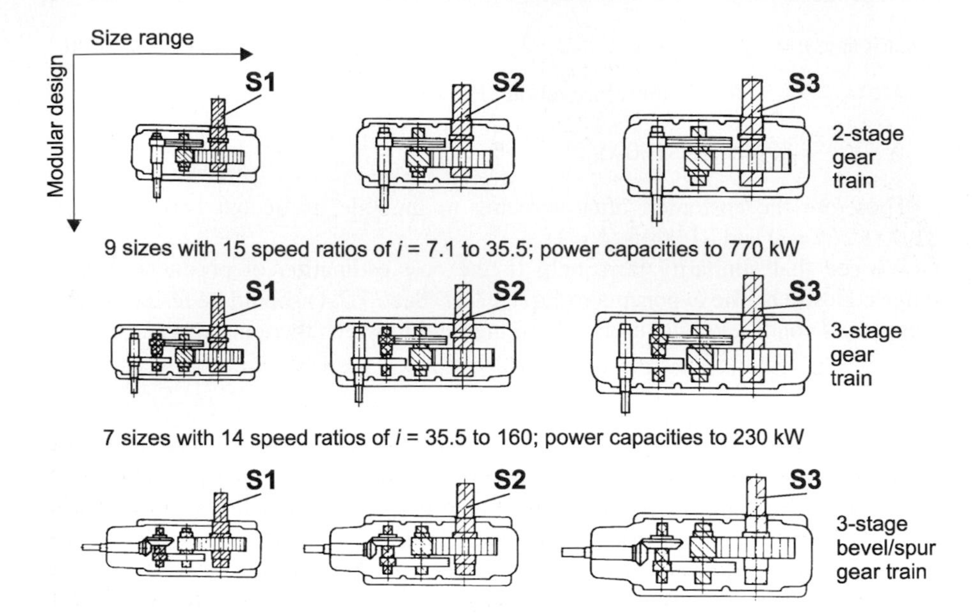

Fig. 7.12-26. Redurex gear train using the size range/modularity principle (from Flender)

- Very often a **modular design** is associated with a **size range**, that is, the elements fulfilling the same function are made in different sizes (**Fig. 7.12-26**).
- A **transmission in modular design** is shown in the figure**,** which has three different gear housings as the basic elements (building blocks) Moreover, they are made in different sizes as a size range (manufacturer's modular design). From left to right, the housings and gears increase in size, in the corresponding stages of a size range. Going from the top toward the bottom, the identically hatched modular elements (gears, shafts) are installed in the different types of transmissions. Top: two-stage spur gear train; middle: three-stage spur gear train; bottom: three-stage bevel/spur gear train. So, for example, from the top toward the bottom, the same driven shaft S1 with the gear is used in all three types of transmissions as a modular design element. The same is the case with the larger modular design element, the driven shaft S2, and so forth.

b) The **advantages** of modular design are comparable with those of a size range:

- **Cost reduction** of the modular design in comparison with a great number of specific single products with the desired function. An individual function can cause higher costs than a special product with this function. The emphasis is only on the cost reduction of the combination system. The **cost reduction for the customers** is achieved, as the example of the handyman's tool set shows. They must buy the basis element (drive motor) only once and the special elements (for example for sawing, grinding, milling) can be added by the customers. In addition,

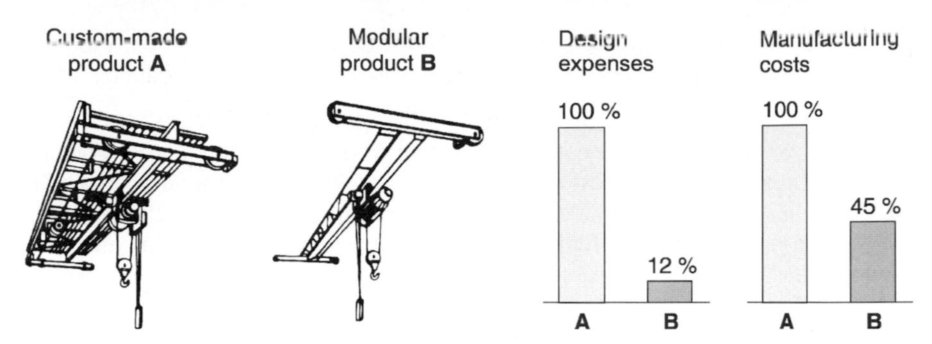

Fig. 7.12-27. Advantages of modular design for industrial cranes [Pei67]

there are cost advantages in maintenance, with the price of spare parts. The **cost reduction for the manufacturer** arises in that individual elements can be made in much greater quantities than with special products, since they are installed in several basis elements, as the transmission example shows in Fig. 7.12-26. **Figure 7.12-27** shows the cost reduction achieved in the case of an industrial crane made as a modular design. The most important elements are produced in larger quantities. Thus the one-time processing costs per piece are reduced in production, storage, purchasing, customer service, and in all overhead cost departments (e. g., design, sales). **Figure 7.12-28** shows the modular design of a conveying system with hanging rails. Of note are the accessory parts (a) and the combination example (b). It is an open modular design that permits the use of as many elements as desired.

- A shortened **delivery time is** partly attributed to the reduced processing time of a order in all overhead cost departments, and also to parts that govern the delivery time frequently being supplied ready from stock. With computerization of both bid release and order handling and scheduling, the delivery time can be shortened even more. Some technical and price details can be settled on a laptop, directly at the customer's facility.

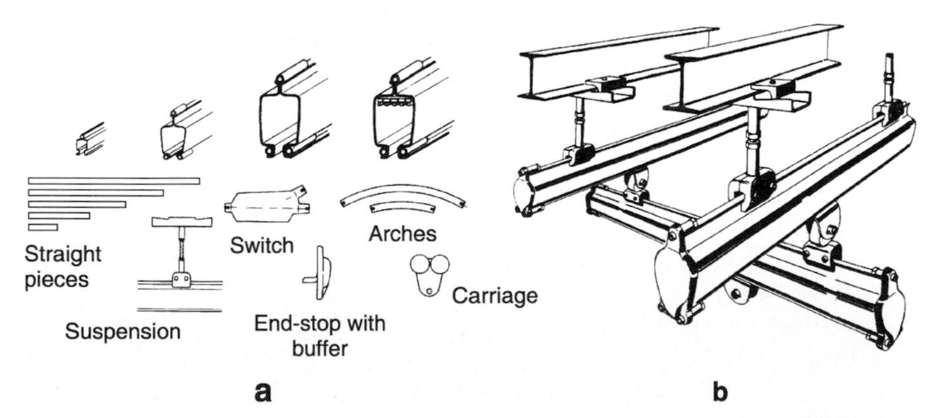

Fig. 7.12-28. Open modular design of a conveying system (manufacturer's illustration: Demag, Duisburg). a: elements, b: combination example [Pei67]

- There is an **increase in the quality** of the elements and of the whole modular design, since there is operational experience from previous products; spare parts are available and easily interchangeable.

The above **advantages** are equally important for manufacturers and customers. It is presumed that the biggest part of the cost savings to the manufacturers is passed on in lower prices to the customers. For the **customers** there is, in addition, the advantage that they are offered greater flexibility in being able to buy elements (functions) later on (for example, agricultural machinery modular designs, handyman's tool sets, and machine tools). For the **manufacturers,** the final advantage is that they develop long-term ties with the customers [Ben90, Pei67].

c) Disadvantages of modular design are:

- **Specific customer wishes** sometimes cannot be fulfilled. The customer must then resort to special designs.
- The modular design is often **technically** not as **appropriate** as a special product is since the elements must satisfy the overall system. Thus, for example, for a handyman's set without speed control, the drive motor basis element often has too high a rotation speed for drilling but too low for grinding. Because of the joints at the elements' interfaces, the weight, volume, or also manufacturing costs are often higher than for a special machine made in similar numbers. What matters is only that the overall system has lower costs. If the demand for certain functions climbs (with the handyman's tool sets, for example, circular saws or surface grinders), it can be that special machines become better and more economical in large quantities. This has happened with many functions of the handyman system.
- The **time required** for the customer for switching from one function to another (e. g., handyman modular set), is expensive for an industrial customer (**customer modular design**). Craftsmen and industrial companies consider this in the purchase of a modular product. No time is used for switching modules for the **manufacturer's modular design,** when the manufacturers only use a modular design in the production of their own products**.** They thus supply the customer ready-to-use products, assembled from elements. An example of this is in the truck and automobile production (platform strategy).
- The manufacturers become less flexible regarding **market wishes**. They seek to justify the **introductory costs** (which are often **considerable)** by sales to match the expense, and will consider product changes only over the long term. In this respect, the conception of a modular design is risky for the manufacturer and requires care in order clarification and embodiment. The aim manufacturer should strive for is therefore a **high flexibility** for products, supplied with **low variant variety**.
- To reduce the considerable **development risk for the manufacturer,** it is better to not look at a complex product in total as a modular design. Instead, realize it only as **local modular designs** within the backdrop of an overall concept [Koh96].

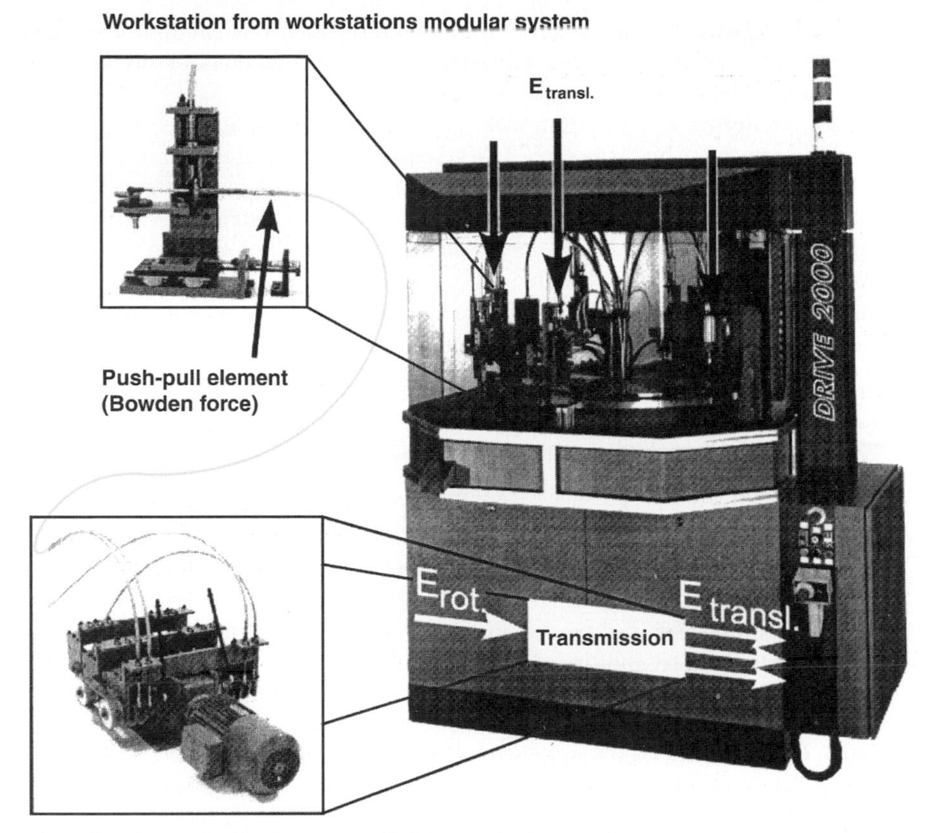

Fig. 7.12-29. Local modular systems for an automated assembly machine. *Top:* configuring the workstations, *Bottom:* configuring the cam disk transmissions [Koh96]

Figure 7.12-29 shows this strategy with the example of an automated assembly system circular table for small hardware items. With the local modular design of an adaptable cam disk transmission (bottom left in the figure) the desired work movements are generated that are directed with push-pull elements to the workstations (top left in the figure). The workstations also represent a local modular design.

7.12.6.2 Design (morphology) of modular products

If it is desired to create a modular design or to redesign an existing one, it is necessary first to have an **overall view** about what is already there, what is desired, and what possibilities there are for achieving that. For this purpose, **clarifying ideas** are often quite helpful, depending on the situation, that provide answers to the following **questions**:

- Which **functions** are implemented, with which function carriers up to now, and how frequently? How can one reasonably integrate functions or divide them up?
- What is **structure** of the system now, or the new one should have? What actually is meant here by "structure"? A system consists of elements (referred to here as elements or building blocks), that relate to each other in some fashion. That adds up to a structure, called the **modular design structure**. The **interfaces** between the individual elements depend very much on the structure.
- Which **type of modular design** is it actually? Who is supposed to deal with that? Is the modular system is supposed to be expandable or not? Which elements or building blocks are possible and reasonable?

Let's begin with the last question. First, we note that for all of the following alternatives there are also hybrid forms.

a) Types of modular designs [Pah97, Kol94]

- **The manufacturer's modular designs** are put together by the manufacturer. The users generally cannot modify them afterwards (e. g., automobile or truck modular designs, gear box designs).
- **The user's modular designs**, on the other hand, are put together by the customer, depending on the desired overall function (for example, handyman's tool sets, kitchen appliances, attachments for agricultural tractors).
- **Closed modular designs** are characterized by a predefined design program with a set number of total functions (e. g., a handyman's set).
- **Open modular designs** are open-ended in their possible combinations (e. g., kitchen furnishings, scaffolding systems, Lego and similar building-block toys). There is only a plan with building example, with applications of use.
- **Structure-bound modular designs** have alternative elements that are meant only for specific places in the modular product (e. g., alternative automobile engines or seats). They have a defined place in the structure.
- **Modular or free-module products,** on the other hand, have elements with different functions that can be arranged as needed on location (e. g., house entry intercoms, electronic surveillance and control systems, etc.)
- We **further** distinguish technical and natural modular systems (chemical system of the elements, cells, plant components), physical and non-physical modular systems (alphabet characters, words, music notes). Also concrete and abstract modular systems (e. g., potential modular design structure), and overall and local modular systems (as already mentioned in Fig. 7.12-29).
- **Combination of different modular designs**
  Modular designs more often exist unconsciously in designs than is acknowledged and discussed in the literature. The use of standard and purchase parts (Sect. 7.12.4.1) very often involves modular designs. The final manufacturer is like a customer of **modular designs from** many different **suppliers** (manufacturers of bearings, fasteners, drives, frames, etc.) who combines these with his own design parts and assemblies. These can arise from a company-specific

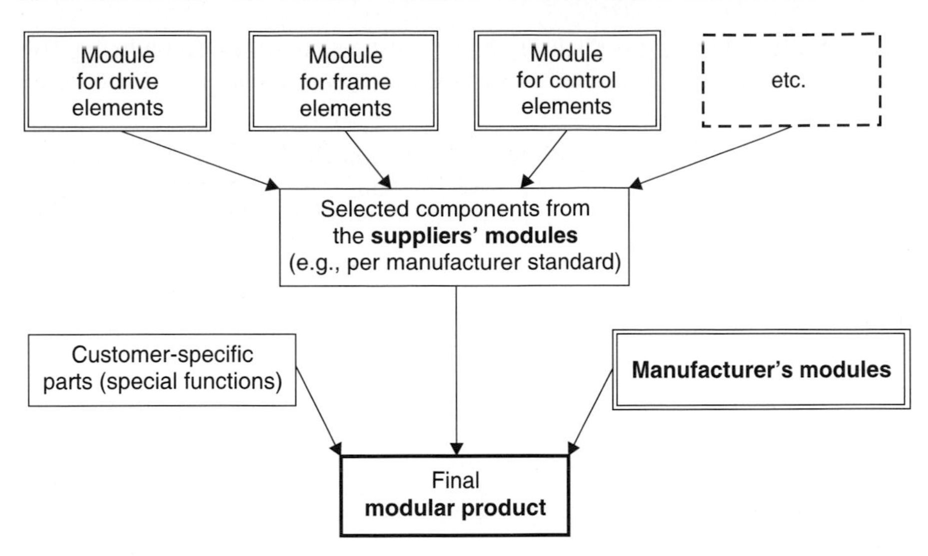

Fig. 7.12-30. Modular design product as a combination of supplier's and manufacturer's modular designs

modular design or only for a particular order as "special parts" (special function); see **Fig. 7.12-30**. The customer does not notice any of that. According to product category, numbers, and company strategy, the most diverse types of modular designs can arise in this way. Sect. 7.12.6.8 presents an example of such a combination of modular systems.

An important strategic question for the company is how far to take the **use of suppliers' modules**. Some companies develop extensive modular systems of their own and rely little on external modular designs. Other companies, such as in the transportation field addressed in Sect. 7.12.6.8, employ only suppliers' modular designs and put together their products from those. There are still others who utilize their own elements only for a few strategically important functions. Important aspects for the decision on this question are:

- Quantities produced
- Company size (design capacity)
- Availability of modular designs and their elements
- Company facilities and strategy (production capabilities)
- Selection of the market segment

b) Types of elements

- Corresponding to the variety of different modular designs, there are also different elements (e. g., physical versus non-physical; abstract versus concrete; technical versus natural).
- **Concrete elements** can be parts and assemblies as well as stand-alone products (e. g., flange/bearing system/electric motor).

- Important for the machine industry is their distinction according to the type of **energy**: mechanical, hydrostatic, hydraulic, electrical, electronic, and software elements.

Their function, shape and their mutual **compatibility** at the interfaces with regard to shape, material, energy and signal exchange are all important. The elements can be named after the subfunctions they realize. In this case, there are the **"must"** and **"can" elements.**

c) The modular product structure

This indicates the relationships and interfaces between the individual **modular products** (e. g., the grinder, drill, and saw configurations for the handyman). By a modular product, we mean the concrete result for a specific application. These modular products have, for their part, a structure relating their elements.

Directed "Gozinto graphs" are suited for the structure description. They have nodes (= object), these connecting arc-shaped lines (= existence relationship) and numbers on these arches, that specify the necessary amounts (e. g., from elements) [Koh96].

d) Functions

They are especially important because modular designs are usually meant to fulfill different functions, which are to be realized with few elements, but produced in the largest possible quantities.

Depending on whether we are dealing with open or closed modular designs or customer's or manufacturer's modular designs, different functional terms are used.

The terms from design methodology are generally applicable [Ehr95, Pah97]:

- The solution-neutral formulation of the desired (planned) purpose of a technical device is the **Total Function** TF. It can be subdivided into **subfunctions**.
- The **Main Function** indicates the main purpose. It is supported by **secondary functions.**

For closed modular designs (example: handyman's modular set), the additional following distinctions may be made:

- The **base function** BF (corresponding to the basis element) is the subfunction that is part of every total function to be fulfilled (e. g., power transmission, switch, motor and gearbox for the handyman's tool set). This is a so-called "**must**" function.
- The **special function,** SF, (corresponding to a special element) is the subfunction that is characteristic for the fulfillment of the respective total function (e. g., drill chuck and drill, sawing attachment, grinding stone with the handyman's modules). This is a so-called "**can**" function.

7.12.6.3 Developing modular designs

a) Fundamental procedure

- **The aim** of the design and development of a modular product is to recognize the subfunctions (elements) most frequently to be realized for the overall system (the modular product to be delivered) and to bring these together. Of these elements, as large quantities as possible should be produced. On the other hand, subfunctions (elements) that are seldom needed should be taken out so as not to burden a frequently needed total function with their costs.
- Since the **development effort** and the **risk** can be considerable, the work must begin systematically and not simply "get started somehow".

- According to **Fig. 7.12-31**, beginning with the **planning phase** at **A,** the existing situation in the business field (product area) is analyzed (A1), for which the methods from Sect. 7.12.3 can be used. Next, the business field with its potentials should be planned (A2). Only then does the usual task clarification for the technical details as well as for the target costs (A3) take place.

  The **conceptualization** of the modular product (**B, modular structure**) and **C, developing individual elements and modular products** (total functions and function structure; see Sect. 7.12.6.2d) follow this planning. These should occur in an interdisciplinary **team** (for example, occasionally management, design and development, marketing, sales, production, quality assurance, cost control). In **Fig. 7.12-31**, a handyman's modular design to be designed afresh is taken as a basis so that the example is easily understood. It has been consciously simplified.

  The procedure under **C** is well known: all modular design products 1-*n* (total function (e. g., drill, grinder, jigsaw) are developed in the concretization sequence of clarifying the task, conceptualizing, embodiment, and detail design (Fig. 4.4-2). In the Fig. these are shown as shifted in time. This is a compromise from practical experience for limiting the development effort at the beginning. For example, the modular products 1 and 2 are developed first, (here "drilling and grinding"). Then, with the experience gained with these and depending on market behavior, modular products 3 (e. g., circular saws) through *n* (e. g., jigsaws) are developed. It is often also done like this in practice, on more complex problems than the handyman's tool set. The concern here is to not make the mistake (as seen in practice) of having only the modular designs 1 and 2 in mind at the beginning, and neglecting further possibilities 3 to *n* in the modular design concept. This results in faulty compromises later. Staying with the example: The circular saw blade might interfere with the electric motor, for example, since there is too little space on the motor spindle. On the other hand, the rotation speed of the electric motor might be too high for the proposed jigsaw.

  The overall modular design and the essential elements must both **carry out an integrated "clarification of the task"** and the **conceptualization** at the beginning, in a team, as mentioned above. Later, the embodiment and detail design of the individual modular design products 1 to *n* can be carried out.

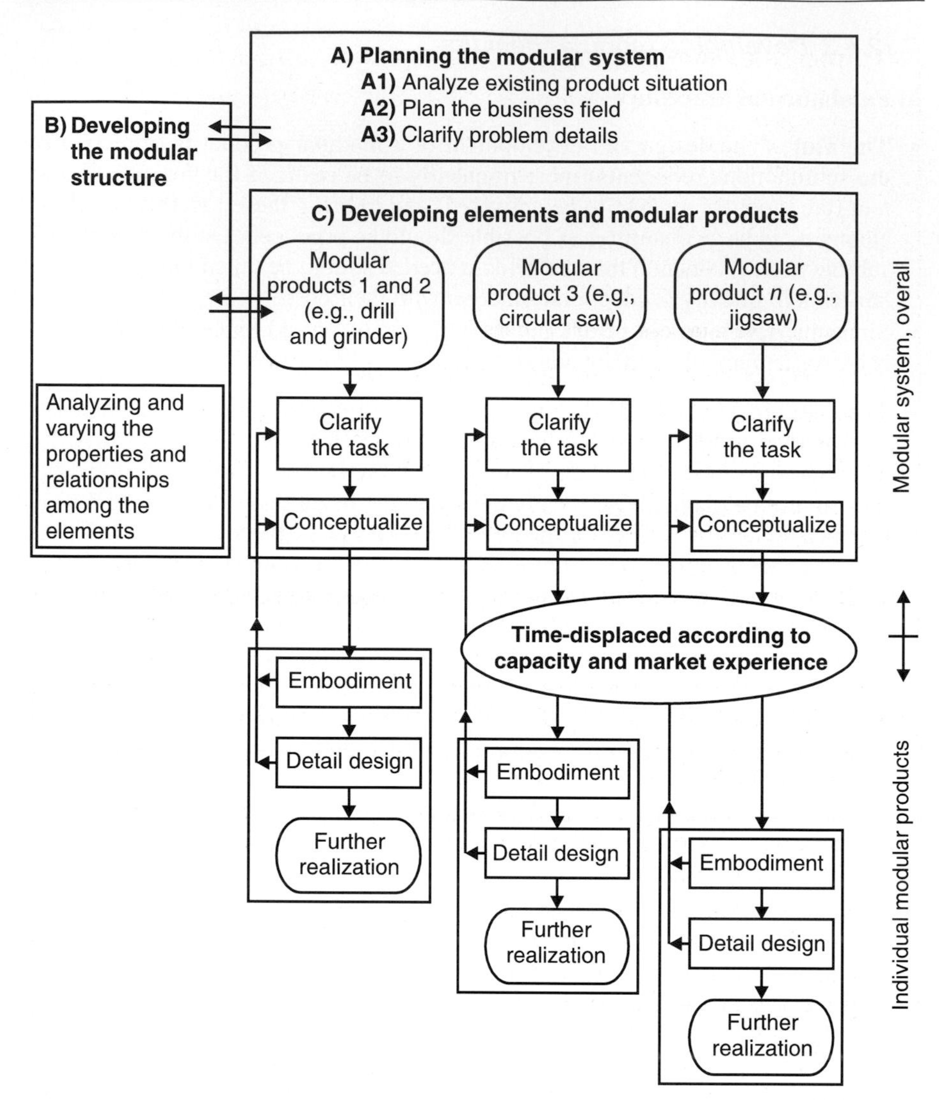

Fig. 7.12-31. Procedure during development of a modular system (similar to [Koh96])

b) Procedure in detail

In accordance with Fig. 7.12-31, first the above team **plans the modular design (Step A)**. For this purpose, the **Checklist A1** can be helpful to analyze the existing situation (**Fig. 7.12-32**).

The next step then is to plan the business field with questions according to the **Checklist A2** (**Fig. 7.12-33**).

A1 Analyze existing situation in product area (see also Sec. 7.12.3)

- Which product and part variants were delivered, to whom, how often, and when?
  a) From own company
  b) Thus far directly from competitors (competition analysis)
  c) "What works and why? What does not work and why not?"
  d) Lot sizes? Warehouse controls.
- Which (technical) problems appear? (complaints, supply time, customer service, ...)
- How high were the costs, profits/losses?
- How large is the market? The market share? Trends?
- Standardization:
  How similar are the variants regarding shapes and materials?
  What is the degree of standardization of the supplied product? (Sec. 7.12.4.1)
- What knowledge is gained from the analysis for further planning and task clarification? What is unclear? What are the risks/chances?

Fig. 7.12-32. Checklist A1 for the analysis of the existing situation in the product area (for Fig. 7.12-31)

A2 Planning the business field for a modular system

- How much market share as the aim?
  In which fields/countries? Till when?
- Which technologies (overall functions, modular products), which prices, which quality, what delivery schedules, which service steps are necessary?
  (Customer surveys needed)
- What is the potential for standardization?
  Modular products, size gradation, size range)
- What quantities sold? In which periods? At what prices? What are the cost targets that can be derived from this?
- What are the realistic quantities (lot sizes) for the most important modular products and elements, based on a preliminary design of the modular system?
  What are the cost potentials from that?
  (Value from Figure 7.5-5 of 15-25 % cost reduction for doubling of quantities)
- Estimate target prices and costs for modular products and important elements.
- What type of modular system would be most suitable (Sec. 7.12.6.2a)?
- What are the development expenses? Personnel capacity? Schedules?
  Development and marketing costs? Is investment a must?
- How and by whom and till when should the project be organized?

Fig. 7.12-33. Checklist A2 for planning the business field for a modular design (for Fig. 7.12-31)

It follows from checklist A2, that a provisional concept idea for the whole modular design must be available in a first version.

c) As preparatory work for the development of modular products and their elements (Fig. 7.12-31, Section C) and in parallel to that the modular structure (B), **an overall, technical, and economical task clarification** (A3) must be carried out. Task clarification for the individual modular design products 1 to *n* follows this.

This will be pursued here shortly for the modular design example of handyman's tools. A market and competition analysis shows that following **total functions** can be sold in a specific target market:

- total function TF1 "Drill" 100 000 times for price X1
- total function TF2 "Grinder" 90 000 times for price X2
- total function TF3 "Circular saw" 40 000 times for price X3
- total function TF4 "Jigsaw" 5 000 times for price X4
- etc.

In addition, accessories functions (designated as the function carriers: drill stand, handgrip, grinding wheel, etc.) occur with certain frequency. Additional technical requirements must still be determined.

It is clear however, that the **type** of **modular design** (Sect. 7.12.6.2a) dealt with here is a customer's modular design and a closed structure modular design. The customer is provided a precise technical description of which element to position where, so that the total function of the modular product (= element configuration) can be realized.

d) After that the **concept** for the whole **modular design** is developed, for which a **function structure** (and its variants) can be of help.

- **Setting up simplified function structures**
  Total functions are subdivided into subfunctions, so that as few subfunctions as possible occur, but are repeated. Since there usually is some idea of what the elements might be, a symbolic modular structure with labeled elements (e. g., as blocks) can be set up and variants for that considered. That has the advantage that the variants are easy to visualize. However, there is also the disadvantage that certain concrete elements become fixed in the thinking.
- **Conceptualizing in regard to solutions for subfunctions**
  If the solutions of the subfunctions are not known, the next step is solution search followed by evaluation and selection. Therefore, we seek suitable elements. Then different concept variants can be formed through combination. The selection is aided by a rough cost calculation (Chap. 9).
- **Embodiment and detail design of production documents**
  Because of higher development costs for more complex modular designs, not all modular design products (total functions) are addressed now, only some important ones (see above, under **a)** and Fig. 7.12-31).

For detail design and documentation, attention is paid to part numbering corresponding to the modular design logic, since that facilitates a computerized processing of the system. For the selection of elements and their combination leading to the modular product (with the possible bid preparation and job processing), software decision tables can be used. For the variant parts list that displays the product structure, a drawing numbering system with parallel coding is expedient (identification or counter and classifying number).

- For **fast bid preparation,** using a **computer-based configuration system** on a laptop directly at the customer's facility has proved itself in practice. It must be worked on jointly by design and marketing. The individual elements and modular design products relevant for the customers are described with their technical performance parameters. The sale prices and manufacturing costs are assigned. Thus, the success of standardization that has been developed is facilitated for sales, thereby also motivating them. It is better to avoid customer-specific special solutions, or else sell these at accordingly higher prices.

7.12.6.4 Modularization

To optimize the processes in design and development, procurement, production, and assembly, as well as customer service, products are divided into **modules.** For each module, clear responsibilities are set. Interfaces are thus kept small in the organizational sequence and in the function and product structures.

Usually it is a question of complex assemblies (sometimes enhanced with specific individual parts) that can be tested by themselves, and are also clearly accountable from quality viewpoint. For the formation of modules and for working with them, the definition of clear interfaces is necessary.

The aim is to reduce the complexity in products and processes through skillful distribution of the tasks for each organization and the persons responsible.

Examples of this are the so-called system suppliers. They are responsible for the design, development, and delivery of subsystems of a product. Such systems for automobiles can be the complete passenger compartment, the sunroof, or the front axle. For example, the first model of the Smart automobile was assembled in the final assembly from seven modules. The module suppliers have their respective system responsibilities.

According to Piller [Pil99], we can distinguish among the following types of the modularization:

Generic modularization
Putting together a product using always the same number of standard parts, that may have different features, on a fixed platform.

Quantitative modularization
Putting together products with many different built-in generic parts (standard parts) on a basic module.

Individual modularization
Putting together products from modules in fixed or variable numbers, that are in part customer-specific, in part come from a standard series, on basis of a root product.

Free modularization
Free combination of standardized and individual modules without the necessity of a standardized root product.

7.12.6.5 Use of platforms

If the same parts are planned and standardized for several products, we can speak of platforms [Kru98, Ker99]. Unlike the usual modular designs where the elements are defined once and then used unchanged, with platforms it is necessary and possible to adapt to the different modifications. Platforms minimize the costs of development, licensing, tooling, etc. that are often high. With platforms we can define particular parts such as specific sensors for a whole engine family; complex assemblies such as the basic assembly of the automobile body; or complex subsystems such as the automobile seat. Corresponding examples are found in many fields where products are made in series. What is important is that the platform should not be visible to the customer; otherwise, the individuality of the products would be lost. In the VW Company there is a so-called "hat" that is put on top of a platform. It refers to the outer skin of the respective vehicle. The hat is visible to the customer and is put on the platform. This means that 1 700 M$ in development and production costs were saved [Wil02]. **Figure 7.12-34** shows the various portions of the platform components in vehicles of the VW Company [Kru98].

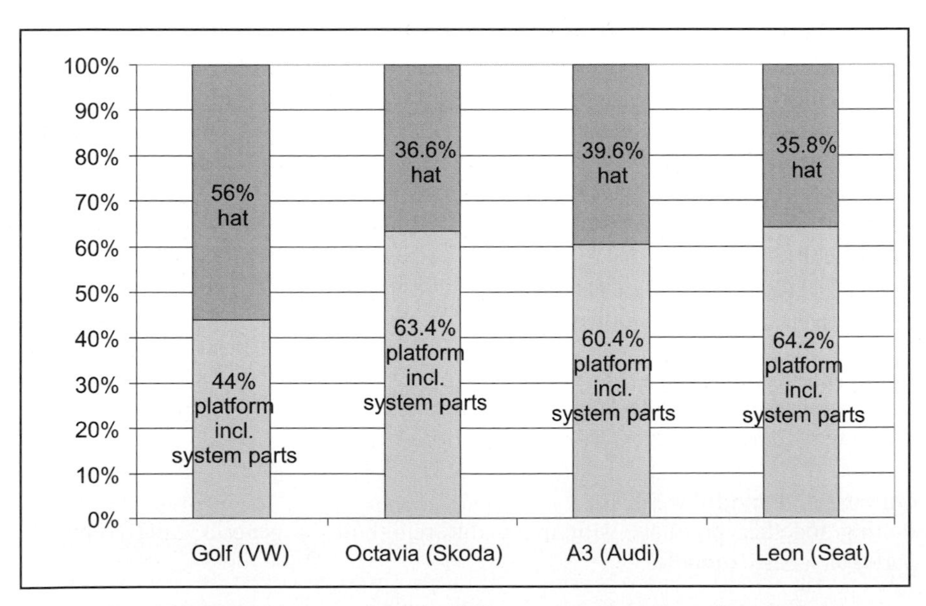

Fig. 7.12-34. Platform portions for different car models (from Kruschwitz)

Automobile example: Many discussions and publications have revealed that some vehicles with different design and brand names frequently use about two-thirds of parts that are similar.

Major appliance example: Products such as refrigerators, stoves, etc., are frequently produced under different brand names and with different designs from one manufacturer, with a very high proportion of similar parts.

Advantages:

- The production quantities increase through standardization, which can lead to considerable cost reductions.
- Development times decrease since the platform solutions can be applied in several products simultaneously. Innovations are thus transferred at an accelerated rate.
- The logistic processes in the company are unburdened by the reduction of materials flow. Parts and assemblies of all of a car manufacturer's models, for example, are supplied from the same sites.

Disadvantages:

- The standard platform can become a brake on innovation, since at times the importance of a larger product family must be considered.
- Inept advertising or publication of information that describes the platform design of different products can give rise to a cannibalization of the different products among each other.

7.12.6.6 Basic solutions, standardization

The preceding sections describe measures for directly limiting product and part variety. The measures described in the following act very early during product development to prevent the escalation of this variety. Thus, not only are the manufacturing costs lowered, but also the design and development costs.

In working with basic solutions the solution concept as such is set. The dimensioning occurs in the detail designing according to the individual requirements. Occasionally, detailed instructions are also given for the development process, which, for example, can include specific production standards, etc. With that the standardization can be a preliminary stage for the application of parametrics.

Figure 7.12-35 shows the use of the “gas spring basic solution” as solution for the function “dampened smooth motion control”.

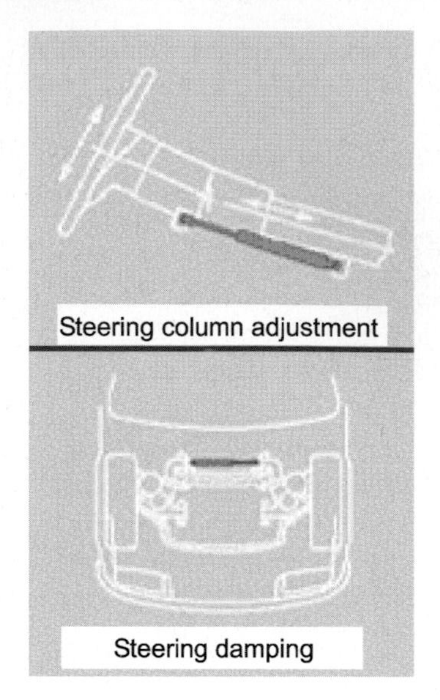

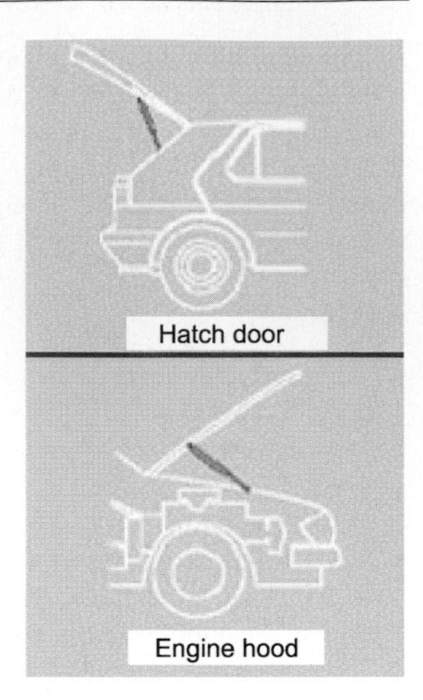

Fig. 7.12-35. Gas spring as a basic solution for the same function in different applications

Advantages:

- Well-proved principle, thus the reliability increases
- Shortening of the development time
- Need for tests or simulations reduced

Disadvantages:

- Fixation on available solutions, which hampers innovation
- Sub-optimal solutions

7.12.6.7 Parametrics, design logic

There are products made in single-unit production with known principles that are customer-specific, variant designs (see Sect. 4.5.2). In such a case the "inflexible" product in size ranges, as shown in Figs. 7.12-20 and 7.12-26, can be disadvantageous and sub-optimal with regard to costs. That happens when a CAD program exists for automated design according to the design logic. For clarification, the following **example of a large turbine transmission housing** is given:

The geometry of a steel housing for a high-speed gear train (turbine transmission, see Figs. 7.12-18 and 7.12-24) depends only on a few key parameters. These are for example, the center distance, the tooth width, the speed ratio, the peripheral tooth speed, and the journal bearing size. Earlier, the housing was cast in GCI (see Fig. 7.13-14) as a product in size ranges, with predefined center distance (increasing in steps) and housing width. For customer-specific requirements (e. g., speed ratio, rotational speed, power), technically less acceptable compromises had to be found in the inflexible "corset" of the size range, center distances, and housing widths, corresponding to Fig. 7.12-21. It is clear that the housings that were used for a specific application were usually too large and expensive. Therefore, the creative idea was to replace the cast housing with welded steel housing, which is produced for each customer's specifications (i. e., differential rather than integral design, similar to Fig.7.12-15, lower part). Such transmissions can be designed automatically with a CAD variant program. First, the gear-set details are calculated and the gear diameters, widths, the center distance, and the shaft diameters are established. On this basis, the geometry of the welded gear housing can be completely generated by the CAD variant program [Fig88, Lin92, Mer96]. If needed, the program can also calculate the manufacturing costs. The housing is now the smallest possible size for the customer-specific gear set. It is so to speak, a tailor-made suit that was made automatically. It is done this way in practice. Possibly, even the data for automatically cutting the housing parts from steel sheets is generated at the same time.

So, what about the **management** of **variants?** Fundamentally, more instead of fewer variants are generated! However, there are no variants that increase the design (process) costs. There is need to store only the customer-specific program versions or drawings; everything else is automated. Of course, some special efficiency-related features of large turbine transmissions are added. The size-dependent manufacturing cost portion of the housings is about 40 to 60% (Fig. 7.13-16, right). Costs are therefore proportional to the third power of the dimensions. Every decrease in size therefore acts to lower the costs significantly. Since the production – (including casting) involves only very small numbers, there is little quantity-dependent decrease in costs (Fig. 7.7-1). That is a chief argument for making a product in size ranges. Producing and storing expensive molding patterns is avoided by using welded design.

7.12.6.8 Example of a modular design in the storage and handling area

In the storage and handling field (moving, storage, and handling of materials), special solutions are developed for each customer. For example, in parts assembly, clothes manufacture, warehousing, etc., every plant design must be adapted to the customer-specific products, the throughput quantities, the space limitations, etc. Such plants can have from 500 to one million parts. The number of different parts can range from 100 to 1 000. To cope with these quantities and the variety of parts constructively, organizationally, within a given time, while continually being aware of costs, modular designs are applied in many different forms.

Technical task statement / field of application of the modular design

For production, assembly, and storage of parts of all types, materials flow systems are needed that are adapted to each task statement and limiting conditions. The

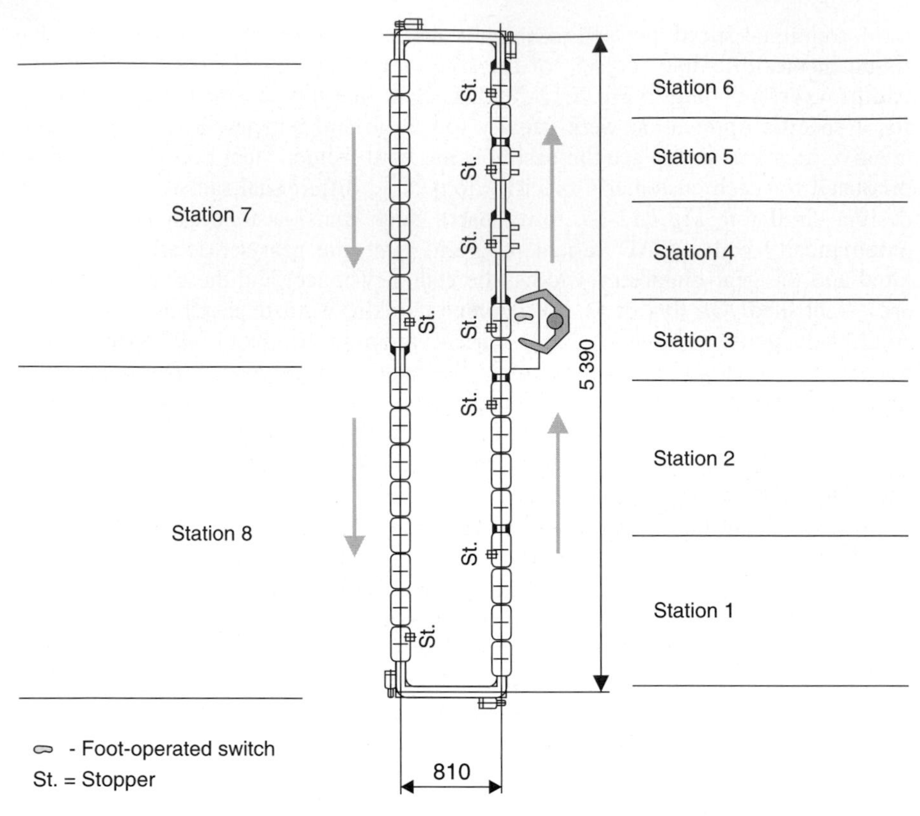

Fig. 7.12-36. Example of an assembly area

part transport system presented here[14] is suitable (**Fig. 7.12-36**, **Fig. 7.12-37**) for parts with a weight of up to 20 kg. The most important functions of the materials flow system plant are transport, stowing, buffering, storage, distribution, feeding, warehousing, supplying, handling and positioning. The essential elements are frames, guides, drives, etc. These plants are marketed worldwide and assembled finally as a functioning plant only at the customer's location.

In order to allow a configuration that is flexible, the system manufacturer resorts to purchase parts (supplier's modular designs) and own elements, as far as possible. Only very specific parts, such as the interface between the transport system and the individual product, are made job-specific (Fig. 7.12-30). For frame parts, fasteners, etc., available purchase parts are used. These are interchangeable and represent an open modular design.

On top of that are arranged the actual company-specific elements from the SHS modular system (storage and handling system; manufacturer's modular system). Parts at the interface to the customer product are sometimes job-specific.

[14] Storage and handling system of Dr. Hafner Montage- und Produktionssysteme GmbH, Kaufbeuren.

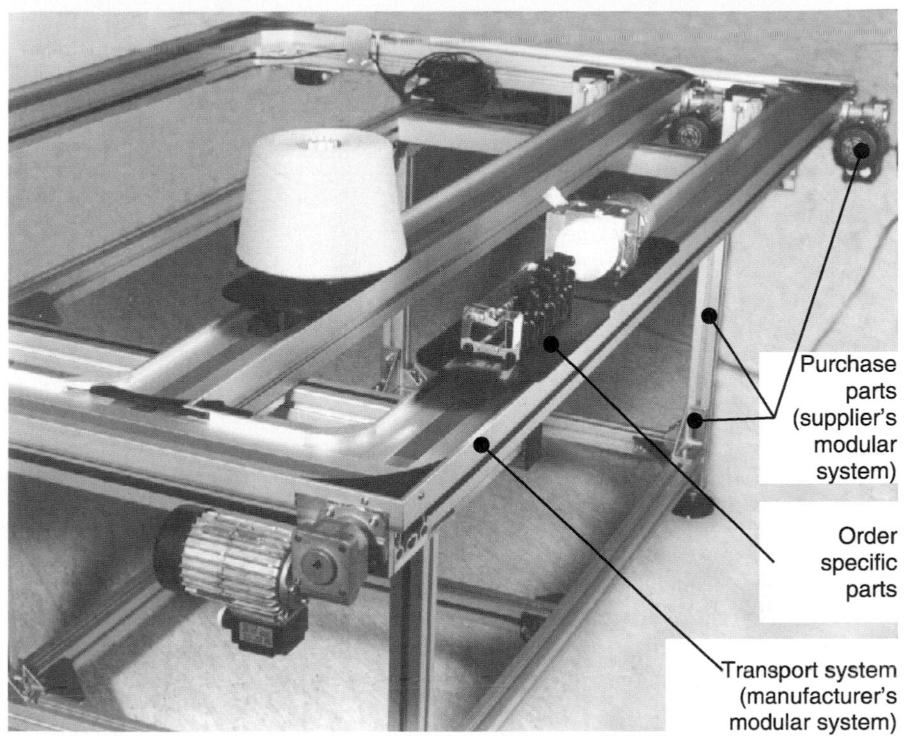

Fig. 7.12-37. Combining different modular designs in one plant

This cross-linked modular design has following **advantages**:

- For **secondary functions** (e. g., the frame), low-cost **purchase parts** (sections from a supplier's modular design) are used as elements (costs, reliability, delivery time, spare parts, etc.)
- For **primary functions** (e. g., the transport, positioning), elements from the SHS modular system are used. Thus, the project planning and design expenditure are kept within limits.
- Only at the **interfaces** to the specific customer product are **special parts** (e. g., a holder of a special shape) used, if necessary. There is unavoidable design expense only for these.

A computer-based selection and project planning system supports the project planning and design functions. **Figure 7.12-38** shows an excerpt from the results of this system. A CAD layout of the plant forms the basis of a quotation. In this layout the essential elements such as rails, switches, drives, etc., are determined according to type and quantity (detailed quantity schedule). From the parts list of the layout the data are delivered to the calculation program; this identifies the manufacturing costs per piece and the assembly time for each part. Added up,

Bid calculation						
Custemer: Exampel				Date: 97/02/03		
Proj. No.: Sample 01				Prepard by: J. Expert		
Pos. No.	**Qtty.**	**Item**	***MC*/pc**	**h-assembly**	**S *MC***	**S h-assembly**
1 032	15	Arc 90	150	0.15	2 250	2.25
1 033	10	Arc 60 right	130	0.15	1 300	1.50
1 034	8	Arc 60 left	130	0.15	1 040	1.20
1 036	15	Arc 45 right	100	0.15	1 500	2.25
1 037	12	Arc 45 left	100	0.15	1 200	1.80
1 038	100	Rail	50	0.05	5 000	5.00
1 956	1 000	Coupler	0.2	200		
Summe:		Layout			15 555	102.25

Fig. 7.12-38. Result of the selection and project planning system

along with the appropriate surcharges, they show the quotation price and provide the basis for ordering and assembly planning.

The **necessity and advantages** of such computer-based systems are outlined in the following; the numbers are only provisional values. The expense that had to be invested in the preparation of the system was beneficial, because:

- The **bid preparation costs** are **decreased** and the **bid accuracy** is **increased.** That is shown by the following estimate:

  Typically, one order results from about every ten bids made. The **bid preparation** for a plant having about 300 000 parts and about 150 000 $ in costs (1 part 0.5 $) takes about **one day with a computerized system** and about **two weeks without such a system.** Thus, **with** a computerized system an employee can prepare about 200 bids per year resulting in 20 orders per year; **without** the system, only 20 bids per year can be developed, resulting in perhaps two orders per year. If the costs for this employee average 50 000 $ per year, then the bid preparation costs in the first case are divided into 20 orders: 2 500 $ per 150 000 $ order. However, in the second case they are divided only into 2 orders: 25 000 $ per 150 000 $ order. This correlation leads to the practice of bids often not being developed in detail, but being only roughly estimated. The results are, as with the $/kg cost calculation of cast parts, that only **those** bids produce orders in which the costs were estimated to be too low. That leads to economic difficulties not just for the individual manufacturer but also for the whole industrial sector The greater the bid scatter is in a given line of business, the greater the drop in the average sales price.
- **Order processing** is **faster** and has fewer **mistakes.** With bid preparation that is computer-based, a detailed quantity schedule is developed (Fig. 7.12-38). That has advantages not only during the cost calculation for the bid, but if the order is received the necessary parts can be ordered and supplied very quickly (Fig. 6.2-2).

- There is **reliable installation** and **startup** at the plant. The detailed quantity schedule is also an important prerequisite for the later smooth assembly and startup of the plant, which is otherwise often delayed by overlooked parts [Hub95a].

7.12.6.9 Example of modular design for sports cars

The Porsche Company was in a crisis during 1991–1992 that threatened its very existence [Gut98]. Sales Figs dropped and the business results were negative, mainly caused by the collapse of worldwide automobile sales and the devaluation of U.S. currency. In addition, the company's model policies did not satisfy the changing market needs, and its production and cost structures were no longer competitive.

Porsche had produced three models that were totally independent of each other in technical details:

- 911: 6-cylinder, air cooled, rear engine
- 944: 4-cylinder, water-cooled, front engine
- 928: 8-cylinder, air cooled, front engine

With regard to road performance, quality profile and price, the models did not distinguish themselves clearly enough. The missing standardization between the size ranges gave rise to high costs in production, purchase and logistics.

The decision was made to completely redesign two model series (the Porsche 911 and Porsche Boxster), with a concurrent conversion to a comprehensive similar-part strategy.

The following demands formed the basis of the new design:

- Optimal coordination of the package concepts and interior dimensions to each other
- Dimensional coordination of the body shell
- Stylistic harmony between the two series
- Coordination of production sequences for both series

As example of the results, we mention here only two points:

Two vehicle concepts but one front-wheel drive car (Fig. 7.12-39)

Features that establish concept and styling, such as the position of the windshield base, the pedal area and the constructively demanding firewall area, were identical for both vehicles. Clearly defined interfaces to the interior and to the rear of the car assured the model-specific parts were securely connected.

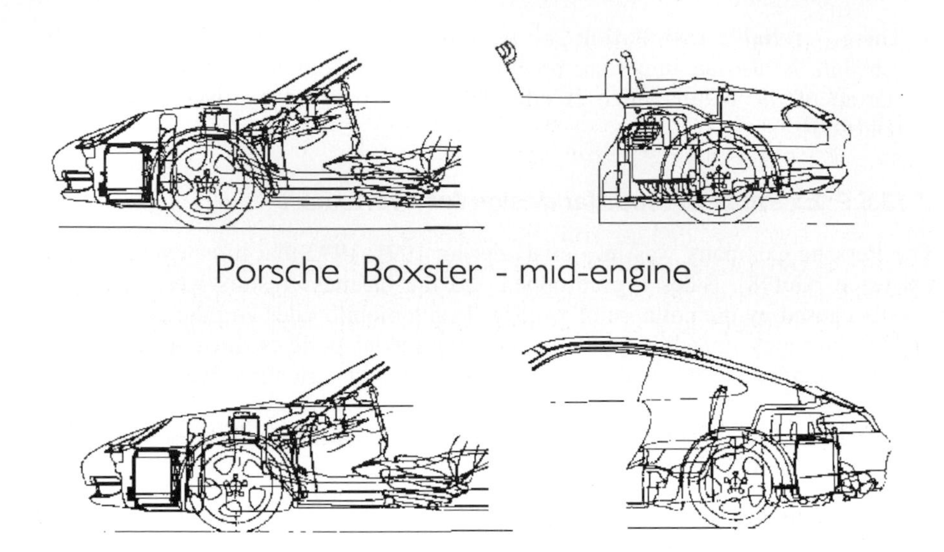

Fig. 7.12-39. Two vehicle concepts but one front-wheel drive car

One door for 3 cars (Fig. 7.12-40)

Door shell systems with the associated parts such as hinges, window lifters, and door locks are some of the most expensive assemblies on the body. Porsche developed a commonality in the door openings, seams, hinge posts, and the shape of the front fender from the bumper to the door opening. The door thus developed can be used in the Boxster, the 911 Carrera Coupé, and the 911 Carrera Cabrio.

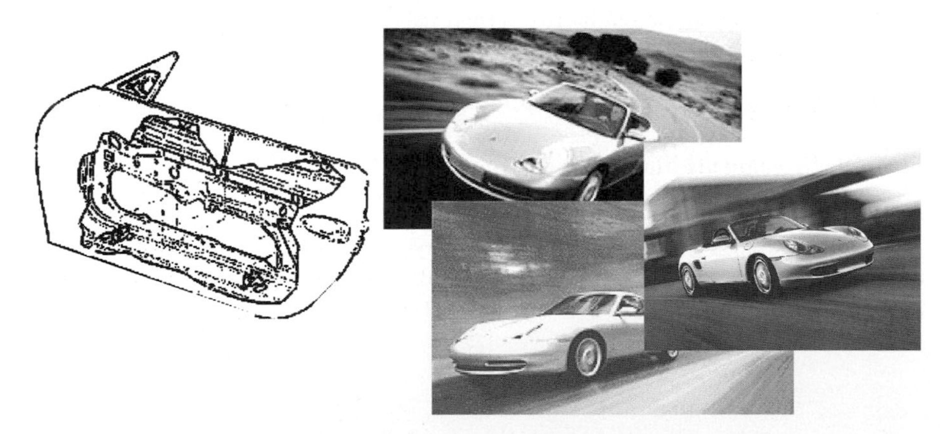

Fig. 7.12-40. One door for three cars

Further examples of the use of similar parts are:

- Cockpit: A common basis cockpit module
- Wheel suspension: Modular design of axle assemblies
- Engine: Conceptually identical engines

The nearly parallel design and development of the two product lines occurred under consistent compliance with the same-part strategy. Through that, many synergy effects became apparent – the economical Boxster line could be made profitable. In spite of the vast conceptual differences, Porsche succeeded with this design in fashioning 43% of the all parts as same parts, thus lowering costs considerably.

7.12.6.10 Example of a modular design / size-range system for tractors

a) State of the art

Agricultural tractors are a good example of cost-sensitive products that are made in medium-sized series; they are needed for food production in almost all countries of the world.

About two-thirds of all tractors made in Germany are exported. Thus, the requirements that are very different around the globe result in great technical complexities for the manufacturers and too large a model/variant variety. That led to the primary concept of the "standard tractor" [Ren99].

Due to their power range of about 10 to 220 kW, the requirements on the tractors are the following:

- Provide traction on farmland.
- Carry and drive many different devices.
- Be suitable for highway driving (50 km/h).
- Have automatic regulation and control (CAN bus network for the internal and external processes (ISO 11783), such as drive-by-wire and GPS positioning [Ren02a].
- Meet safety, comfort, and environmental requirements.

Contributions to variant management in the tractor industry are the following: In 1974 Welschof described for the first time the family principle [Wel74]. Renius refined this in 1999 [Ren99] and proposed that since the tractor technology worldwide is always becoming more diverse, the technology could be divided into grades I through V that, in turn, have seven different components each (**Fig. 7.12-41**).

Components:	Power			1 Drive			2 Engine					3 Transmission					4 PTO			5 Hydraulics				6 Cabin			7 Electronics		
Technology grade	low	medium (~40-80 kW)	high	only rear	all-wheel opt.	all-wheel stand.	1-cylinder	2-cylinder	3-cylinder	4-cylinder	6-cylinder	very simple	simple	semi-automatic	automatic	continuously variable	540 RPM	540 and 1000 RPM	3 or 4 speeds	rear lift	remote control	rear & front lifts	load sensing	no cabin	simple cabin	comfortable cabin	none	some	extensive
I	X			X			X	X		X		X					X			X				X			X		
II	X	X			X			X	X	X	X		X					X		X	X			X	(X)		X	(X)	
III		X	(X)		(X)	X			X	X	X			X				(X)	X	X	X	X			X	(X)		X	(X)
IV		X	X			X			(X)	X	X			X	X				X	X	X	X	X			X			X
V		X	X			X				X	X				X	X			X	X	X	X	X			X			X

Fig. 7.12-41. Modularity principle for tractors: Proposal for division of the worldwide demand of tractors into 5 technology grades by means of the chief components

Each grade has a typical performance bandwidth (small, medium, large). It increases from technology grade I toward V, to higher values. There are additional differences for the seven tractor components: chassis, engine, transmission, power take-off (PTO), hydraulics, cabin, and electronics. The technology grades are typical for the economic level of the corresponding markets: In developing countries, grades I and II are represented; in the industrialized countries, grades III, IV, and V are utilized [Ren02b]. The **cost structure** for technology grades IV and V is approximately as follows: Transmission and rear axle, 30%; engine, 17%; cabin, 14%; electrical equipment, 10%.

b) Terms used for modularity principle

The aim of a low-cost tractor system must be to cover a wide range of tractor models with the smallest possible parts variety.

To narrow down the variety, the following terms are to be distinguished:

- **Product line** (or the product program of the manufacturer) which consists of several **families** (for example, four) in highly developed markets. This is composed of perhaps three to five **types** each, for which the power rating is regarded as a characteristic feature of a model [Ren99]. Special designs such as, for example, viticulture tractors or tractors for municipal use are excluded here.
- A **family** or size range (also called a **series** in the tractor industry) refers to several types that are very similar to each other, with many similar parts and usually a uniform wheelbase. **Figure 7.12-42** shows a family structure as a typical goal for a "Tractor Product Line – Europe 2002" [Ren02b]. Series 1 has a wheelbase of 2 150 mm, Series 2 of 2 400 mm, etc. Actual industry programs always deviate a little from such targets because of existing product lines.
- A **type** (or tractor model) is determined mainly by an engine with a specific power rating. For example, Series 1 (with a 2 150 mm wheel base) consists of models with three-cylinder engines of 30 to 55 kW.
- Within a type, there are still more **variants**. These arise from specific customer preferences such as, for example, all-wheel drive instead of rear-wheel drive, adjustable wheel track, and with or without a cab.

Tractor Series	1	2	3	4
Typical wheel base [mm]	2150	2400	2750	2950
Engine output [kW]	30 - 55	60 - 90	90 - 125	135 - 210
Engine speed [RPM]	2 100 - 2 500			
Diesel engine	3 Cyl. Displ. ~ 3L	4 Cyl. Displ. ~ 4L	6 Cyl. Displ. ~ 6L	6 Cyl. Displ. > 7L
	turbo-chargers and coolers very common			
Range of functions	medium	very large (variants)		large
Level of comfort	medium	high	very high	

Fig. 7.12-42. Modularity principle for tractors: Several product families/series form a product line – example "Standard tractors – Europe 2002"

c) Modularity rating

For cost-based evaluation, Renius [Ren02b] proposed a modular design rating that refers to parts:

$$\text{Modularity rating} = \frac{\text{number of different parts}}{\text{number of models produced}} \leq \text{limit}$$

In the interest of lower costs, this rating should be as low as possible. Target values for tractors are not generally known [Ren02a] because they depend on company-specific factors. The manufacturer must define what counts as a part (e. g., own parts, purchase parts, manufacturer's own standardized parts or assemblies; however, not standard parts such as screws, retaining rings, etc.) This strategy increases the quantities per part, lowers the manufacturing costs, can put pressure on the purchase prices, and reduces the logistics and assembly costs. Also consequently, the process costs decline in purchasing, production, and customer service. The philosophy resembles the platform strategy of automobile manufacture described above, but was consistently applied in the tractor industry many years earlier [Wel74].

d) Practical procedure for limiting part count

The aim is to try to make the investment-intensive parts and assemblies similar of same, such as large castings, sheet metal and forged parts, etc., within a tractor family.

The following are **especially uniform within families**: engine concept, basic measurements (e. g., wheel base), gearbox modules, chassis and body (sheet metal parts), cabin, piping/wiring systems (hydraulics, fuel, air, air-conditioning equipment, electrical equipment, electronics, etc.), and fuel tanks. From a **modular design viewpoint,** the aim is to use same parts in the families. Examples are uniform cabins (the human being is always the same), uniform electronic parts and software, uniform device interfaces (e. g., "rear-end interface"). This principle is especially marked between adjacent families (however, very little between Series 1 and 2, because of the technology leap). Some companies augmented Series 2 at the upper end by the use of six-cylinder engines. They broke through the scheme of Fig. 7.12-42 because the customers favored under-loaded six-cylinder engines as opposed to highly overloaded four-cylinder engines, in spite of higher costs.

It would be ideal to cover all five technology grades mentioned above with only one product line according to Fig. 7.12-42. This would allow the entire international market to be covered by a single modular design. Unfortunately, experience shows that is not possible [Ren02b] because of the much too large a range of the necessary technology and the corresponding revenue differences.

What is, however, attainable is a realization of technology grades III, IV, and V with one well-planned product family or series, particularly in the more developed countries.

An example of that is the Series 6 000 from John Deere, developed and produced mainly in Germany; it is assigned to Series 2 with an additional six-cylinder engine (Fig. 7.12-42). In 2002, it was available worldwide in the Series variants 6005 (III), 6010 (IV), and 6020 (IV) [Ren02c]. **Figure 7.12-43** shows Model 6420 as an example.

Fig. 7.12-43. Typical European medium-class standard tractor in 2002, with variants corresponding to technology grades III, IV and V; from J. Deere

7.12.7 Summary

In the preceding sections a whole range of possibilities were shown for limiting the variant variety to product and part level and to process level. To satisfy the virtually infinite number of customer wishes with the smallest possible number of variants, it is necessary to attempt a variety of measures, not just a single measure. The discussion shows, in part, different perspectives and focal points that will be summarized here.

The left part of **Fig. 7.12-44** shows how, without variant management, the individual customer preferences are realized by 100% individual modules (assemblies, parts, etc.), in addition to the usual standard parts. The right side of the figure shows how the number of customer choices can be kept within limits by using different standardization approaches that were shown in the preceding sections.

Different combinations of the standardization approaches crystallize out, depending on the market and the product. Some examples (two very different ones from automobile manufacturing and two from machine and plant engineering) should illustrate this:

- A **Formula 1 racing car** certainly contains some standard and purchase parts, but will be predominantly made of individually designed and produced parts. For such a product the demand for optimal performance is certainly foremost, despite other demands and possible cost savings from the use of standard parts.

  On the other hand, for **high-volume automobiles** many customer wishes have to be satisfied with low costs if possible. Therefore, such a car is largely produced from especially developed parts that are, for all practical purposes, standard parts because of the large quantities made. Special wishes that were not preplanned are either not filled, or filled only at high costs. High-volume manufacturers set up their own subsidiaries for the fulfillment of these special wishes.

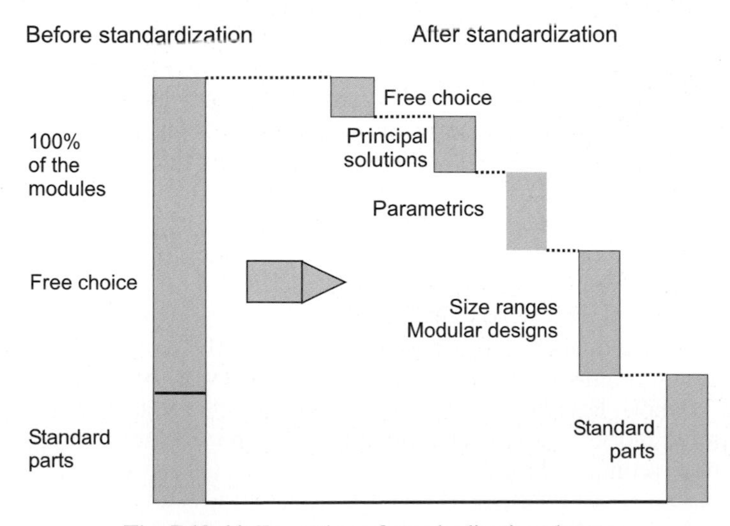

Fig. 7.12-44. Formation of standardization classes

- **Assembly plants for medium volumes** (Sect. 12.6.8) are indeed focused on specific customers, but mainly for reasons of standard items, cost, availability, and reliability. Merely unavoidable adaptive or interface modules are specifically resolved as parametric parts or type or individual parts.
- Between the extreme cases discussed above are the so-called model solutions (e. g., specific **gear trains**). Thus, high-speed turbine transmissions are designed totally for specific customers to achieve an exact speed ratio (Sec 7.12.6.7). Model solutions are the basis of customer-specific solutions, which have defined the solution principle but not the individual design.

The few examples discussed above show that the situation must be specifically tested in each individual case to find the optimal solution for now and for the future. Due to market shifts, new technical advances, etc., decisions made earlier must always be given a second look. Essentially, the aim is always to strive for the reuse of products, parts, and processes by applying the options shown here.

7.13 Results of a cost benchmarking project

7.13.1 Overview and procedure

Benchmarking is the comparison of features (processes and products) with the best in the world, including those made by competitors. The aim is to learn and to introduce timely measures to become better than the competition. Benchmarking projects can be carried out in a partnership and by mutual agreement with other companies, or internally and hidden [Cam94, Mer94, Pie95]. For technical product comparisons, one can buy competitive products and examine them according to all relevant technical properties.[15] It is frequently said, however, that this is not possible in the area of the costs. For cost benchmarking, one must analyze the causal processes and the corresponding affecting variables (cost drivers). The necessary information can be collected from internal and external sources (e. g., from common suppliers and customers, in a surprising variety of forms (personal information, brochures, expositions, analyses, publications, etc.). They are arranged, systematized, and assembled like a puzzle. Subsequently, new data are brought forth. Benchmarking is an iterative process. Kreuz [Kre97] provides seven steps on how to proceed (**Fig. 7.13-1**). For details, the reader is referred to the cited literature.

A Massachusetts Institute of Technology study in 1991 [Wom91] showed what decisive and international effects benchmarking can have, pertaining to Japanese, American, and European car manufacturing. Under the rubric of "Lean Production", a rethinking process was also introduced in the German industry that has had effects on the triad of costs, times, and quality, and thus also on the profits of the companies. Similarly, McKinsey has carried out several benchmarking studies [Romm93, Klu94, Rom95]. The MIT study compares the strategies and measures of successful and less successful companies with the catchwords of "simplicity and emphasis setting" [Ehr06].

---

[15] Case studies on product benchmarking in [Sab97].

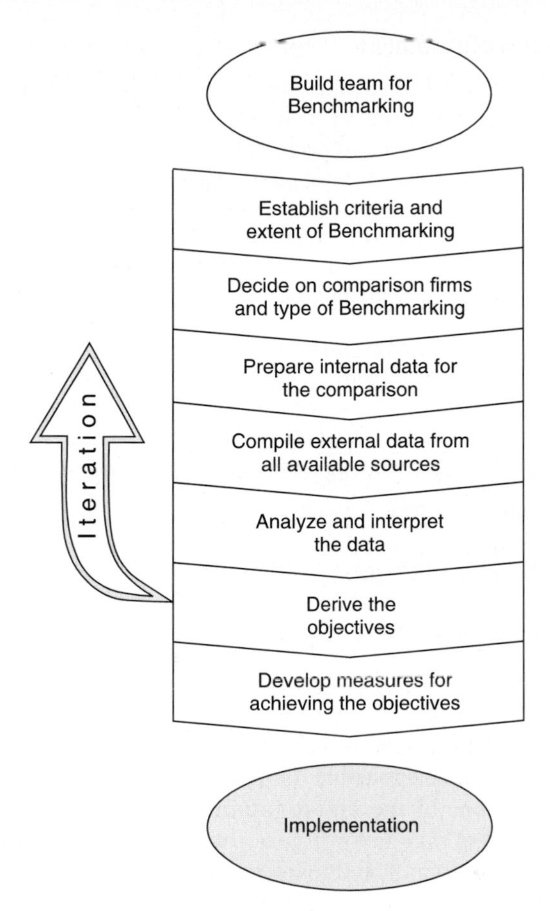

Fig. 7.13-1. The benchmarking procedure [Kre97]

In the following the process of benchmarking is not reviewed. Rather, the results from a long-term benchmarking research project are presented.

7.13.2 Costs benchmarking in the gear transmission industry

In the period from 1978 to 1994, the Gear Drives Research Association FVA, together with the Institute of Mechanical Design of the Munich University of Technology analyzed the costs of single- and two-stage stationary gear trains [Bru93]. This involved five research projects, with the participation of 8–15 companies. It dealt with strongly object-related benchmarking, which necessitated comparing the supporting processes. As examples, only some results will be described here. As will become evident, many of the patterns recognized here are generally valid for the whole machine industry.

a) Purpose and accomplishment

One of the aims of the investigation was to recognize the essential parameters affecting the manufacturing costs of gear drives and their parts, and to be able to improve them as regards design and production technology.

A second aim was the evaluation of each company's internal processes and facilities relative to the direct competition. Consequently, the company's strengths and weaknesses could become apparent.

In detail, following were accomplished:

- A ranking of the **cost parameters** and the dependence of the manufacturing costs, in particular, on size of the unit and quantities produced (Sects. 7.6 and 7.7; Figs. 7.6-3, 7.7-1, 7.7-2).
- **Cost structures** of the gear drives and/or their parts and their dependence on size and quantities (Sect. 7.7, Figs. 7.7-3 through 7.7-5).
- **Rules** and **measures** for lowering costs.
- **Quick cost calculation procedure** for manufacturing costs, related to the desired transmitted torque [$/Nm].

The **procedure** in all five projects was essentially identical:

- **Concept and description** of the project (e. g., cost analysis of gears). Search for interested FVA member companies.
- Formation of a **working group** from the participating companies, which consisted mainly of parties responsible for production planning and cost analysis.
- Deciding on the **scope of the investigation**, i. e., the variants whose costs had to be calculated. This may involve gears of 50 to 2 500 mm diameter, straight and single-helical, quenched and tempered and/or hardened, made from two to three different materials. Value was placed on a technically clear and binding comparison basis for all. The participating companies had to have production experience for all variants.
- Development of **drawings** for the parts whose costs had to be calculated (unfinished and finished parts), and a guideline for a uniform calculations sheet, which corresponds to the work schedule.
- Shipping the documents to the companies. These documents were routed **anonymously** after completion to the Institute for evaluation, via the distribution point of FVA in Frankfurt. Absolute secrecy was an assumption for the projects. This was achieved by coding the companies (Code No.).
- **Presentation** of the company data that were evaluated, to the members of the working group and **arrangements for further investigations**. For example, the setup and production times as well as the machine tool types with their place cost unit rates were evaluated for each individual work step. The costs could be calculated from that. The technical data were imported according to agreed-upon calculation methods.

The last three operation steps were performed repeatedly for enhancing the scope of the investigation and for covering the database with different components.

b) Inter-company spread of the manufacturing costs

Amazingly, the large spread of the manufacturing costs among the participating companies was generally from 1:2 to 1:4, although technically everything was the same.

This is shown with the example of gears in **Figure 7.13-2** (see also Sects. 9.3.7.1 and 9.3.7.2).

The large spread is primarily in the different time standards (for example, feed rates, setup times, cutting speeds, etc.) attributed to production planning and not to different workloads in the companies, as is often presumed. There are apparently different internal guidelines for time standards for otherwise identical technical situations. The following conclusions may be drawn from this:

- In the single-unit and limited-lot production there is still considerable potential for rationalization for purely production technology reasons that constitute between 20 and 30% of the production costs depending on the situation in the company.
- Even with standard materials, purchasing can save at least another 10%; for forgings, where there are even more potential savings.
- Depending on size, some manufacturers are more expensive than others. Some companies make small gears at a low cost, but not large gears, and vice versa. This results from the convergence of many factors and often explains the typical strengths and weaknesses in the market. This **size dependence** of the calculated manufacturing costs on gears is presented more exactly in **Fig. 7.13-3** for four companies relative to the mean value of the competitors. Again, there are more expensive companies (B) and companies with lower costs (M). Furthermore, there are companies whose average costs differ with the part size (A, I). That can be important for long-term shifts in the sales program (Sect. 9.3.7.2).

In spite of the large number of participating companies and the great spread of the parameters, it was possible to form conclusions and develop rules for low-cost design that are valid industry-wide (e. g., Sect. 7.7).

Some years later the above investigation into gear costs was repeated. It was found that the spread between the companies was even greater regarding production times and costs of materials.

Thus, the conclusion was that large production and cost differences between competing companies portray a stable situation. Naturally, it may be that individual companies became more economical while others deteriorated. This large spread in times and costs, with technically the same components and/or processes, occurred with all investigations on gear drive components. These are described later, in Sects. 7.13.3 through 7.13.7.

Meanwhile, a project on the comparison of process costs was carried out [Gle97]. This investigation showed that with overhead costs in the indirect fields (design, sales, etc.) there is also a similar spread as described above [Bro98].

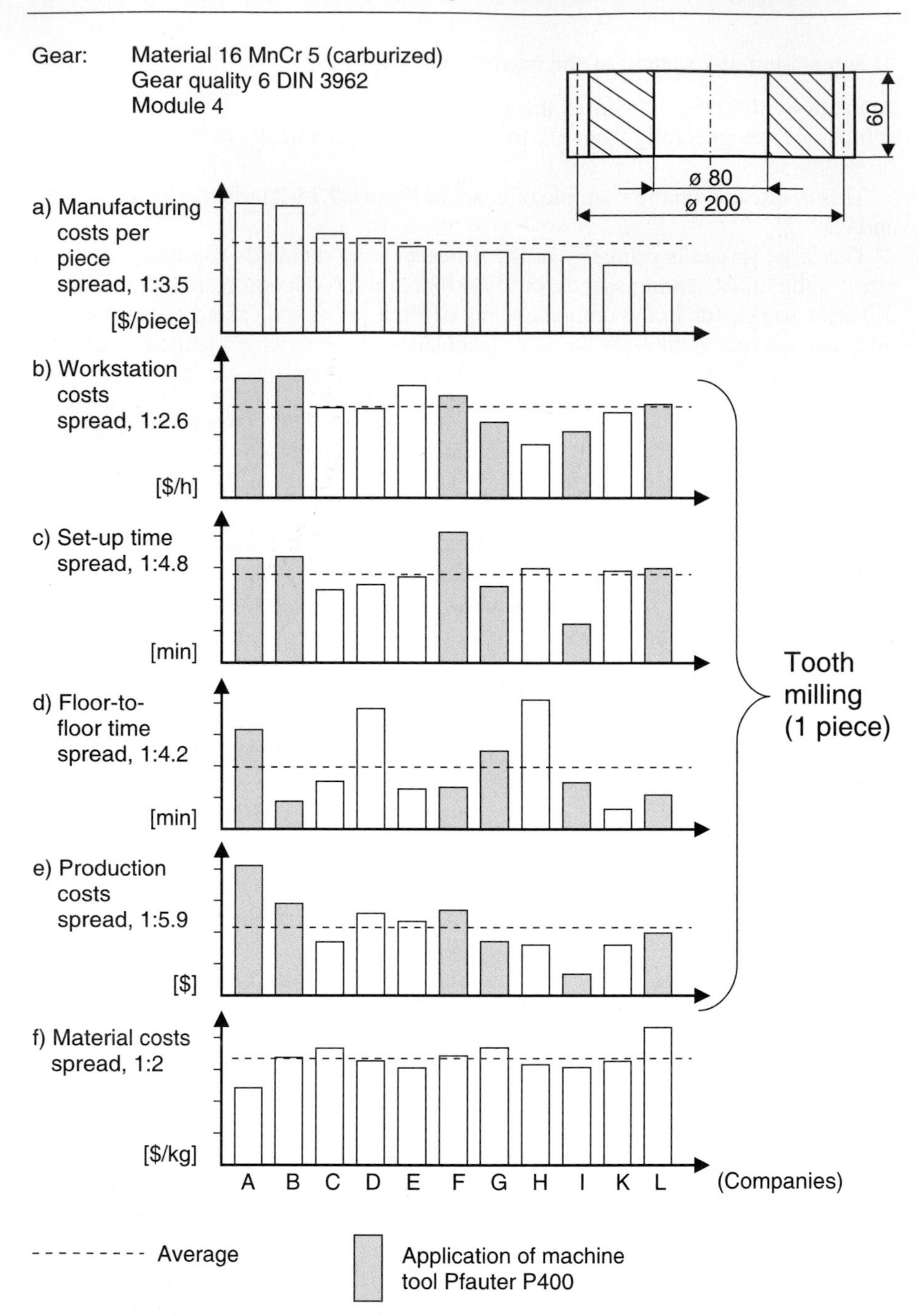

Fig. 7.13-2. Costs of gears calculated by eleven companies of the Gear Drives Research Association (FVA) (same drawing, same work schedule, same calculation sheet; shaded: same machine tool – Pfauter P400; A to L: encoded companies [Fis83])

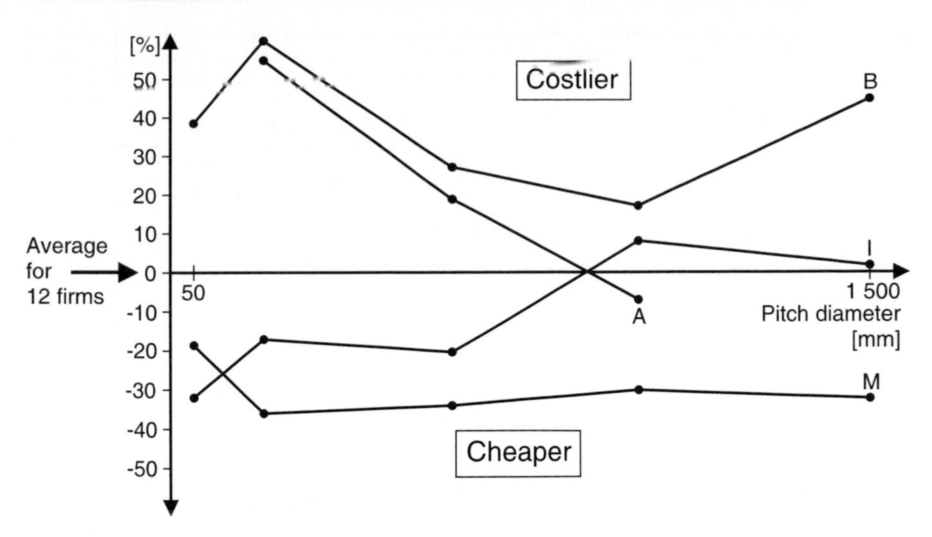

Fig. 7.13-3. Calculated manufacturing costs of gears of four selected companies A, B, I and M relative to the calculated mean value for all twelve companies [Fis83]

7.13.3 Gears

The parameters examined for the cost analysis of gears are apparent from **Fig. 7.13-4**. Pitch diameter range is 50–1 500 mm (in individual cases, up to 2 500 mm) spur gearing and helical gearing with 20-degree helix angle, quenched and tempered, as well as case-hardened and gas-nitrided [Fis83].

The following findings came out of this study:

- The **size** (gear diameter) is a leading parameter for the manufacturing costs (Sect. 7.6, Fig. 7.6-3). For small gears there is little change since the dominant cost elements, setup and idle time costs, do not depend significantly on size. For large gears the manufacturing costs change with nearly the third power of the dimensions since the material and production-time cost elements, including the weight-dependent heat treatment costs, replace the above portions. The costs per unit of transmitted torque were highest for small sizes and dropped to a nearly fixed value for large gears.
- The **number (lot size)** is another important parameter for manufacturing costs. Due to the high share of setup costs for small gears, the manufacturing costs fall off quickly with increasing lot size. They approach the cost structure of large gears, so that their relationships and rules apply (Sect. 7.5, see also Figs. 7.7-1, 7.7-2, 7.7-5, and 7.7-6).
- The **width/diameter ratio** of the gear sets is of little importance, if manufacturing costs per unit of transmitted torque are considered (Sect. 7.6.2).
- The **module** has considerable influence on the costs, but only for gears with low tooth numbers.

Parameter	Range investigated	Effect on *MC* for pitch-ø			Comments	Figure or Sec.
		ø<200	ø≈500	ø>1m		
Size pitch-ø	d_1 = 50 to 1 500 gear 16 MnCr 5	med.	high	very high	φ_L = linear scale for geom. similar size change	7.6-3 7.7-5
Quantity (lot size)	1 to 100 (dep. on size)	high	medium	low		7.7-1 7.7-2 7.7-5
Speed ratio	i = 1.12 to 6.3				For i = 6, the gear sets the *MC*	7.13-6
Width/ diameter	b/d_1 = = 0.3 to 1.2	medium/low			Depends on speed ratio	Sec. 7.6.2
Gear design	Solid, welded, cast gear			med./ high	Welded gear: method of costing for heat treatment important	-
Material and heat treatm.	Case hard./ quench-temp. gas-nitrided	high	very high	very high	Cost influence for same size: medium to low	7.7-3 7.13-5 7.13-16
Gear quality	4 to 8	med./ high	medium	med./ low	Behavior very company-specific	Sec. 7.11.6
Module/ no. teeth	m = 2 to 20 mm	med.	low	low	For < 35 teeth: effect medium to high	-
Helix angle	$\beta = 0^{\circ}; 20^{\circ}$	low	medium	low	For wide gears: medium to high	-
Other quantities						
Pattern cost share (wood patterns)						7.13-11 7.13-12
Integral / Differential design pinion						7.12-15
Cost structures: - by cost categories - by production operations						7.7-3 7.7-4
Manufacturing costs spread						7.13-2 7.13-3

Fig. 7.13-4. Parameters investigated in the cost analysis of the FVA gears [Fis83] and references to additional Fig. and section numbers in this book (Explanation: "Effect small": less than 10%; "Effect medium": 10–20%; "Effect high": more than 20%)

- The **helix angle** has no significant effect on the manufacturing costs per unit of transmitted torque. Helical gears, however, transmit the torque at a lower cost for the same tooth stress as spur gears. The thrust bearings that might be needed and their assembly costs are not considered here.
- Higher **gear tooth quality**, particularly for small gears, leads to higher manufacturing costs. For large gears, the influence is less, in particular because of the high weight-dependent portions of cost. Here there are significant company-specific cost differences (Sect. 7.11.6).

- For **small gears, material** is of secondary importance provided the gears are made in small lots. On the other hand, for **large gears** and with large lot sizes, a manufacturer should aim for a material of lowest possible cost, with high load capacity. The heat treatment costs (annealing, hardening, etc.) behave in the same way as the material costs, since they are traditionally calculated in \$/kg. Both elements can be summarized as "weight-dependent" costs.
- **Heat treatment** or **hardening** are common with the materials suitable for that, and are of great importance for costs since they directly influence the size. As **Fig. 7.13-5** shows, the manufacturing costs per unit of transmitted torque decrease for case-hardened as opposed to quenched and tempered gears, to approximately 30%. On the other hand, gas-nitrided gears have a lower load capacity and require a slightly more expensive material, 42 CrMo 4 instead of 16 MnCr 5. Therefore, the costs drop in this case only to 45%. It should be noted that case-hardening works all the better when the torque, and thus the size, is larger.
- The **gear wheel shape** (solid, welded or cast gear) is of interest only for pitch diameters above 1 000 mm to 1 500 mm (with single-unit production). For diameters of around 2 000 mm, welded or cast designs are more economical, particularly vis-a-vis solid gears, since they have high portions of size-dependent costs (design variants, Fig. 7.13-17).

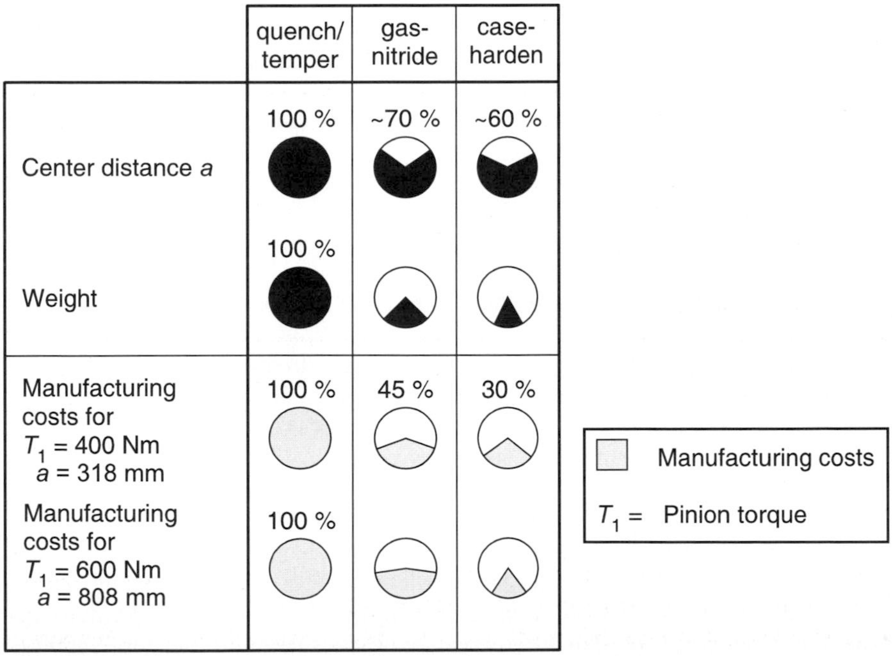

	quench/ temper	gas-nitride	case-harden
Center distance *a*	100 %	~70 %	~60 %
Weight	100 %		
Manufacturing costs for T_1 = 400 Nm, a = 318 mm	100 %	45 %	30 %
Manufacturing costs for T_1 = 600 Nm, a = 808 mm	100 %		

Manufacturing costs
T_1 = Pinion torque

Fig. 7.13-5. Comparison of manufacturing costs of gear sets with the same Hertzian pressure of safety, for speed ratio $i = 3.55$ (materials: 42 CrMo 4 quenched/tempered, 42 CrMo 4 gas-nitrided, 16 MnCr 5 case-hardened, lot size 4, load capacity proof according to DIN 3990, [Fis83])

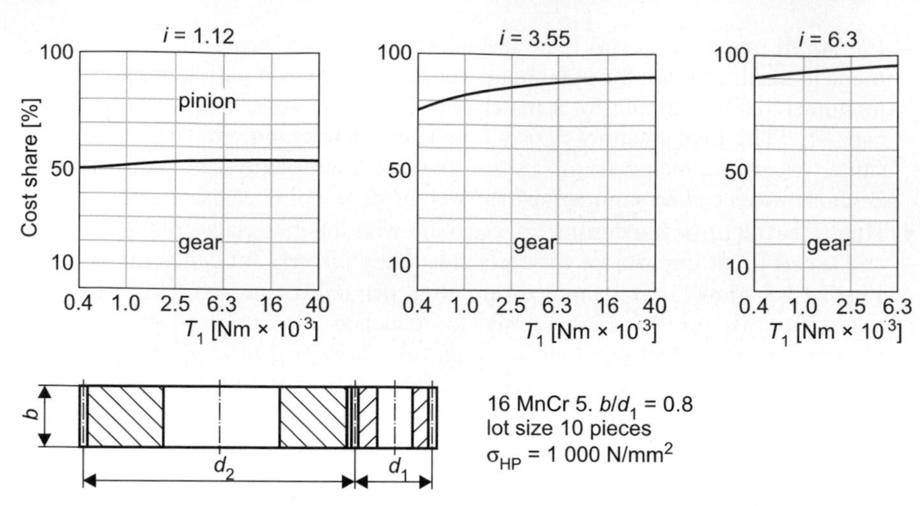

Fig. 7.13-6. Cost shares of gear and pinion in the manufacturing costs of gear sets ($b/d_1 = 0.8$, 16 MnCr 5, lot size = 10 [Fis83])

- The essential cost points may be recognized from the **manufacturing cost structures.** These are subdivided according to production costs either from setup times, from total time units, or from costs of materials (Fig. 7.7-3), or also according to costs for production operations / material costs (Fig. 7.7-4). They are of outstanding help for design and they change little over the long-term. In **Fig. 7.13-6**, the relationship of the costs of pinion to gear is portrayed. It is clear that for a speed ratio near 1:1, the gear and pinion each has a 50% share of the cost of the gear set. However, with $i = 3.55$, the gear brings about 80–90% of the costs, so one can concentrate mainly on the gear concerning the cost estimate for a gear drive.

7.13.4 Comparison of welded and cast gear housings

Welded and cast gear housings for two-stage gear drives shown in **Fig. 7.13-7** were analyzed for manufacturing costs [Haf87].

The figure also shows an averaged cost structure of the whole gear drive. While the gear set (gears and shafts) causes approximately 43% of the manufacturing costs of a gear drive, the housing with 36% has the second-highest contribution. The cost analysis of the housings is also especially beneficial because there is considerably greater freedom in design for the housing than for the gears and shafts. The housings were designed according to the criteria of the participating companies, as shown in **Fig. 7.13-8** (welded housing) and **Fig. 7.13-9** (cast housing). From these the enormous size differences are obvious (weight: 40 kg to 7 300 kg). **Figure 7.13-10** gives an overview of the investigation. There is a high probability that the basic cost results are also valid for other welded or cast housings.

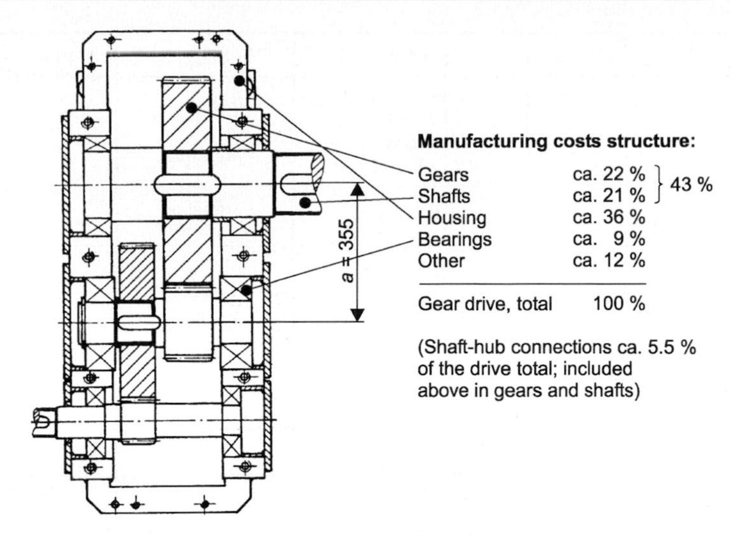

Fig. 7.13-7. Two-stage gear drive, with center distance *a*, part-based cost structure [Haf87]

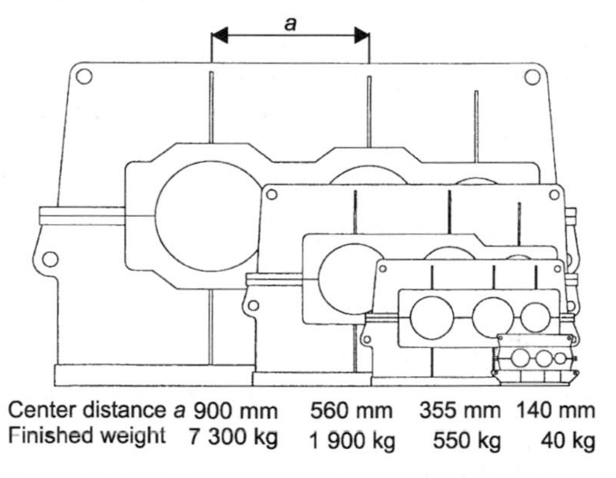

Fig. 7.13-8. Size range: welded housing [Haf87]

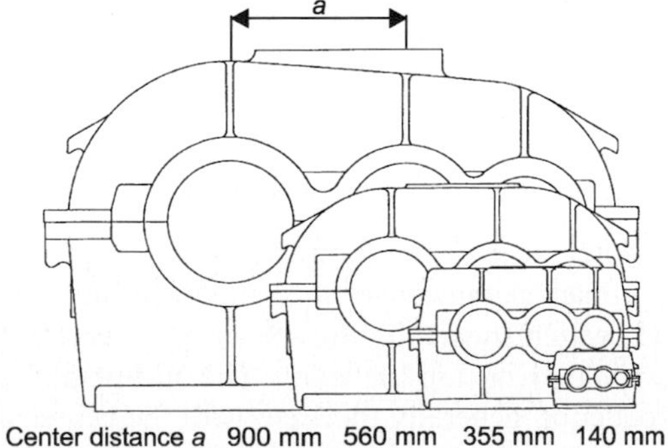

Fig. 7.13-9. Size range: cast housing [Haf87]

Parameter	Range	Prod. type	Figure No.
Size	Center distance a = 140 ... 900 mm Finished weight 40 ... 7 300 kg	Casting Weldment	7.13-11
Quantities (lot size)	1 ... 50	Casting Weldment	
Width/length of housing	0.3 ... 0.56 (for housing from GG 25, a = 355 mm)	Casting	7.13-14
Materials	for Casting: GG 25, GGG 40, GS 45, GAl Weldment: St 37, St 52	—	
Casting process	Sand cast - manual	—	
Patterns in various qualities	Wood (H1, H2, H3, S1, S2); Styropor	—	7.13-12
Welding process	Inert gas weldment E/MAG	—	
Housing machining	conventional milling, turning, boring	Casting Weldment	
Other items			
Cost structures Limiting quantities		Casting Weldment	7.13-13 7.13-15 7.13-16
Type of design		Casting Weldment	7.13-14 7.13-30

Fig. 7.13-10. Parameters and ranges of the investigations for the FVA cost analysis: welded/cast housings [Haf87]

Designers generally have **three questions** (Fig. 7.11-2):

- Which production process (welding or casting) is appropriate?
- If this decision is made, how does one design the housing appropriate to the task, and at low cost?
- How much of the production is in-house or outsourced? (Complete housing from the supplier or only the unfinished housing, preprocessed, or all in own shops?)

The following **findings** came out of this investigation:

- The **size** (here, center distance or indirectly, the weight) is, as for the gears (Sect. 7.13.3), a primary parameter of the manufacturing costs. That is evident from **Fig. 7.13-11**, which shows the growth laws for cost elements of cast and welded housings plotted on log-log axes. The **unfinished housing costs** for both types of production generally increase with the center distance a (in general, the dimensions) at more than a quadratic rate. That is, a housing twice as big costs at least four times as much. The costs for **machining** the feet and joint

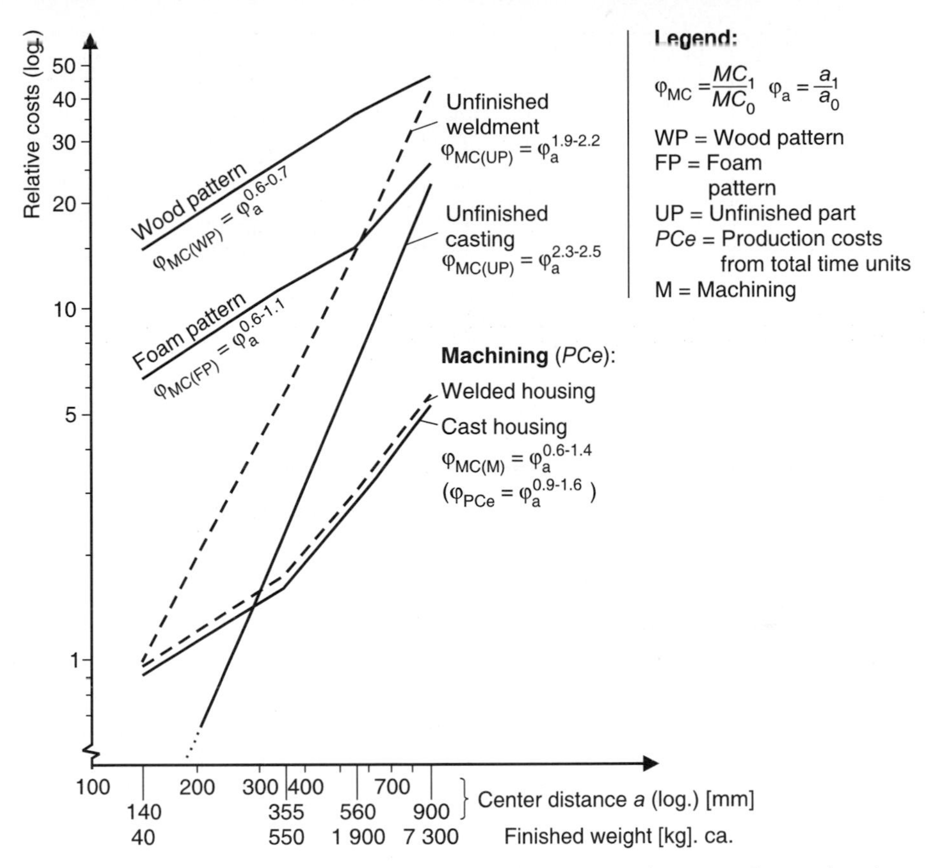

Fig. 7.13-11. Cost growth laws (average values) for cost elements of gear housings (lot size = 1) referred to the center distance; φ_a = step size [Haf87]

surfaces as well as the bearing surfaces (*PCe*: production costs from total time units) increase approximately linearly with the size. Machining is faster for large dimensions because of switching to larger, more expensive machines. The **pattern costs,** on the other hand, increase less steeply, which comes from the fact that the larger the product is, the increasingly smaller part the pattern costs play in the manufacturing costs.

That is also clear with the example of gears (Sect. 7.13.3) in **Fig. 7.13-12.** Even for casting one piece ($N_{tot} = 1$), the pattern costs for a gear of 2 000 mm diameter are only about 7% of the manufacturing costs (i.e., one-half), compared to a gear of 1 m diameter. Besides, in the figure the heavy influence of the quantities cast can be seen. With ten pieces cast, the pattern cost portion is only about 0.5% of the manufacturing costs of the gears with 2-m diameters, and can therefore be neglected. A main reason for that is the explosion of the material costs that grow very nearly proportional to the weight (almost to the third power of the linear dimensions). Figure 7.11-10 shows this with the example of all types of cast steel parts, which vis-a-vis GCI generally have two to

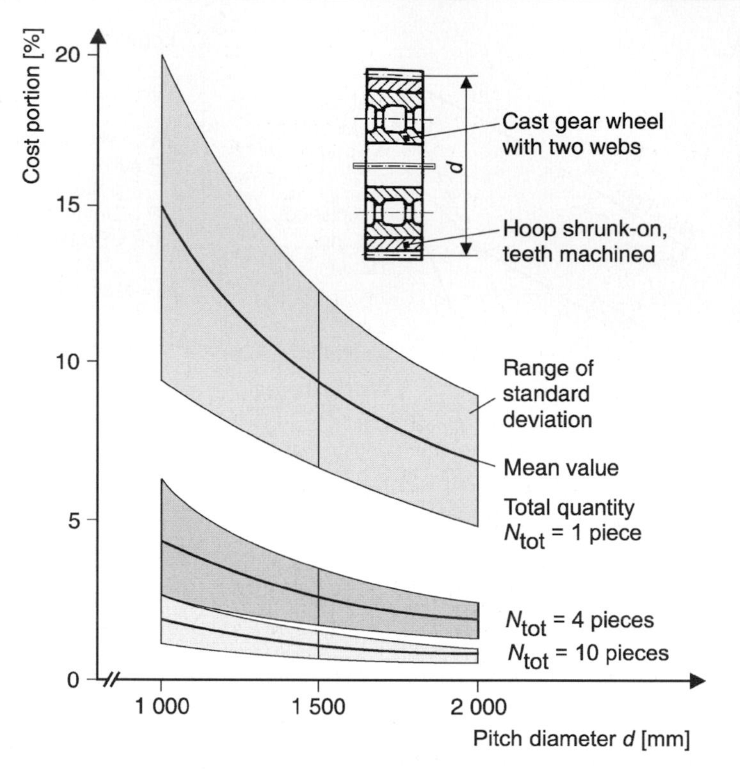

Fig. 7.13-12. Pattern cost share of the manufacturing costs for gears (N_{tot} = total number) [Fis83]

three times higher material prices per unit volume or weight. In addition, from Fig. 7.13-16, in the case of welded housings, the great influence of material costs is clear.

- The **number (lot size)** is an additional leading parameter for the manufacturing costs, particularly for castings, because introductory costs, (advance costs) for preparing the patterns are divided by the produced quantity. This is most effective going from count one on to two to five (Fig. 7.11-8). In contrast to that, the number has hardly any influence with a specified welding process (manual MAG welding), at most on the training-effect. The situation is different if welding is done with fixtures because fixture costs are advance costs, just as with patterns.
- Hence the **limiting values** (i. e., the numbers at which a casting becomes more promising than a weldment) are of great importance for the designer's question of casting versus weldment? **Figure 7.13-13** shows that the size also has a considerable influence on the limiting number. For small gear housings, a casting can be more economical than a weldment for 10 to 45 pieces; for large castings the crossover point is reached at 2 to 8 pieces. Also, here again there is large scatter that is company-specific! These results contradict the widespread rule, "under three to five pieces, use welding". The size influence can be explained

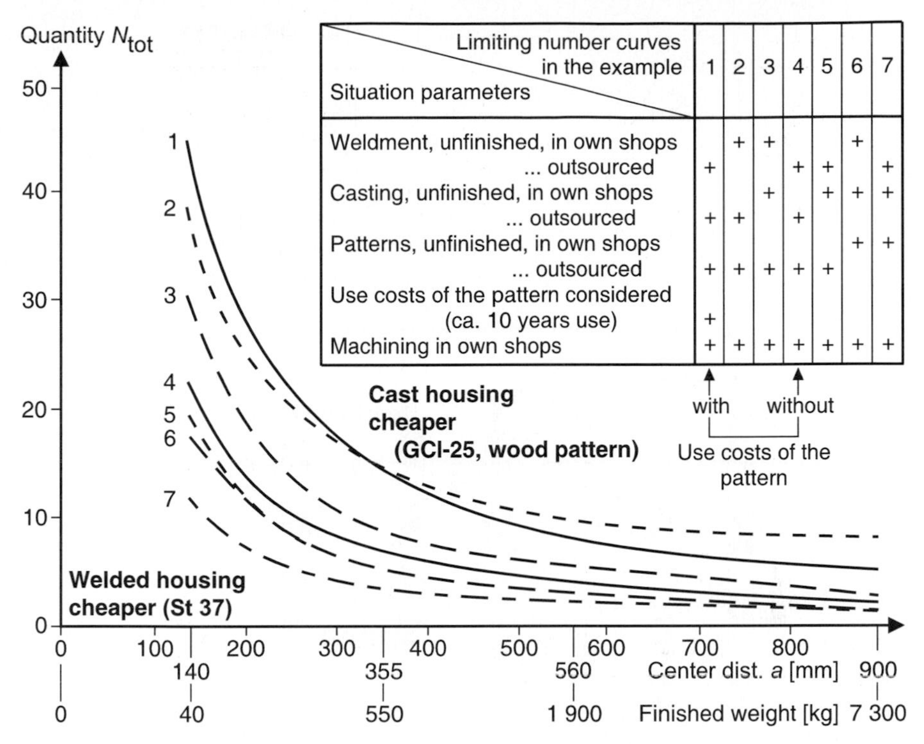

Situation parameters \ Limiting number curves in the example	1	2	3	4	5	6	7
Weldment, unfinished, in own shops		+	+			+	
... outsourced	+			+	+		+
Casting, unfinished, in own shops			+		+	+	+
... outsourced	+	+		+			
Patterns, unfinished, in own shops						+	+
... outsourced	+	+	+	+	+		
Use costs of the pattern considered (ca. 10 years use)	+						
Machining in own shops	+	+	+	+	+	+	+

Fig. 7.13-13. Crossover quantities for cast and welded designs, for single-unit production of housings [Haf87]

by the small cost growth of the patterns (Figs. 7.13-11 and 7.13-12). Note from Fig. 7.13-13 that the limiting number curve moves up if the use costs of the wood pattern (storage and repair costs) are included. In addition, there is a shift if parts and manufacturing are in-house or provided by suppliers. The figure shows that from the cost viewpoint alone, the decision about weldment versus casting is complex (see Sect. 7.11.1c).

- The influence of the **width/length ratio,** *b/l*, of a gear housing is of interest in this respect. From the investigation, no conclusions could be drawn whether it is better to build long and slender or more compact, cubic-shaped units (Sect. 7.6.2a). **Figure 7.13-14** makes it clear that compact housings have about 13% lower manufacturing costs than long, flat units. The same tendency is shown by welded housings. This is probably also valid as a rule for other types of housings. In gear design, it has long been known that a decrease in the center distance lowers costs (Fig. 7.6-1). One must consider, however, the unequal load sharing over the width of the gear, shaft bending, and the space required by the rolling bearings.

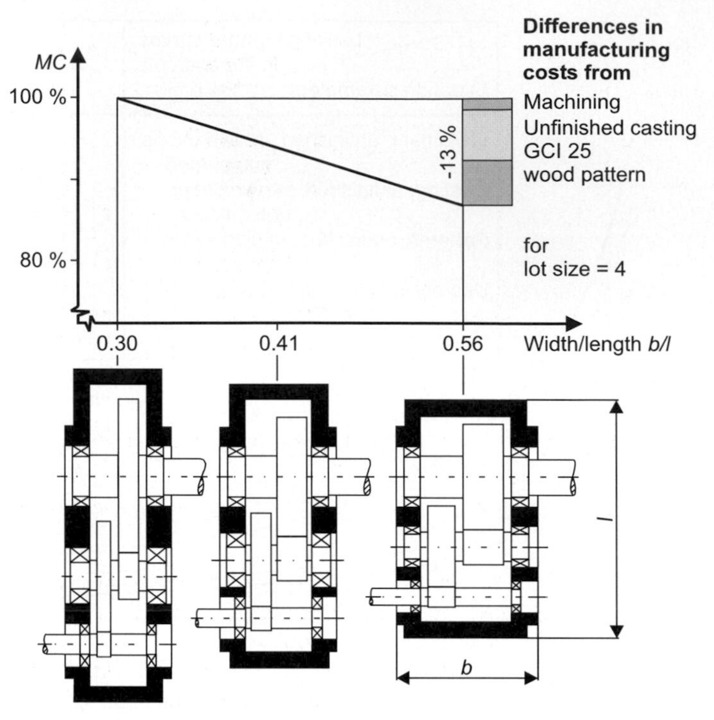

Fig. 7.13-14. Influence of the width/length ratio on the manufacturing costs of cast housings with information on the causal cost portions [Haf87]

- For **lowering the costs of the unfinished part, machining and the pattern,** the important items in the cost structure depend on the size of the unit and the quantity, as shown in **Fig. 7.13-15** for cast and welded housings.
  - With increasing size, for all housing designs it is the **unfinished part costs** that come to dominate. Accordingly, attention should be paid to keeping material costs low (relative costs of unfinished part GCI 25 : NCI 40 : CS45 : Cal 2 = 1 : 1.5 : 2.5 : 1.3. Cast aluminum (CAl) is therefore relatively low-cost [Haf87]). It is thus better to use thinner walls with ribs. For welded housings, strive for few parts, few seams, and few scraps (see also Sect. 7.11.5.2).
  - The **machining costs** are important for small welded and cast housings in large quantities. Therefore aim for small, few and properly placed machined faces.
  - For small cast housings in small quantities, **pattern costs** are dominant. Patterns are economical if they are built from few simple basic shapes and have few partitions, cores and inserts (Fig. 7.11-11a).

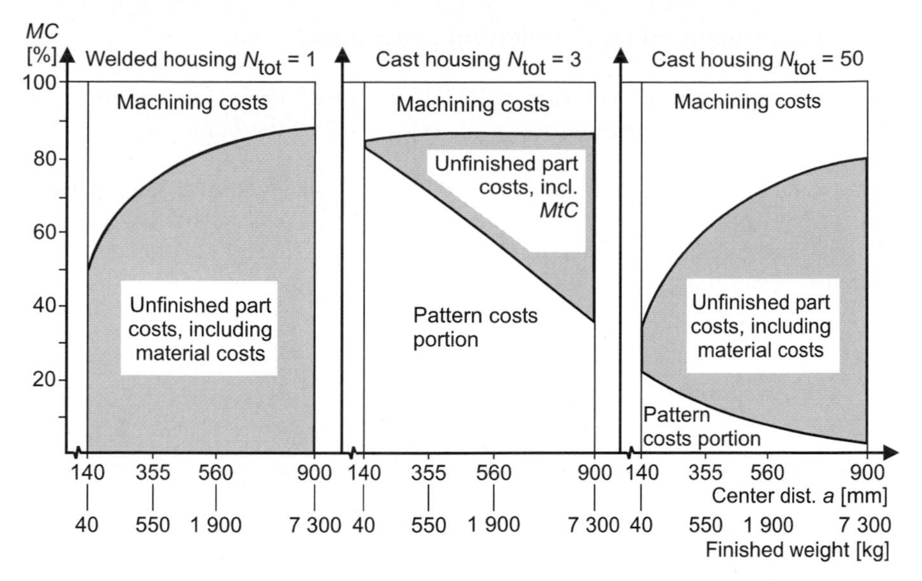

Fig. 7.13-15. Manufacturing cost structures of welded and cast gear housings depend on size of the unit and number N_{tot} (total quantity produced) [Haf87]

Figure 7.13-16 shows the cost structures for welded housings. The material costs, which increase with the size, far exceed the costs of the individual production steps. It is also clear that the actual welding constitutes only a small part of the total manufacturing costs (see also Fig. 10.2-2).

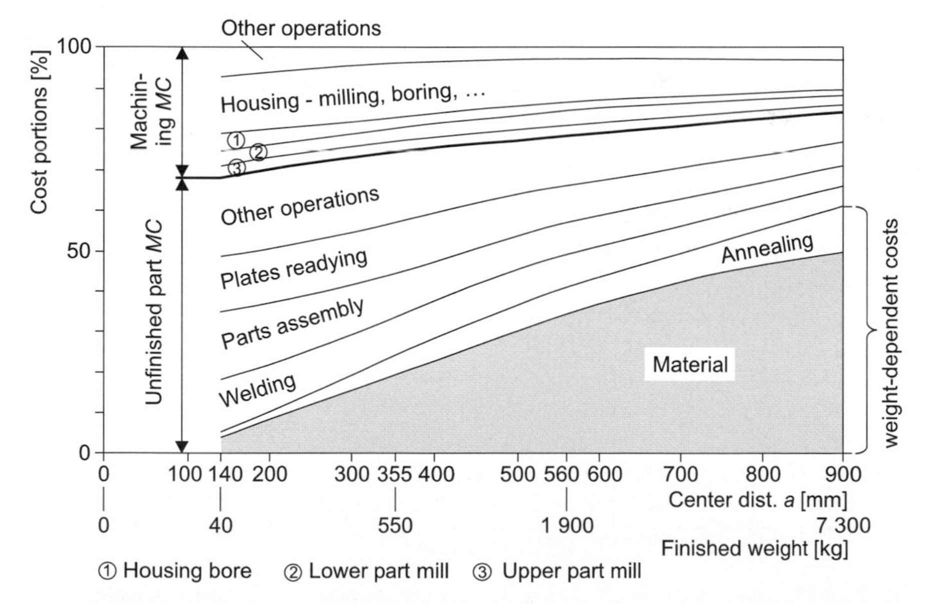

Fig. 7.13-16. Cost structure of welded housings, with in-house production of the unfinished part and machining (lot size = 4) [Haf87]

7.13.5 Heat treatment and hardening procedures

In gear making heat treatments are applied on parts for stress-relief and normalizing. Case-hardening and gas-nitriding are often used as the hardening processes [Bru94].

The costs for that may amount to 20–30% of the manufacturing costs, particularly for big components (e. g., gears). In practice, they are generally charged proportional to the component weight. As shown in **Fig. 7.13-17**, the rate ($/kg) is frequently reduced in steps with increasing weight.

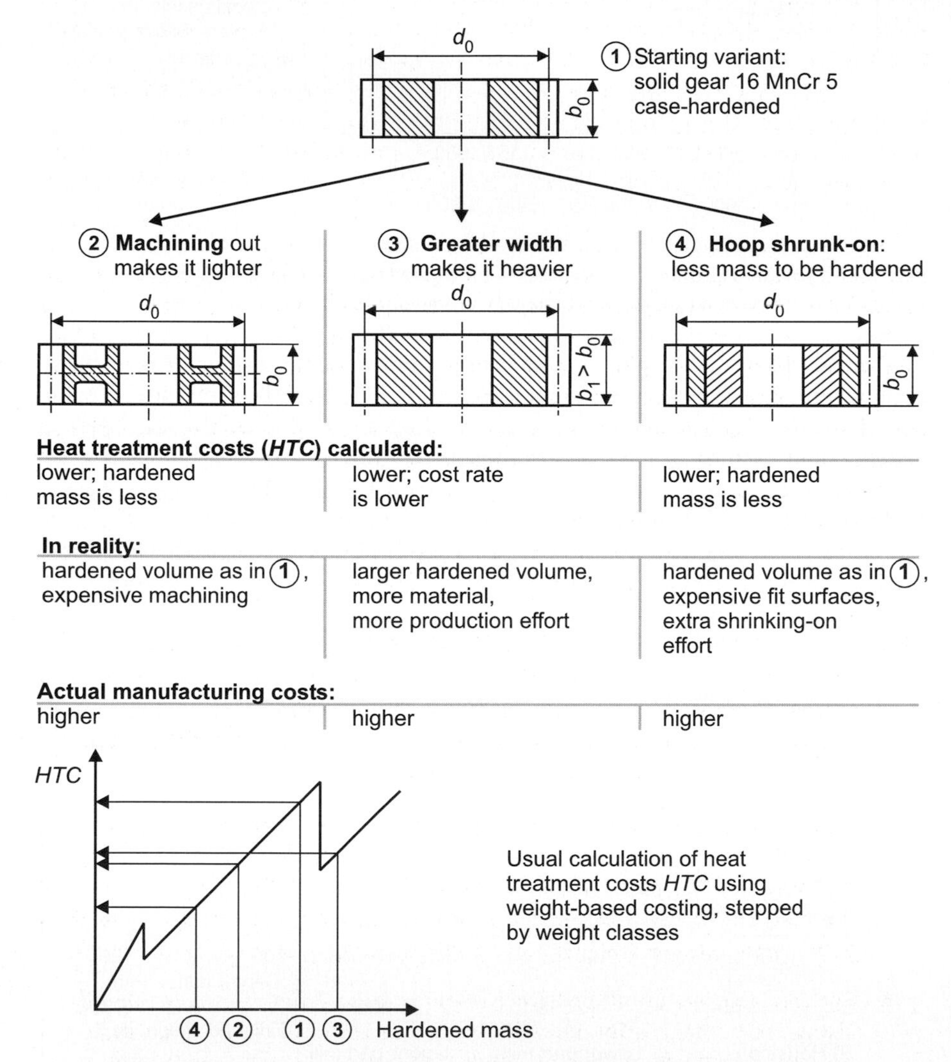

Fig. 7.13-17. Incorrect decisions during the design of gears due to cost calculations not based on cause [Bru94] (see also Fig. 8.4-8)

A cost-conscious designer will attempt to design the components for low costs according to this cost accounting. This is shown in the figure with the example of a case-hardened gear about 1 500 mm in diameter. Knowing that for the initial gear ① lower heat treatment costs are charged if it is lighter, the designer decided to reduce the gear weight from 4 050 kg by about 1 600 kg by machining. That reduced the computed manufacturing costs for case ② by 8%. In reality, the hardness costs were virtually the same as for ① because the furnace size and the heating time are the decisive parameters for these costs. The manufacturing costs actually increased about 3% due to the machining costs, if the same hardness costs are used. In a similar manner, the designer could also have implemented cases ③ (less wide, in order to use the benefit of the step change) and ④ (the shrunk-on hoop with machined teeth has a lower hardened weight than the whole gear).

The inference is that **only cause-based cost accounting can be of real help for low-cost design.** That is similarly valid for not allowing credit for design standardization, since the overhead costs are not assigned cause-based to the manufacturing costs (see Sect. 7.12.4.1, and variant management Sect. 7.12; also Sect. 8.4.3, Fig. 8.4-8.)

In general, for make the calculation of heat treatment costs more cause-based, there is a simple explanation: **Heat treatment costs** are up to 70% fixed costs, which have to do little with the weight of the component to be hardened. They include personnel costs, maintenance costs, depreciation of the furnaces, etc. Only 5–10% of the costs are related to the component weight, on which the usual weight-cost calculation is based (auxiliary material costs, additional energy costs). Therefore, for a heat treatment manager it is a matter of achieving the highest throughput of parts to be handled, for the available equipment. Then the costs per part become small. The degree of furnace utilization is therefore definitive; hence, the importance of the newly developed furnace hourly rates [Bru94]. It is thus important to use time and not the weight as a primary cost parameter, as for production, (reference parameter, see Fig. 8.4-10) [Som92].

This gives rise to the following rules for the designer:

➔ If **heat treatment** is done **in one's own shop,** do not trust weight-cost accounting. Before cost-raising measures, as described above, are carried out, consult with the heat treatment manager and cost control.

➔ If **heat treatment is outsourced** to suppliers, then design according to the probable cost accounting of the supplier (see analogous cases of procuring cast and welded parts from suppliers: Sect. 7.11.2.2b). The tendency, therefore, is for low weight, if weight-cost accounting dominates. Watch your own additional costs (see above in the example for machining costs.)

➔ For the decision on alternative hardening procedures, the **costs** should always be compared **for the whole process chain** (e. g., costs for material, pre-treatment, hardening, and post-processing such as grinding). Furthermore, consider loading (e. g., transferable torque).

7.13.6 Shaft-hub connections

As shown in **Fig. 7.13-18**, eight different shaft-hub connections were analyzed, with diameters in the range of 20–500 mm and in lot sizes of 1–100 pieces [Kit90, Ehr91]. As already mentioned in Sect. 7.13.2b, with the nine participating companies also, the spread in **manufacturing costs** (calculated as full costs) was greater as the sizes decreased. For example, the scatter in production costs of a 20 mm toothed coupling (DIN 5480) ranged widely. For the process step "cut internal teeth in hub" (gear shaping or broaching), it was as much as 1 : 8, the place cost unit rates 1 : 3, the total time units 1 : 29 (!), and for setup times as much as 1 : 5. For diameters of 500 mm, the production cost spread was only about 1 : 2. Here again the main causes are the assumed times for production and idle times and the setup times. In spite of that, the results are largely reliable. The costs accounted for here include the costs for machining (shaft, hub), for additional parts, and joining. The spread of the technical data is also considerable (e. g., 1 : 4 for the calculated transmitted torques); see Fig. 7.13-22, [Kit90].

a) The findings showed the following:

The **type of shaft-hub connections** has, next to the size (not reviewed here), considerable influence on the manufacturing costs, as shown in Fig. 7.13-18. That shows the importance of the main cost influence, the concept or function principle (see Sect. 7.3). Strange to say, according to company inquiries, about 60% of all shaft-hub connections are made as feather-keyed connections, even though their manufacturing costs for the same torque are double to eightfold those of round or

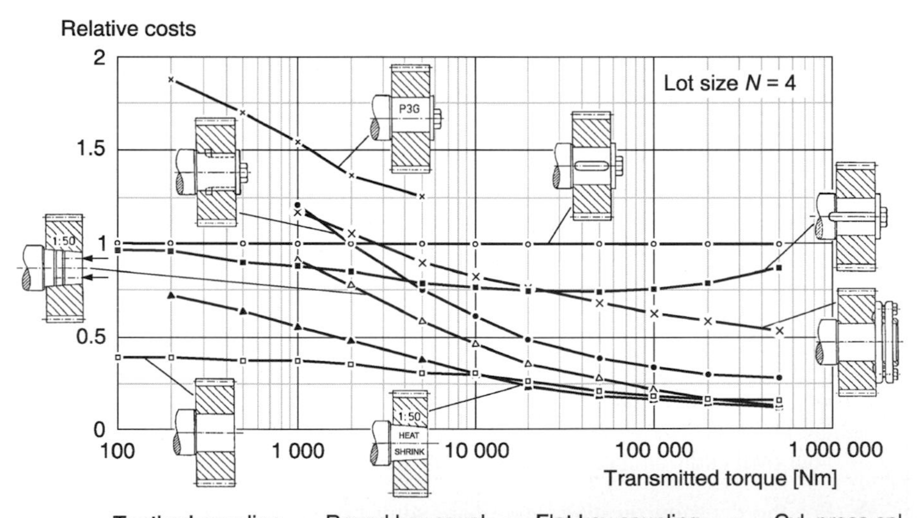

Fig. 7.13-18. Relative costs of shaft-hub connections are dependent on torque. Round keyed connection set to 1, width/diameter = 1. Relative costs consist of production costs for machining (shaft, hub), costs of additional parts and for joining [Kit90]

conical pressed connections. In the figure, round keyed connections are set to 1 (**one** key per connection) (DIN 6885A). We see that relative to this, most shaft-hub connections become more economical, the higher the transmitted torque is. Of course, keyed connections can be dismantled more easily, but the team members from industry knew that this is the essential basis for the choice of keys only in a few cases. It seems to be an old engineering tradition to favor form-based connections ("You see the function"). The conical oil-pressure connection is a low-cost alternative if the ability to be dissembled is required. However, a fixture is needed for that.

➔ Friction **press fits** are preferable if assembly or disassembly conditions allow it. Avoid keyed connections.

➔ Design the effective key length to be large.

➔ Rectangular **keys** are better than round-end keys.

➔ Key lengths of more than 1.5 times the diameter do not transmit much higher torque.

While the transmitted torque has been used as a comparison criterion for the costs up to now, **Fig. 7.13-19** shows the shaft-hub connection costs relative to a keyed connection (set = 1, DIN 6885A) if the **diameter** is **specified.** In addition, here the cylindrical press fit is generally only half as expensive.

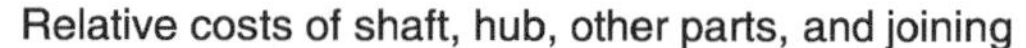

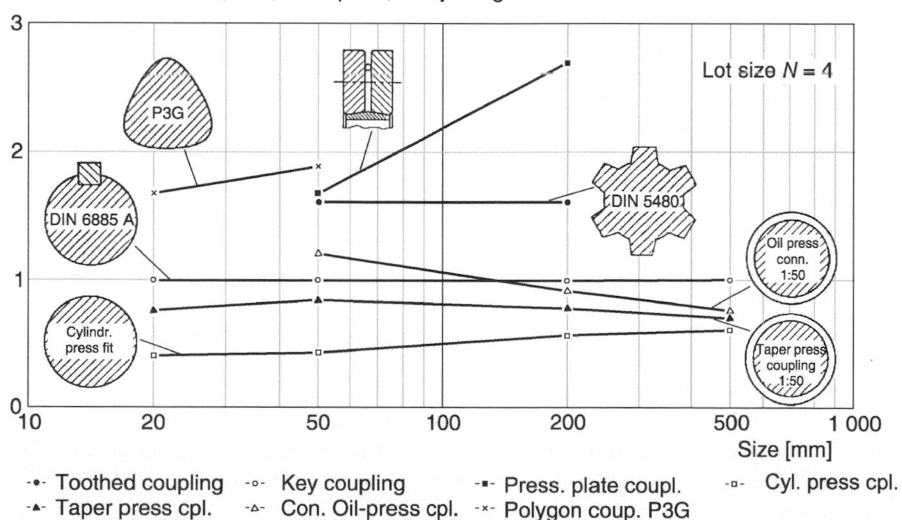

Fig. 7.13-19. Relative costs of shaft-hub connections referred to keyed connection DIN 6885A (set to 1) [Kit90]

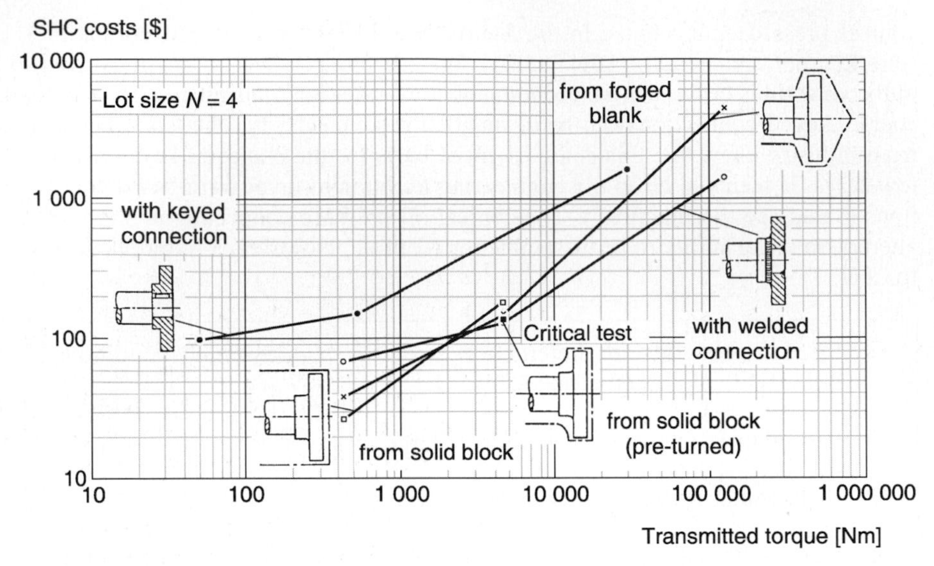

Fig. 7.13-20. Costs of shaft-hub connections of flanged shaft variants as function of the transmitted torque [Kit90]

The shaft-hub connection costs for **flanged shafts** are indicated in **Fig. 7.13-20**. Alternatives are primarily the one-piece flanged shaft (integral design, from-the-solid, forged) and the two-part shaft (differential design; with keyed connection, also welded flange).

Here also, keyed connections are quite expensive. For smaller torques, flanged shafts from-the-solid are preferable; welded flanges are better for larger torques. The length of the shaft in this case amounted to 5.5 times that of a substitute diameter of a comparable two-part variant and about 1.7 times that of a flange outer diameter.

➔ **Flanged shafts smaller than 40 mm** in diameter should be built as integral design (from-the-solid) (see also Fig. 7.12-15).

➔ Flanged **shafts larger than 50 mm** in diameter should be built as welded design.

Since the **selection of shaft-hub connections is** not only concerned with costs, but also, for example, with the ability to be dissembled and self-centering, **evaluation profiles** are presented in **Fig. 7.13-21** for six shaft-hub connections that involve five additional criteria. The larger the inner white area is, the more preferable is the particular connection (the assembly expense is found separately, although it is also contained in the shaft-hub connection costs).

Figure 9.3-3 will show how **quick cost calculations** may be used for the selection of shaft-hub connections.

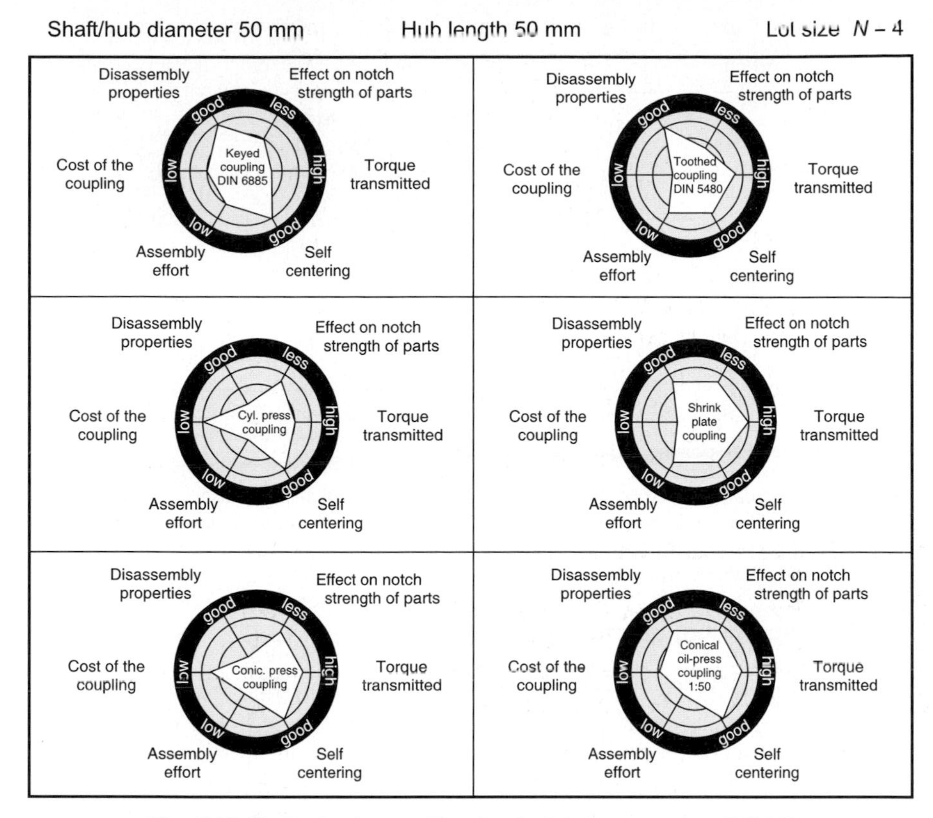

Fig. 7.13-21. Evaluation profiles for shaft-hub connections [Kit90]

b) Industry-wide comparison of technical design calculations

An industrial survey on keyed connections concerning transmitted torques (**Fig. 7.13-22**) and cylindrical press fits was carried out by the Institute. This survey was to find how much the size and thus the costs of shaft-hub connections depend on the technical design features.

The results were startling. It turned out that the **spread** in the **technical design** of six or seven companies is of almost the same order of magnitude as the spread in the production times and costs. This was despite the fact that the companies' technical data for the calculations (loading situation, technical drawing, material, heat treatment) were completely identical. The cost spread has therefore as its basis the spread in design and production. This is true in spite of the generally accessible results from technical investigations [Kol84]!

For **keyed connections** (Fig. 7.13-22) the transmitted torques have a spread of 1 : 4, the permissible pressures on the key, of 1 : 3.5. Within a company there were similar spreads if different persons did the same calculations. With **press fits,** the spread is even larger, of nearly 1:10 [Kit90].

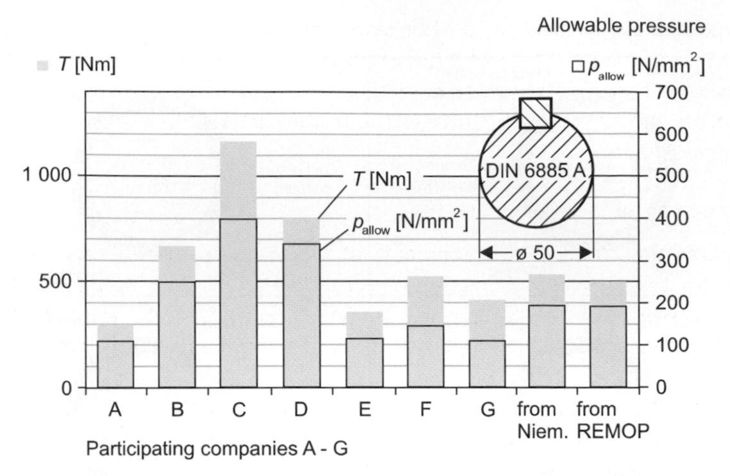

Fig. 7.13-22. Spread of company information on the transmitted torque T and Surface pressure p of keyed connections, according to DIN 6885 Form A ($b/d = 1$; shaft material C 45 N; hub material 1 MnCr 5 E) [Kit90]

7.13.7 Assembly of gear drives

Assembly is the **repository of lapses and mistakes** of all the preceding processes in design, production, procurement and planning. Therefore is the least rationalized and is difficult to preplan. For this reason, the assembly of two-stage gear drives (Fig. 7.13-7) and some two-stage bevel gear drives was analyzed for 5 to 12 medium-sized companies with single-unit and limited-lot production. The aim was to work on cost reduction measures using the knowledge gained on the important components of assembly costs and their main parameters, [Hub95a, Hub95b]. Here also, many findings are typical for the entire machine industry.

a) The gear drives studied

Two kinds of gear drives were investigated: two-stage spur gear and two-stage bevel gear drives. These were in sizes from 140 mm center distance (90 kg) to 900 mm (18 300 kg). The lot sizes were one and five (some were 20 and 100). **Assembly** referred to the assembly of the actual gear drive and of the oiling system and measuring and control instruments. This also includes the coating of a few teeth with a thin layer of marking compound or lacquer and a test run to check for the full contact over the tooth width and finally, acceptance by the customer.

The **situation of** the companies was as follows: The assembly costs portion of the manufacturing costs of the gear drives was 4–30% (average 10%). The assembly duration totaled 0.5–30 days (average of 5 days). The actual assembly time was 10–60 hours. The waiting time and delays amounted to 0.5–30 days, which led to delivery delays of up to 50 calendar days. This resulted in contract penalties.

We see the basic fact about assembly in custom-built single-unit and limited-lot production. With few custom-built variants (most where gear drives in size ranges or modular designs), the share of defective gear drive assemblies was only 5–15% of the total. However, with single-unit production of customer-specific special solutions, it grew to 15% to more than 50%.

Therefore the investigation was split into the analysis of satisfactory and deficient assembly processes.

b) Satisfactory gear drive assembly

Here the main parameters were the size of the unit and, to a certain degree, also the lot size. The spread in the assembly costs was 1 : 9. The lot size change from one to five pieces caused a decrease of only about 10% of the entire assembly costs, since the setup expenses at the assembly stations were rather small compared to those of mechanical processing.

Figure 7.13-23 shows that assembly of parts dominates, with 40–45% of the assembly costs, while the setup constitutes only 5–8%. In the parts assembly, in turn, the dominant factors were the mounting of the gears on the shafts, the assembly of the bearings, and the tuning the bearing play. The entire assembly costs, *AsC*, increase with center distance a of similar gear drives, approximately with a power of 1.3 to 1.9 of the linear ratio $\varphi_L = a_1/a_0$. Here 1.3 holds for small gear drives and 1.9 for large ones. Therefore, the assembly costs of variant 1 compared to variant 0 are: $AsC_1/AsC_0 \sim \varphi_L^{1.1,\ldots1.9}$.

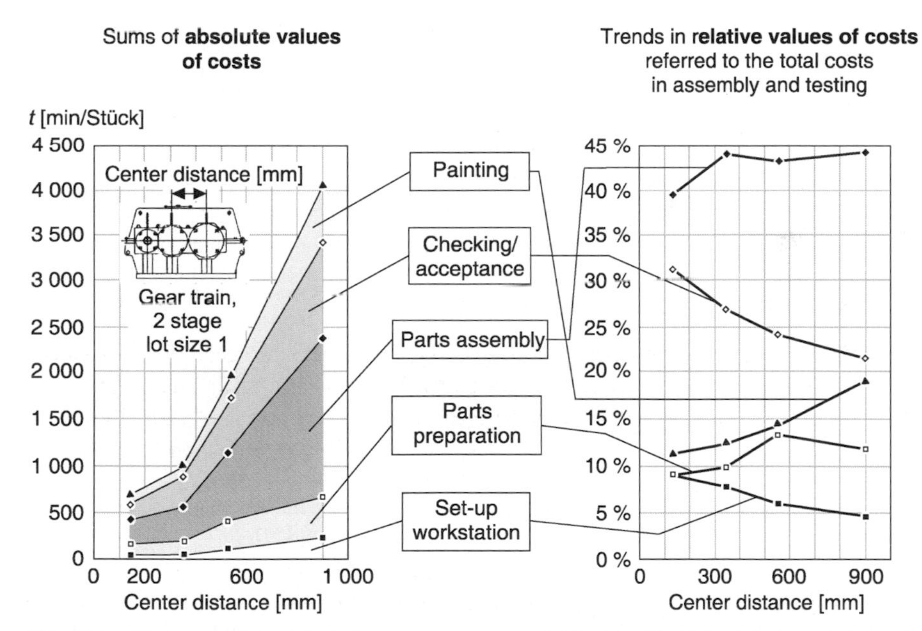

Fig. 7.13-23. Costs in the assembly and testing activities as functions of size of the unit (assembly lot size: 1 piece) [Hub95a]

c) Deficient gear drive assembly

As mentioned above, a considerable portion of all assembly activity results in defects; this causes most of the avoidable costs. First, an on-time start of assembly is delayed in many cases because of parts not being available: "The assemblers are chasing the missing parts". When at last the assembly begins, the items listed in **Fig. 7.13-24** show up, primarily from the outside, to cause problems in assembly.

The dominant mistakes are from design and part production, with each accounting for a third. The more customer-specific adaptive design there is, the higher is the portion of design defects. Most mistakes are primitive and would not have been allowed to occur in the case of orderly work. They can be classified as personal mistakes (shortcomings) and interface errors (incomplete flow of information). Corresponding measures for reducing costs can be formulated.

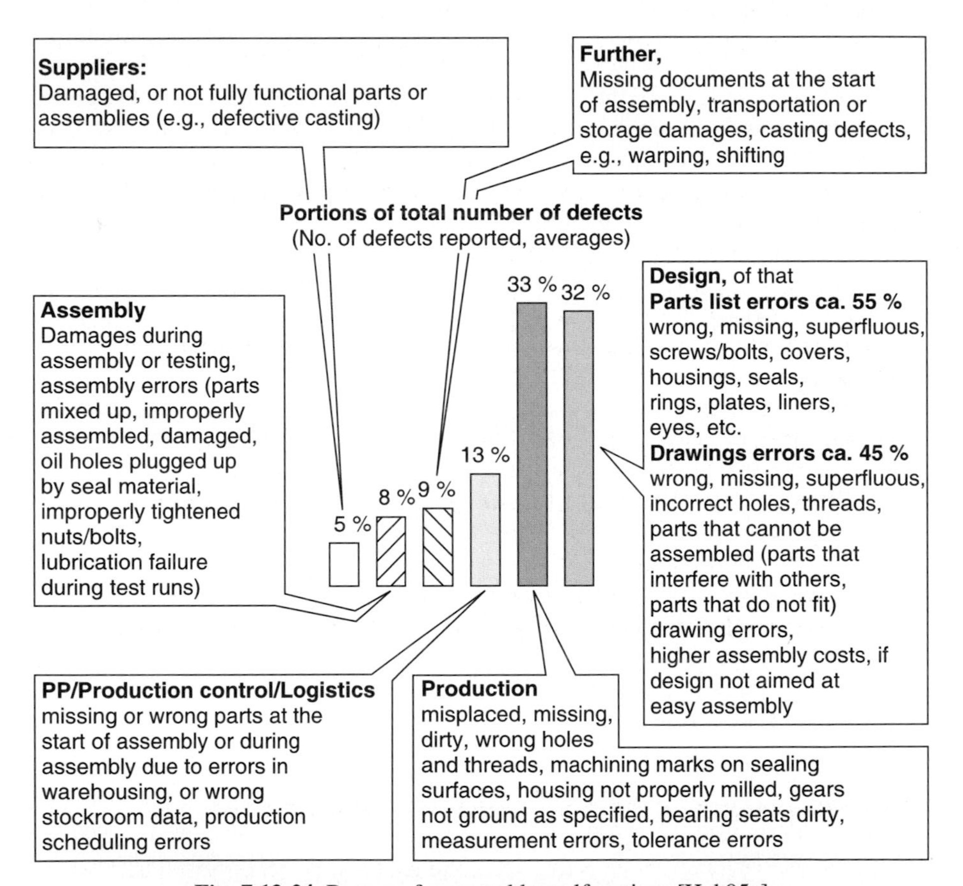

Fig. 7.13-24. Reasons for assembly malfunctions [Hub95a]

d) Cost reduction measures

From the above knowledge, the priority is in getting on from the defective to more satisfactory assembly processes and guaranteeing part availability at the planned assembly start.

It is advisable, first, to carry out a **company-specific analysis of the assembly situation** according to time and costs, so that the **essential weak points** may be recognized (similar to [Hub95a]). This is in any case a necessity for quality management. Such an investigation uncovers like a magnifying glass, most mistakes and lapses in the whole product realization process. As a result, not only the costs of fixing the mistakes, but also the much more painful delays in delivery of the product (loss of image, penalty for poor performance, loss of follow-up jobs) can be brought under control. This is particularly true of systematical mistakes, since accidental mistakes cannot be completely prevented.

Such an investigation will result in a **faults documentation**, which helps in the following organizational steps:

- **Integrated organizational** measures can help in time to safeguard the part availability:
  - **Assembly preplanning** (delivery and accessibility condition of in-stock parts) together with all the affected departments of production, procurement, warehouse, design, quality assurance.
  - **Product realization in teamwork** (Simultaneous Engineering) in which assembly must be included early on.
  - **Production and cost advice** with emphasis on assembly requirements.
  - **Spatial proximity** between assembly and design, if possible.
- **Measures in design:**
  - **More time** for the persons responsible for a careful final check of drawings and parts lists (perhaps with a checklist of frequent mistakes).
  - Information on **Design for Assembly (DFA)** [Hub95a]; attention to rules (Figs. 7.11-52 and 7.11-53a-e).
  - Measures for **motivation improvement**.
- **Measures in part production and procurement:**
  - Making employees more **quality conscious** and improve the possibilities of self-checking.
  - Close **collaboration** between quality assurance (also metrology) and procurement and production.

7.13.8 The complete gear drive, and a cost reduction example

In the preceding sections, 7.13.3 to 7.13.7, the analyses of the essential gear drive components and processes were presented, and the possibilities for cost reduction were shown. To complete the discussion, the following **overview for the whole gear drive** (Fig. 7.13-7) is provided.

a) Manufacturing costs for the entire gear drive, depending on size of the unit (weight) and lot size – rules for cost reduction

Figure 7.13-25 shows to the left the steep rise in the manufacturing costs of the unit with the size of the unit and, to the right, the fast drop in the limited-lot range of 1–5 pieces. Costs of the gear set and the housing are dominant.

Why these changes come about becomes clear from the manufacturing cost structures shown in **Fig. 7.13-26**. On the left, it is apparent that the steep rise in the nearly volume-proportional material and heat treatment costs replaces the part of the production costs from setup times (setup costs). Large gear drives therefore have high materials costs; small units have high setup costs (see also Fig. 7.6-3).

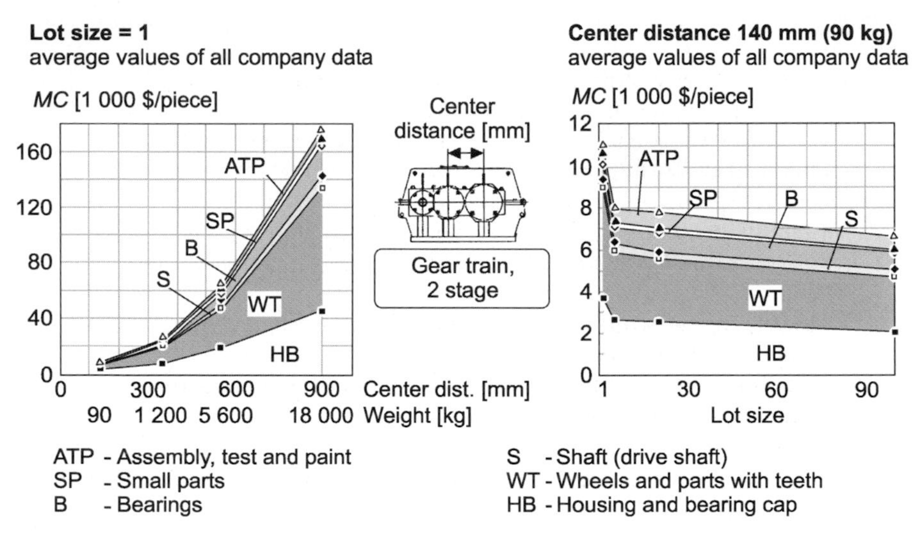

Fig. 7.13-25. Manufacturing costs of gear drives as function of size of the unit and the lot size [Hub95a]

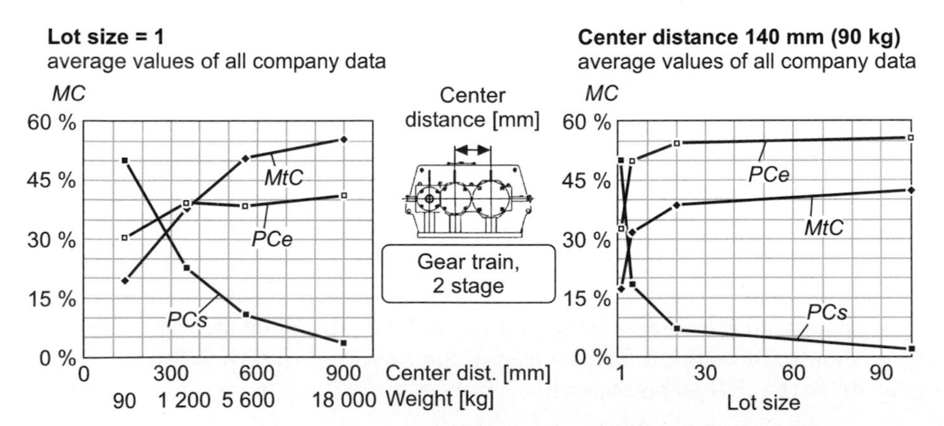

Fig. 7.13-26. Behavior of the portions of production costs from total time units *PCe*, costs of materials including heat treatment *MtC,* and set-up costs *PCs* [Hub95a]

To the right in the Fig., we see the steep drop in setup costs with the lot size; these get divided directly by the quantity. The production costs from total time units, *PCe,* are 55% of the *MC* compared to the materials costs, with approximately 40% of the MC. This is due to the relatively small size of the gear drive (center distance 140 mm; 90 kg). As seen at the left, at the center distance of 900 mm (18 000 kg), the situation is reversed. The costs of materials are 55% and the production costs from total time units, *PCe,* approximately 40%. The larger and heavier a product is, the more dominant are the costs of materials (see Fig. 7.7-3).

The following **rules for the lowering costs** can be derived from these cost procedures (see also the rules at the end of Sect. 7.7.3):

➔For **small** sizes (center distance):

- **Reduce production costs from total time units.** They represent the dominant part at the manufacturing costs. For example, by:
  - Use of more rational, efficient **production processes** (particularly with **increasing lot size**).
  - Use of **economical production equipment** (avoid "large" machine tools, e. g., for drilling, milling, grinding, planing or gear-cutting).
  - Use of **rational production sequences** (reducing the number the processing steps – comprehensive processing).
  - **Reducing the volume of material cut** (make the unfinished part outline as close as possible to the finished part outline (however, see Fig. 7.12-15); use little extra material; allow coarser tolerances; optimize rough and fine machining.
  - Finish processing **in one set-up**.
  - Providing **more favorable machining conditions** (material choice).
- **Lower production costs from set-up times** (in particular for small lot sizes). For example, by:
  - Planning on **fewer production steps** and less switching of machines.
  - **Testing of alternative production sequences** and machines for efficiency (for high set-up costs).
  - **Reducing the number of the different parts** (same parts, repeat parts, products in size ranges, modular designs); thus increasing the quantity and lot sizes.
  - With **increasing lot size** and for small product size, design for **economic material use**. Steps for that are the same as given below for large sizes.

➔For **large** unit sizes (center distances) and for **small sizes**, that are produced in **large lots:**

- **Lowering the cost of materials** (Sect. 7.9.2). They represent the largest part of the manufacturing costs in these cases. This is achieved, e. g., by:
  - Small design (optimize spatial arrangement) depending on the case, and lightweight design (Sect. 7.9.2.2).
  - Integral or differential design (Sect. 7.12.4.3).

- Decrease or optimization of the waste or scraps (for example through optimization of shape or the burning plan for plates).
- Use of lower cost material (perhaps even stronger, easier to machine).
- Use of same parts, part families, products in size ranges, modular designs, etc. (Sect. 7.12).

• If for large unit sizes the **production costs** also need to be lowered, the first priority for this should be given to the **gear set**. See above for the needed measures.

Naturally these cost-cutting measures, which are derived from data acquired from different companies, must be checked again for a company-specific application.

b) Cost reduction example

Starting from an existing design of a two-stage gear drive (**previous design, Fig. 7.13-27**, on the left), a new design is to be generated by using the above findings, for a manufacturing cost reduction of 10%. The cost structure of the previous design is shown **in Fig. 7.13-28**, on the left.

b1) The following requirements were revealed for the **problem**:

- Cost target: 10% lower manufacturing costs.
- The redesign should be like modular design.
- Housing to be welded design.
- Input and output shafts (with keys) unchanged; customer choice.
  $d_{in} = 7.0$ mm, with key DIN 6885 Sheet 1,
  $d_{out} = 160$ mm, with key DIN 6885 Sheet 1.
- Lot size 4.
- Input from: electric motor, 250 kW, 1 500 rpm, clockwise.
- Output: 120 rpm ± 1% to conveyor belt, in the open, clockwise.
- Operation: 10 h/day; utilization factor $K_A{}^* = 1.25$.
- Bearing life: 10 000 h.
- Safety factor against pitting failure $S_H \geq 1.5$.
- Safety factor against tooth breakage $S_F \geq 2.0$.

$^*K_A$ = utilization factor, according to DIN 3990; multiplies torque for calculating endurance strength (= 1.25 for moderate oscillation)

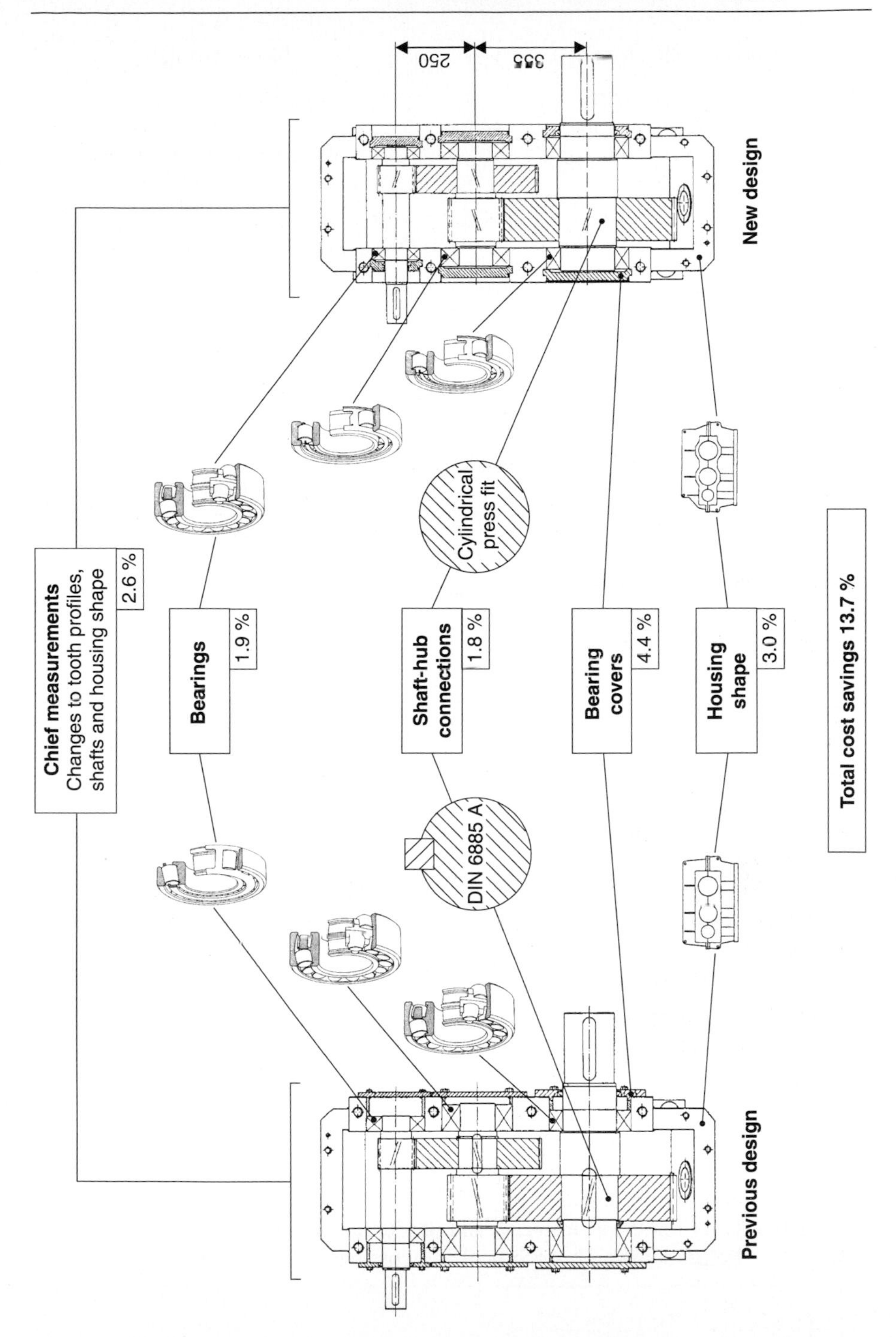

Fig. 7.13-27. Effectiveness of the measures for the cost reduction at a gear drive [Kit90]

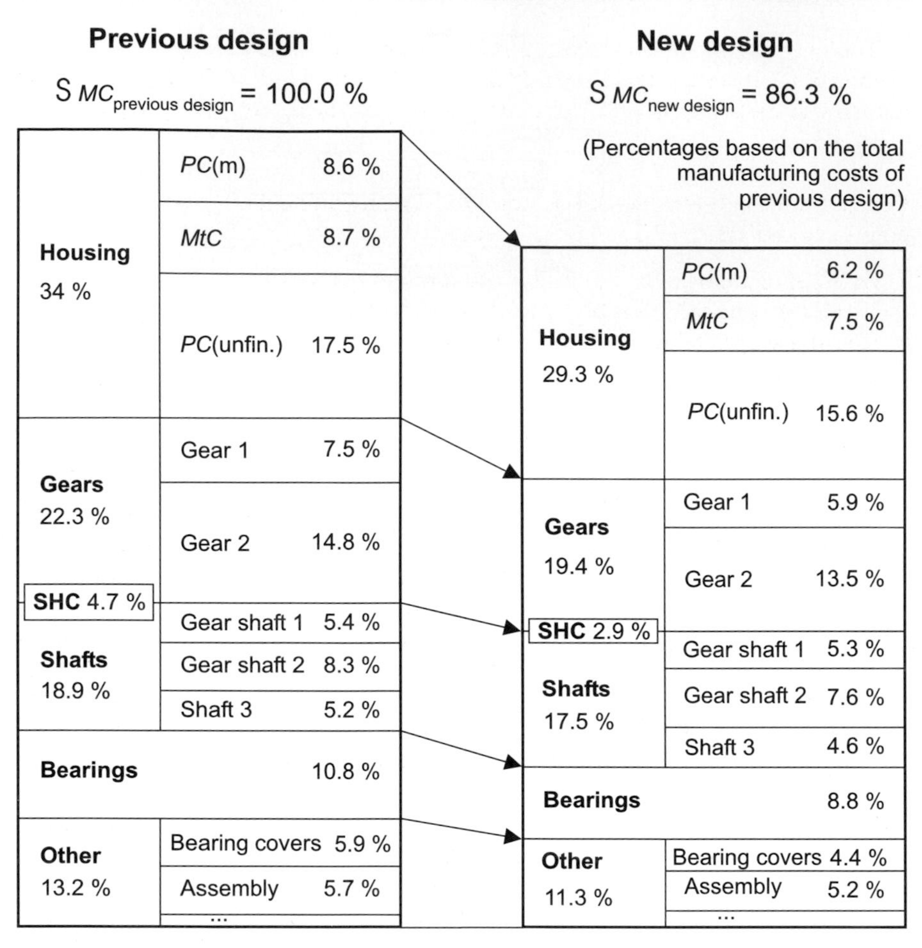

Fig. 7.13-28. Comparison of cost structures of the previous and the new design for gear drives [Kit90]

b2) The **procedure** followed the **procedure cycle** from Fig. 4.5-7, with iterations (see analogous Fig. 10.1-12).

I1	Clarify the task
I2	Analyze the previous design
I3	Emphasize cost lowering as well as solution freedoms and potentials
II	Solutions search
II3 to II5	Look for shape, material, and production variants
III	**Select solution** by technical and cost-based evaluation

The task clarification (Subsection I1) and the analysis of the previous design (Subsection I2) were carried out already, as at b1) above or in Fig. 7.13-28 to the left, so that to that no new information is to be generated. The results on the other hand are:

For **I3**: **Solution freedoms and potentials,** corresponding to the known cost focal points of gears and housings, are as follows:

- Division of the overall speed ratio into the 2 stages
- Tooth data (module, number of teeth, tooth width)
- Housing shape and its width to length ratio

and in addition, in secondary portions:

- Type and dimensions of bearings, shaft-hub connections, bearing caps.

For **II & III: Search for and select solutions**

- **Figure 7.13-29** shows some **shape-based solution variants** that were suggested during teamwork (shaft-hub connections, housing shape, bearing caps). In the same way, information is generated for cost-based evaluation.
- The variation and evaluation of the center **distances**, a (standard number series R40), the **speed ratio distribution**, and **gear widths** showed most favorable values for the center distance ratio $a_1/a_2 = 250/355$ mm and the speed ratio $i_1 = 1:3.75$ in the first high-speed stage. In the process, instead of the tapered-roller and spherical-roller bearings used earlier, lower-cost cylindrical roller bearings could sometimes be substituted. Since these could only take small axial loads, a small helix angle had to be aimed for. This itself brought about a cost reduction of 1.9% (Fig. 7.13-27, right).
- In a similar manner, further solutions were investigated. The **housing shape design**, particularly the side walls, brought about a production cost reduction as well as a reduction in materials cost. The amount of material needed for the housing was lowered from 994 kg to 868 kg (−12.7%).

b3) Result

As it can be seen from Fig. 7.13-28 on the right, with a reduction of 13.7% the cost target could be clearly exceeded. The procedure during the cost reduction was analogous to that shown in Fig. 10.1-12. Because of the given restrictions, the individual elements in cost reduction are not from the large cost units, as would otherwise be expected, but in the small items such as bearings, bearing caps, and shaft-hub connections.

Finally, the force-fit was chosen instead of keyed connections (Fig. 7.13-18), and the bearing caps were set in instead of bolted on. We also see that many small improvements can lead to the overall success. The new design was also evaluated and accepted by the company representatives, and is therefore of practical utility.

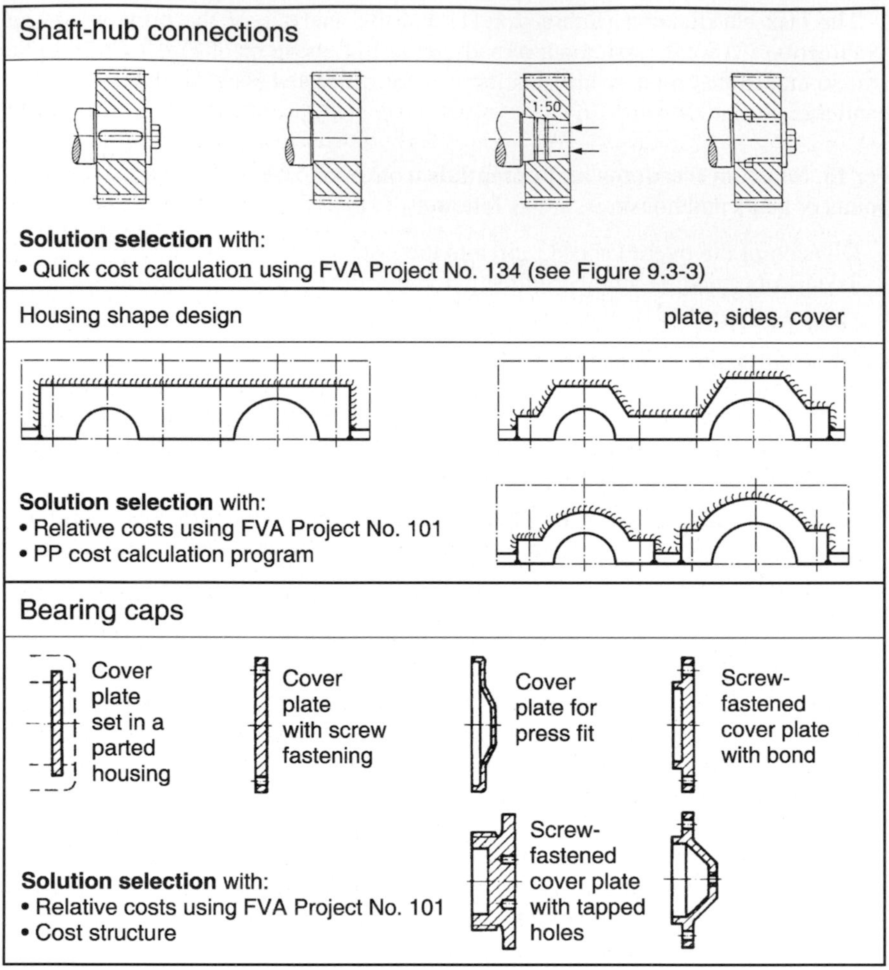

Fig. 7.13-29. Solution variants of gear drive parts and information for solution selection [Kit90]

7.14 Influence of product disposal on manufacturing costs

7.14.1 Motivation for disposal-oriented product development

In Germany, manufacturers can be mandated by law to **take back** their products that have been used and retired, fully or in part, at their expense. The intent is to decrease the waste stream, and is based on Germany's Recycling and Waste Law [Bun94]. The responsibility of the producers for their products thus spans their use and is extended all the way to their disposal. The **disposal costs** must be calculated after considering the appropriate regulations [Bun92] – as a component of the lifecycle costs – for economic control of products as well as for setting the price (Fig. 5.1-5). In addition to ascertaining and allowing for these costs, it is important for producers to see how the disposal costs can be minimized.

Although the Recycling and Waste Law has not yet been enacted, manufacturers in Germany are already taking back and disposing of products. This is often implemented as a test of the impending take-back regulations. Many manufacturers are testing their products' disposal properties and costs, in cooperation with salvage operators. Independent of the legal status, the take-back of old appliances and gadgets in the consumer goods field is offered to the final owners against delivery charges (e. g., take-back of appliances by Siemens). In the capital goods field, a free-of-cost take-back of old products frequently means a competitive advantage because of the direct customer-manufacturer contact.

Measures for the reduction in **disposal costs** are most effective in the early stages of product definition. In many cases, the **manufacturing costs** of the products can also be lowered in the process. The **parameters** for lowering disposal costs affect two junctures of the product lifecycle:

- During development, the **product structure** and **geometry** as well as the **materials** choice determine a portion of the disposal costs.
- After taking a product out of service the other portion of the disposal costs arises by the **choice of the logistics, the disposal process chain** and the **specific disposal procedures.**

Measures for lowering the disposal costs can be applied [Bri95] at these points. However, the significant portion of these costs is determined during product definition [VDI93a]. Knowledge of **disposal requirements** is especially necessary for affecting the disposal costs (**Fig. 7.14-1**).

The difficulty in determining the disposal properties during design is primarily due to the long time lapse between this ascertainment and the disposal itself. Unlike production-oriented design, here the feedbacks are more time-delayed, since the service life of the product lies between its design and the disposal. Calling in a **disposal specialist** in early product design stages does not therefore guarantee the disposability of a product at the time it is taken out of service. This is somewhat different than calling in a production specialist regarding producibility at the time of production.

In addition, the **disposal market** is still a political and vague market where there are great variations by region and time regarding costs of and revenues from salvaged materials. In addition, in many fields the disposal technology has not matured at the same rate, as is the case with standard production processes. Thus, for growth of the disposal market for machines, appliances, etc., which depends on government regulations, a strong development of disposal processes may be expected.

Nevertheless, the manufacturer of durable goods needs to respond to the possible take-back laws, analyze the contemporary state of the disposal technology and redesign the **products** for their eventual **disposal**. A procedure with this aim is proposed [Phl97] in the following. Likewise, the manufacturer should plan **disposal logistics** early in product development. This can make the disposal costs much lower than if it is considered only later.

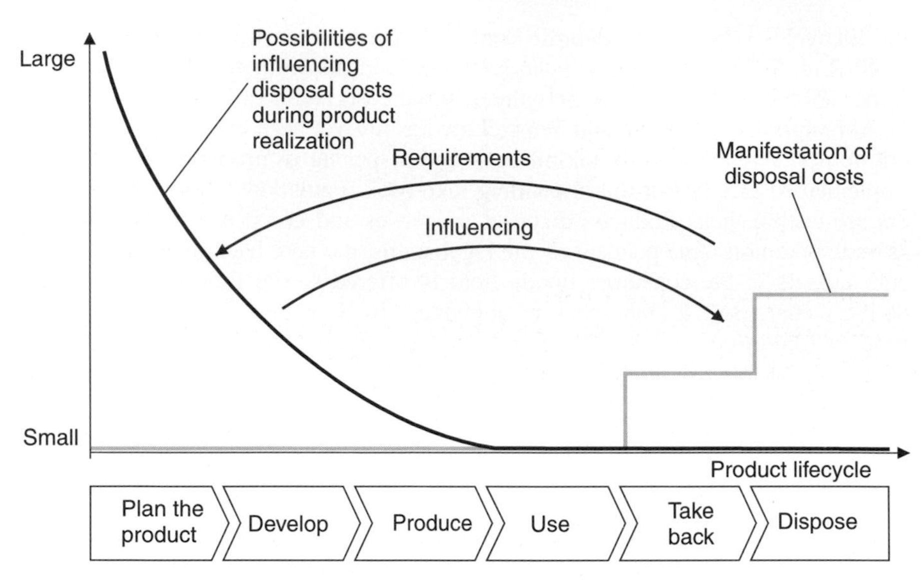

Fig. 7.14-1. Product design can strongly influence the disposal costs. However, the requirements of utilization processes for disposal must be known

7.14.2 Product development procedure for lowering disposal costs

In developing products that have low disposal costs, the emphasis is on the analysis of the disposal properties of products and the identification of the need for action. Synthesis steps can be taken on the basis of a careful analysis without specific tools and only with the help of the familiar design methodology. The design of products with low disposal costs follows the procedure cycle (Figs. 4.4-1 & 4.5-7). The steps of the procedure cycle are described as follows [Phl97; Phl99]:

I Clarifying the task

- Determine the **disposal strategy**
  For setting the disposal strategy the different possible disposal paths are clarified for a product for their relevancy to disposal costs (e. g., separation by dismantling and/or shredding and subsequent separation). For that, a market analysis with disposal tests can be carried out.
- Clarify the **requirements**
  The demands on a product with low disposal costs depend on the disposal paths that are most economical at the end of its lifecycle. A product that is to be disposed of later through shredding and separation techniques must be designed differently than one which is largely to be dismantled. The demands on a product designed for dismantling are presented in Sect. 7.14.3.
- **Analysis of the present state of a forerunner product**
  A intellectual and test-oriented analysis of a previous product should look at, for example, disassembly times and cost and with regard to the revenues from

the recycled material or dump costs. This helps to clarify which requirements are met and where there are still weaknesses (material value analysis, jointing analysis, disassembly analysis).

- **Evaluation of weak points**
  Regarding the evaluation of weak points, these are classified relative to their priority and the expense of the change to identify the action requirements.

II Search for solutions

- **Achieving this on the product**
  The treatment of the selected weak points of the product for creating variants that are more readily disposed of is on should be based on the results of analysis, aided by design methodology. Additionally, the synthesis can be supported by design guidelines or annotated variational features, etc. [Bri95; Phl97.]

III Select the solution

- **Evaluation and selection of product variants**
  The redesigned products must be analyzed, as described above, intellectually and test-oriented, to verify to what extent the weak points were eliminated.

7.14.3 Example of an adaptive design for lowering disposal costs

The Siemens model TC 22 coffee percolator (**Fig. 7.14-2**) was developed under the specifications of easy disassembly and a reduction in materials used (disassembly-based disposal path). The aim of product design was to separate the device into its constituents by a hammer blow on the heating plate and to recover polypropylene (68%), wiring (10%), steel (10%) and aluminum components (6%). As part of the

Fig. 7.14-2 Coffee percolator, used as example

research project, this percolator was subjected to a disposal analysis and redesign that are discussed in the following [Pro96].

I Clarify the task

I.1 Decide on the disposal strategy

The disposal strategies for the percolator (manual disassembly and mechanical preparation) were examined by means of crushers and cutters, as well as various sorting-operations. The disassembly domain is described below.

I.2 Clarify the requirements

A product is considered to be disassembly-oriented, if it can be **easily separated**, **quickly disassembled** and its disassembly properties are obvious and unambiguous (associativity) [Phl97].

a) We define the **separability** of a product as the number of needed destructive and nondestructive separation steps that lead to sorting of the product parts into easily marketable constituents. Requirements concerning separability can be formulated generally as follows:

- **Minimization of the number of the disassembly steps**, through minimizing of material variety, part count, disassembly operations and a consolidated arrangement of hazardous materials.
- **Keeping the contamination** of the targeted components **within bounds**.
- **Minimization of waste material and hazardous constituents.**

b) We define the **ability to be dissembled** as how easily the various assemblies, parts and material connections can be accessed and loosened in the course of disassembly. Requirements concerning the ability to be dissembled can be described as follows:

- **Connections** to be loosened and **product parts** to be taken out must be **easily accessible**. For that axial, straight accessibility and short disassembly paths are needed. The connections to be loosened must be reachable with the requisite tools, or appropriate handling surfaces should be provided.
- The disassembly of the device, up to the separation into its constituents, needs good, **unambiguous** ways of **loosening** the **connections.** After taking the product out of service the opening of the connections should be feasible single-handed, and without tools.
- The disassembly procedure should be planned this way during product design, i. e., to **minimize activities and movements during disassembly** in separating of connections and the removing the parts. Alternating directions and movements should be avoided during disassembly, as well as changing the tools used for disassembly. Aim for an easy handling of the object being disassembled.
- Any **danger to the workers during disassembly** from the product is entirely unacceptable.

c) Associativity refers to the recognition of the decisive properties of a product, module, or part by a user unfamiliar with the products regarding their handling (necessary analysis steps, association of the constituents, etc.)

- **Connections** that are to be loosened for the desired separation must be placed so they can be found quickly by a worker unfamiliar with the products, and the parts to be loosened must be properly arranged and located. The separation process for connections should also be easy to recognize. It should be possible to loosen the connections with standard tools.
- To help sort the parts to be separated, **all materials, hazardous** or otherwise, should be unambiguously and **clearly marked** as assembled, and conform to standards. To avoid mistakes in sorting, make similar parts from the same materials.

Demands on a disassembly-oriented design are not always the same as the demands regarding design for low-cost disassembly, since the disassembly costs always depend on the desired disassembly strategy and are specific to the product and the user. Demands on the design for low-cost disassembly must therefore be clarified in a way that is specific to the company.

I.3 Analysis of the present state of the percolator

We can now proceed to realize the above requirements with an adaptive design for low-cost disassembly of the percolator as follows. During the adaptive design of the coffee percolator, the concept was given. Only embodiment and a detail design could be adapted to the disassembly requirements.

The methods for analyzing existing products for their suitability for disassembly and the cost relevance of the weak points involve three steps:

- Verifying the demands on the disassembly-oriented design under laboratory conditions.
- Practice testing of the disassembly operations.
- Scenario formation of future utilization areas and practical methods.

From the analyses, **weak points** could be identified relative to the above requirements, although the device was already designed for disassembly.

During the **structure analysis** of the **connections,** the screw joint at the base with the housing, for example, was seen to be deficient. The screw is poorly accessible because of its location deep in the device. The shape of the screw head (a countersunk lock screw) demands a special-purpose tool (**Figs 7.14-3** and 7.14-4).

In the next step, the **practice test in the disassembly operations** was examined first, to see if the identified weak points present a problem in disassembly as it is practiced today. Secondly an attempt was made to get a realistic grasp of activity profile and working times during disassembly, to see if any conclusions could be drawn regarding further weak points.

During the practice test of disassembly operations it was tested whether the weak points recognized earlier in fact also present a problem. The activity process and corresponding times (costs) were noted in order to discover further weak points.

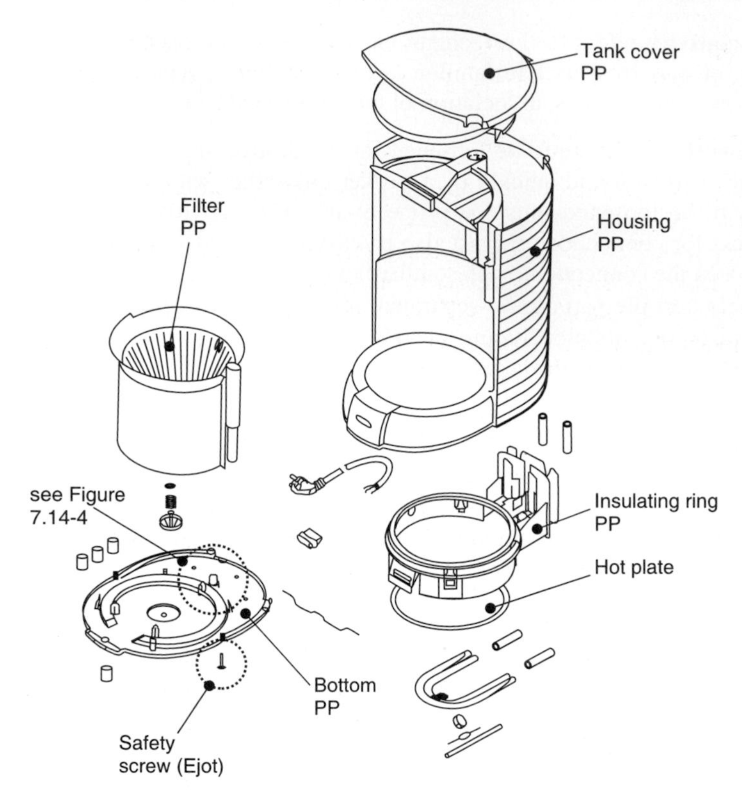

Fig. 7.14-3 Coffee maker parts and materials (*PP: polypropylene*) (from Siemens)

The important thing was to carry out the test with several coffee makers, to explore the **training-effect** (decrease of idle times for orientation and tool search, Sect. 7.5.2). This was accomplished by simulating a **low-volume disassembly**.

For a complete disassembly of the coffee maker, today about 70 seconds are needed to recover polypropylene (PP), polymer mixtures, wiring and steel, and relay scrap. This costs about 0.46 $ if we take into account the revenue from the recovered constituents and the costs. The screw joint at the base was confirmed by the disassembly test as a weak point. The worker needed on the average 7 seconds for loosening this connection before the device could be taken apart with a hammer blow to the heating plate.

Subsequently, using a disassembly cost tree [Phl97], the **optimized disassembly minimum** of 7 s was found under the present conditions. A limited disassembly of the tank cover (46 g polypropylene) and filter (121 g polypropylene) was found to be the most economical. The rest of the device was relegated to a polymer mixture fraction (to the dump!), which would cost in total only about 0.30 $ to dispose.

A user survey showed that as of today a rise in the revenue of about 0.08 $/kg can be expected for polypropylene and about 0.15 $/kg for pure material parts. With conditions remaining otherwise the same, with optimized disassembly time

(0.3 s), the housing and as a result a pure polypropylene constituent of 566 g also can be recovered. This disassembly would cause 0.26 $ in disposal costs.

I.4 Evaluation of weak points

- Weak points that are recognized at the **desk** (in the lab) and can be eliminated at no cost should be eliminated.
- Weak points that during the **practice test** prove to be a danger for the worker (e. g., danger of injury) or the environment are to be eliminated, as are important deficits in associativity and cost focal points. The screw joint of the base and housing in the percolator was recognized as a weak point, resulting in a call to action.

II Searching for solutions, and III selecting a solution

For the screw joint between the base and housing, which was identified as the weak point, a variant was pursued. A snap ring joint as shown in **Fig. 7.14-4** replaced the screw joint. This avoided the need for unscrewing; a hammer blow on the heating plate can loosen it.

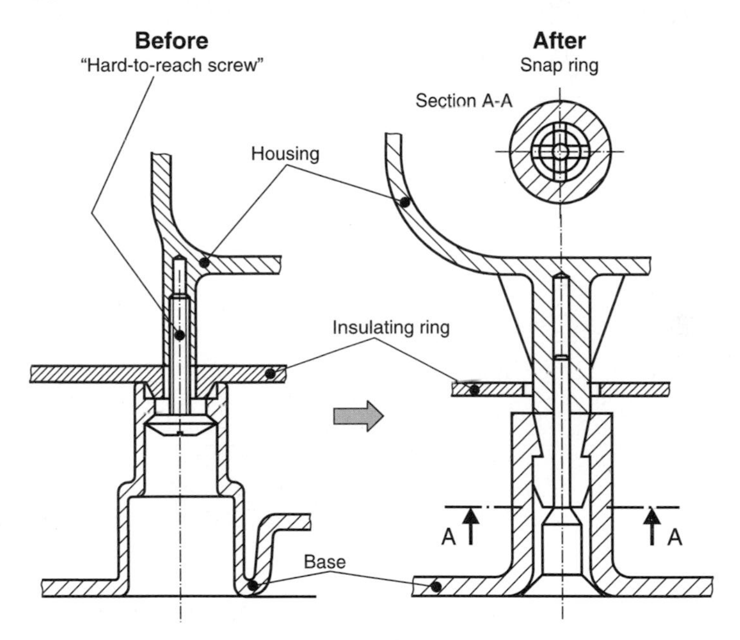

Fig. 7.14-4 Previous solution (*left*) and a redesign of the base/housing connection (*right*)

To sum up, this solution has the following positive aspects:

- A decrease in the variety of connections, and thus the saving of a disassembly tool.
- The elimination of a disassembly step, since the snap joint is taken apart by a hammer blow on the heating plate.

- Eliminating a tool change in the disassembly sequence.
- Eliminating the change from fine to gross motor activities, with a saving of 5 seconds in the disassembly time corresponds to a disassembly cost savings of 8%. The economical disassembly minimum would remain the same with this solution.
- The housing could also be recovered economically at a polypropylene price of 0.08 $/kg.
- The disposal costs fall from 0.29 $/device to 0.27 $/device (at 0.08 $/kg for polypropylene).

7.14.4 Some simple rules for lowering the disposal costs

By analyzing many devices as part of a research project for the design for low-cost disposal [Phl97], the following rules were identified:

➔ Reduce the **material variety** in products to a few, with little loss in the value of materials that are utilized more (for example St, PP, etc.)

➔ The parts made from these materials should be quickly and unambiguously recognizable and taken apart (no permanent connections, no laminations, no surface deposition, etc.)

➔ Avoid materials that cause problems with reuse, or plan these to be easily separable (e. g., do not combine Fe with Cu).

➔ If **dangerous** and **hazardous materials** are unavoidable, provide clear statements about them and design them to be easily removable.

➔ **Valuable** reusable **components** (e. g., electric motors, sensors, etc.) or materials should be able to be dismantled first, quickly and easily.

8 Fundamentals of Cost Accounting for Product Development

The knowledge of costs and their origin is a prerequisite for cost management. Cost accounting represents how cost originate. Therefore, the basics of cost accounting [Wöh96, Sch96] are presented here in view of their importance in product development. For a deeper understanding of the material, suitable literature for engineers is [VDI90, War80, War90] recommended.

The terms and perspectives of cost accounting are presented according to the common business practices. Every company is different, has different emphases, etc., and can use its own costing model. In practice, terms such as cost allocation can vary in the details from what is given here. The basic scheme, however, is the same.

In the foreground is the calculation of the costs for realizing a product in a company-defined frame. To consider the influence of costs in the entire company, wide-ranging deliberations are necessary. The costs to be considered in each case and the costing procedures to be employed depend on the decision situation.

The central problem of cost accounting is to assign the cost (cause-based) to the cost units (causation principle). As shown below, this is very expensive and sometimes nearly impossible. Since cost accounting itself must not be too expensive (economic efficiency principle), in view of this reality it must charge the costs with simplified procedures.

After an overview of costs and cost accounting, cost unit accounting with differentiating overhead costing is central since it represents the common costing procedure in the machine industry. The limits of differentiating overhead costing for cost management are shown, and newer approaches are presented for costing that is more cause-based, such as process costing. In conclusion, direct costing is explained.

8.1 The origination of manufacturing costs

Costs arising in the economic activities of a company can be represented much simplified, as a "black box" (**Fig. 8.1-1**). The company produces **outputs** (products, services) and conveys these to the market. For that, it receives **earnings**. For product realization, it also consumes many **outputs** (production factors such human labor, resources, materials) which give rise to **costs**. The economic aim of the company consists of safeguarding and maximizing the difference between the

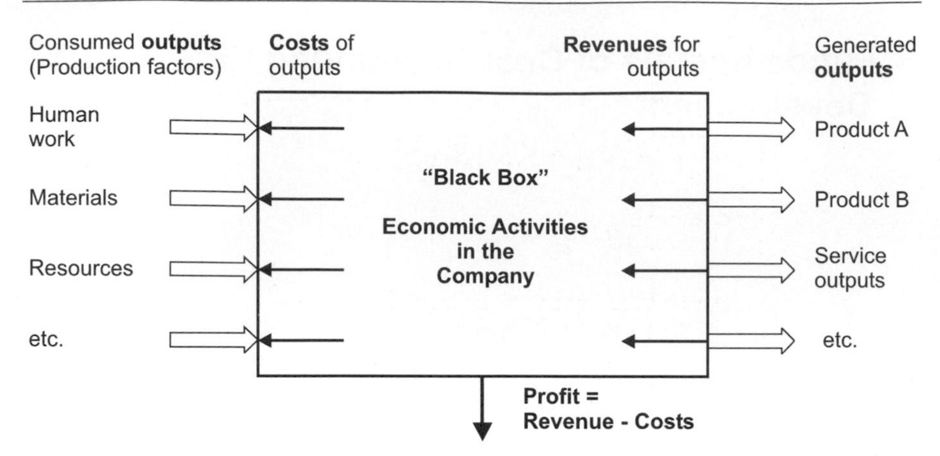

Fig. 8.1-1. The company as a "black box" for the performance activities and for the corresponding cash flow (economic activity)

revenues and the costs (i. e., the **profit**) on a long-term basis. Profit and its growth are the best measures of the success of a company from the economic viewpoint.

Thus, costs arise in the realization of a product, which are recorded more or less in detail, and processed in the costing department for different purposes. Cost accounting is the model of company activities from the cost viewpoint. It describes the activities in the black box shown above. The business and financial accounting (Fig. 8.3-1) record the cash flows of the company with its environment, that is, the inputs and outputs of the black box. As with any other model, this does not exactly represent the reality. There are some consequences from that for cost management (Sect. 8.4.4).

8.2 Cost terms for product manufacture

The terms for costs and cost information are defined and explained in DIN 32 990 [DIN89a] and in the guideline VDI 2234 [VDI90]. Here they are reproduced only as far as they are absolutely necessary for cost management. The terms and perspectives of cost accounting are presented as the usual scheme that is the basis of every company's internal costing. An investigation [Lan96] showed that the extent and content of cost accounting differ widely because every company, more or less, uses its own costing scheme. In particular, smaller companies often have only incomplete cost accounting (**Fig. 8.2-1**). Thus the terms, procedures, etc. can vary in practice from the following description.

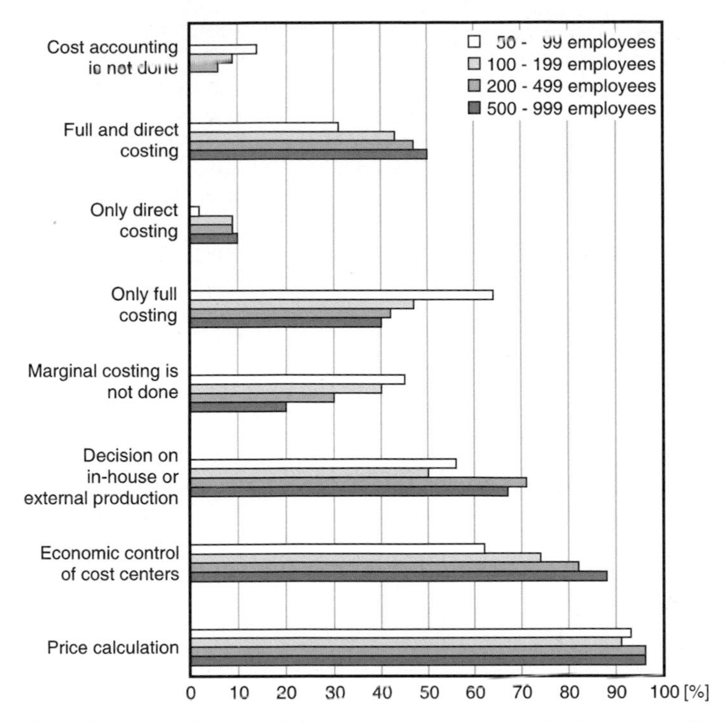

Fig. 8.2-1. Content and extent of the company cost accounting in practice [Lan96]

8.2.1 Definition and organization of the costs

First of all, we give a short review of the essential terms. In Sect. 8.3, the individual terms are dealt with in detail.

> ➔ Definition [DIN89a]: **Costs** are the consumption assessed in money, of production factors and external services as well as sales, for the realization and depreciation of goods and/or services.
> Production factors are resources, materials, human labor, etc.

Industrial management refers to the term **costs** strictly as the company's outputs and distinguishes that from **expenses** that exceed the costs for foundations or others involved in product realization. In addition, costs do not lead inevitably to **expenses**, as one sees in the example of accounting writing-off for acquired production machines (Sect. 8.4.5) or interest on the company's own capital.

The costs in a company can range from thousands to several billions of dollars, according to company size. To make these "cost blocks" manageable, costs are classified in different ways depending on the purpose of cost accounting (**Fig. 8.2-2** and Fig. 2.1-2). The most important division perspectives are:

Classification of costs			Costs for ..... (examples)		
			Transport activities in turning shop	Round steel for production of product XYZ	PC in design office
Questions, aims of cost accounting	Cost type?	How caused?	Personnel (wages)	Materials (Material costs)	Accounting costs (write-off)
	Cost center?	Where created?	Turning shop	Materials management	Design
	Cost unit?	Assign costs to one cost unit (product)?	**No:** Overhead costs with usual time recording system	**Yes:** Direct costs	**No:** Overhead costs with usual cost accounting system
	Fixed or variable?	Dependent on the degree of activity	**Partially:** e.g., 70 % fixed, 30 % variable; sometimes workers assigned to other work stations	**Yes:** 100 % variable	**No:** 100 % fixed, if no other application in the time span under consideration

Fig. 8.2-2. Overview of classification possibilities for costs

- **Types of costs** (e. g., for material, personnel, capital, etc.; Sect. 8.3.1).
- **Cost centers** (e. g., departments – purchasing, manufacturing, design, etc.; Sect. 8.3.2).
- **Cost units** (e. g., product, machine part; Sect. 8.3.3).
- **Accounting** (direct costs, overhead costs; Sect. 8.3.2).
- **Dependence on the degree of activity** (fixed and variable costs; Sect. 8.5).

If all costs are given as **one** cost value, it is called **full costing** (Sect. 8.4). Often this value is not expressive enough. It is complemented by the values of **direct costing**, which divides the full costs according to fixed and variable costs (Sect. 8.5).

Of significance is the classification of costs according to the date of the calculation. The costs known after the event, accruing in a given period, are called the **actual costs**. They can vary from case to case for the same courses of events (e. g., through waste). Therefore, we calculate **normal costs** for simplification (from the average of the previous periods). Finally, such average values are assumed (including expected changes) as **standard costs,** to be achieved for a future period.

8.2.2 The terms sales price, total (factory) costs and manufacturing costs

Figures 8.4-2 and 8.4-9 show how manufacturing costs result from material and production costs and how the total costs ensue by the addition of other costs (administration, sales, etc.). From the total costs, **a calculated net sales price** can be calculated by the addition of a desired profit. Since one must reckon on reductions (discounts, rebates) in the course of order negotiations, as well as with augmentation for risks and commissions, these are included here and a **calculated gross sales price** is reached. With this, an offer can be made. However, what actual **price market** will be attained is a completely different question.

No close correlation essentially exists between costs and the market price (Sect. 4.4.3). A **price accepted by the customers** corresponds to their idea of the benefit of the product, and depends strongly on the price quotations of the competitors. The costs for manufacturing the product do not interest the customers. For the manufacturer there is, however, a one-sided connection between price and costs: The price can be higher than the costs but it must not fall below the total costs in the long run. Where this low limit lies is different with each competitor due to the design of the product, the production technology, and the capacity utilization or rationalization measures carried out earlier (Sect. 7.13, Fig. 7.13-2, Fig. 7.13-3). It shows up only in times of crisis due to severe competition. Lapses during "fat years" concerning improvement of the product quality, decrease of the delivery time, increase of market share, reduction of the variable and fixed costs become the touchstone of survival in "lean times".

From this point of view we see that **cost data may be disclosed only in-house,** but be **kept** absolutely **secret from the outside** (i. e., the competition or the customer). The competition can derive inferences on production advantages, etc., and the customer will accept prices only just above the costs. It is perhaps different for dedicated suppliers (e. g., in the automotive industry), who often have to disclose their internal cost accounting to the customer.

Finally, it should be mentioned that prices become costs when the company buys products. For example, the customer's material direct costs arise from the vendor's prices for the production material.

8.3 Cost accounting in a company

The accounting method in a company comprises the domains shown in **Fig. 8.3-1.** The cost accounting group occupies an important position since the master data are processed here for the other departments as well as for different management decisions. Only this is considered in what follows.

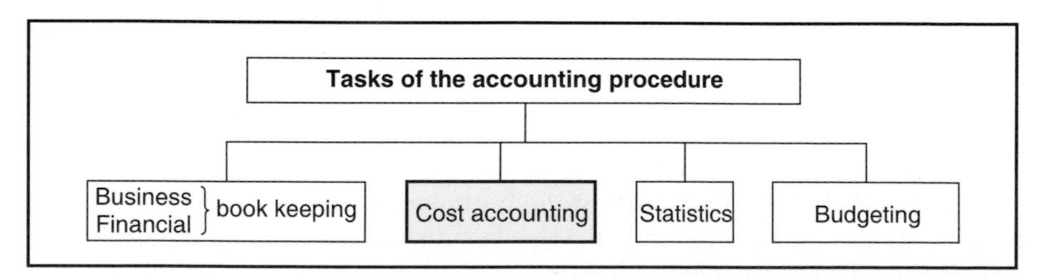

Fig. 8.3-1. Areas and tasks of the company-internal accounting

Cost accounting has three essential goals [VDI90]:

- The **ongoing control** of **the efficiency** of the company activities by comparing the accruing costs with the realized output of the company.
- Determination of the **probable costs** of processing an order as a basis for preparation of the quotation price (preliminary calculation).
- Determination of the **real accrued costs** of processing an order (follow-up calculation).

Unfortunately, the costing group seldom deals with the **goal of the early support of cost-effective decisions in product development**, so that technical personnel must resort to self-help in many cases (Chap. 9). However, the self-help **must** be based on the costing group and be coordinated with it.

Cost accounting records, divides, and examines accumulated costs (Fig. 8.2-2) with the following questions in view:

- **Which** costs resulted? **Cost types** (e. g., material, labor costs)
- **Where** did costs result? **Cost centers** (e. g., turning, milling, sales, etc.)
- **For what** did costs result? **Cost units** (e. g., ship transmission, lever)

The accumulating costs are therefore separated according to cost types and allocated to cost centers and cost units. Finally, the costs so charged are evaluated for those task areas shown in Fig. 8.3-1.

8.3.1 Cost type accounting

➔ Definition: **Cost type** is the designation for costs with the same features; cost types are usually identified according to the type of production factors consumed [DIN89a] (**Fig. 8.3-2**).

8.3.2 Cost center accounting

➔ Definition: A **cost center** is an operational area of cost generation, demarcated according to specific criteria.

Such criteria can be functional, accounting, organizational, and spatial. Important cost centers are departments, shops, machines, machine groups and, possibly, individual workstations (also called cost stations in this respect) [DIN89a].

Costs are recorded with the aid of the operational accounting sheet distributed to the different cost centers (**Fig. 8.3-3**). The distribution usually occurs in cost centers according to the following plan:

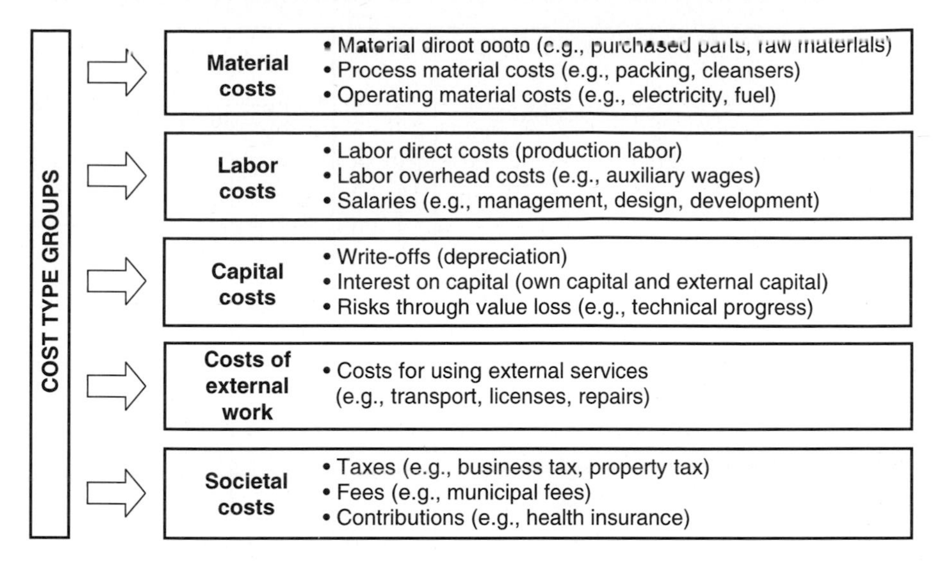

Fig. 8.3-2. Production factors "consumed", classified into cost types

- General cost centers (e. g., facilities management or energy supply) whose services are used by all or by almost all other cost centers.
- Production auxiliary cost centers (e. g., workshop, production planning).
- Production main cost centers (e. g., machine shop, assembly, and individual machines).
- Materials cost centers (e. g., purchasing, material testing, warehouse).
- Administration cost centers (e. g., management, administration).
- Sales and distribution cost centers (e. g., sales, advertising).

The number of cost centers depends on the requirements in the company. Every cost center should have clearly marked limits so that profitability studies for the individual cost centers may be made.

A part of the costs that occur in the company can be assigned directly to the cost units (products, services); these are called direct costs. They do not need to be charged by detouring around the cost centers.

➔ Definition: **Direct costs** are all costs that can be **directly assigned to an allocation object** or are assigned in a specific application case.

The most important allocation objects are **cost units**. In this context there is a distinction between, for example, material direct costs (costs for materials, purchase parts, packing material, etc.) and production labor costs.

Frequently, direct costs are determined as the product of the **amount** (e. g., number, weight, man-hours) and the **value** per unit (e. g., price charged, wage rate) [DIN89a].

Step		Types of costs \ Cost centers	Aux. cost centers: General aux. cost center (e.g., general energy supply)	Aux. cost centers: Aux. production cost center	Main cost centers – Main production cost centers: Turning shop	Main production cost centers: Assembly	Main production cost centers: Sum	Materials cost center	Administration cost center	Sales/marketing cost center
Step 1 Costs acquired	DC	1.1 Material direct costs *MtDC*			150	100	250			
		1.2 Production labor cost *PLC*			60	32	92			
	Overhead costs	1.3 Overhead wages	2	10	21	19	40	2		
		1.4 Salaries	8	16	6	6	12	8	25	27
		1.5 Personnel extra costs	4	8	3	3	6	4	12	10
		1.6 Material overhead costs	9	3	7	8	15	1	1	1
		1.7 Energy (outside purchase)	2	4	14	10	24			
		1.8 Maintenance and repair (outsourced)	5	8	10	9	19	2		
		1.9 Taxes, insurance, fees, rent, contributions	6	8	10	10	20	3	13	12
		1.10 Advertizing, publicity		1					2	4
		1.11 Customer service, commissions								30
		1.12 Depreciation	8	11	25	26	51	4	10	15
		1.13 Accounting interests and risks	2	1	4	3	7	4		
Step 2 Costs distribution		2.1 Sum 1.3 ... 1.13	46	70	100	94	194	28	63	99
		2.2 Apportion general aux. cost center (Key: e.g., installed capacity)		6	13	13	26	1	5	8
		2.3 Apportion aux. Production cost center (Key: *PLC* (1.2))			50	26	76			
Step 3 Determine *OC* surcharge rates		3.1 Sum overhead costs *OC* (2.1...2.3)			163	133	296	29	68	107
		3.2 Production labor costs *PLC* (1.2)			60	32	92			
		3.3 Production overhead costs surch. rate *POCS*			270%	420%				
		3.4 Production costs *PC* = *PLC* + *POC*					388			
		3.5 Material direct costs *MtDC* (1.1)					250	250		
		3.6 Material overhead costs surch. rate *MtOCS*					29	12%		
		3.7 Manufacturing costs *MC* = *PC* + *MtDC* + *MtOC*					667		667	667
		3.8 Administration overhead costs surcharge rate *AOCS*					68		10%	
		3.9 Sales overhead costs surcharge rate *SOCS*					107			16%
		3.10 Total costs (without *DDC*)					842			

Values in 1000 $

Production overhead costs surcharge rate (line 3.3)

$$POCS = \frac{\text{Production overhead costs}}{\text{Production labor costs}} \cdot 100\ \%$$

Material overhead costs surcharge rate (line 3.6)

$$MtOCS = \frac{\text{Material overhead costs}}{\text{Material direct costs}} \cdot 100\ \%$$

Administration overhead costs surcharge rate (line 3.8)

$$AOCS = \frac{\text{Administration overhead costs}}{\text{Manufacturing costs}} \cdot 100\ \%$$

Sales overhead costs surcharge rate (line 3.9)

$$SOCS = \frac{\text{Sales overhead costs}}{\text{Manufacturing costs}} \cdot 100\ \%$$

Fig. 8.3-3. Simplified operational accounting sheet and calculation of overhead cost surcharge rates

Costs that cannot be assigned directly to specific cost units are lumped together as overhead costs. According to [VDM95] they have reached over 50% of the total costs in one machine manufacturing company. Typical overhead costs are costs of the administration and salaries that cannot be assigned directly to, for example, a crankshaft or a gear.

➔ Definition: **Overhead costs** are all costs that **cannot be directly assigned to an allocation object** or to an application [DIN89a].

Important overhead cost types are, in general, salaries, auxiliary wages, auxiliary and operating material costs, accounting write-offs, and interest. "Accounting" costs do not accrue directly, but they should allow a substantial capital preservation.

Overhead costs are charged indirectly in full costing with formulas (overhead cost surcharge rates) on allocation objects, in particular, cost units (Sect. 8.4).

The surcharge rates are set with the **operational accounting sheet**, which is also used for other tasks (e. g., efficiency control; Fig. 8.3-3).

The fundamental structure of the operational accounting sheet is identical in all industrial enterprises. In it, the costs of the different cost centers are organized according to cost types, in tabular form. If direct costing (Sect. 8.5) is used, then costs are divided into fixed and variable costs in the operational accounting sheet.

Setting up the operational accounting sheet occurs in several steps:

Step 1: Acquiring the costs

The numbers for the direct and overhead costs are acquired from bookkeeping and distributed, based on their origin, to the cost centers. The direct costs serve only as a basis for the overhead cost surcharge rates, which are calculated later.

Step 2: Distributing the costs

First, the costs of the general auxiliary cost center are moved to the other cost centers (distribution key, for example, with energy supply, according to installed capacity/cost center, line 2.2). Then the costs of the production auxiliary cost center are moved to the production main cost centers (distribution key, for example, production labor costs, in line 2.3). The sum of lines 2.1 through 2.3 shows the overhead costs for each main cost center (line 3.1).

Step 3: Determining the overhead cost surcharge rates

Now the overhead cost surcharge rates are found, e. g., the manufacturing overhead cost surcharge rates by division of line 3.1 by line 3.2, etc.

Furthermore, in steps not shown here, the planned costs are compared with the actual accrued costs and company's operational values are found [VDI90].

8.3.3 Cost unit accounting

Cost unit accounting assigns costs to the individual product (cost unit). This cost calculation has to fulfill the following tasks:

- Determination of the manufacturing and total costs per cost unit (product) as a basis for price setting.
- Success verification of the products for the production program planning.
- Comparison calculations for different products at different times and different production processes.

According to the timing of the calculation, we distinguish between **preliminary** and **follow-up cost calculations**. The preliminary calculation occurs **before** product realization and is based on the preplanning of the production. The follow-up calculation occurs during the actual manufacturing and calculates actual costs. With longer projects, intermediate calculations are carried out during production. If this occurs continuously during the development, it is called **concurrent calculation** (Chap. 9).

The terms used in practice and their assignment to the chronological course of the product creation process are different. To avoid misunderstandings, one should check how terms are used in practice and how they are used here. The cost calculation process has been simplified here and proceeds as follows:

Upon an inquiry, the **quote** is generated. If it results in an order, it is further developed. After extensive preparation and completion of the design documents, the **preliminary calculations** take place with work schedule preparation. After manufacturing, the **follow-up calculations** are carried out. The contents and extents of the individual steps can be different. Time overlap is possible since certain parts may be already produced while others have still not been designed.

8.4 Cost calculation procedures

We categorize **division costing** and **overhead costing** as essential procedures. The division calculation is used mainly in uniform mass production, and is very simple. A company which produces, for example, only one type of jacks can divide all costs accrued during an accounting period by the number made during this period and thus obtain the per-unit costs. An expanded form of the division calculation is the computation with equivalence digits [DIN89b].

The more different the products are in a company, and more variable the quantities produced are, the more significant the problem of cause-based cost allocation becomes. In these cases, overhead costing is utilized. We distinguish according to the number of the reference quantities between the summary (Sect. 8.4.1) and the differentiating overhead costing (Sect. 8.4.2). It is the most widespread procedure of cost unit accounting in the machine industry.

Special forms of differentiating overhead costing are workstation costing and machine hourly-rate accounting. These are based on a more detailed separation of the cost centers, down to individual workstations (e. g., machines) and to a more differentiated recording of the overhead costs (Fig. 8.4-9). Thus they are more precise and their use is increasing.

8.4.1 Summary overhead costing

With summary overhead costing, the whole company is considered as a single cost center, and the overhead costs are added as a block onto the direct costs as a reference parameter. The calculation with only one reference parameter is simple, but is beset with problems as shown below. Therefore, summary overhead costing, like the division costing, may be used only for companies with very similar products.

Material direct costs *MtDC*, **or** the production labor costs *PLC* can be used as a **reference parameters.** The differences that result from the choice of a reference parameter are explained in the following example:

A company that makes welded assemblies has total costs *TC* of 1 000 000 $ in an accounting period, of which 400 000 $ are material direct costs *MtDC*, and 100 000 $ are production labor costs *PLC*. According to the reference parameter used, those result in overhead costs *OC* and overhead cost surcharges *OCS*, as depicted in **Fig. 8.4-1**. In practice, one cannot make this comparison because when a given reference parameter (e. g., the material direct costs) is chosen, only this can be used in the calculation. The different options are not practiced in parallel.

a) **Establishing overhead cost surcharge rate**

		Calculation with Reference Parameter **MtDC**	Calculation with Reference Parameter **PLC**
Total costs	*TC*	1 000 000	1 000 000
Direct costs	*DC*	400 000	100 000
Overhead costs	*OC = TC - DC*	600 000	900 000
OCS	*OC/DC*	**150 %**	**900 %**

b) **Calculation: Machine frame 1**

		with Reference Parameter **MtDC**	with Reference Parameter **PLC**
Direct costs	*DC*	2 000	1 000
Overhead costs	*OC = DC × OCS*	3 000	9 000
Total costs	*TC*	**5 000**	**10 000**

c) **Calculation: Machine frame 2**

		with Reference Parameter **MtDC**	with Reference Parameter **PLC**
Direct costs	*DC*	2 000	2 000
Overhead costs	*OC = DC × OCS*	3 000	18 000
Total costs	*TC*	**5 000**	**20 000**

Fig. 8.4-1. Example for the summary overhead costing: Different reference parameters result in different surcharge rates and total costs

First, with the same total costs in the company there will be different overhead costs. That results in different overhead cost surcharge rates. These then lead to completely different total costs in the costing of new products. **Figure 8.4-1a** shows the costs accrued in one accounting period and the calculated overhead cost surcharge rates. With these, the product (cost unit) calculation is then carried out.

Cost calculations for machine frame 1, for which the direct costs are 2 000 $ *MtDC* and 1 000 $ *PLC*, are as shown in **Fig. 8.4-1b** and depend on the chosen reference parameter for the overhead and total costs.

If another machine frame 2 is chosen, which has the same material direct costs of 2 000 $ (same weight) but double the production labor costs of 2 000 $ (more parts and welds), the results are as shown in **Fig. 8.4-1c**. According to the reference parameter used, the second machine frame may cost the same as the first frame (reference parameter *MtDC*), or twice as much (reference parameter *PLC*)! The summary overhead costing is therefore not suitable for these machine frames, since they have such different cost structures.

8.4.2 Differentiating overhead costing

Differentiating (or differentiated) overhead costing is widespread in the machine industry. For this calculation procedure, the company is divided into several cost centers to which the overhead costs are assigned over several reference parameters. For example, we may categorize:

- Material overhead costs — *MtOC*;
- Production overhead costs — *POC*;
- Administrative overhead costs — *AOC*;
- Sales overhead costs — *SOC*.

Reference parameters, for charging the overhead costs, are commonly:

- Material direct costs — *MtDC*;
- Production labor costs (also production direct costs) — *PLC*;
- Manufacturing costs — *MC*.

The costs, according to **Fig. 8.4-2** and Fig. 8.4-9, are found as follows:

Material costs *MtC*

The costs of materials are made up of the material direct and overhead costs.

$$MtC = MtDC + MtOC$$

They make up 15–60% of the total costs in the machine industry (on the average, 37.8% [VDM06]). The material costs are therefore an important cost element in the company as well as in the products (Sect. 7.9.1). The term material costs includes all costs of all purchased goods (i. e., costs of raw material, semi-finished products, standard and purchased parts, parts externally produced, etc.). They also

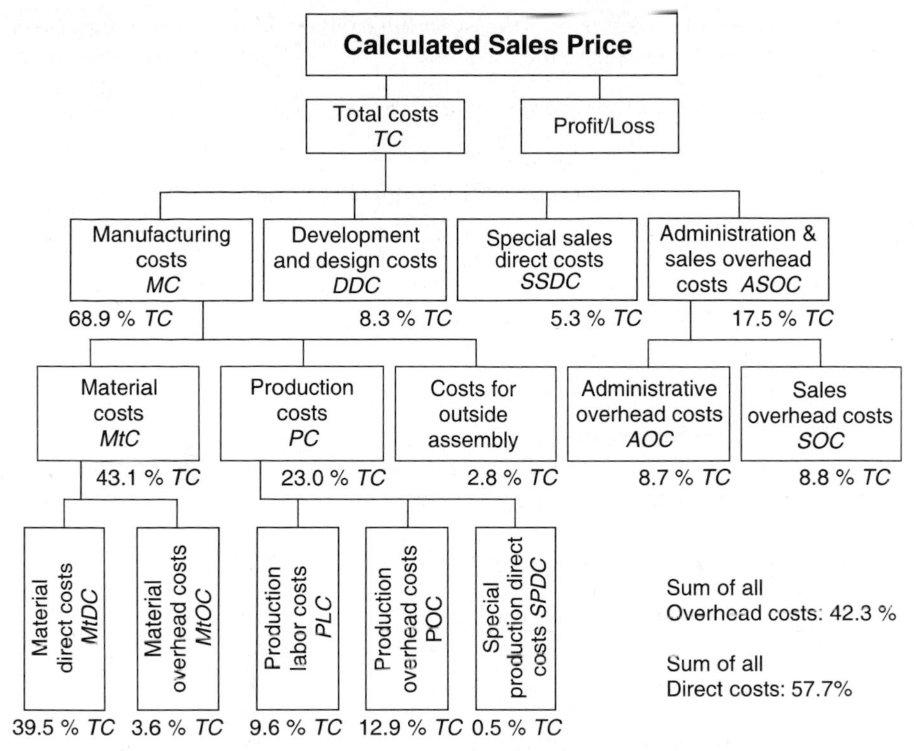

Fig. 8.4-2. Calculation scheme in the machine industry (differentiating overhead costing); percent details according to VDMA, 2002 [VDM06]

include, for example, the supplier's production labor, administration costs, profits, etc. In the case of a large portion of externally supplied parts, it is important for cost management to recognize or estimate how the costs are put together by the supplier to take measures for lowering the costs. This is best achieved in cooperation with the supplier (Sect. 7.10.4).

Material direct costs *MtDC*

The material direct costs are found by multiplying the quantity of material used with the value per unit quantity (generally the cost price/unit). The material quantities used for the cost unit are obtained from requisition slips, the cost prices from the bills paid, and additional procurement costs (transportation, etc.).

$$MtDC = \text{quantity} \cdot \frac{\text{value}}{\text{unit}}$$

Material overhead costs *MtOC*

Material overhead costs are computed from the material direct costs by using the material overhead cost surcharge rate. The material overhead costs contain the costs of space for the warehouse, interest on the capital tied up in inventory, costs of material control according to the order, etc.

$$MtOC = MtDC \cdot \frac{MtOCS}{100} \%$$

$MtOCS$ = Material overhead cost surcharge rate (usually 5–10–20%)

Production costs *PC*

Production costs are the sum of production labor and overhead costs.

$$PC = PLC + POC$$

The production labor costs are comparatively low, on average 8.5% of the total costs. In contrast, the production overheads make up a larger part, on average 18.5% of the total costs. Production costs include the costs of all production steps, as well as of the assembly. Instead of production labor costs, we may also work with production direct costs *PDC*.

Production labor costs *PLC*

The production labor costs are found from the labor rate and the time/piece

$$PLC = plc \cdot t$$

plc = Labor wage rate [$/min or $/h]
t = Time per piece [min or h].

The time per piece (Fig. 7.6-2) is acquired from the worker's work orders, either as listing of the actual times (follow-up calculation), or from the standard times from production planning (preliminary calculation).

Production overhead costs *POC*

Production overhead costs are computed from the production labor costs by using the production overhead cost surcharge rate.

$$POC = PLC \cdot \frac{POCS}{100} \%$$

$POCS$ = Production overhead cost surcharge rate (usually 200–500%)

The production overhead cost surcharge rate is determined with the operational accounting sheet for each cost center. The production overhead costs contain the costs for machines, shops, supervisors, etc.

Special production direct costs *SPDC*

Special production direct costs are costs for fixtures, models, etc., which are provided only for the production of one cost unit. Tools used more generally are considered under production overhead costs.

Outside assembly

Figure 8.4-2 mentions the costs of assembly from external sources, according to [VDM95]. In the usual scheme of the differentiating overhead costing, they are not listed separately. How far they and other cost types are broken down and determined, should be decided specific to the company (see also Fig. 8.4-9, where the sales direct costs *SDC* are identified separately.)

Manufacturing costs *MC*

The manufacturing costs are the sum of material and production costs.

$$MC = MtC + PC$$

The manufacturing costs *MC* are, on average 68.6% of the total costs. In practice, these costs are also designated as *MC1*. In *MC2* the design and development costs *DDC* are also included.

Total costs *TC*

Total costs are the sum of the manufacturing, design and development, as well as administration and sales overhead costs.

$$TC = MC + DDC + ASOC + SSDC$$

In addition, special sales direct costs *SSDC* should be considered where appropriate. For the above conditions, the design and development costs are typically 3–25% (average 8.6%) of the total costs, and the administration and sales overhead costs are 15–20% of the total costs. In practice, even additional costs can be charged as appropriate (e. g., production risk, warranty costs, etc.). Such unusual features should be decided specific to each company. The example in Sect. 8.4.3 shows how differentiating overhead costing is carried out.

The scheme and the method of calculating overhead costing appear to be simple. For cost management, however, the following **four problems** arise:

1. Costs of the product as a sum of the costs of many parts

The costs for a product arise from adding up the costs of its many parts (an automobile, a specialty machine, consist of 10 000 or more parts), its materials costs and the costs of the many production steps, including the assembly. This **problem of quantity** is indeed manageable with the use of computers, but the gathering and preparation of data for transparency in cost management are expensive.

2. Cost calculation prior to product realization – preliminary calculation

Another problem is the **point in time of the calculation**. Afterwards, it takes "only hard work" to put together the accrued costs from the materials requisition and wage sheets. However, to determine the cost targets, the quotation price, and the concurrent cost control during the development of a product, the costs must be calculated on the basis of documents that are still not available or are incomplete (drawings, work schedules, parts lists). That is difficult, and the results are uncertain (Sect. 9.1).

3. Updating the cost data

Because cost data change with time (wage increases, changes in material prices, etc.), they must be kept up-to-date. The continuous updating of the data and also the calculation with current, past, and future cost rates at the preliminary and follow-up calculation are in theory solved with the computer by the use of flexible marginal costing [Mül93]. In business practice, there are still considerable gaps in this method.

4. Cost influence of quantities, sizes, standardization, etc.

Overhead costing charges the overhead costs proportional to the production labor and material direct costs for a product. This is valid, however, only in few cases. As is clear from examples in Sect. 8.4.3, design, administration, and sales costs depend on the novelty and the complexity of the specific order, and not absolutely on the often very low material direct costs and production labor costs (Sect. 7.12). With process costing, we attempt to assign the costs to other effective variables – cost drivers – based more on how they are caused (Sect. 8.4.6).

Other problems of overhead costing are dealt with in Sect. 8.4.4. Differentiating overhead costing indeed considerably reduces the problems of summary overhead costing shown in Sect. 8.4.1. However, in principle, they still persist.

8.4.3 Examples of real cost generation and overhead costing

Possible incorrect decisions based on overhead costing are shown with the examples **a** through **d** below. They especially affect the costs with different quantities, sizes, variant and special designs (**a**, **b** and **d**), and very definitely the constructive measures (**c**). The numbers in the examples are greatly simplified to make the relationships comprehensible. On the other hand, the circumstances are realistic (see also Fig. 7.13-17).

Example a: Special design of a mixer cover

This example of a welded cover for a stationary concrete mixer shows how and where during the production costs really arise and are charged, based on the usual overhead costing.

The cover (**Fig. 8.4-3**) is a special design for a customer whose wish must be fulfilled. One suspects, however, that the agreed-upon price of 1 600 $ hardly covers the actual (that is, resulting) costs.

How the costs came about is as follows: The call from a customer brought about the sequence of events. A design engineer had to drive to the customer to obtain the measurements of the cover directly from the concrete mixer. The customer asked the design engineer: “What will the cover probably cost?” The design

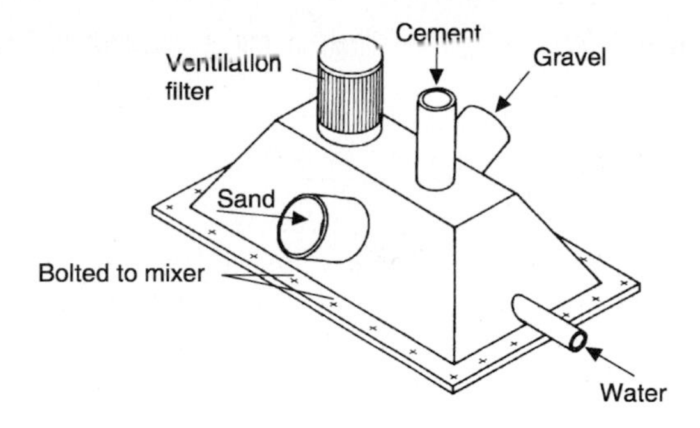

Fig. 8.4-3. Cover for a stationary concrete mixer

engineer thought about it: The weight of the cover is about 300 kg; the usual kg-price (weight cost unit rate) is approximately 6 $/kg; therefore, he answered: “The cover will likely cost about 1 800 $”.

The design engineer drives back, prepares a preliminary design, and gives this to the production planning and/or to costing. The latter does a preliminary calculation as differentiating overhead costing, with workstation cost rates (**Fig. 8.4-4**) by means of the preliminary quantity schedule (needed material quantities and production times). The sales department, which also received the estimated value of 1 800 $ mentioned by the design engineer, is informed of the result (total cost

Material direct costs		[kg]	[$/kg]	*MtDC*		
Sheet metal		250	1	250		
Pipes & small parts				50	**300**	***MtDC***
Material overhead costs	*MtOCS* = 10% (of *MtDC*)				**30**	***MtOC***
Material costs					**330**	***MtC***
Production costs	t_s[h]	t_e[h]	[$/h]	*PC*		
Sawing		0.5	40	20		
Welding	0.5	2.5	60	180		
Mech. working	0.5	5.0	60	330		
Painting		0.5	80	40	**570**	***PC***
Manufacturing costs *(MtC + PC)*					**900**	***MC***
Design and development costs						
DDOCS = 10% (of *MC)*					**90**	***DDC***
Administration and sales overhead costs						
ASOCS = 25% (of *MC*)					**225**	***ASOC***
Total costs					**1 215**	***TC***
Sales price					**1 600**	**Earnings**
Profit				**24 % of earnings**	**385**	**Profit**

Items without designation, in $

Fig. 8.4-4. Preliminary calculation for the mixer cover (differentiating overhead costing with workstation cost rates), according to Fig. 8.4-3

Material direct costs			[kg]	[$/kg]		*MtDC*		
Sheet metal			260	1		260		
Pipes & small parts						90	**350**	***MtDC***
Material overhead costs	*MtOCS* = 10% (of *MtDC*)						**35**	***MtOC***
Material costs							**385**	***MtC***
Production costs	t_s[h]	t_e[h]	[$/h]	*PCs*	*PCe*	*PC*		
Sawing		0.5	40	0	20	20		
Welding	0.5	3.25	60	30	195	225		
Mech. working	0.5	5.0	60	30	300	330		
Painting		0.5	80	0	40	40		
			Sum:	60	555		**615**	***PC***
Manufacturing costs *(MtC + PC)*							**1 000**	***MC***
Design and development costs *DDOCS* = 10% (of *MC*)							**100**	***DDC***
Administration and sales overhead costs *ASOCS* = 25% (of *MC*)							**250**	***ASOC***
Total costs							**1 350**	***TC***
Sales price							**1 600**	**Earnings**
Profit					**16 % of earnings**		**250**	**Profit**

Items without designation, in $

Fig. 8.4-5. Follow-up calculation of the mixer cover (actual costs charged)

1 215 $). They offer the cover to the customer for a price of 1 800 $. The customer negotiates and agrees upon a price of 1 600 $.

The customer is satisfied because he received a 12.5% discount. Marketing is pleased because they expect a profit of 385 $ (24% of the price).

After the cover is completed, a follow-up calculation with the actual quantity schedule is carried out (**Fig. 8.4-5**). From that, the total costs come out not 1 215 $, but as 1 350 $; thus, the profit turns out to be not 385 $, but 250 $ (about 15%). With that, one would be satisfied in the normal case and the difference looked upon as the usual inaccuracy between the bid and follow-up calculations. The value of 6 $/kg for an approximate price-calculation is also confirmed.

It could now be concluded that the special design of such dust covers represents a good business opportunity that can be pursued based on overhead costing. The real situation shows, however, such a big loss that for similar cases in the future a considerably higher price must be charged.

In **Fig. 8.4-6**, the actual sequence of steps of the order with the real cost generation is considered in process costing (Sect. 8.4.6). For simplification, all work considered up to now in overhead cost departments (e. g., purchasing, design) is assessed with a workstation cost rate of 50 $/h. It is further assumed that with this workstation cost rate, all other costs (e. g., security, management) will be charged, based on cause.

A look at the actual steps in the process shows a completely different picture than the one we saw with overhead costing:

- The **material direct costs** are cause-based (quantity and value) schedules and do not change.
- If we consider the expenditure for this order in detail, we see that the usual **material overhead surcharge rate** of 10% on the *MtDC* does not suffice for this relatively small order. Because of the time required in the order and receiving departments, it costs 60 $ instead of the 35 $ that was charged.
- The **production costs** are entered, referring to (production) process that are largely cause-based.
- The **design costs** are very much higher for this special design than that found from the average value of the design and development overhead-cost surcharge rate of 10%. They amount to 17.2 h design time of 960 $ (including travel costs) instead of 90 $ charged in the preliminary cost calculation or 100 $ for the follow-up calculation.
- For the **administration** and **sales overhead costs** there is a change from 225 $ or 250 $, to 300 $.
- In the **total result** the process costing identifies the real costs of 2 285 $ for this job instead of 1 215 $ or 1 350 $. Instead of a profit of 250 $, the job causes the loss of –685 $ = –43% of the earnings (Fig. 8.4-6).

The example clarifies the problematic nature of cost accounting: It should be cause-based as far as possible – that is a main basic principle, the causation principle. On the other hand, the expenditure for cost accounting should also not become too large (economic efficiency principle).

In this example, only the production and material direct costs were entered in overhead costing directly and, thus, cause-based, for the product. Starting from that, all other costs were charged with overhead cost surcharge rates. These arise in proportion to the costs accrued in an earlier accounting period (e. g., a year) to the production costs at a cost center (Sect. 8.3.2). That is simple and is correct on average over the accounting period. However, special features of individual jobs such as special designs or the influence of quantities and sizes on the costs are not accounted for correctly; rather, are "obscured". Furthermore, production and material costs are size-dependent; administration, marketing, design and development costs, on the other hand, are not. Therefore, in case of doubt the designer should resort to production and material (direct) costs and consider the resolution of overhead costs with caution. However, according to Fig. 8.4-2, those costs are half of the total costs.

Material direct costs				[kg]	[$/kg]	*MtDC*		
Sheet metal				260	1	260		
Pipes & small parts						90	**350**	***MtDC***
Material overhead costs				*t* [h]	[$/h]	*MtOC*		
Ordering				0.6	50	30		
Material receiving				0.6	50	30	**60**	***MtOC***
Material costs							**410**	***MtC***
Production costs	t_s[h]	t_e[h]	[$/h]	*PCs*	*PCe*	*PC*		
Sawing		0.5	40	0	20	20		
Welding	0.5	3.25	60	30	195	225		
Mech. working	0.5	5.0	60	30	300	330		
Painting		0.5	80	0	40	40		
			Sum:	60	555		**615**	***PC***
Manufacturing costs *(MtC + PC)*							**1 025**	***MC***
Design and development costs				t_{ek} [h]	[$/h]	*DDC*		
Telephone conversation				0.2	50	10		
Travel to customer site				8.0	50	400		
Travel costs						100		
Proposal				3.0	50	150		
Detail design				6.0	50	300	**960**	***DDC***
Administration and sales overhead costs				*t* [h]	[$/h]	*ASOC*		
Telephone ordering, order preparation, etc.				2.0	50	100		
Shipment				1.0	50	50		
Bookkeeping				2.0	50	100		
Processing account				1.0	50	50	**300**	***ASOC***
Total costs							**2 285**	***TC***
Sales price							**1 600**	**Earnings**
Profit						**-43% of earnings**	**-685**	**Loss!**

Items without designation, in $

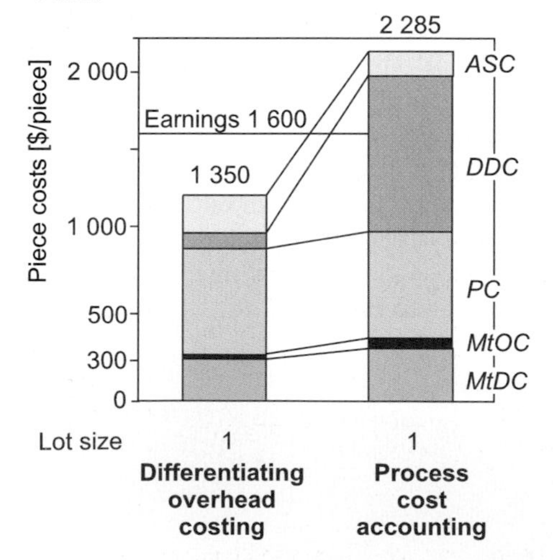

Fig. 8.4-6. Process costing for the mixer cover ("real" costs)

Example b: Special design of three identical mixer covers

If the **order is for three identical covers** instead of one, and the earnings per piece remain the same 1 600 $, the real costs and the costs calculated with overhead costing and profits and/or losses change (**Fig. 8.4-7**) substantially.

Follow-up calculation for **one** cover, with lot sizes 1 and 3

	Differentiating overhead costing		Process cost accounting	
Lot size	1 piece	3 pieces	1 piece	3 pieces
MtDC	350	350	350	350
MtOC			60	20
MtOCS = 10 % (of *MtDC*)	35	35		
MtC	385	385	410	370
PC *PCs*	60	20	60	20
PCe	555	555	555	555
PC	615	575	615	575
MC	1 000	960	1 025	945
DDC			960	320
DDOCS = 10 % (of *MC*)	100	96		
ASC			300	100
ASOCS = 25 % (of *MC*)	250	240		
TC	1 350	1 296	2 285	1 365
Earnings	1 600	1 600	1 600	1 600
Profit	250	304	-685	235
Earnings yield	16 %	19 %	-43 %	15 %
Cost reduction 1-3		**-54**		**-920**
		-4 %		**-40 %**

Items without designation, in $

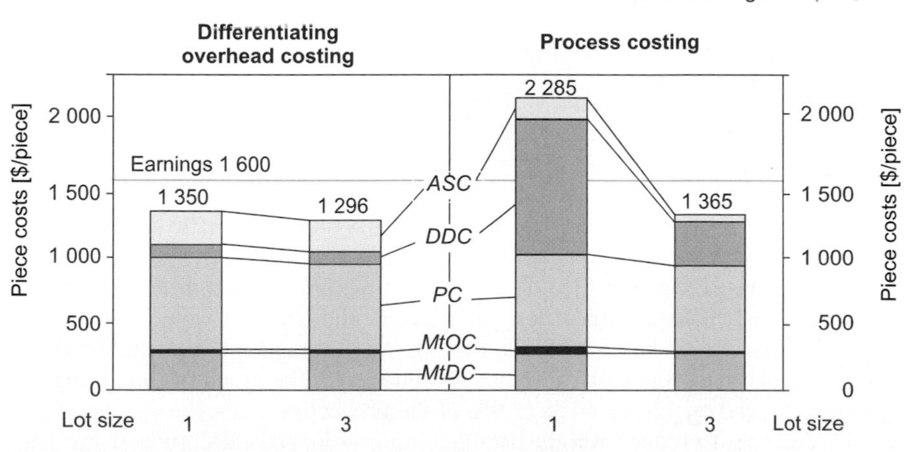

Fig. 8.4-7. Costs calculated with overhead costing, and "real" costs (process costing) for the production of three identical mixer covers in one lot

With **process costing** the real costs arising in materials management (*MtOC* = 60 $), design (*DDC* = 960 $), administration and sales (*ASC* = 300 $), remain fixed for the job regardless of whether one or three pieces are involved. Per piece, they then amount to only another 440 $ (*MtOC* = 20 $, *DDC* = 320 $, *ASC* = 100 $), instead of 1 320 $, as it was up to now for single-unit production. That already shows a saving of 880 $/piece. Furthermore, setup times are available during the production for welding and mechanical work of 30 min each (Fig. 8.4-6). With a lot size of three, every piece takes only 10 min of that and saves about 40 $/piece, as opposed to single-unit production. Together, this results in a saving of 920 $/piece (–40%) if we calculate the real total costs.

With differentiating **overhead costing,** the production costs are lowered due only to the consideration of the decrease of setup times by 40 $/piece.

Thus the design and development overhead costs also reduce automatically by 4 $/piece, found with the manufacturing costs as a basis, the administration and sales overhead costs by 10 $/piece. The total calculated costs reduce in total only by 54 $/piece (–4%).

Thus if the earnings remain constant at 1 600 $ for the production of three pieces, reckoned with the real costs, there is a profit of 235 $/piece (i. e., approximately 15% of the sales price). On the other hand, in the case of single-unit production we see a loss of 685 $ (43% of the sales price). Due to the high costs, which occur only once, quantity has a drastic influence in such a case. With the usual overhead costing, on the other hand, the improvement in the profit is only small, from 250 $/piece (16%) as stated previously, to 304 $/piece (19%) now.

We see that overhead costing cannot make any appropriate statement about the real costs and their generation. This includes cases of mixed production between standard or routine tasks and special designs, as well as of single-unit production and smaller lot sizes. **Thus, it does not provide a reasonable starting point for cost management.** Instead, one proceeds to use-based process costing (Sect. 8.4.6) or attempts to at least charge the design and development costs as direct costs.

Example c: Weight-based costing of heat treatment

Another example of **results of cost accounting** that are not cause-based is repeated in **Fig. 8.4-8** (see also Sect. 7.13.5, Fig. 7.13-17). It is usual to charge the costs proportional to the weight of the parts (weight-based costing, Sect. 9.3.2.1) put into the furnace for heat treatment (annealing, stress relieving, hardening, etc.). In spite of having the same external measurements (decisive for the furnace size), light, thin-walled parts are assigned lower costs than heavy parts. The weight of the gear shown in Fig. 8.4-8 (to the right) was therefore reduced by 1 650 kg by machining out material. According to the calculation, this would save about 3 950 $ in heat treatment costs. Since this saving was much higher than the 650 $ costs for machining, the cost-conscious designer could be proud of the savings of 3 300 $ achieved by this, equal to 13.9% of the production costs.

The decision to reduce weight by machining is in general correct if the heat treatment is done by a supplier who uses weight-based costing, not carried out in one's own shop. Then the heat treatment costs for the customer do in fact decrease with weight.

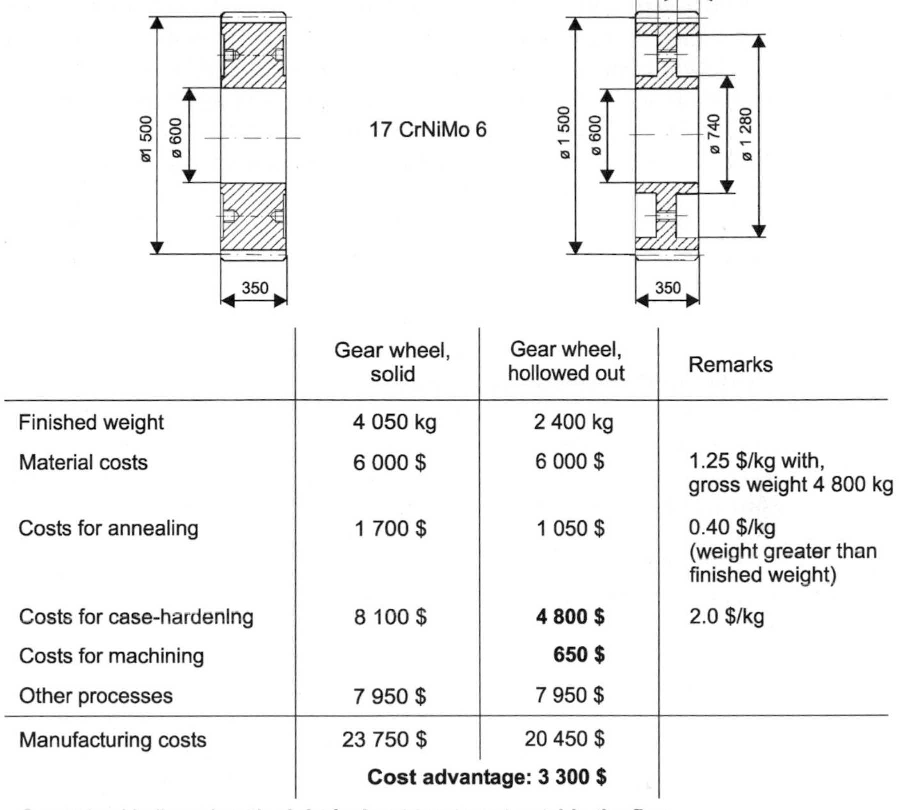

	Gear wheel, solid	Gear wheel, hollowed out	Remarks
Finished weight	4 050 kg	2 400 kg	
Material costs	6 000 $	6 000 $	1.25 $/kg with, gross weight 4 800 kg
Costs for annealing	1 700 $	1 050 $	0.40 $/kg (weight greater than finished weight)
Costs for case-hardening	8 100 $	**4 800 $**	2.0 $/kg
Costs for machining		**650 $**	
Other processes	7 950 $	7 950 $	
Manufacturing costs	23 750 $	20 450 $	
	Cost advantage: 3 300 $		

Gear wheel hollowed out is **right** for **heat treatment outside the firm,** if weight-based costing is done there.
It is **wrong** for **heat treatment in the company's own shop.**

Fig. 8.4-8. Cost comparison of differently shaped gears (see also Fig. 7.13-17)

However, if the heat treatment is carried out in one's own shop, the gear becomes more expensive due to the machining costs, since the required size of the furnace and the heat treatment time are the same. Here the heat treatment costs depend basically on the largest measurements, which are identical, and are not determined by the dimensions of the machine part (Sect. 7.13.5 [Bru94]).

Example d: Influence of quantities and dimensions on the costs

To show the influence of the quantities produced and dimensions on the real costs and the costs charged, we carry out further calculations with the numbers of the gear of the preceding example that was not hollowed out.

The gear's production costs are 23 750 $. If we accept, as in example **a**, Sect. 8.4.3, that the design and development costs are charged at a surcharge rate of 10% on the *MC* and a design hourly costs 50 $, a design cost of 2 375 $ was charged for the gear. That corresponds to 47.5 h design time.

For depiction of the influence of the **quantity**, it is assumed that there is an order for a limited-lot production of five identical gears. The manufacturing cost/piece would drop only slightly (e. g., around 5%; see Figs. 7.7-1 and 7.7-2) to about 22 500 $. The design costs went up 10% on these manufacturing costs (2 250 $/piece) and amount to 11 250 $/order. The real design costs (design time) will not increase, however, as compared to single-unit production.

For looking at the influence of the **dimensions**, it is assumed that the same gear with size ratio $\varphi_L = 0.5$ is manufactured in lot size = 1. The diameter then is 750 mm (all other dimensions corresponding), the weight approximately 500 kg, and the manufacturing costs about 4 000 $ (Sect. 9.3.4). The design costs are again 10% of that (i. e., amount to 400 $), which corresponds to 8 h design time. In comparison to the larger gear, with 47.5 h, this is much too low! In reality, **on the average** over an accounting period the surcharge rate for the design costs as 10% of the manufacturing costs is **correct**, but **wrong in individual cases**! The behavior of the administration and sales overhead costs is similar.

If the costs calculated in this way work their way into the prices, one does not need to wonder when a lucrative serial production job is lost to a competitor who calculates costs more correctly (Fig. 7.5-1; Sects. 7.12.5 and 7.12.6).

8.4.4 Disadvantages of overhead costing

Most commonly in companies with single-unit and limited-lot production, making products varying in complexity and sizes, calculated costs and the costs that are really generated, differ from each other (Sect. 8.4.3). As mentioned, in the machine industry approximately 50% of the total costs are overhead costs ([VDM95], Fig. 8.4-2) and therefore are not assigned to a cost unit directly. This throws light on the possible subsidization of certain products. The following are the **consequences for design**:

a) As opposed to a program of simple, standardized products, complicated products (e. g., **special designs**) require large time expenditure in many overhead cost departments, (e. g., in design, production planning, purchasing, and sales). This does not become obvious in cost accounting (Chap. 6, Sect. 8.4.3, example **a**). If these costs are not considered in the price, a danger exists that increasingly more orders are accepted for special designs, which apparently produce profit, but in reality must be subsidized by the standard products.

The standard products put fewer burdens on the overhead cost departments, since they have already been launched and most of the drawings, work schedules, and fixtures are at hand. Despite that, they get burdened with too high overhead costs due to the overhead cost surcharges and therefore often fare badly – seemingly at a loss – although in reality they create profits. The danger exists that such products will disappear when cheaper competition arrives. The latter are often small companies whose overhead costs are small since they do not have expensive development, production planning, and marketing departments. With that, the company is pushed more and more into a technologically demanding market niche and must add qualified personnel in its development and production departments. Thus, the overhead costs keep rising further. The company could unknowingly have made good profits with the old standard products.

Thus, it is essential that at least the costs of design and production planning be assigned to the product from the appropriate time sheets. This is indeed carried out already in many companies. It is necessary, however, not simply to write down the hours for checking (as it is largely customary). The hours charged should also be evaluated and with future orders or bids the costs calculated accordingly and the costs reflected in the prices!

b) If **single-unit and limited-lot production change** for a product type, the cost reduction that is calculated with overhead costing would be too small for products made in larger quantities. The time needed in design, production planning, marketing, and purchasing does not much depend on the quantities. In spite of that, cost calculations continue to be made with constant overhead cost surcharges for the second piece, third piece, etc. This is also valid for similar products with varying weights.

The influence of quantities is often considered in the preliminary calculation only for the production setup times, which is divided over the larger number. It is assumed that the production processes do not change. In the case of larger products, depending on the setup time, that results in reductions in total cost of only a few percent in going from one piece to two pieces/lot, while in reality there is a much greater reduction (examples **b** and **d**, Sect. 8.4.3). The greatest disadvantage of concealing the influence of quantities in design lies in that the efforts toward similar parts, part families, product size ranges and modular design (i. e., internal standardization) are not heeded or do not become apparent (Sects. 7.12.5.1 and 7.12.6). These measures aim at increasing the number for parts in spite of single-unit production of complete products. If the design engineer knew what considerable effect it has to design only one instead of two different covers, for example, much more attention would be paid to that.

Sometimes well thought out modular designs do not find a ready market only because the economic advantages to be achieved were concealed by overhead costing, and therefore could not be passed on to the market. Then special designs prevail, although this is technically and economically absurd. Product standardization in the above sense is carried out basically to reduce costs due to greater quantities produced (Fig. 7.12-4, Fig. 7.12-13, Sect. 7.12). If, however, quantities are not sufficiently heeded in cost accounting, there is no stimulus for the designers to push for internal standardization.

c) Cost calculation procedures greatly simplified for expense reasons (e. g., the weight-based costing) cannot be used as a means for cost management. They lead at times to wrong conclusions (examples **a** and **d**, Sect. 8.4.3) or do not sufficiently consider constructive influences.

In the **following decisions,** full costing with differentiating **overhead costing** is **inadequate** as a cost accounting procedure:

- Cost calculation of products for mixed production (e. g., in large/small quantities, large/small dimensions, high/low complexity, and standard/special designs).
- Cost calculation of products with capital-intensive production, since the narrow labor cost reference basis shows large scatter in manufacturing costs.
- Assessment of the appropriate use of available resources.
- Assessment of in-house versus external production.
- Assessment of investment of production machinery.

d) Distortion of the influence of quantities, novelty, complexity, and size by the usual differentiating overhead costing **also shapes** the mentality of the **sale**. Companies lose conventional standard products while favoring unidentified expensive, custom-built special designs. With that, the lot sizes for parts and assemblies shrink, and the design expenditure climbs. The company becomes the technically demanding, small and expensive maverick. The result is that the cost structure of the company (Chap. 2; Fig. 6.1-1) becomes increasingly less favorable.

These remarks are not intended to question the company's cost accounting. The latter is simplified so as not to be too expensive and must support other tasks such as wage accounting, etc. Cost management must work with the existing cost accounting in the company and be based on their data (Chaps. 3 and 4). However, limits and possible errors in the cost accounting should be known and considered accordingly in these decisions (Sect. 8.4.6). **In the collaboration among marketing, design, production planning, and costing, the decisions at times must be different from the "cent-precise" data provided by costing** (example **c**, Sect. 8.4.3).

8.4.5 Workstation costing

Differentiating overhead costing is being increasingly replaced by workstation costing because of the disadvantages described above. It is mostly employed, however, only in production. Therefore, the problems with the charging of typical overhead cost departments remain as before.

A workstation cost rate includes the production costs referred to the man-hours (or the duration of machine utilization) for the specific place of work. The rate can refer here to a machine or manual work place. It contains the wages and wage incidental costs, machine costs (accounting write-off, interest, space, energy, repair costs), tool costs, and remaining production overhead costs (e. g., for production control, quality control, supervisor, etc.). The special direct costs of production are not included. In some companies, instead of workstation cost rates, machine hourly rates are used that contain only the machine costs. Then the wage and wage incidental costs and the remaining production overhead costs are considered separately. **Figure 8.4-9** shows a comparison of the different cost calculation procedures.

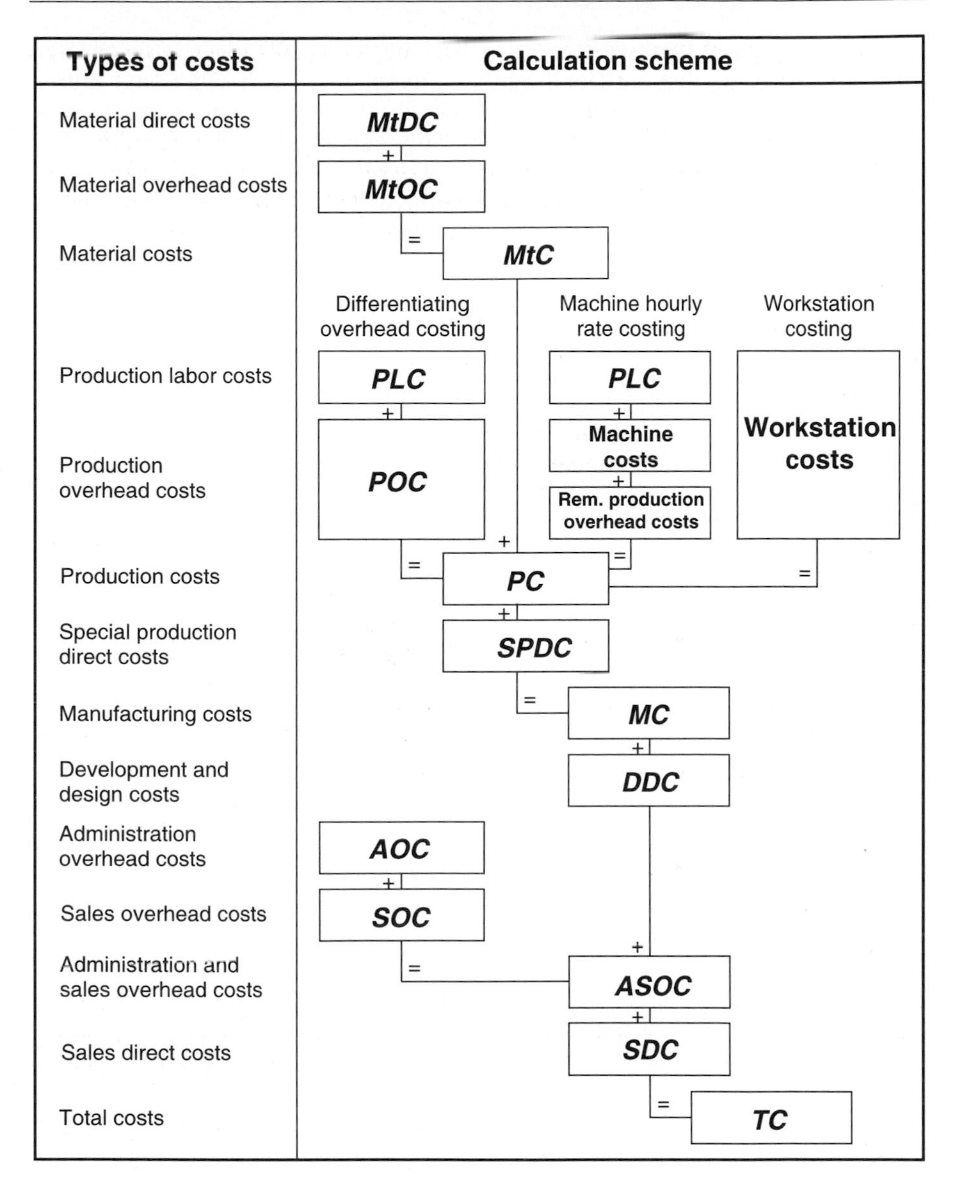

Fig. 8.4-9. Differentiating overhead costing, machine hourly rate, and workstation costing

The **advantage of workstation costing** as opposed to the usual type of differentiating overhead costing lies in that specific machines or machine groups are treated as individual cost centers. It therefore provides cost calculation based more on use; the calculation, however, becomes more expensive. This principle of cost accounting corresponds basically to differentiating overhead costing; the difference is in the considerably larger number of the cost centers. In some companies, a hybrid type of overhead costing and workstation costing is used.

The advantage of workstation costing becomes especially clear in companies that make products that are significantly different from each other, and utilize machines and workstations with correspondingly different expenses during production. Furthermore, using workstation costing reduces accounting variations. In overhead costing, the production labor costs are the basis for the production overhead surcharge, which is often 8–10 times as high. Thus, variations in production labor costs greatly affect the calculated production costs. Workstation costing therefore goes from the value base (the production labor costs) over to a quantity base (the production time).

The following two machines provide an example of a more use-based costing by using workstation cost unit rate (see also Fig. 8.5-1):

Workstation cost unit rate:		
	Vertical mill	60 $/h
	Boring mill	70 $/h
Overhead costing:		
	Average costs of the cost center (for both machines)	65 $/h

If these two machines are combined in a cost center for differentiating overhead costing, only the average costs of 65 $/h are reckoned. The vertical mill is valued too high, the boring mill too low. If quotation prices for wage work are calculated from that, the turning work is quoted too expensive with the result that no orders come in, while the boring mill, which is quoted too low, is overloaded. It is assumed that the competitors in the market do not estimate just as poorly. Analogous cost distortion also results for products made in-house that use the machines differently.

8.4.6 Process costing

In the past, solutions for problems of overhead costs have appeared from different authors. These have come to be known as Process Costing, Events Costing, Activity-based Costing, Transaction Accounting, Cost-driver Accounting, Resource-oriented Cost Calculation, and the like [Ber95, Bru94, Buc99, Eve97a, Hor91, Hun97, Fis93, Mül93, Schh93, Schm96, Sto99].

Process costing was developed to increase cost transparency in the indirect fields (design and development, production planning, marketing etc.) whose costs used to be charged as overhead costs as a lump sum. It starts with a relatively simple fact: During the product creation process, a great number of activities and processes that incur costs are necessary for the individual cost unit. For greater cost transparency in these fields, process costing tries to record and assign the planning, operation, checking, and coordination tasks pending in the indirect fields, based on cost units, similar to workstation costing in the production field. Process costing is not a new cost accounting system; rather, it uses the traditional cost types and cost centers accounting. Process costing serves as a supplement to, not a substitute for, cost accounting in the indirect field.

Field	Product realization process/part process	Reference quantity (cost unit)	Cost set
Indirectly productive field	Obtain order Prepare drawings Order material Plan production sequence	Number: orders Number: drawings Number: positions Number of planned operating cycles	Costs/order Costs/drawing Costs/position Costs/planned operating cycle
Directly productive field	Material	Weight	Costs/weight unit
	Turning Boring Tooth milling	Production time Production time Production time	Costs/production time unit " "
	Heat treatment	Treatment time	Costs/treatment time unit
	Sand blasting Polish bored surface Polish milled teeth	Production time Production time Production time	Costs/production time unit " "
Indirectly productive field	Delivery Prepare bill Register payment	Number: positions Number: positions Number: Entry postings	Costs/position Costs/position Costs/entry posting

Fig. 8.4-10. Largely use-based calculation, with process costing in indirect fields, and workstation costing in production [Bru94]

By the application of process costing, the costs arising in the indirect fields are also assigned through the choice of suitable reference parameters (cost drivers) and process cost unit rates to the individual products, largely use-based.

Figure 8.4-10 shows how in production and materials management (a directly productive field) the costs are charged on reference parameters on a quantity basis such as weight, production time, and in differentiating overhead costing, use-based (Sect. 8.4.3). The production process is represented correctly by cost accounting: It is virtually always process costing. On the other hand, the assignment of the overhead cost departments (the indirectly productive fields) is not about overhead cost surcharges with process costing, but carried out on reference parameters. Reference parameters are, for example, the number of the positions in purchasing or production planning, the number of the working steps, etc. In Sect. 8.4.3, a type of process costing was shown to determine the real costs of a concrete mixer cover.

Process costing has been introduced in the machine industry only with caution and hesitancy because of the computation expense, and then only in individual fields (development, production planning).

8.5 Direct costing

In full costing described up to now, **all** accrued costs are assigned to the individual cost units. Direct costs are assigned directly, and overhead costs assigned by using surcharges of the cost units.

Direct costing separates the fixed and variable costs and considers first **only the variable costs**. Usually the variable costs are assumed proportional to the quantity produced (proportional costs).

> ➔ Definition: **Fixed costs** are costs whose value is **independent** of the appearance of a specific cost-controlling variable [DIN89a].
>
> ➔ Definition: **Variable costs** are costs whose value is **dependent** on the appearance of a specific cost-controlling variable.
> Note: Cost-controlling variables include the activity, the order and contract quantity, the lot size, as well as the length of the planning or accounting period.
> Important variable cost types are the material direct costs and the production labor costs [DIN89a].

For distribution of fixed and variable costs, the facts about the **period under consideration** are important. Almost all costs are fixed only over a short duration; the material is ordered, wages and salaries must be paid, etc. Over the long-term, all costs are variable. The company may go out of business. Accounting periods of one quarter to one full year are common.

The **direct costs** are taken usually as **variable costs. Overhead costs** consist of both **variable and fixed costs**. They are determined when one ascertains the fixed and variable cost elements of each cost type in a parallel cost calculation as shown in Fig. 8.4-9.

The differentiation between fixed and variable costs is of great significance. If no new investments are made, then fixed costs remain and cannot be influenced. Also, in cost management we must compare alternative solutions only regarding variable costs.

8.5.1 Application of direct costing

By the separation of the full costs into variable and fixed costs, direct costing is suited for decisions of various kinds:

- **Help in decision-making for negotiating orders and contracts**
  Direct costing should be applied, besides full costing, if we need to decide whether orders whose market price is below the total costs are to be accepted. If there were a loss using full costing, we would not accept the order. If through other orders the fixed costs for an accounting period are already covered, and

the achievable price is above the variable costs, the difference between achievable price and variable costs **(margin)** can lead to more profit. We would then approve the order as long as it can be managed capacity-wise and is strategically sensible (Sect. 8.5.2).

- **Help in decision-making for the use of free capacity**
  If the capacity utilization of a company or a cost center decreases or is low, the full costs of a cost unit (product) itself increase, since the fixed costs must be spread over a smaller count. One would then refuse orders that do not cover the increased full costs, although really only more orders are needed. In this case, direct costing can give information on the lower limits of the price (limiting costs, for example, only for materials and wages), which is reasonable for a short-term price policy. Naturally, the orders must not continue to be below the full costs.
- **Help in decision-making for the product program**
  With direct costing, the optimal production program can be planned better [War90]. One can more easily ascertain what margin for the fixed costs of a company the different products supply (Sect. 8.5.2). It is absolutely justified and common in this case to sell newly designed products for a short time below full cost until they are accepted in the market. Then a drop in the full cost causes greater capacity utilization and/or quantities, below the price at which it can penetrate the market.
- **Help in decision-making for the crucial launch numbers**
  With the help of direct costing, the crucial quantities at which to bring out the product (the breakeven-point) can be ascertained (Fig. 6.1-2, Sect. 8.5.2).
- **Help in decision-making for choosing the production process**
  The problems are explained with an example. For a machining job the alternatives are a vertical mill and a boring mill. It is assumed that the production time and quality are the same for both machines. The workstation costs are known only as full costs so it is expedient to choose the vertical mill because it has about 70 $/h – 60 $/h = 10 $/h lower workstation cost rate (**Fig. 8.5-1**). However, if the workstation costs are known to be variable costs, the boring mill is more favorable since it has about 30 $/h – 25 $/h = 5 $/h lower workstation cost rate of the variable costs. Therefore, depending on the cost information, one reaches opposite decisions (Sect. 7.11.1a).
- For the last point this is generally true: For **alternative production machines** and processes **available in the company, variable costs** must be used for comparison since fixed costs accrue anyhow and are therefore not affected by the decision. Let us suppose the decision is made to procure a new machine (e. g., a boring mill) since the available machine is overloaded with orders. It is also advisable to use the available machine at full capacity the first, even though it is not advantageous from the viewpoint of full costing (here the vertical mill). A new machine is procured only if this is also used to capacity (precondition: the company wants to continue with in-house production).

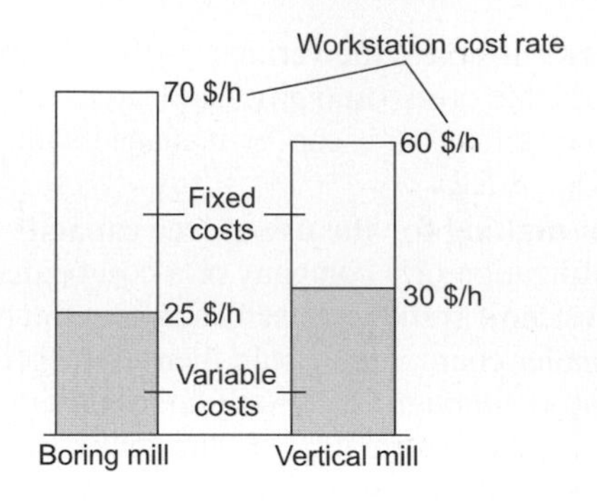

Costs considered for choice of process	Decision with identical production time and quality	
a) only workstation costs calculated as full costs	Boring mill at 70 $/h too expensive	Vertical mill costs only (60/70) · 100 = 86%; is preferred
b) only variable costs known	Boring mill costs only (25/30) · 100 = 83%; is preferred	Vertical mill at 30 $/h too expensive
c) variable and fixed costs known, even so, machine used to capacity	Boring mill is preferred, due to lowest variable costs. Procure new machine under constraint before overloading: Decision for vertical mill	→ Deploy only if boring mill overloaded

Fig. 8.5-1. Different decisions on the production process, depending on which costs are compared

- **Help in forming the make or buy decision**
  From this example we see that both cost accounting procedures, full costing and direct costing, are equally necessary and justified. Depending on the type of decision, one or other procedure is to be used. In management literature the term "crucial, or decision-relevant costs" [Sei90] is also used for this (Sect. 7.10.3).

8.5.2 Marginal costing

Marginal costing is used if we apply direct costing with regard to earnings to the profit or loss account. The margin is the difference between the earnings and the variable costs:

Margin = Earnings – Variable costs

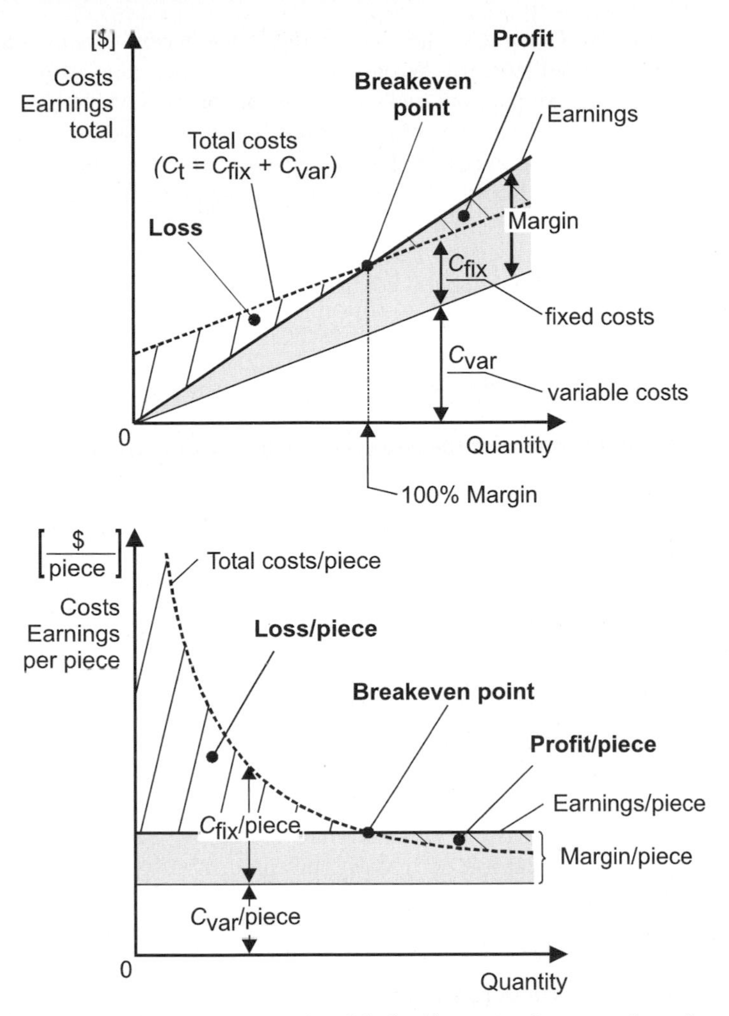

Fig. 8.5-2. Profit threshold diagrams (simplified). Above, total costs and earnings. Below, costs and earnings per piece

It covers the fixed costs wholly or partially. If it is 100%, the entire fixed costs are covered. If it is less than 100%, there is loss. Similarly, an amount above 100% represents profit (**Fig. 8.5-2**).

One can therefore apply the term margin to the whole company or to individual products. In the short-term, one can also set a price below the total costs for an individual product, that is, allow a margin smaller than 100% if one recoups that later or earns margin with other products far above 100%. However, it is risky if rival businesses do not calculate their costs properly or offer products for a long period below total costs for market-policy reasons, or force other companies down to the lower market price. So the existence of not only one's own company but also of many companies in the same line can be endangered since the losses are suffered continually. Generally, the market does not permit raising the prices again to the level needed [Vec86]. Wrong or inaccurate cost calculation with large scatter with different competitors causes a drop in the achievable prices in the market. The company that receives orders regularly is the one whose prices are by chance low due to calculation scatter. In this respect, a line of business has interest in a precise and use-based cost calculation of all companies (avoidance of bankruptcy losses).

The following limits use-based calculation: "A product gives rise only to its variable costs and not the fixed, but it contributes to the covering of the fixed costs of the company".

Marginal costing has the following tasks:

1. Showing the dependence of the costs on the degree of activity

The most important purpose of marginal costing is to show the dependence of the costs on the degree of activity (utilization of production, quantities produced). The profit threshold diagrams in Fig. 8.5-2 show that the profit threshold (breakeven point) is reached at a specific degree of activity or only with a specific quantity. The upper part of the figure that shows the entire costs for all orders; the variable costs increase proportionally with the quantities produced. The fixed costs are independent of that. The total costs are the sum of the two; this straight line intersects the earnings straight line at the breakeven point. The lower part of the figure displays the same information on a per-piece basis. After that the total costs become increasingly smaller per piece with increasing quantities. This dependence is a cause of decrease due to quantity (Sect. 7.5).

Here only the simple basic correlations are presented. In reality the variable costs do not increase strictly proportionally, but proceed progressively or degressively in parts. Also, the fixed costs are not invariant over the entire range, but show jumps, since above a certain number additional resources must be procured.

2. Assessment of the product program

With the help of marginal costing we can judge correctly, which products generate which share of the profit or loss and how the future planning should be conducted.

This will be clarified with the following example. To improve understanding, very simple numbers and the following assumptions were chosen:

- A company produces in each case 1 000 pieces of the products, A and B, in one accounting period.
- Summary overhead costing.
- Overhead cost surcharge rate 300%.
- The variable costs result from the direct costs and one-third of the overhead costs.
- The fixed costs are constant.
- Per piece, the following costs arise:

Product	A	B
Earnings	550	300
Direct costs DC	100	100
Overhead costs $OC = DC \cdot 300\%\ OCS$	300	300
Total costs TC	400	400
Profit/Loss	+150	−100

The full costing for the company in one accounting period and the profit threshold diagrams for all products are shown in **Fig. 8.5-3**.

Products	**A**	**B**	**Total**
Quantity	1 000	1 000	
Earnings	550 000	300 000	850 000
Direct costs	100 000	100 000	200 000
Overhead costs (300% *OCS* on *DC*)	300 000	300 000	600 000
Total costs	400 000	400 000	800 000
Profit	150 000	-100 000	**50 000**

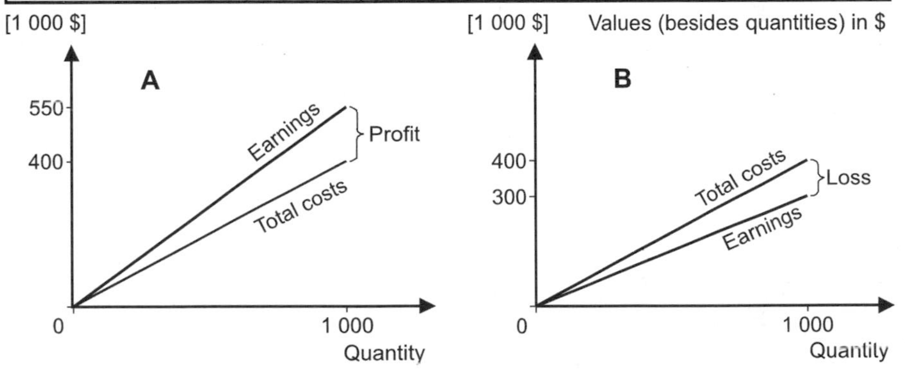

Fig. 8.5-3. Cost of products A and B with full costing

With full costing, in the production of 1 000 pieces of the product A there is a profit of 150 000 $ and for 1 000 pieces of product B a loss of 100 000 $ (Fig. 8.5-3). In all, there is a profit of 50 000 $ for the company. In view of the full costing results, management considers stopping the production for product B, without further changes in the company, to increase the profit (the fixed costs remain constant!). According to full costing, the profit is higher (150 000 $). In reality, however, there is a loss, as identified by direct costing (**Fig. 8.5-4**). We see from direct costing that product B supplied a margin of 100 000 $. Without the production of B this would disappear, and there would be a loss of 50 000 $ for the entire company instead of the 50 000 $ profit.

Products	A	B	Total
Quantity Earnings	1 000 550 000	1 000 300 000	 850 000
Variable costs (= *DC* + 1/3 *OC*) Margin	200 000 350 000	200 000 100 000	400 000 450 000
Company's fixed costs (2/3 *OC*) Profit			400 000 **50 000**

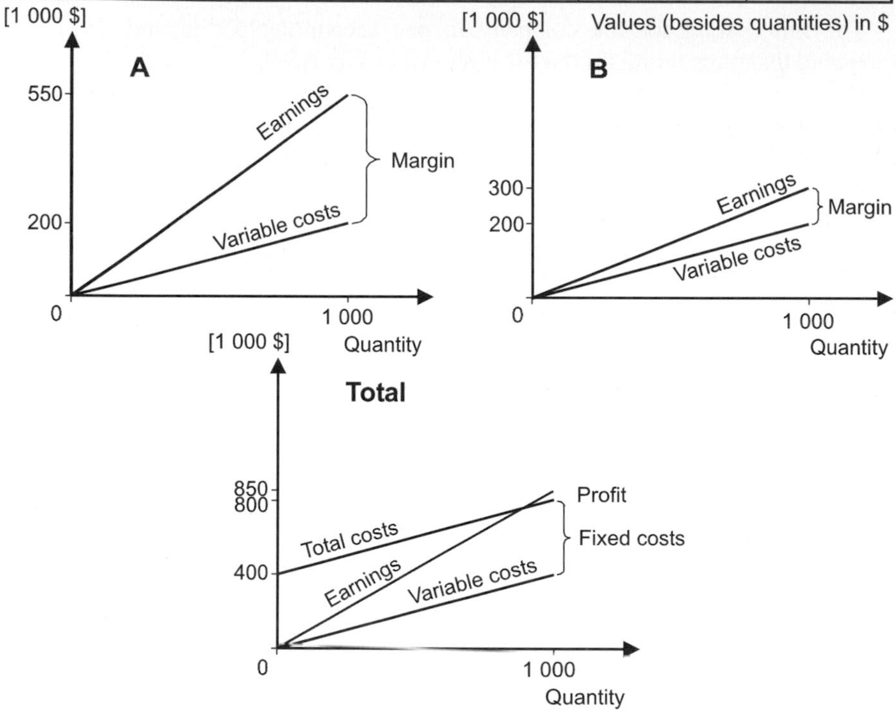

Fig. 8.5-4. Cost of products A and B with marginal costing

The necessary measures should be the following:

- Do not stop production of B.
- On a long-term basis, look for a successor product for B with higher earnings (e. g., new concept).
- Reduce variable costs of the products (e. g., by low-cost design, and value analysis).
- Sell and produce products A and B in greater numbers, provided that it can be assumed that the fixed costs of the company do not increase.

In practice, sometimes all three measures are applied simultaneously. We see the big effect of sales promotion but that it is not always necessary and sufficient to apply constructive measures alone. It is a basic entrepreneurial principle to use the full capacity of a company if possible, even if orders were at a level that would

Additional order for product B	A	B	Total
Quantity Earnings	1 000 550 000	1 100 330 000	 880 000
Variable costs (= *DC* + 1/3 *OC*) Margin	200 000 350 000	220 000 110 000	420 000 460 000
Company's fixed costs (2/3 *OC*) Profit			400 000 **60 000**

Values (besides quantities) in $

Fig. 8.5-5. Additional order for product B

hardly make any profit according to full costing. Nevertheless, a profit can still arise from the given margin, as the following extension of the example shows.

For product B an additional order of 100 pieces can again be accepted with earnings of 300 $/piece. With full costing one would not accept the order because there is an additional loss of 100 $/piece or a total of 10 000 $. In reality, the company's profit increases due to this additional order by the entire margin of this order (i. e., by 10 000 $), since the fixed costs of the company were already covered (**Figs. 8.5-5** and 8.5-6)!

8.5.3 Limit costing

In practice, costs do not exhibit an ideal linear course as was shown in Sect. 8.5.2. The variable costs may change in steps linearly, degressively or progressively, and with the fixed costs, sudden changes can occur. Therefore, we define limit costs as:

$$\text{Limit costs} = \frac{\text{additional costs}}{\text{additional production unit}}$$

They can be different according to shape of the cost curve and the segment of utilization under consideration. In practice, they are often equated with the variable cost/piece. The fixed costs do not find any entry in the limit costing. They are taken as a block into the income account. In addition, limit costing is a form of direct costing **(Fig. 8.5-6)**.

Limit costs are compared with sales earnings. The earnings surplus remaining serves to partially or completely cover the fixed costs or, in addition, shows a profit. A comparison of full costing with limit costing emerges from Fig. 8.5-6, in which the examples from Fig. 8.5-3 and Fig. 8.5-4 are shown graphically as "flow streams". With full costing (Fig. 8.5-6, top), every product accounts for itself. The "earnings stream" of the products A and B fills the separate "cost pots" A and B. At A, there is a profit overflow. With B, the "cost pot" is not filled; there is a loss. With limit costing (Fig. 8.5-6, bottom), the "fixed cost pot" of the company is strictly separated from the "pots" for the variable costs of the products. The earnings resulting for the products in the observation period first fill the "pots" of the variable costs of the products. The surplus fills in the income account as a margin the "pot" of the fixed costs of the company. If it does not become "full", a loss results; if it "overflows", there is a profit for the company.

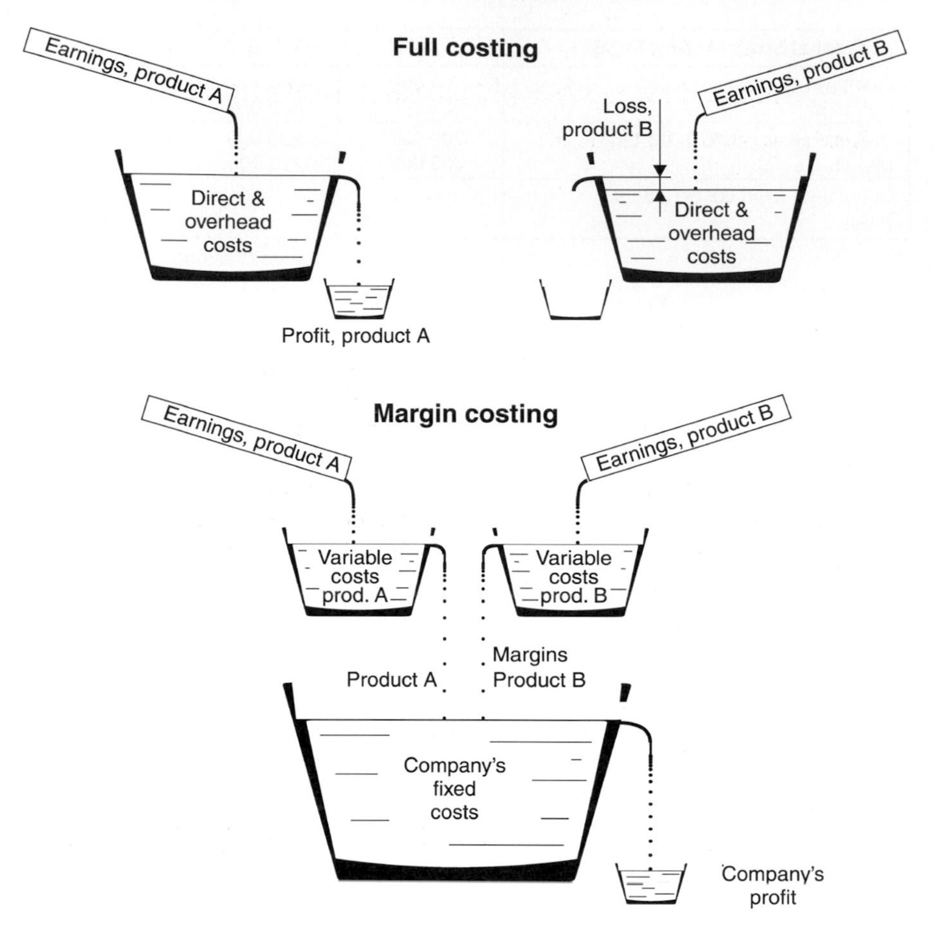

Fig. 8.5-6. Explanation of full and direct costing (marginal costing) (according to the examples in Fig. 8.5-3 and Fig. 8.5-4)

The advantage of this view is a clear separation between the costs that are directly necessary for production and those that are not directly related to production. In the case of a loss, management knows where measures are to be applied first: to the fixed costs, not to the proportional ones!

When looking at direct or limit costing, it must not be forgotten that over the long-term the entire costs (full costs) of the company must of course be covered! Therefore, in bad times one can attempt to increase the sold quantities through price cuts. With that, however, the profit threshold also moves to higher quantities, and it is questionable whether through the markdown so much more can really be sold. Furthermore, perhaps additional orders at prices that are too low (small margins) are accepted in good times because other orders have already covered the fixed costs. For a short period of time, that adds more margin, and also to profits. The problem, however, is that these measures lead to long-term price cuts for all products in the market, and in bad times an increase in prices is not feasible [Vec86].

9 Early Identification of Costs during Product Development – Development-Concurrent Cost Calculations

An essential element of cost management is a running cost calculation, concurrent with the product development process. In this way an early identification and influencing of costs can begin, as far as possible, right at the decision time (short-loop feedback control system, see Fig. 4.4-2). Thus at the beginning of a product development process – starting from the setting of the cost target – costs of the new product can be entered into the company costing structure. This allows a continuous cost comparison and stimulates cost-cutting measures in the case of deviations from the targets.

Conventional cost accounting is often not suitable for this task. Therefore, so-called quick (or, short) cost calculation procedures have been developed which make use of important parameters involved in product development. After an overview, the essential procedures for short calculation will be introduced in this chapter. Finally, computer-integrated calculation will be introduced, which is of increasing importance.

9.1 Overview

9.1.1 Aims of cost calculation concurrent with product development

Cost management requires a running calculation[16], concurrent with the product development process. During the usual preliminary calculation **on completion of the design**, the costs are determined on the basis of generally **complete** design documents containing parts lists and work schedules, which at times have to be reworked (Fig. 4.4-2). Here lies the main problem of the running calculation or early cost identification: We want to determine costs in the development process **quickly** and **early**, although the documents are **not** yet **complete** and the product is not yet firm in its details. Ideally, we would like to know the costs when only the requirements are clear and hardly anything is known of the new product [Bec94, Bec96, Ehr96, Kön95].

During the calculation concurrent with product development, two distinct supplementary tasks have to be differentiated:

---

[16] Unless otherwise stated, the term 'calculation' in this context implies cost calculation.

a) Search for costs of the whole product during the product development

In the case of complex products, it is a problem just to keep in view the manufacturing costs of the whole product during the development process (at times, over several years), and to collect these and continually to compare them to the target costs (Sects. 4.8.3.2 and 7.12.6.8). The costs of a complex product result from a sum of the costs of its many parts, which in turn are of different categories (costs of materials, costs of individual processes). This depends not so much on the accuracy of the data at the beginning of the product development process (Sect. 9.3.7.3), as that nothing is forgotten and any deviations from the company's standard are recognized and considered.

One problem here is that the available cost data are of variable quality. So, for purchased parts, assemblies, etc., exact costs are known or can be confirmed, but for new parts only estimates or only costs for prototypes are known. Also, the costs are not known for parts that are later mass-produced. The quality of the data (exact, estimated, etc.) should be noted in order to enable an estimate of the accuracy and risks during product development [Lin01]. Here test, tool, or model costs, and at times disposal costs for existing materials or tools no longer required due to the changes, must also be accounted for.

Furthermore, costs exist in very different degrees of detail. For large purchased assemblies the purchase price is shown as materials cost. Small, low-cost hardware items produced in-house also show up in material and production costs. Here a sensible compromise must be found in the level of detail. In the structure and establishment of the oversight of the whole product, it is absolutely necessary to accommodate all participants in the task execution and in the calculation procedures used, from the time of bidding on an order up to the cost calculations. It is a typical task of project management.

b) Calculation of individual assemblies and parts

As a subtask during the cost search of a whole product, the costs of individual important assemblies, parts, or even important work processes that are redesigned must be reestablished during product development.

In the simplest case, using the costs of an existing similar product for the new product solves the problem. By adapting the costs to the new product, this procedure often suffices in the machine industry, particularly in the early stages of the development process.

It takes more effort to determine the costs with formulas established statistically from the existing product and part spectrum, or with cost growth laws (Sect. 9.3). In exceptional cases, a precise calculation is also necessary. The computer-integrated calculation described in Sect. 9.4 forms a borderline case of the early cost identification. It is largely built, particularly in the case of designs of variants (Sect. 4.5.2), on complete drawings. However, it shortens the frequently long period between detail design and costing with work scheduling, etc., and consequently helps in early cost identification.

Many different costing procedures that can be used early in the process are known under different names (short cost calculation, quick cost calculation; cost function etc.) [Bec96, Bro96, Ehr85, Ger94, VDI87, Hor96]. Before we present individual procedures, we will discuss some important points of view.

Since the product cannot be described in detail yet and the design-concurrent calculation is supposed to be simpler than the usual preliminary calculation, the **results** are of **necessity less accurate** than the results of the preliminary or follow-up calculation. How much inaccuracy is permissible depends on the intended purpose of the design-concurrent calculation.

Intended aims can be, for example:

- Supporting calculations for quotation on a bid
- Continuous checking **during** product development to ensure that the cost target is attained (this is the most important purpose)
- Variant comparison, also of competing products (benchmarking)
- Recognizing potentials for cost reduction
- Checking and early estimating of supplier costs, bids
- Establishing cost structures, relative costs, and rules
- Time guidelines for jobs during the work schedule preparation
- Industry-wide checks of the type of costing and its accuracy.

The **necessary accuracy** must therefore always be established specific to the process and the company. Here the problem is presented only for the purposes of comparing bid calculation and comparison of variants.

If the price is determined for **bid calculation** to consider the probable manufacturing costs, a high level of accuracy is necessary. If the quotation price is too high, the order does not materialize. If it is too low and one gets the job, the price cannot be subsequently raised, and losses ensue (Sect. 8.5.2) [Eve77]. What is important to know is that the costs are only one aspect in the determination of prices. We should consider not only "Bottom up = How much **will** the product cost?" but also "Top down = How much **may** the product cost?" (Fig. 4.5-3).

For **variant comparison**, the accuracy needs to be only so high that it is possible to decide between the variants with confidence. During concept development the costs of the variants are often so far apart that from experience a cost estimate suffices.

The **possible accuracy** of cost determination depends, as does the determination of all other properties, on the **degree of familiarity** of the product. Hence the costs of a repeated part are exactly known, while those of a new design (Sect. 4.5.2) are known only with less accuracy. The accuracy improves in the course of product realization, as the degree of familiarity with the product improves. In **Fig. 9.1-1** [Bro96] (see Fig. 4.8-3 too) an example is used to show that during planning, the costs of a new design are known with an accuracy of ±20%; during the preliminary calculation, with ±5% accuracy; and are exactly obtainable at the follow-up calculation (Sect. 9.3.7).

Since new designs consist of a large number of known parts and assemblies whose costs are known, the inaccuracy is less for the whole product than for individual new parts. Figure 9.1-1a shows that for totally new designs (100% new parts), there is a ±20% assumed inaccuracy at the planning stage; this reduces to an

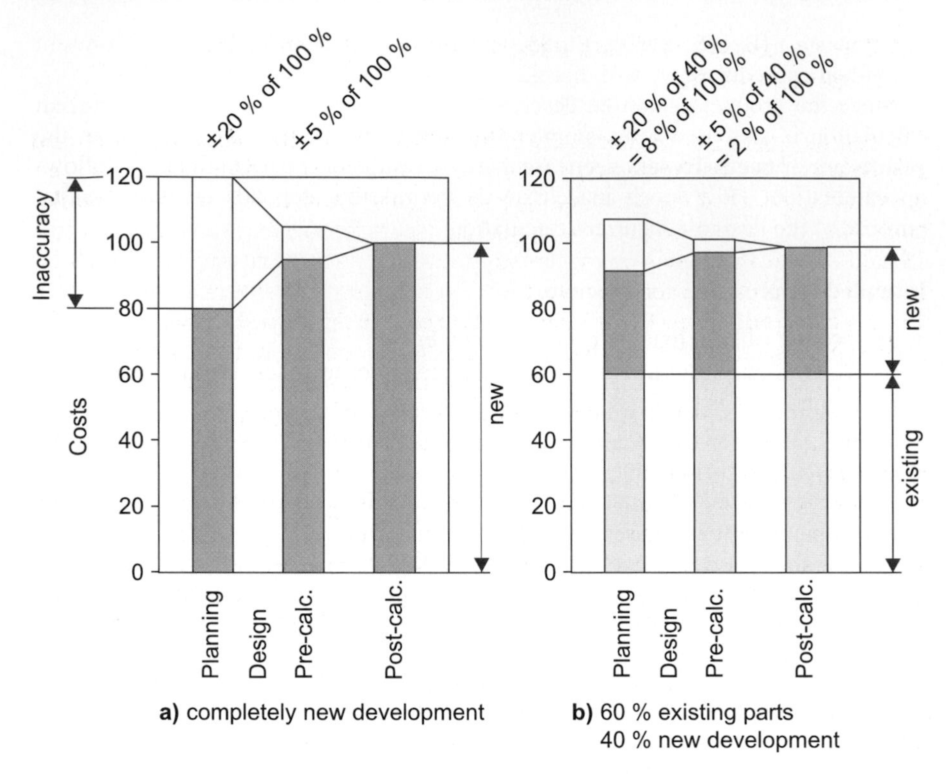

Fig. 9.1-1. Accuracy of cost statements depends on the degree of familiarity with the product

inaccuracy of ±5% at the preliminary calculation. Figure 9.1-1b shows a design with ±60% of the parts already existing and ±40% that are designed anew. That reduces the inaccuracies (a) at the planning stage to $(0.20)\cdot(0.40)=0.08=8\%$, and (b) at preliminary calculation to $(0.05)\cdot(0.40)=0.02=2\%$. At the follow-up calculation stage, the inaccuracies have been eliminated.

In contrast to what is, in principle, less accuracy (better: more uncertainty) of the cost statements at the beginning of a project, is the demand or necessity of an exact determination of the cost target. In the case of strength, multiplying the calculated stress with a safety factor that (as every technician knows) is in fact a "risk factor" solves the problem of the uncertainty. That is not possible in the case of costs since the market (or an already set sales price) force a cost target generally below the scatter range shown in Fig. 9.1-1. It is the task of cost management to attain the cost target through appropriate measures.

Another aspect is the **object** for which **calculations** are made. Are the costs to be determined for a complex product, for a component, or only for a design space? Corresponding to the object, the data must be distributed and tracked according to assemblies and parts, or according to costs of materials, costs of individual production steps, and assembly.

9.1.2 Progress of cost calculation concurrent with product development

For cost management it is necessary to have **the costs in the same structure and breakdown** from the **beginning** of the product development, when the cost targets are established, as at the later stages of **preliminary and follow-up cost calculations**. Only then are continuous control and comparison of the values possible [Sau86]. The realization of this apparently natural demand often runs into difficulties in practice because the necessary cost data are not often available in a suitable form. Getting **continuity** (Fig. 10.1-8) and **transparency** in this regard is a prerequisite for successful cost management. That is also valid for processes (see Sect. 4.8.3.2).

In the case of bigger projects, such as the development of an automobile, specific programs for planning and following through of the projects are employed. For more modest products in the machine industry, the use of **spreadsheet analysis** has proved itself in getting an overview of the frequently very extensive "number cemetery" and to keep to the project deadlines. If an overview is provided, they can be quickly updated and evaluated. They also allow a parallel calculation with several columns for costs of the old product, target costs, and actual costs of the new product at specific times. With that, comparisons and graphic evaluations are easy to make (Sect. 10.1, Fig. 10.1-8). Such a table also generally promotes cost transparency for product developers.

The **basic procedure** is described briefly in the following [Ehr87a, Grä98, Grö91, Mei92, Hein95, Bot96, Ehr96, Bin97, Rei97, Sch95, Sch98a, Stö98, War80, Wel98]. What is important is to always adapt the procedure to the given situation and to note that it is a longer process, with iterations and learning effects.

The costing of one or several similar products (sometimes also of a competing product) is used as the starting point [Roma93, Sau86]. Either the starting data are simply assumed or adapted by extrapolating to the current product. The cost targets are entered in a further column for the assembly or part. By comparing the actual with the target costs the necessary starting points for cost reduction are recognized. During the preparation of the table, ideas or actions (cost reduction potentials) with which one can influence the costs, also become evident. The actions are noted down, as well as the party responsible for their realization and the expected cost savings, and the development process is begun. At later set points of time, the state of development is determined, the goal and actual states are compared, and deviations or aims not yet attained are ascertained. If necessary, new measures are introduced and again responsible parties determined, etc.

9.1.3 Procedure for quick cost calculation

Just as there are many procedures for early identification of properties (e.g., strength, deformation, etc.) there are different procedures of quick cost calculation. Their applicability and accuracy depends on the determination of cost-relevant data for the specific stage of development. The accuracy of the statements improves with increasing concretization of the product. This statement is valid, in principle, regardless of the methods that are employed. Examples that lead to good results for known solution models already in the concept phase [Ste92] do not contradict this.

They are tacitly limited to known designs. Not only the concept but also the existing solutions are considered.

In practice, the **following procedures have** proved to be **useful:**

- Cost estimation (Sect. 9.2)
- Search or similarity calculation (i. e., search and adoption of the costs of similar products (Sect. 9.3.1)
- Determinations of the costs with a basic parameter (e. g., the weight, which represents the main effect on the costs; Sect. 9.3.2)
- Determination of the costs by design equations (Sect. 9.3.3)
- Quick cost calculation with several parameters (Sect. 9.3.4)
- Calculation using cost growth laws (Sects. 9.3.5, 7.7, 7.12.5)
- Computer-integrated calculation (Sect. 9.4)

Here only the basic structure of the procedures is shown. All procedures must be adapted to the products and the operational factors. **A direct adoption is not possible!**

Some additional programs for **project cost calculation** will be mentioned here. Within the U.S. space program commercially marketed software programs were developed, with which product development, manufacturing, and lifecycle costs for larger development projects could be estimated [Das88, Rec97]. The PERT system is suitable for both time and cost control. It is based on the network planning technique. In contrast, the PRICE system is oriented more toward the costs of a product and is used for aerospace projects as well as in the machine industry [Rec97]. An extensive subdivision of the development process is essential for the use of these programs, where a variety of cost-relevant data is entered for the individual elements. The programs are largely based on the principle of subdividing the product or the process. For these parts, cost and/or time relevant factors are provided and summarized in formulas. The programs are calibrated for the specific circumstances (i. e., the factors and powers in the formulas must be adapted to the respective circumstances by comparison with similar known objects). An analogy also exists with the subdividing estimation (Sect. 9.2) and the quick cost calculation formulas (Sect. 9.3).

9.1.4 Possibilities for reducing the effort

The procedure of concurrent calculation shown here requires effort, but is necessary for cost management because at the time of the design decisions the corresponding cost information must also be available. In practice, this results in a decrease in expense in the case of complex products. This is because **not all parts** or assemblies of a product are new, and therefore their costs do not have to be calculated again (Fig. 9.1-1). From Fig. 7.12-5, a product contains the following types of parts whose costs are known or must be determined:

- Same parts: costs known
- Repeated parts: costs known
- Standard parts, purchased parts: costs known, obtain quotation

- Similar parts: estimate based on earlier calculation, or calculated with similarity principles (Sect. 9.3)
- New parts: new calculation or qualified estimate (Sect. 9.2).

If the **number of new parts** is kept **small** (Fig. 7.12-6), the introductory costs, the project time decrease, the reliability, and the product delivery readiness improve (Fig. 7.12-4). In addition, the product cost can also be calculated faster and more precisely.

With reduced in-house production, the calculation effort is also less because, as experience shows, the suppliers **provide price quotations relatively fast** (Sect. 7.10.2).

In addition, **cost calculation with differential costs** [Ger94, Rau78, VDI95] reduces the effort. The solution S_1 and the solution variants S_2, S_3 differ from each other mostly only in sub-fields. Accordingly, the costs also differ only in these sub-fields. The procedure is then as follows: For solution S_1, the total costs CS_1 are ascertained. For the variants S_2, S_3, etc., only the costs that are different from that of the solution S_1, i. e., ΔCS_2, ΔCS_3 are determined, and so forth. The costs of the solution variants that result are:

$$CS_1 = CS_1,\ CS_2 = CS_1 \pm \Delta CS_2,\ CS_3 = CS_1 \pm \Delta CS_3$$

Less effort is required since the costs for **not all parts of a product have to be calculated precisely** (Sect. 9.3.7, Fig. 9.3-10).

9.2 Cost estimation

The estimation of manufacturing costs goes faster than calculation; it is, however, less accurate, so it is often omitted for lack of confidence. Under certain conditions and if applied systematically, cost estimation is sufficiently exact. The estimation must be based on experiences with similar situations, parts and processes and not on a vague feeling. The **estimation accuracy** can be **increased** by the following four measures:

1. Distributive estimation

Due to error compensation (averaging out random errors of many individual estimates), the accuracy of the overall result is higher than that of the individual estimates (Sect. 9.3.7.3). It therefore pays to estimate the costs of many parts or assemblies separately if possible. In the case of “A” parts, also estimate the material and production costs separately, and perhaps separately for each production process (see Sect. 4.6.2 for Pareto analysis.) The costs of individual parts are also easier and more exactly assessable than are the costs of the complete product. Knowledge of cost growth laws also helps (Sects. 7.5, 7.7.1 and 9.3.5). With the increase in accuracy, the estimate becomes easier to justify.

2. Estimation through several persons

In the same way, accuracy improves if several competent persons estimate the costs independently of each other. After discussion with others, a mean value is reached. Technical and accounting knowledge and corresponding professional experience are necessary for this purpose. It is of value to call a brief meeting of specialists (maximum 1 h) who know about the technology and the costs. A well-prepared statement of the task is necessary: Concepts, cost data with flipcharts and/or projected on a screen, discussion, agreement on steps, and documentation of the results (Sect. 9.1.2).

3. Combination of estimation and precise cost computation

In general, we will calculate cost-decisive "A" parts by preliminary calculation, by comparison with formerly produced parts (Fig. 9.1-1), or through quotations from the suppliers, more precisely than "B" and "C" parts. The relative total error e_{tot} is small if the cost of the major portion of the product is determined precisely (Sect. 9.3.7, Fig. 9.3-10).

4. Comparative estimation

Estimation results are improved if one uses certain bases, such as costs of similar parts (average values – cost per unit weight, average cost/part), see Sect. 9.3.1. This has been proven in the electronics industry and in the specialty machine industry where there is a large number of components in a product. We determine the costs of the product by adding up the average values of the cost/part or station. For a circuit, layout, etc., the number of the available components, multiplied by the average cost per component, gives a reliable cost estimate. However, one should use not just an average value for all parts of the machine, but several average values for specific part types or processes. For example, for a machine tool use the cost per meter guide length, cost per station, etc.

As advantageous the estimation is regarding speed and low effort, nevertheless the following **disadvantages** remain:

- Recognized mistakes cannot be utilized sufficiently to improve the estimate. In addition, estimation remains a matter of "feeling".
- The results are largely person-dependent and rationally difficult to understand. If the estimator is no longer available, there is no longer a continuity in the result.
- Cost estimation cannot be taught in a short period.

The estimation results must be held on to and be compared with the actual costs later and the deviations reviewed. Thus, a continuous improvement of the results is achieved.

9.3 Quick cost calculation

Quick cost calculation is a simplified method of costing for a defined cost driver, as shown in the standard, DIN 32 990 [DIN89a]. Other terms such as short, similarity and equivalence-figure calculation are used [Mei92]. The most common definition of quick cost calculation is all the methods by which costs are made accessible in the respective realization stages of the product (e. g., in design or project planning). As opposed to the procedures of operations scheduling and costing, which are based on complete product documentation, these are reduced to the most important and known parameters. Typical production parameters such as feed and cutting speed are generally excluded, since as a rule they are still unknown during design.

In the preparation of quick calculations, the validity domains and the limiting conditions (Fig. 7.1-3) should be indicated. If these are missing, gross mistakes can result during their use. Furthermore, attention should be paid to who is supposed to make use of the results. A production and cost consultant has considerably more production and cost knowledge than a design engineer. Accordingly, the quick cost calculation can then be more extensive and more precise.

For the realization of quick cost calculations, there are three possibilities [Kie79, Pic89, Rei97, Stö97, Stö98]:

1. Comparison calculation of constructive variants

Design variants or alternatives are calculated conventionally. If specific design-relevant parameters (e. g., in the case of shaft-hub connections, the type, width, diameter) are varied during the calculation, diagrams with these variables as input variables are obtained (Sect. 7.13.6 [Kit90]).

2. Analysis of documents of the preliminary calculation

Through analysis of the formulas for operations scheduling and preliminary calculation, essential cost parameters can be recognized. These, as well as the fundamental formula and calculation structure, are important for the setup of cause-driven quick cost calculation formulas. The derivation of similarity principles is also mainly through the analysis of time and cost documents (Sect. 9.3.4).

3. Statistical evaluation of costs and time units

From existing work schedules with preliminary or follow-up calculated times, or cost calculation results as well as design dimensions taken from shop drawings are initially tied together with hypothetically formulated functions. Through statistical evaluations, the optimal coefficients and exponents of the formula (cost function) are then determined (Sect. 9.3.4).

9.3.1 Search calculation; similarity calculation

A relatively simple and rapid means for determining the costs of new products is to compare them with the costs of existing products; this is frequently done in practice. Comparison is done quickly and is reliable if the compared objects do not differ too much and if clear and current data are available. A precondition for such comparison is the search for and finding of similar existing objects. Because companies often have comprehensive master data records with tens of thousands of parts, this search is not simple.

Hillebrand [Hil86, Hil90] shows computer-aided possibilities of searching for similar parts according to arbitrary features and with a mathematically established similarity measure. If several similar parts are found, the costs of the new part can be determined very closely through interpolation by using a cost function. The features that can be searched for must be determined appropriately and encoded beforehand. The computer-aided similar-part search can also be part of a Cost Information System (Sect. 9.4.2). Newer programs can search only for terms in arbitrary files without having to be classified [Mül94] (Sec.7.12.3.1c and 7.12.4.1).

The **quick cost calculation with part classification systems** works on a similar basis. Goetze [Göt78] starts from the finding that the shape of a part is particularly responsible for its production costs, and that a shape-descriptive classification system for cost estimating would be suitable. He supplements Opitz's classification system [Opi66] with additional key points that enable an assignment to constructively determined cost-effective parameters.

9.3.2 Determination of costs based on one parameter

Often a single parameter defines the product so adequately that it can also be used for cost computation. This procedure is very simple. However, it gives acceptable results only when the new product is **very similar** to the comparison products, both **constructively** and from a **production** viewpoint. In the following, three well-known procedures are described.

9.3.2.1 Weight-based costing

With this method, the manufacturing costs MC_0 of a product 0 are referenced to its weight W_0, which yields a "weight-cost ratio" MC_w:

$$MC_w = \frac{MC_0}{W_0} \left[\frac{\$}{\text{kg}}\right]$$

The manufacturing costs MC_i of a similar product i are then calculated by multiplying its weight W_i by the weight-cost ratio MC_w. It is assumed that the **costs** are directly **proportional to the weight (Fig. 9.3-1)**. Thus, the costs per unit weight MC_w are fixed:

$$\frac{MC_i}{MC_0} = \frac{W_i}{W_0}$$

$$MC_i = MC_0 \cdot \frac{W_i}{W_0} = W_i \cdot MC_w$$

Assumptions for a satisfactory accuracy of this method are:

- **Same type of products** (same design, same production, same materials, same quantity and dimensions range).
- **No significant extrapolation** beyond the given range.
- The larger the products are (i. e., the **material cost is a large share**), the more precise the weight-cost calculation becomes (Fig. 7.9-2). This is also valid for small products that are mass-produced, and all products for which the material cost portion is high (Sect. 7.7, Fig. 7.7-6).

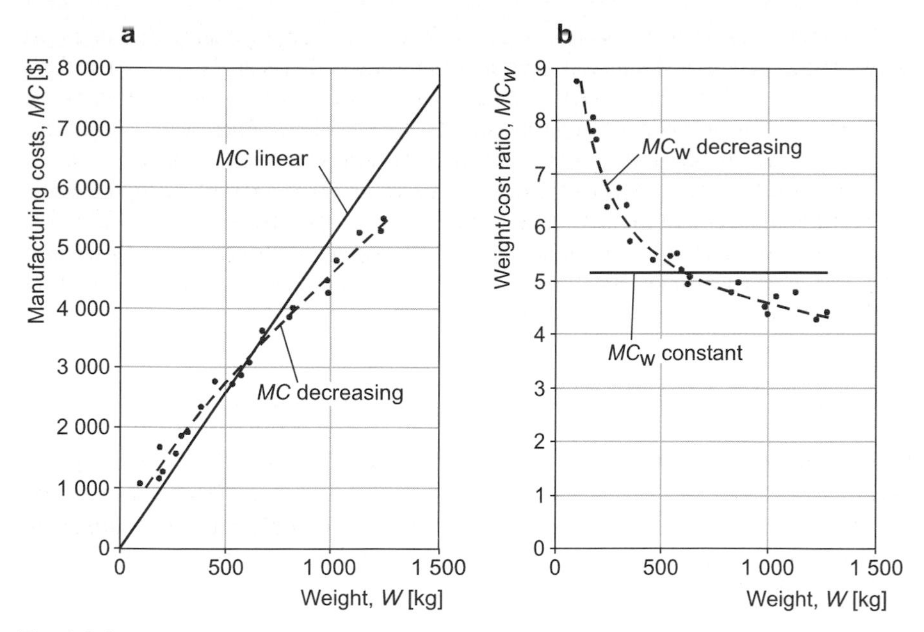

Fig. 9.3-1. Weight costs a) linear and decreasing behavior of the manufacturing costs *MC* b) constant and decreasing behavior of the manufacturing costs per unit weight MC_w

The **procedure** is **improved** when the weight-cost ratio MC_w is taken not as fixed, but decreasing as the weight increases (i. e., the costs increase not proportionally, but decreasingly with weight). See Fig. 9.3-1, right; for reasons, see the text in Fig. 7.3-2. For a new product *i*, the weight W_i is determined first, then from the manufacturing costs/weight curve MC_w the corresponding weight-cost ratio MC_{wi} are determined. With that, the new manufacturing costs MC_i are calculated (example, Sect. 10.3.3):

$$MC_i = W_i \cdot MC_{wi} \left[\text{kg} \cdot \frac{\$}{\text{kg}} \right]$$

It is also advantageous to specify not just one weight-cost ratio (e. g., for welded parts) but several (e. g., for small, medium and large, simple and complicated parts, etc.). In [Ruc82], standard values are given (e. g., for welded designs) for shop production costs in production hours per ton [h/t], classified according to dimension and product type.

9.3.2.2 Material cost method

This method (which is described in detail in the standard VDI 2225 [VDI97]) begins with the premise that for a specific product type, the manufacturing costs have a fixed relationship of material to manufacturing costs. Thus, the cost structure from *MtC* and *PC* is fixed. The cost of materials of the new product can usually be determined quite rapidly from the parts list and the material purchase prices, or from material relative costs. The manufacturing costs are then calculated from the known ratio of material cost (material cost portion) *mtc* as follows:

$$mtc = \frac{MtC}{MC} = \text{const, and known from existing similar products.}$$

The costs of materials MC_{new} of the new product are determined from the parts list (or the material volume is found from the drawing) and is multiplied with the purchase prices and, at times, with material relative costs (Fig. 7.9-10). The manufacturing costs of the new product MC_{new} are then found from

$$MC_{\text{new}} = \frac{MtC_{\text{new}}}{mtc}.$$

The same conditions are valid for achieving a satisfactory accuracy as in the case of the weight-cost method.

9.3.2.3 Quick cost calculation with performance governing parameters

Instead of weight, the manufacturing costs can also depend on parameters that indicate performance for complete products such as compressors, mills, filters, or dryers. These might be power requirements, suction output, filter surface, or dryer surface. Costs are graphically depicted with these parameters. The resulting curves are often straight lines with suitable axes (e. g., log-log). From that, the costs can be read off for other performance data.

With an example from [Göt78], **Fig. 9.3-2** shows how the costs depend on the numbers of pole-pairs in electric motors. The updating occurs through new entries or with inflation rates. This type of quick cost calculation is especially useful for project planning since the relationship between the required technical performance and costs (or price) are immediately available.

9.3.3 Design equations

Design equations were proposed by Kesselring [Kes54], as explained in [VDI97, Ger94]. In a design equation the correlation between the costs of a product and the essential technical parameters are expressed in a closed-form equation.

The advantage of a design equation is that it shows the essential relationships and leads immediately (both qualitatively as well as quantitatively) to products that have the lowest cost or are the lightest, smallest, or produce the least loss, etc. The disadvantage is that it is very difficult to set up useful design equations for complex products, since extensive simplifications must usually be made, or the relationships are very extensive and complex. We can avoid a closed-form equation relating technical parameters and costs by using a computer program instead. It is then possible (for example, for design of variants) to summarize all technical and economical relationships in this program and determine the optimal solution by using optimization programs [Fig88].

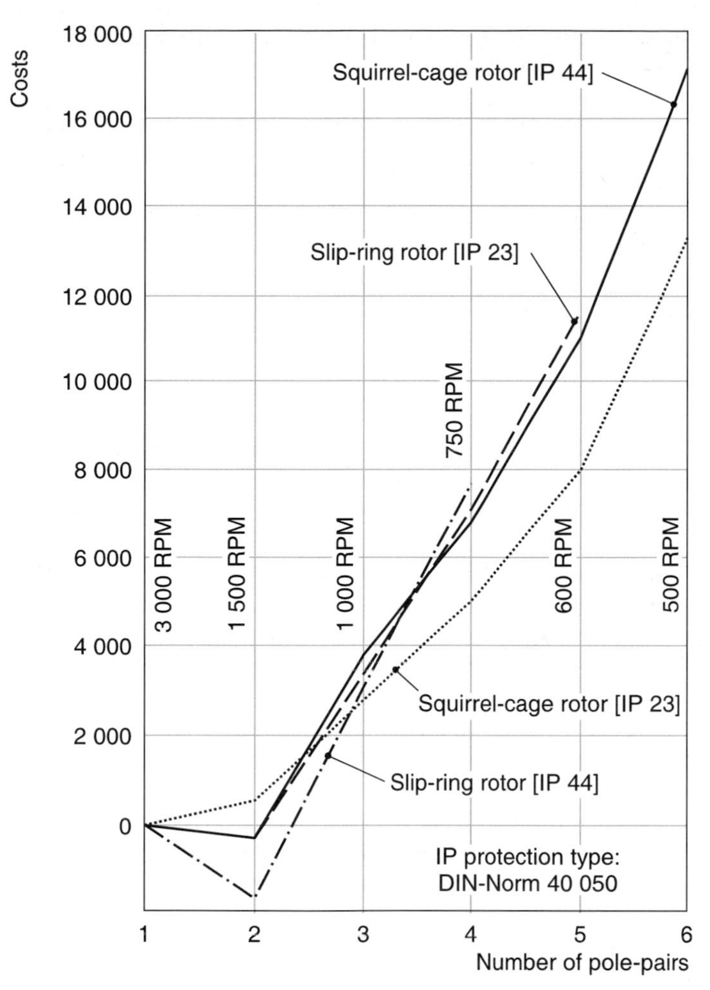

Fig. 9.3-2. Cost vs. number of pole-pairs for different speeds, for three-phase motors with 45 kW rated output [Göt78]

9.3.4 Quick cost calculation formulas with several parameters

There are machine parts or production processes, for which the manufacturing costs or production times cannot be expressed with physically derivable relationships. This is the case, for example, with cast parts for which a variety of empirical or estimated quantities can be used (Sect. 7.11.2.2). In such cases, the actual time units of a variety of parts are used, and these times are coupled with pertinent parameters (e. g., geometry, tolerances, or material types). We search for a **statistical connection** between the parameters and the production times or cost. The usual procedure for that is regression analysis (Sect. 9.3.4.1). Since this approach consists only of a linear combination of variables, the more generally used procedures are mathematical optimization (Sect. 9.3.4.3) and cost calculation with neural networks (Sect. 9.3.4.4).

9.3.4.1 Development of quick cost calculation formulas with regression analysis

If Y is the object entity (e. g., costs or time) and X_1, X_2, ... X_n are the independent variables or quantities, then regression analysis is expressed as a multiple **linear regression** equation of the form:

$$Y = a + b_1 \cdot X_1 + b_2 \cdot X_2 + ... + b_n \cdot X_n$$

Also, a product-exponent approach (Figs 9.3-4a, 9.3-8a) which implies taking logarithms of the independent variables can be employed as a special case. It leads to the **nonlinear regression** equation:

$$Y = a \cdot X_1^{e1} \cdot X_2^{e2} \cdot ... \cdot X_n^{en}$$

With the regression calculation, the factors a, b_1 to b_n or the exponents e_1 to e_n are determined so that the sum of the squares of the deviations of the actual values from the values calculated with the formula becomes a minimum.

Just as with cost accounting for different spectra of parts, mixed approaches of combined linear and nonlinear terms are frequently used. A simple example is the determination of manufacturing costs from the material and production costs. The material and production costs themselves result from the multiplication of the amount (kg, min) with the ratios ($/kg, $/min), which are then added. With regression analysis this problem can be solved by writing not just one equation for the whole part spectrum and all types of costs, but rather several equations (e. g. for material and production costs) are compiled separately. If the formula statement is obtained with mathematical optimization processes (Sect. 9.3.4.3), then arbitrarily mixed formula statements are also possible (Fig. 9.3-4b). In no case do the main variables (significant variables) or the suitable formula statements appear by themselves. However, in general it is difficult to find the variables and to set up suitable formula statements.

Regression analysis with one variable is also the basis for the above-mentioned procedures of cost computation using the weight and one parameter (Sects. 9.3.1 and 9.3.2). In addition, **weighted regression** [Ste95] can be used as the computation process for improving (increasing the accuracy) the quick cost calculation. The essential distinguishing feature in this case is that the regression formula must be determined anew for every inquiry. From the present database, only data that have the most similarity with the inquiry data are used for the pertinent calculation. In addition, the values that lie nearby are weighed more heavily in the calculation than those farther away. The increase in accuracy requires a longer computing time.

With the help of statistical methods during the preparation of quick cost calculation formulas,

- the main variables are determined from a large number of given variables;
- quantitatively, the factors and exponents of these variables are defined in a formula statement; and
- different formula statements are evaluated relative to each other (Fig. 9.3-4).

Because it is a statistical approach, certain conclusions may be drawn:

- **Sufficient actual data records** must be available (i. e., drawings of parts, the corresponding work schedules with times and/or costs).
- Only the **available part spectrum** is represented according to its **type** in the equation. The equation is valid only within the limits of the given part spectrum. In extrapolating, we must be aware of the uncertainties. If a formula is set up for relatively simple parts, for example, cast parts without special core work, it is not meaningful for complicated parts.
- There is **no totally correct formula** for the calculation of times and costs, but there is always one that better approximates the data records used. It is left to the ability of the responsible person to select the data records that are representative of the totality of parts being considered.
- The **recording and evaluation** of the data records is **extensive**. For the actual application (the calculation with the chosen formula), a simple calculator is enough.
- Statistically developed formulas, as opposed to those derived analytically (Sect. 9.3.5), have the disadvantage in that they are **difficult to interpret**. From the formula setup, no direct conclusions regarding the design can be drawn (Sect. 9.3.4.3).

Statistical procedures can be used where not only analytical calculation is not feasible, but also for parts or production processes that are very complicated and too extensive for calculation. Here the quick cost calculation formula approximates the time or cost accounting in a manner satisfactory for most purposes, so that the calculation can be more rapid or done with less effort.

For the basics and the execution of regression analysis, refer to the literature (e. g., [Eve77, Mag82]). Extensive software for regression analysis is available for PCs, for example, as spreadsheet programs.

It should be observed that during regression analysis the statistical quantities (e. g., standard deviation and certainty measure) are to be calculated by comparison with the entry data. With an examination verified in practice, we need not go back to these data because the values underlying the calculation would only confirm the calculated values. Rather, data not used for the regression analysis should be chosen for checking.

In [Ehr85], a series of additional examples of regression analyses and quick cost calculations are shown.

9.3.4.2 Example of a quick cost calculation with several independent variables

The following example shows a quick cost calculation for shaft-hub connections, which is elaborated on by variation, calculation, and a mixture of statistical evaluation and cost growth laws [Kit90] (Sect. 7.13.6).

To be able to approximately calculate the costs of **shaft-hub connections (SHCs)** in-house, the **quick cost calculation** depicted in **Fig. 9.3-3** was developed as an example. Starting from an SHC cost basis DfC_0 (for 50 mm shaft diameters, lot size = 1), and by using several constants A_x and exponents E_x, the manufacturing costs can be calculated if the shaft diameters and the lot sizes are altered.

Quick cost calculation formulas
for SHC costs of:

Shaft-hub connection of shaft and gear

in size-range design with b:d = 1:1 (geometrically similar main dimensions, comparable production and shape)

b = hub width; d = bore diameter

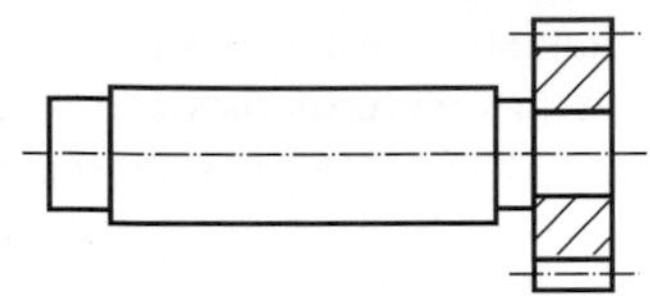

Equation:

$$DfC_i = DfC_0 \cdot (A_0 + A_1 \cdot (\frac{d_i}{d_0})^{E_1} + (A_2 + A_3 \cdot (\frac{d_i}{d_0})^{E_2})/n_i)$$

DfC = SHC costs as differential costs for a basis variant without SHC

Goal:
- SHC costs DfC_i from manufacturing costs

Range of independent variables:
- Connection diameter: d_i = 20-500 mm (max.) as per connection type see picture below
- Lot size n_i = 1-100 pieces

Validity range of goals:
- Cost comparison basis: 15 May 1986
- the data basis are manufacturing costs for smooth shaft (C 45) and hub with outside teeth (16 MnCr 5 or for d = 500 mm: 17 CrNiMo 6) fastened with a shaft-hub connection
- Ratio of connection hub width to bore diameter, b:d = 1:1

Parameter:

Average values	SHC cost basis d_0 = 50 mm, n_0 = 1	Constants				Expo-nents		Average calculation error
Connection type	DfC_0 [MU]	A_0	A_1	A_2	A_3	E_1	E_2	[%]
Feather key fastening	237.5	0.135	0.099	0.741	0.035	2.3	2.0	3.6
Cylindrical press fit	102.4	0.117	0.128	0.723	0.012	2.4	2.0	3.8
Toothed shaft-hub connection	413.2	0.119	0.068	0.801	0.039	2.5	2.3	2.0
Taper press fit	183.0	0.131	0.119	0.749	0.019	2.2	2.0	4.0
Tapered oil press fit	246.6	0.215	0.120	0.669	0009	2.1	2.2	1.9
Tension-stack fastening	230.3	0.378	0.204	0.456	0.023	2.6	2.5	4.6
Polygon fas-tening P3G	527.5	0.133	0.092	0.569	0.207	2.3	1.2	1.4

Fig. 9.3-3. Quick cost calculation for SHC costs for shaft and gear [Kit90]

Since the cost comparison basis is 1986, the DfC_0 must be multiplied by an inflation rate (available from VDMA), or the calculated costs DfC_i must be compared with in-house cost calculations. The structure of quick cost calculation might have long-term validity.

An **example calculation** will illustrate the procedure.

Task

For a shaft-hub connection for a shaft and gear, the available space is specified (i. e., joint diameter d and joint length b). Among the selection possibilities are a feather key connection and a cylindrical press fit; both transmit the required torque. The lowest-cost shaft-hub connection will be used in an application. The known data are the production lot size ($N_i = 3$ pieces), the bore diameter ($d_i = 150$ mm) and the hub width ($b = 150$ mm, $b{:}d = 1{:}1$).

With the formulas shown in Fig. 9.3-3 and the corresponding constants and exponents, the calculation formulas can be prepared. Only the bore diameter ($d_i = 150$ mm) and the lot size ($N_i = 3$) are still to be substituted, as given above ($d_0 = 50$ mm).

As calculation formula for the feather key connection (FK) is:

$$DfC\left(\text{FK}\right) = 237.5 \cdot \begin{pmatrix} 0.135 + 0.099 \cdot \left(d_i / 50 \right)^{2.3} \\ + \left(0.741 + 0.035 \cdot \left(d_i / 50 \right)^{2.0} \right) / N_i \end{pmatrix}$$

and for the cylindrical press fit (CP):

$$DfC\left(\text{CP}\right) = 102.4 \cdot \begin{pmatrix} 0.117 + 0.128 \cdot \left(d_i / 50 \right)^{2.4} \\ + \left(0.723 + 0.012 \cdot \left(d_i / 50 \right)^{2.0} \right) / N_i \end{pmatrix}$$

Result

The calculated SHC costs are, for

- feather key connection, DIN 6885A: 410 MU
- cylindrical press fit: 223 MU.

The cylindrical press fit is therefore approximately 43% more economical than the feather key connection and, thus, is by far the lowest-cost shaft-hub connection (Figs 7.13-18 and 7.13-19).

9.3.4.3 Development of quick cost calculation formulas with optimization processes

Regression analysis determines the factors in **one** calculation cycle of the equation. In the first go-around, a formula useful for quick cost calculation is usually not established; rather, several different approaches must be set up and their suitability tested. Mathematical optimization processes can replace this iterative procedure. Baumann [Bau82] used this procedure the first time for the determination of cost functions, and it is explained here briefly with an example (**Fig. 9.3-4**):

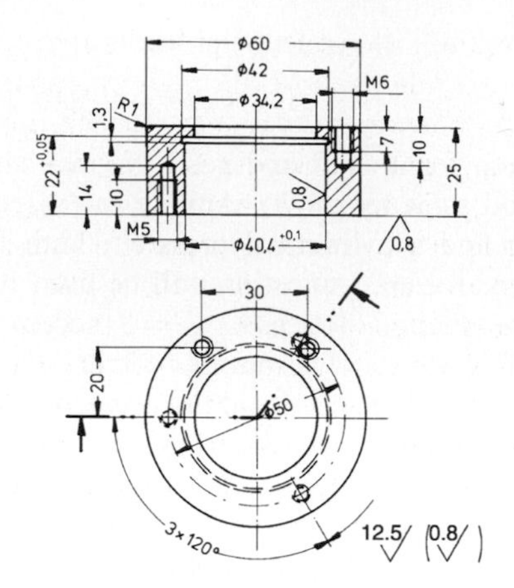

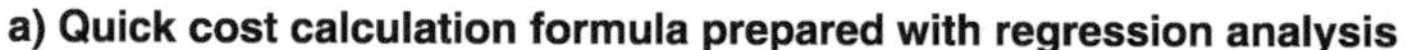

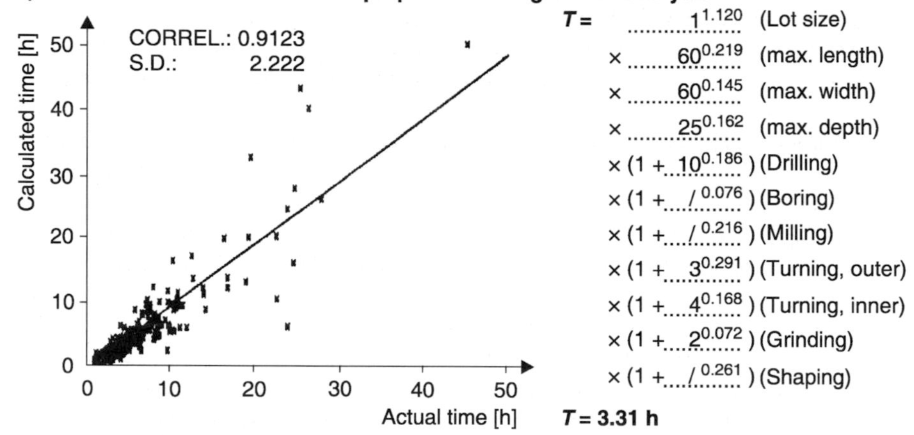

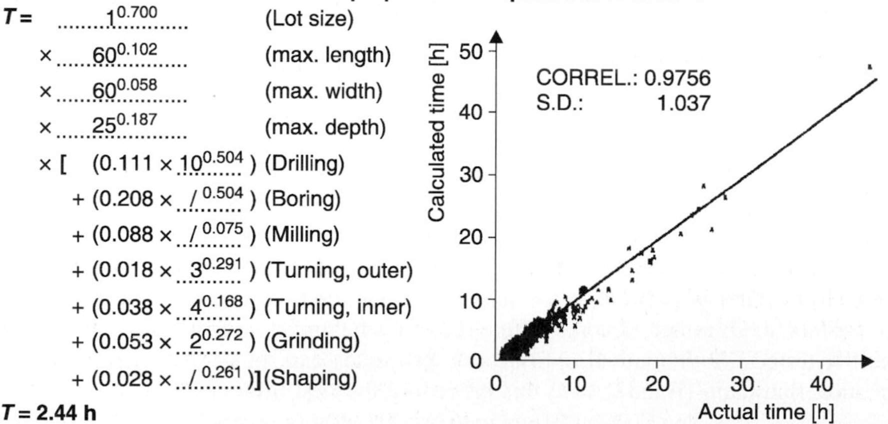

Fig. 9.3-4. Comparison of quick cost calculation formulas using the example of a ring [Bau82] ("/" means: no value for boring, milling, shaping)

The production time units Y_j for a large number of fixture parts are known (by multiplication by the current hourly wage, we find the manufacturing costs). The quantities X_1, X_2..., X_i are independent variables for the production time, for example, lot size, maximum measurements, etc. The factors to be determined are a_i and the exponents e_i for the independent variables. Thus, the production times calculated from the formula show minimal deviations from the known values Y_j for the entirety of the parts (see notation in Sect. 9.3.4.1).

The factors and exponents are now changed systematically (optimization strategy), starting from an initial value, until the sum of the deviations reaches a minimum. If, for example, the lot size X_1 has a large exponent e_1, the manufacturing time for large parts becomes too long, since in the formula the lot size dominates relative to other variables. It is just the opposite for very small parts; the deviations would therefore be high. This formulation is then rejected relative to other formulations. According to the optimization strategy, a new formulation is created. With that, the production times of all parts and their deviations are calculated again until a satisfactory formulation is attained.

Figure 9.3-4 shows a comparison of the possible formula setups between the regression and the optimization calculations. With the same variables a product formulation is possible only for regression analysis (Fig. 9.3-4a) (additive formulation, transformed logarithmically). On the other hand, the optimization calculation (Fig. 9.3-4b) allows a combination of product and linear formulations, which corresponds more to the cost build-up, as resulting from the statistical coefficients of measure (correlation and standard deviation). Furthermore, the frequently necessary use of $(1 + X_i)$ as the variable is seen, so that with the absence of the variable (e. g., the part is not milled) the overall result is still meaningful and does not become zero.

The example also shows that the application of the quick cost calculation formula can be relatively simple. The formula can be stored in a programmable calculator. The independent variables are read off either from the drawing (for example, 60 mm for the maximum length and width, here the diameter) or counted (e. g., 10 for the number of drilled holes) and used in the formula. The result is found quickly. The material costs are determined separately from the volume of the blank piece.

9.3.4.4 Use of neural networks for determining cost

From some authors the use of neural networks for the determination of costs has been investigated [Büt95, Sch92, Wol94]. The task of a neural network is to determine the desired output data with a given number of input data. That is achieved when the entry values in the "neurons" are weighted for so long and formed until a satisfactory result is achieved. For a special problem, the network must have access to a suitable number of input and output neurons in the input and output positions (**Fig. 9.3-5**).

Between the input and output states, an arbitrary number of neurons can exist in hidden states. The connections between the neurons of a network are arbitrary. The information that is given at the entry neurons (e. g., gear diameter, width etc.) is converted into an internal representation by weighting and arbitrarily selectable functions. The network computes the result, for example, the manufacturing time, for a fitting input pattern by the applying the formulas for all its neurons from

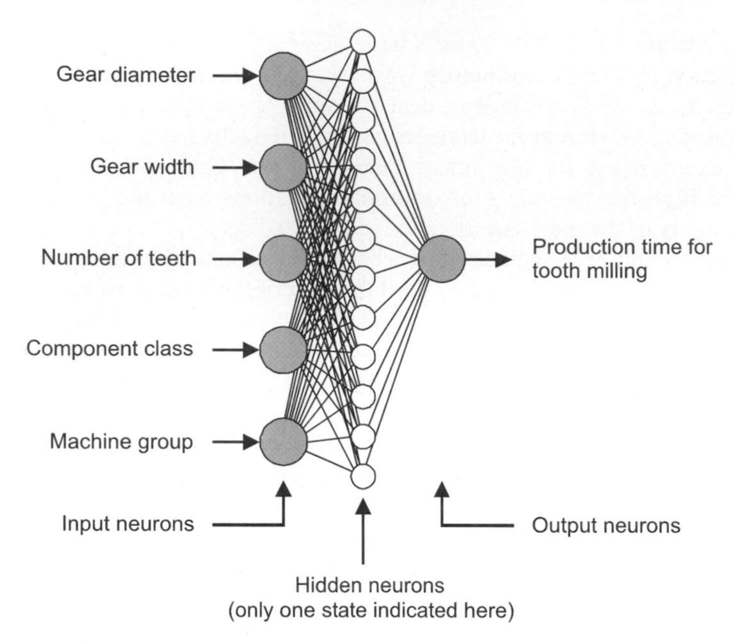

Fig. 9.3-5. Basic structure of a multi-layered neural network [Wol94]

state to state (i. e., referring to Fig. 9.3-5, from left to right). The output signals subsequently represent a result of the internal representation.

The neural network cannot solve an unfamiliar problem by itself. The network must be "trained" with known data. Known data are entered for this purpose, just as they are in the case of regression analysis. Unlike in regression, the factors of a formula provided by the person responsible are not determined in a calculation process and then outputted. Rather, the internal weighting and connections of the neurons are adjusted in many calculation cycles such that the answers of the network lie within the assumed error tolerance. The individual weighting and connection of the neurons that take place remain concealed from the user. The number of needed calculation cycles depends on the desired tolerance and thus the "training time".

Herein lies the difference from the procedures described up to now, which work with an explicitly known formula. In particular, this has disadvantages for low-cost design. If, for example, in the case of multiple regression analysis the influence of the part weight on the costs is found to be negative, one can see that it is an accidental statistical result [Bor97, Gau98]. This is similar to the statement that "the number of the births is equal to the number of storks", which has nothing to do with reality. In neural networks, such an absurd connection cannot to be recognized.

On the other hand, the advantages of neural networks as opposed to regression analysis and the optimization calculation for quick cost calculation formulas is their outstanding adaptability for complicated, non-linear data that cannot often be achieved by other methods. For a detailed description and application of the procedure, refer to the literature [Büt95, Bec96, Sch92, Wol94].

9.3.4.5 Use of fuzzy logic for determining cost

References [End00, Lei01, Sch01] describe the use of fuzzy logic for determining cost. For the very reason that fuzzy logic incorporates vague information, the authors feel that it is suitable for determining costs in the concept phase of the development process, when information is still sparse. As in the preceding sections, with fuzzy logic the cost information is also determined for a new product from costs of existing products. These must exist, prepared in a suitable form. Corresponding product models were developed for this purpose, which presume a methodical task execution. Consequently, this procedure becomes quite expansive.

9.3.5 Quick cost calculation with cost growth laws

A cost growth law (also called similarity law) is the relationship of the costs of products similar to each other. For that, geometrical, material, shape, and production similarities must exist, which is usually the case for systems in size ranges (Sect. 7.12.5). In general, production costs are proportional to production times. Therefore, the relationship can also contain times, which has advantages for the updating of the information. By similarity, we mean (Sect. 7.12.5.3):

- **Geometric similarity**, where the products have the same proportions, but differ only by the size ratio (linear scale, magnification factor)

  $$\varphi_L = \frac{\text{length}_1}{\text{length}_0} = \frac{L_1}{L_0}$$

  (Pantograph magnification). Materials and production must also be identical
- **Geometric semi-similarity**, where certain dimensions change with different size ratios. Therefore, for example, for a roll the diameter can change with the ratio $\varphi_D = D_1/D_0$ and the roller width with a different ratio $\varphi_B = B_1/B_0$.

In the contrast to that, there are also products and parts that are not similar geometrically, but are still similar as regards the production technology. They form a production family and are manufactured with same resources (machine tools, fixtures, tools, etc.; see Sect. 7.12.3).

The preliminary design, the manufacturing costs of which are known, is designated as a **basic embodiment** (MC_0); the embodiment, which is to be created concerning the manufacturing costs, is the **follow-up**, or the **next embodiment** (MC_1). The cost growth law in the case of geometrical similarity like this, is:

$$\varphi_{MC} = \frac{MC_1}{MC_0} = f\ (\varphi_L\ ),$$

in the case of semi-similarity, e. g., of rolls:

$$\varphi_{MC} = \frac{MC_1}{MC_0} = f\ (\varphi_D, \varphi_B\ ).$$

The **purpose of the application of cost growth laws** is the quick calculation of the costs for larger or smaller size follow-up embodiments, starting from the technical and cost data of the basic embodiment. It is not necessary to first produce the follow-up embodiments, make drawings, and then to calculate. Rather, we can do this now, after the design and costing of the basic embodiment "at the desk", and thus save time.

The **advantage** is that we already recognize while designing a product in a size range, how the cost structures change with their most important cost elements. This is accomplished through different cost growth laws of different production processes or material costs (Fig. 7.6-3, Eq. (7.7/1)). Therefore, we are better able to keep costs in mind while designing. Furthermore, it is beneficial that the fundamental relationships are in part valid industry-wide and hardly need to be updated. The adaptation occurs over the company-specific and currently calculated basic embodiment.

The relationships contained in the cost growth laws have great importance in practice for the different rate of growth of the material costs and the costs of different production processes (Fig. 9.3-7). If they are known, cost estimates are easy to uphold (Sect. 9.2) and measures for low cost design can be derived (Sect. 7.12.5.5).

The **disadvantage**, as opposed to quick cost calculations which are statistically established (Sect. 9.3.4), is that we must at first derive a cost growth law (by statistical evaluation or through derivation from production-time formulas), and then form the basic embodiment and calculate its costs. Furthermore, the number of true geometrically (semi-) similar products is not large (Sect. 7.12.5.4).

We can categorize different types of cost growth laws:

- **Summary cost growth laws** of components (groups, machines) present the costs for these directly, without dealing with individual production operations. Since their accuracy is limited, they are employed for recognizing essential cost dependencies for products in size ranges and to derive rules [Ehr79, Fis83, Pah84, Die88, Kön95]. An example of a summary cost growth law is Eq. (7.7/1) that was established by statistical evaluation of a large spectrum of machined parts:

$$MC_{1n} = \frac{PCs_{01}}{n} \cdot \varphi_L^{0.5} + PCe_0 \cdot \varphi_L^2 + MtC_0 \cdot \varphi_L^3 \left[\frac{\$}{\text{piece}}\right] \quad (7.7/1)$$

  A calculation example is shown in Sect. 7.7.2. However, it should be checked with specific parts as examples. As far as possible, the exponents and the structure of the equation should be adapted specifically to the company and to the pertinent part. For quick cost estimates, we can employ Eq. (7.7/1), or the basic cost growth as a function of φ_L^x, depicted in **Fig. 9.3-6**. From the log-log representation (Fig. 9.3-6a) values can be read off more simply, and it shows how the cost growth behaves above and below $\varphi_L = 1$. In the linear representation (Fig. 9.3-6b) the exponential increase of the costs is recognized better for φ_L^2 and φ_L^3.
- **Differentiated cost growth laws** use the growth laws of the production times for the production processes individually for a part (**Fig. 9.3-7**) [Fis83, Pah84]. The relatively small deviation for the preliminary costing from accounting based on use and error compensation is advantageous. The times or costs of every production process and every component are determined (Fig. 7.13-11

and 10.3-7). A disadvantage is the higher preparation expense for the cost growth laws and more calculation effort, which can be reduced, however, by the use of computer programs.

The calculations, with summary and differentiated cost growth laws, are described in Sect. 10.3.4.

Examples for the derivation of differentiated cost growth laws

For a specific production process (e. g., turning), the relationships that exist between design parameters (dimensions, material) and the production time units are determined through analysis of the documents from operations scheduling. The terms and contents of the time units are described in [REF71]. These are sometimes used in the companies differently. Here the organization is according to machining, idle and setup times (Fig. 7.6-2).

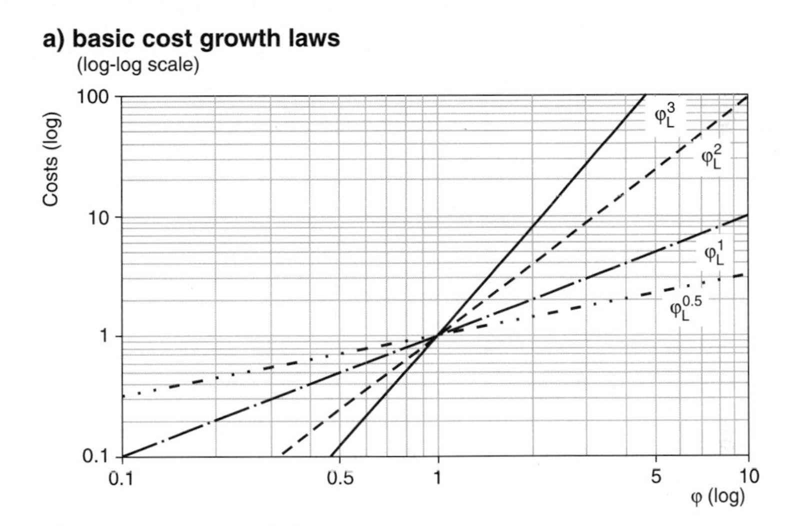

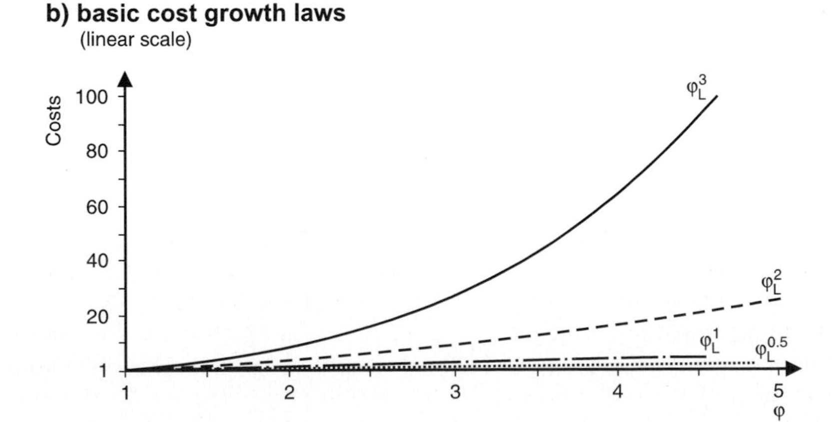

Fig. 9.3-6. Basic cost growth for different exponents x for φ_L

Machine type	Process	Exponent		Attainment certainty
		calculated	rounded	
Universal lathe	External and internal turning, turn screw threads, slice, turn grooves, turn chamfers	2 1 1.5 1	2 1 1 1	++ + + +
Vertical mill	Turn - external & internal	2	2	++
Radial drilling machine	Drill, countersink, cut threads	1	1	0
Drill & milling center	Turning, drilling, milling	1	1	0
Groove milling mach.	Milling key ways	1.2	1	+
Universal grinding machine	External cylindrical grinding	1.8	2	++
Circular saw	Sawing sections	2	2	0
Sheet shears	Shear sheet metal	1.5...1.8	2	+
Trimming machine	Edge/trim sheet metal	1.25	2	+
Press	Straighten sections	1.6...1.7	2	+
Beveling machine	Bevel sheet metal	1	1	++
Gas cutting machine	Gas cut sheet metal	1.25	1	++
MIG- and E-hand welder	I welds, single-, double-vee butts, hollow, fillet welds	2 2.5	2 2	++ ++
Annealing		3	3	++
Sand blasting (calculated according to weight or surface)		2 od. 3	2 od. 3	++
Assembly		1	1	++
Fixture for welding		1	1	++
Hand plastering		1	1	++
Varnishing/painting		2	2	++

Legend:
- ++ Expect good certainty of attaining estimate with rounded exponent
- + Certainty of reaching less than with ++
- 0 Greater scattering possible

Fig. 9.3-7. Exponents for times per unit (total time units) with geometrical similarity, for selected types of machine tools and production processes [Rie82, Pah97]; for the application: see Fig. 9.3-6

a) Machining times t_m:

The machining times can usually be established easily, e. g., during straight turning:

$$t_m = \frac{D \cdot \pi \cdot B \cdot i}{v_c \cdot f}$$

with D = diameter
B = turning length
i = number of cuts
f = advance
v_c = cutting-speed.

With that, the growth law of machining times for turning becomes:

$$\varphi_{\text{tmtu}} = \frac{\varphi_D \cdot \varphi_B \cdot \varphi_i}{\varphi_{vc} \cdot \varphi_f}$$

If the optimal cutting speed for a specific material, the number of cuts as well as the advance, are set as constant:

$$\varphi_{\text{tmtu}} = \varphi_D \cdot \varphi_B = \varphi_L^2$$

That means that the machining times, e. g., for the turning of a similar shaft twice as big are four times as long, and accordingly the corresponding costs (Fig. 9.3-6).

In [Hub95a] (Sect. 7.13.7b) dependence on the order of $\varphi_L^{1.3...1.9}$ for the assembly costs of gears was shown (1.3 for small, 1.9 for large gears).

Growth laws for additional production processes are shown in Fig. 9.3-7, and in references [Fis83, Pah84, Die88] (see also Fig. 7.13-11, Sect. 7.13.7b). How that is accounted for is explained in Sect. 10.3-4.

b) Idle times t_i:

Idle times arise with every part, besides the machining times (e. g., for loading, emptying, measuring, reversing or changing the workpiece). In general, they are not subject to any physically describable relationships and must be determined statistically. As a rule, idle times depend on the size of the workpiece. In [Pah84] a series of idle times were evaluated. In particular, similar assembly activities such as “tighten workpiece”, “change tool”, etc. usually increase proportional to the linear size. Thus, on the average:

$$\varphi_{ti} \approx \varphi_L$$

c) Set-up times t_s:

Setup times are concerned with items like the preparation, installing, and removal of fixtures and tools. They arise once per task. The evaluation by [Lan74] shows that setup times follow:

$$\varphi_{ts} = \varphi_L^{0.2...0.4},$$

and, from [Ehr79, Fis83], on the average:

$$\varphi_{ts} = \varphi_L^{0.5}$$

Depending on the field and on the fraction the setup time is of the total time, it can be set without great error as $\varphi_L^0 = 1$ (i. e., the setup time is fixed and independent of the size). That is the case for a specific production machine in the first approximation. The differentiated similarity viewpoint assumes that the basic and the next embodiment are produced with the same resources and same technological data (e. g., cutting speed, number of cuts, feed). If these change considerably, the corresponding embodiment must be suitably recalculated as a new basic embodiment. The modification of the setup time is then recorded.

9.3.6 Procedure for performing the quick cost calculation

To avoid setbacks in the preparation and application of the quick cost calculation formulas, a systematical procedure should be followed [Bau82, Kie79, Pic89]. The essential steps are:

Step 1: Planning the project and establishing the requirements

The neglect of this step is the cause for many unnecessarily wasted man-years. Projects with six to seven man-years are known to the authors, the results of which were outstanding but were not put into operational use, since little thought was given to the purpose, the integration of the later users, or keeping the plan up-to-date.

Note that by using quick cost calculation along with the usual calculation, a second calculation path is formed. In a company with narrow functional thinking, this can lead to management's disapproval (the technical people are supposed to see to their technology and not cost accounting!). Management should therefore be brought into the picture. This second calculation path leads to the goal faster than the company's cost calculation but does not address aims such as time guidelines, economic consideration of production processes, and the like. We stress once again that with cost management it is not a question of the "cent-precise" accounting of costs during product development, but rather the early identification and influencing of the costs, even if all information is not there yet [Sei97].

This new path needs as much care as conventional costing. The danger always exists that this care is not provided and so the work done thus far becomes useless. Therefore, the preparation, application, and maintenance of quick cost calculation **must** be coordinated with the respective departments! The production personnel (production scheduling, cost control), should prepare and update the quick cost calculation, since they have the experience with preliminary and follow-up costing.

Before planning quick cost calculations, we must consider who might work with them. If it is to be a production and cost advisor, the system can be complicated and thus more precise; if it is for design engineers, then it must be simple and it may contain only those variables known to them. One must consider which means and tools are necessary for the cost information system. Finally, the efforts for preparation, use, maintenance, and updating must be estimated. No unaccountable differences between the results of the quick cost calculation and the conventional calculation must arise; otherwise the quick cost calculation will not be taken seriously.

Step 2: Analysis of the database

If we know which products and which time or cost data are to be evaluated, we must determine whether sufficiently current data records (costs reckoned on a fixed date, times according to the current status of production) are available. There should not be too much data in one field and too little in others. We must pay particular attention to the reliability of the outermost values (e. g., to check very small or large, very simple, or complicated parts).

Step 3: Determine possible effective variables

It is important to find really definitive effective variables for times and costs. If the quick cost calculation is to be used for bid purposes or for designing, it is especially important to not include any typical production parameters that the personnel cannot handle. It is also important that the variables do not depend directly on each other, so that no intercorrelation (e. g., dimension and weight) might occur.

Step 4: Data collection and data verification

Since this step usually represents the largest volume of work, and it is very time consuming to enter forgotten data afterwards, it is recommended that data collection formulas be devised and tested. Supporting staff can then assume further data collecting responsibility. It must finally be checked whether certain types of parts are over-represented while others are hardly accounted for.

If the data have been recorded, gross mistakes can be tested for first (validity check). These are, for example, data outside of the operational range (part weight too large or too small), or orders of magnitude beyond possible limits.

Step 5: Statistical evaluation

Through the person responsible, a cause-based formula statement is constructed, and regression or optimization calculation or neural networks determine the needed factors.

During this first testing of the formulations, one must also examine whether formula statements are more favorable to sub-fields of the entire spectrum than a closed formula. Finally, the outlying data are to be tested. It is recommended that the calculated data and the master data points be plotted graphically and the regression line be drawn in (Figs. 9.3-1, 9.3-4 and 9.3-8). Through analysis of the master data, one recognizes the atypical special features for the entire range. Then other (non-linear) variables can be introduced and thus new relationships are tested (see above). Thus, which variables are significant and which formula statement is optimal are determined iteratively.

Step 6: Presentation of the results

Depending on the subsequent use, a preparation of the formulas thus obtained follows on forms with input and output quantities and programming on desktop computers. The master data sets must be clearly at hand for the later updating processes. It is worthwhile to evaluate the new formulas further (e. g., for rules, cost structures, etc.), and to make the results available to the design engineers as a basis for the target cost oriented design.

Step 7: Implementation in the company

Dealing with quick cost calculation in project planning, design and operations scheduling is a new idea and may be distrusted. Therefore, an implementation phase with training sessions is necessary. It is also important that the scattering of conventional cost computations (which are often significant) are clarified (Sect. 9.3.7.1). Only then will the personnel see that all calculations have deviations and that the possible accuracy is a question of effort and the calculation time (available information), and that the question of the necessary accuracy depends on the intended purpose.

Besides the implementation, the organization for the maintenance of quick cost calculation also needs to be resolved.

9.3.7 Accuracy of quick cost calculations

Accuracy means the absolute or relative deviation of specific data from a reference point. For manufacturing costs, we choose the actual cost of a product as the reference point. Through uncertainties of production (e. g., for waste, refinishing, or unforeseen deviations through material defects), the actual costs show strong scatter for identical products that are produced one right after the other or after long intervals. Therefore, the planned, pre-calculated costs (plan costs) are better used as the reference, which should of course be checked continually with the help of follow-up calculations. The **standard deviation** σ is usually denoted as the statistical measure of accuracy. In the case of normal distributions, 68.3% of the actual values lie in a range of $\pm 1\sigma$ about the values determined with the quick cost calculation formula.

The measure for the accuracy of quick cost calculations is the preliminary calculation. The earlier they are used in the product realization process, the more the results of the quick cost calculation may deviate from those of the preliminary calculation. For them at least, less detailed input parameters are available. Furthermore, the cost differences pending for the decision are usually large, so the calculation deviations also may be greater.

Under accuracy, therefore, is understood the deviation from the values of the preliminary calculation, whereby these are also not exact. The **absolute error** $\boldsymbol{E}$ of the value y computed from the quick cost calculation to the value of the preliminary calculation Y, is:

$$E = y - Y$$

and the **relative error** $\boldsymbol{e}$:

$$e = \frac{y - Y}{Y} 100 \, [\%] \tag{9.3/1}$$

We must distinguish between the relative error of a single quick cost calculation formula (e. g., for times or costs of a single production process or of an individual part), and the relative error from a sum of results of different quick cost calculation formulas. For example, the sum of times or costs of several production processes for a part, for the costs of an assembly, or a whole machine made of several parts calculated with quick cost calculation, or even the sum of the costs over a counting period.

In the first case (for a single estimate), we must reckon with greater relative errors of 5–20%; in the second case, we will be able to expect smaller relative errors, to 10%, because of error averaging (Sect. 9.3.7.3). For accuracy of relative costs, see Sect. 4.6.3.

9.3.7.1 In-house accuracy of preliminary calculation

In discussion of the accuracy of quick cost calculations the question comes up. How exact is the reference point itself (i. e., how exact is the preliminary calculation, within the company, and industry-wide).

The time determination is based partly on the physically based sequence of processes (e. g., machining times during the machining operation), and partly on agreements (a part of the setup, idle, recovery and extra time units). Even with experienced work process schedulers with the same documents in the same company for both setup and machining times for the same work, time variances of about 1:2, or ±35% have been documented [Käs74].

Pacyna [Pac76], for example, found through statistical investigations in the foundry industry that the quotation prices could vary within a foundry around ±20 to ±30% and the mean values of different foundries, in turn, around ±10 to ±20%.

However, the time determinations by operations scheduling (production planning) and the results of the calculation are often accepted without question. Deviations are not admitted since the same persons seldom calculate the costs of the same job again. In total, we must reckon on considerable deviations in in-house costing in the case of individual jobs, unless the work schedule preparation and time determination are carried out on a computer. We should not demand any greater accuracy from quick cost calculations than the operations scheduling itself indicates. As the previous experience shows, the accuracy of quick cost calculations, in part, is very good [Fis83, Pah84]. In positive cases there is only a few percentage points of deviation from that given by the preliminary calculation. Quick cost calculations, which proceed from the follow-up calculation or use a great deal of averaged data from the preliminary calculation, can be even more precise than statements from production planning based on a similarity calculation (**Fig. 9.3-8**). This similarity calculation is often used in production planning practice because it is simply too expensive to plan every part exactly. It therefore reaches back to the accessible calculations of similar parts (Sect. 9.1.3).

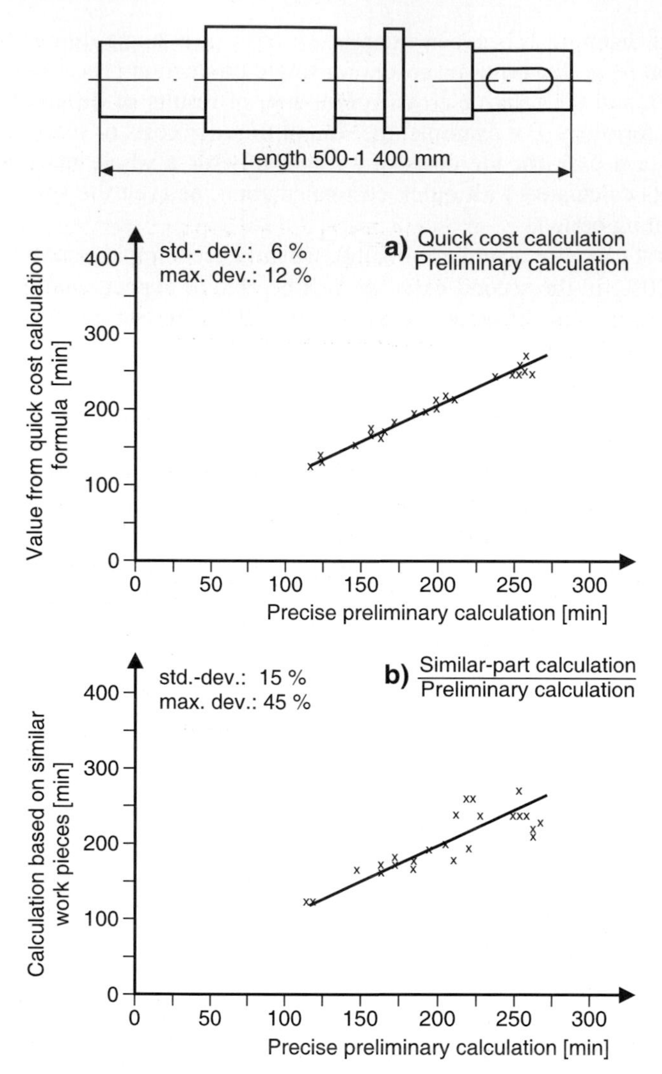

Fig. 9.3-8. Comparison of the deviations between quick cost calculation, similarity calculation of production planning and more precise preliminary calculation [Haf87]

9.3.7.2 Industry-wide accuracy of the preliminary calculation

The deviations of the preliminary calculation from one company to another are larger than those within a company. In different companies the production facilities, utilization, and business calculations are different. The decisive influences for the deviations are not, however, utilization or type of accounting (i. e., company's economical factors), but rather the different time determinations of the operations scheduling for basic and setup times. In the case of the same production machines, the same part and material, as well as the same costing types (the same form

used), deviations of 1:4 come about in the floor-to-floor and machining times. Setup time deviations are even higher (Sect. 7.13.2b).
As Fig. 7.13-3 showed, the calculation differences are size-dependent. For companies that make small gears economically (Company I), it can be more expensive to make large gears relative to the average competitor, or, vice versa (Company A). There are also altogether expensive companies, however; (Company B might be about 30% more expensive than the average) or low-cost companies (Company M might be 30% cheaper than the average). With such deviations it is surprising that the expensive companies still exist in the market. The prerequisite made for this calculation comparison was that only time calculations are made for gears for which there is definitive production experience. Keep in mind, however, that the gears constitute only about one-third of the transmission costs and that a company that is expensive as regards gears, for example, might make gear housings more economically.

It is possible that in the case of series production, the large deviations do not come about in the capital goods industry with single-unit and limited-lot production. There, the entire production can be planned in more detail, and there is an even greater pressure of costs. In the course of globalization of the markets, even higher deviations must be reckoned with in the case of international cost comparisons (benchmarking, Sect. 7.13). Besides the errors in cost accounting, different constraints such as exchange rates, costs from rules and regulations, etc., also distort the comparisons.

9.3.7.3 Compensation of random errors

When time or cost estimated values y_i (for data found from quick cost calculations) are flawed with random errors E, a total value $y_{tot} = y_i$ will show a smaller error E_{tot} than the individual values y_i [Kie82]. This effect occurs because the random errors E_i are distributed on both sides of Y_i and therefore, in part, cancel out (Gaussian error compensation). Such total values y_{tot} come about from:

- Part costs (or time units) as a sum of the costs (or time units) of several production processes or the sum of production and material costs.
- Assembly and/or product costs as a sum of the costs of several parts and/or assemblies.
- Period costs (i. e., the sum of time units or costs of several single events over the period of a month, for example).

With the relative errors e_i (Eq. (9.3/1)) of the individual estimated values y_i, the relative error e_{tot} of the total value y_{tot} becomes:

$$e_{tot} = \sqrt{\frac{\sum_{i=1}^{n} (e_i \cdot y_i)^2}{y_{tot}^2}}$$

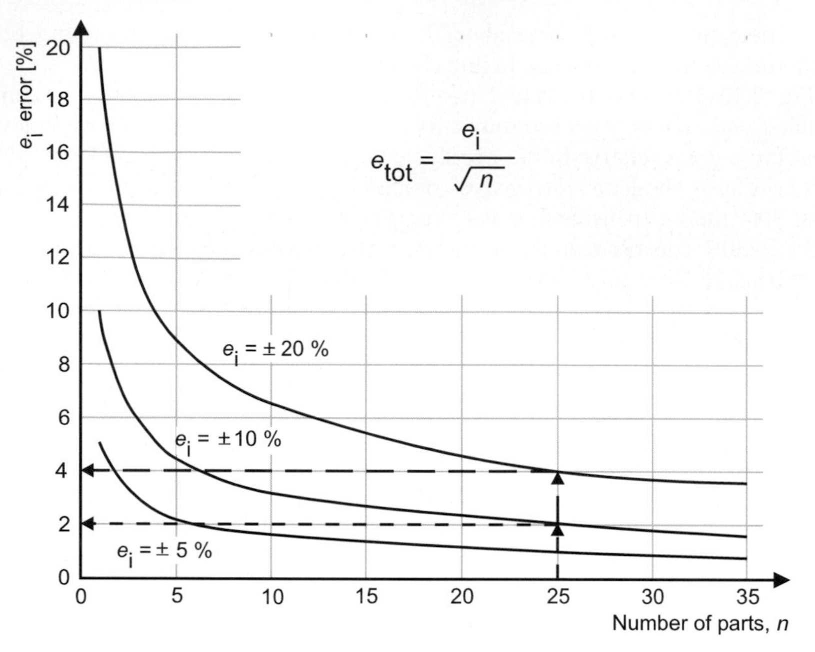

Fig. 9.3-9. Relative error e_{tot} for an assembly made of n parts with same costs and same relative error e_i of the individual estimates

If the **individual estimates** $y_1 = y_2 = \ldots = y_n$ are of **equal size** and they all have the same **relative error** $e_1 = e_2 = \ldots = e_n$, the above equation simplifies to:

$$e_{tot} = \frac{e_i}{\sqrt{n}}$$

This result is shown plotted in **Fig. 9.3-9**. The total error e_{tot} of an assembly made from n equally expensive parts with randomly distributed errors e_1 of the same size, is around n^{-2} smaller than the error of each part. If a product has, for example, $n = 25$ assemblies, for all of which a cost estimate of 100 $ ± 20 $ (e_1 = 20%) is valid, then a cost estimate for the total product is:

$$25 \cdot 100\,\$ = 2\,500\,\$$$

The total error is found from

$$e_{tot} = e_1 / n^{0.5} = 20 / 25^{0.5} = 4\% = \pm 100\,\$$$

If we apply this statement to the known unequal distribution of the individual values from Pareto analysis (Sect. 4.6.2), we have the answer to the question: "How precisely must a single value y_i be determined so that the total value y_{tot} does not exceed a permissible error e_{tot}?" (Fig. 9.3-10).

The numerical values can be taken individually from **Fig. 9.3-10**: For a permissible relative total error e_{tot} = 10% for the production costs of the transmission, the

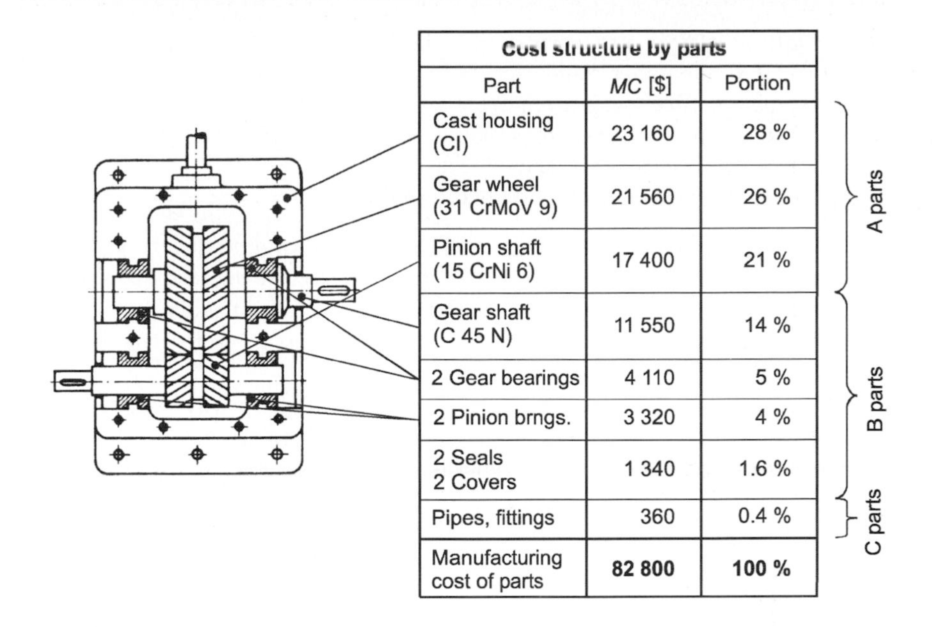

Cost structure by parts			
Part	*MC* [$]	Portion	
Cast housing (CI)	23 160	28 %	A parts
Gear wheel (31 CrMoV 9)	21 560	26 %	A parts
Pinion shaft (15 CrNi 6)	17 400	21 %	A parts
Gear shaft (C 45 N)	11 550	14 %	B parts
2 Gear bearings	4 110	5 %	B parts
2 Pinion brngs.	3 320	4 %	B parts
2 Seals 2 Covers	1 340	1.6 %	B parts
Pipes, fittings	360	0.4 %	C parts
Manufacturing cost of parts	**82 800**	**100 %**	

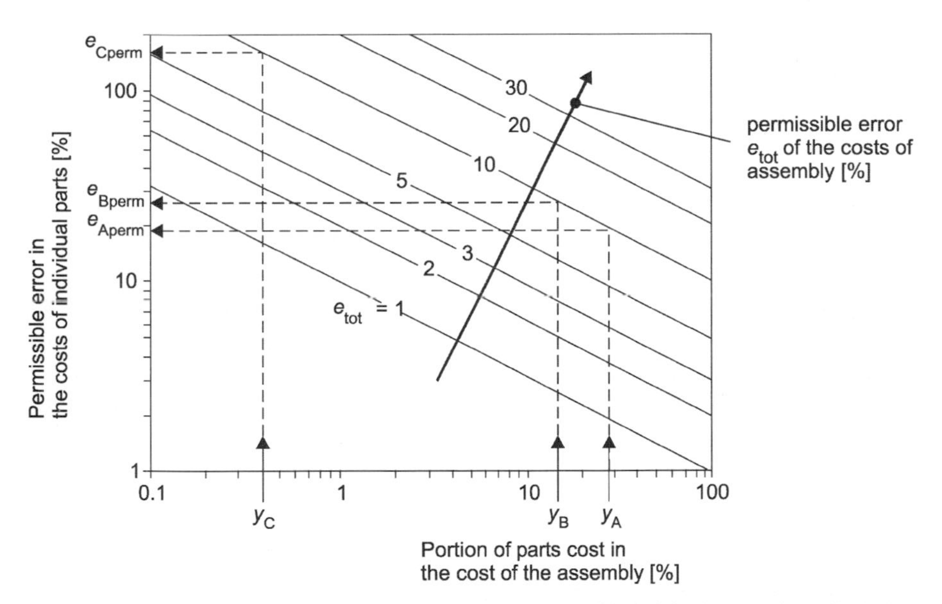

Fig. 9.3-10. Example of permissible relative error e_{iperm} of individual parts, for calculating the permissible error e_{tot} of the assembly [Kie82]

housing with $y_A = 28\%$ cost portion may be calculated with a relative error $e_{Aperm} = 17\%$. On the other hand, for "C" parts with $y_C = 0.4\%$ cost portion, a coarse estimate suffices with a relative error $e_{Cperm} = 170\%$. The different shape of the error curves in Fig. 9.3-9 (hyperbolic) and Fig. 9.3-10 (straight line) is due to the different axis scales: linear and log-log.

The **practical significance** of these relationships is:

- The more detailed one estimates or calculates the cost of a product (e. g., dividing into components or production steps), the more precise the result becomes.
- Individual portions, for example, material costs or costs of individual assemblies, can be estimated or calculated more easily and more exactly than the total value for a product.
- The larger the share the part to be calculated has of the entire product ("A" part), the more precisely it must be calculated in order to achieve a specific accuracy of the overall result. On the other hand, the estimate of "C" parts may be coarse (Fig. 10.1-12).

9.3.7.4 Keeping current

Quick cost calculation formulas can be updated easily with changes in wage or material costs when the formulas are used to calculate not costs directly, but production times or material volume (weights). These need only to be multiplied by the new hourly wages or the new costs of materials. We separate the quick cost calculation like the usual costing, into quantity and value schedules. For a specific period (approximately five years), quick cost calculations that determine costs directly can also be updated with annual inflation rates (Sect. 9.3.4.2).

The company cost accounting method might change (e. g., changing overhead costing to workstation costing), as might the production technology (other production times, more efficient production processes) or the organization (production in production islands). Then the quick cost calculation formula must be re-created, the same as the case with standard costing. This works faster than the initial formula formation if the initially used data records are arranged in a file. These must be fixed by using the new production times. After that, a new formula is prepared with the regression program with the same main variables and same formula approach. Thus a continuous updating is possible in which the new data records are always entered and old ones removed. A new, up-to-date equation is worked out repeatedly, in shorter intervals.

9.4 Computer-integrated cost calculation

The quick cost calculations presented in the preceding sections help in finding the costs of a product in the course of the development process. As shown above, the basis of the quick cost calculation must be the company's own costing. The preparation and use of quick cost calculations can be reduced if the bid and preliminary calculations are set up and maintained so that it can be already carried out during the development process.

A core idea here is that the data found from product development and stored in the CAD system are the input data for work scheduling and costing. If the CAD data can be carried over to a spreadsheet in suitable form, the product developer can run the cost calculation after establishment of the data in the CAD system, by a keystroke. With that, the aimed-for short-loop feedback control system (Fig. 4.4-2) would be attained (at least in the embodiment and detail design stages.)

However, the preliminary calculation and its basis (setting the quantity schedule with work schedule preparation by operations scheduling) are time-consuming due to the planning of numerous work processes with the choice of machines, fixtures, tools, cutting data, etc. Documented experiences of the personnel do not enter in work scheduling. Furthermore, optimal machines are not always available; there is waste, etc. Therefore, work schedule preparation and cost calculation are still frequently carried out manually. Through investigations in the field of the automatic work schedule preparation [Ham93, Die89], improved computer setup and computer support are also becoming more extensive in work scheduling and costing departments. In recent years a series of programs were developed that appeared suitable for finding costs already during the development process [Bul95, Hor96]. However, the market concerning the software is very unclear and is continuously changing. Furthermore, the requirements and conditions in the companies (e. g., costing methods, organization, and type of product and production) are very different. Thus, no complete listing of all software systems available on the market can be given here.

The calculation procedures and the number of cost elements considered account for the essential differences between the systems [Hein95, Hor96, Sche90]. Some systems specialize in manufacturing costs, for example HKB (Sect. 9.4.1), XKIS (Sect. 9.4.2), or DIDACOE [Wie90]. They use the differentiating overhead costing or machine hourly-rate calculation, building on an automatic work schedule preparation. Other systems such as KICK [Fis93, Koc94] and KOMO [Eve90] put the emphasis on registering the costs of indirect fields better than has been done up to now, through process cost principles or "resources". They do not determine the production costs in as much detail as do HKB or XKIS, but use statistically established formulas. Another aspect is coupling the calculation to other programs (e. g., a CAD system). The front load costing [Kok99, Nef00] method uses a probabilistic approach that is based upon probability theory and decision-oriented approach of risk analysis.

Recently, due to increasing computer use and the establishment of software companies, more program systems have been developed; some can even be used directly on the Internet, for example, for turning [Mas01]. The software market is continuously changing. If the need for software arises, a search on the Internet is advisable.

In the following sections, with the help of two specific examples, computer-aided costing is introduced, with increasing levels of integration of calculation and design:

- **Computer integration of work scheduling and costing** (Sect. 9.4.1). Program HKB realizes computer support and combines the individual steps of operations scheduling and costing, which are often carried out in separate departments.
- **Computer integration of CAD, work scheduling and cost calculation** (Sect. 9.4.2). The essential entry data of operations scheduling and cost calculation are already available in the computer database, in the documents constructed during design with the CAD system. The basic idea is to deliver these data directly into a work scheduling and cost calculation program to avoid double work, errors, and delay. However, this theoretically simply idea is not easily implemented in ractice. With the XKIS system, an approach to the solution of these problems is ntroduced; it has been in use in a gear manufacturing company since 1995.

9.4.1 Computer integration of work scheduling and cost calculation

As an example of a work scheduling and costing program, the **System HKB**, marketed commercially for several years, is introduced here [Fer87, Mir91, Hei93b]. The System HKB computes the production costs of individual parts if their basic features such as shape, material, dimensions, tolerances, surface quality, heat and surface treatment, and production processes are known. The costs of materials are determined as usual from the weight and the material cost rate. The production costs are found from the production time (setup, idle, and machining times) and the corresponding machine hourly rate. **Figure 9.4-1** shows with the example of a shaft how the geometry of the individual shape details must be entered. The production time and cost are output according to the shapes.

The program system contains rules and formulas for the computation of production times of the most important production processes for a number of shapes. The basis data necessary for the computation (about materials, fixtures, cost unit rates, etc.) are entered company-specific one time and periodically enhanced and updated. The program is used not only for design-concurrent cost calculation, but also as an economic and administrative computation program in some companies.

From Fig. 9.4-1 it should be recognized that for input into the program there must already be a finished embodiment of the part with dimensions and tolerances. These are entered in corresponding masks (the values in boldface type in Fig. 9.4-1). An interface to the CAD system, CATIA, exists but the identification of individual shape details is time-consuming. The main application of the program is in the embodiment design phase. If the corresponding functions and part classes are already defined and data on existing products are available, then these can be searched and used as a starting point for cost estimates.

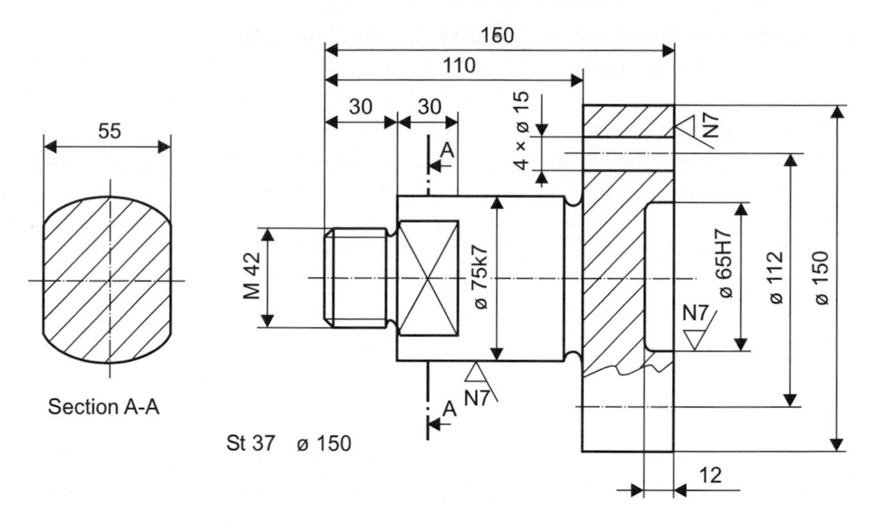

SAMPLE Machine Co.	**Manufacturing Cost Calculation - HKB** Total costs from WS features	Date 87-3-3 Visum HKB

Object:	**F. 11111 shaft**	Quantity: 10
Assembly:		
Comment:	**Docu**	
Type/Macro:	**None_special**	
Raw material:	**Steel_plain_C**	Weight: 20.8 kg
Semi-shape:	**Cylinder** id = **0** l = **150** d = **150** kg = **20.81**	
Manufacture:	**Rolled**	

Workpiece features	t_s	t_e	PC
Starting_piece	0	4	26
Facets_and_Radii n = **3**	3	4	5
Outer_cylinder o = **8** t = **0.50** l = **12** d = **65H7**	10	16	18
Inner_cylinder o = **8** t = **0.50** l = **12** id = **65**	2	3	3
Plan surface n = **1** o = **8** t = **0.50** d = **150** id = **65**	0	0	0
Outer thread i = **30** d = **42**	10	9	10
Level n = **2** p = **1** o = **8** t = **0.50** h = **10** **b = 30** l = **60**	62	3	8
Through/blind hole n = **4** p = **1** t = **0.50** d = **15** l = **40**	35	4	6
Sum (Times in min, costs in Sfr)	121	44	76

Calculation scheme	Direct costs	Overhead	Total
Material costs (incl. 15 % overhead)	23	3	26
Production costs	23	39	62
Special direct costs of production	0	0	0
Business expenses 10 % of production costs	0	6	6
Manufacturing costs	46	48	94
Administration & sales overhead costs 15 %	0	13	13
Total costs	46	61	107

Fig. 9.4-1. Example of cost calculation of a part with HKB

9.4.2 Computer integration of CAD, work scheduling and cost calculation

As the example in **Fig. 9.4-2** shows, it is possible to directly couple CAD systems with programs to the cost information with the **XKIS System**. Programs exist for this purpose for information about:

- **Costs of existing products** (machines, repeated parts) and their cost structures, according to components, types of costs, and production operations. Thus, the focal points for low-cost design are recognized.
- **CAD-to-cost coupling** for the calculations of important, in-house produced new parts ("A" parts). It is understood that immediately after the definition of a part, costs are calculated according to the usual scheme of operations scheduling and costing, whereby production costs and cost structures are worked out, and displayed on the CAD screen for the product developer.
- **Purchased part prices** (e. g., fasteners, material, drive parts, standardized parts) and the costs for externally acquired parts or assemblies (e. g., for cast and welded parts) established with quick cost calculations.
- **Function costs** for recurring functions, for which technically alternative function carriers are available.
- **Collections of data** (e. g., about workstation cost rates, exponents for cost growth laws of specific production sequences).

These programs and information must be installed with care in the CAD procedures and other processes of the company (e. g., work scheduling and costing). The greater the use of computer support in the operations scheduling (CAM), the more integrated CAD/CAM processes will come about. On one hand, parts of the NC data are already in the design. On the other hand, quick cost calculations become superfluous since time guideline (preliminary calculation) programs can be deployed by operations scheduling at the time of the first embodiment drawings.

The System XKIS (Extended Cost Information System: [Sch92, Ste93, Ste95, Rei96a]) supports the product realization process, beginning with the concept development up to the production planning.

Use in the concept phase for similar part search

Before settling on the concept, the feature-based repeating and similar part search acts to avoid overhead costs in the form of unnecessary processes in the company. When existing drawings, parts lists, work schedules and so forth are found for a ew product or a new shape, they do not need to be created all over again.

The feature-based repeating and similar part search is based on the hierarchical modeling of parts and assemblies, with the aid of features as descriptive objects. A **feature** is the object description of an assembly, a part, or a shape with object-specific geometrical and semantic properties in each case. The goal of a search is to find a part or an assembly that conforms to the desired requirements completely or in part.

Next to repeated part use, the repeated use of individual elementary features (e. g., a special gear tooth system) is also of interest for a new part. This is the case if we consider the incurrence of overhead costs of new parts and variants in indirect fields. The object thus found either can be adopted unchanged into the current embodiment, or is simply adapted. Then in the additional steps, available tools, fixtures and NC-programs can be reused.

Calculation for the new embodiments

The basis for both the automatic calculation of the new embodiments and for the repeating and similar part search is the interactive feature identification. That takes place in XKIS starting from a 2-D or 2.5-D drawing in the CAD system CADAM. Here the interactive identification of the features with the mouse is necessary by clicking on the descriptive geometry elements. Alternatively, the features are generated largely automatically in a 3-D model in the CAD system Pro/ENGINEER. By exporting essential geometrical and technological data into a neutral database, this information can be accessed by any system (CAD, NC, CAPP, etc.). The greater information content of 3-D models largely makes an automatic extraction of the feature data possible for models that are built up systematically from the CAD database.

The underlying embodiment does not have to be fully detailed. With a rough-sketched embodiment, the designer can quantify the properties that affect costs and provide these to the system. The extraction of the non-geometrical properties (material, heat treatments, surface quality, etc.) occurs in the CAD system likewise with the aid of selection lists. All necessary data are asked for by the system; these are the same as for the repeated-part search. Building upon these product data that are computer-interpretable, the **automatic rough work-planning** takes place. The basis for that is the manufacturer's detailed production and costing model, as well as the product model and the knowledge base, for which the calculation is carried out.

The **outputs of the results from CAD** are costs or times, as desired, in the form shown in Fig. 9.4-2, above and below.

In the upper portion, the individual machined surfaces of the planet gear are labeled with arrows indicating the costs or times. This assignment clearly shows the details of how costs arise. The costs comprise all total time units, whether they can be influenced or not, to which the machined surfaces and thus the features, are to be assigned.

For a clear representation of all costs that arise during the production of the part, the **cost structure** is shown in the lower part of Fig. 9.4-2. The machines for the various processes are represented with setup and idle times next to the costs or times for the processing of the features-analogous to the shape-related representation. In addition, the costs that are assigned entirely to the part are shown. The analysis of these cost results is a basis for changes to the embodiment that can then be evaluated immediately. Thus, the required short-loop feedback control system (Fig. 4.4-2) is realized virtually without loss of time. Due to the detailed cost statements, the product can be optimized for low cost, keeping in mind the available production facilities. Furthermore, with this tool the production variants for the same component can be compared very rapidly as regards cost and time, with repeatable results and low effort.

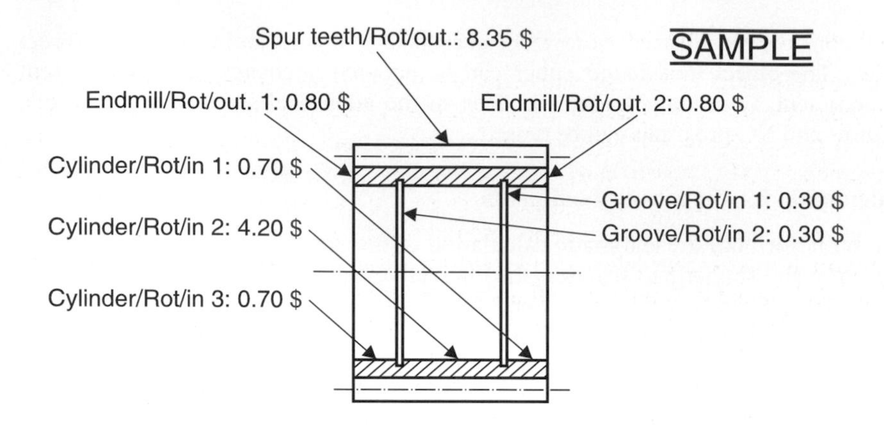

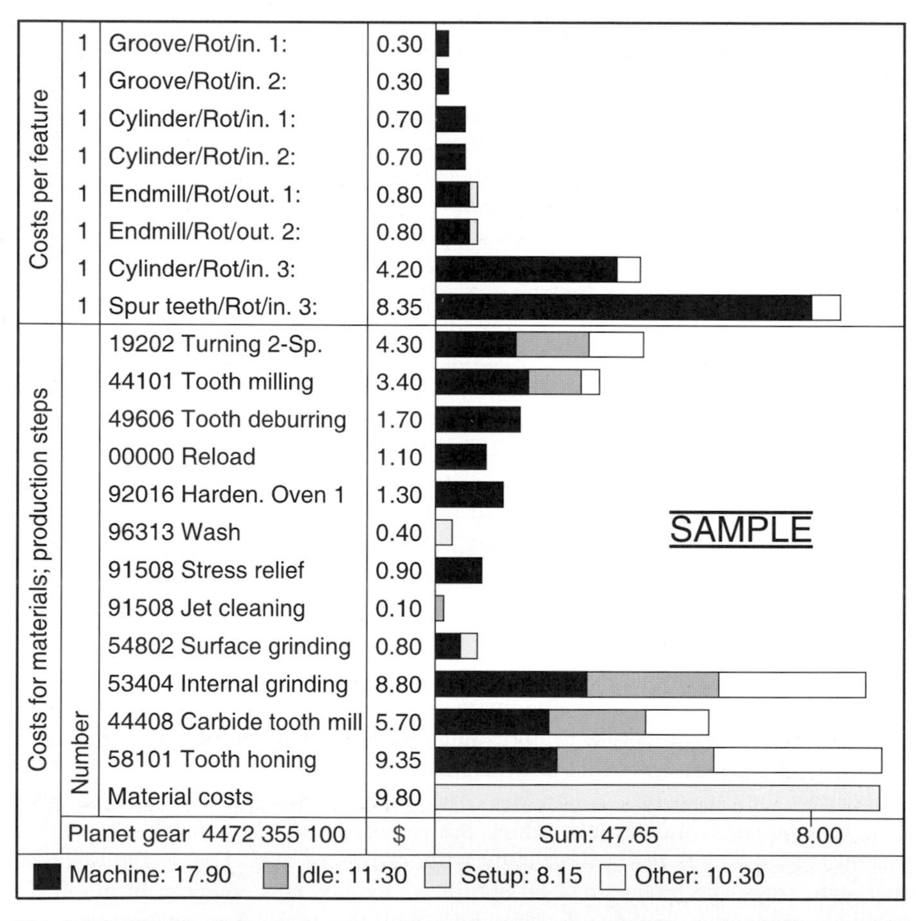

Fig. 9.4-2. Results of the design-concurrent calculation from XKIS on the CAD screen for the example of a planet gear [Ste95]

Development and maintenance of the system

The **development of this system** required a multi-year cooperation of the company and the university [Sch92, Ste93, Ste95, Rei96a]). In addition, a close cooperation of the appropriate fields in the company was necessary (i. e., product development, production planning, production, costing and data processing). On one hand, this was necessary to integrate the system in the affected company processes so that it is accepted by all of them and the results keep on being used (e. g., the approximate work schedule as basis for the later final work schedule). On the other hand, specific components, shapes (e. g., gear teeth), materials, etc., depending on production and lot size, had to be standardized. The interdisciplinary team determined with which machines, cutting speeds, tools, fixtures, etc. the production is optimal. A cost reduction was achieved by this standardization alone.

The **maintenance of the system** through experts is enabled via interfaces to work scheduling and costing. The departments of production planning, time recording, resource design, etc. also have access to the product model in the database, the data and knowledge bases, and the automatically constructed approximate work schedule. From the knowledge thus gathered, instructions are passed on to production planning for optimized production.

The possibilities of actively managing variants with feature-based product modeling and knowledge structuring go considerably farther than that assumed in repeating and similar part search. Beginning with the standardization of the product shape design in the form of features on all hierarchical levels, the production planning can also be standardized and thus the production itself, in the form of the knowledge base for work scheduling. Next to this **preventive management of variants**, a limited revision of the available part spectrum is also possible, based on the feature-based product model. Due to the unmanageable variety of shapes, parts, and assemblies, great importance is attributed to such computer-aided tools in most companies with design tied to overhead costs. Variants of products and parts arise in a company, on one hand, from different product implementations for different customers (externally made variants). On the other hand, different design engineers and work planners produce different documents for no specific reason, even though the components might in fact be identical (internally made variants; Sect. 7.12.1.2). The latter types of variants decrease with the XKIS system.

Benefits of the system

Following are the advantages of this system [Ste95]:

- Management of variants with corresponding reductions in manufacturing and overhead costs.
- An automatic investigation of whether a product can be made with the available resources.
- Optimization of the product shape during product development, resulting in a reduction of manufacturing costs.
- Decrease of expenditure in work planning and product development by decreasing the iterations between product development and production planning.

- Simulation of different production processes for a product.
- Decision documentation.

These advantages cannot be directly calculated, like many rationalization measures in product development. Indeed, it does not show up in the usual cost accounting, and process costing is not yet sufficiently implemented (Sect. 8.4.6). Examples, with advantages and estimated potentials for cost reduction are itemized here:

- **Management of variants**
  A savings potential of 25 000 $ is realistic with complete repeatability, for **one** complex gear part found with repeated part search. This was established from process analyses in the company for resource consumption accruing one-time in the product realization process [Rei96a].
- **Time savings**
  The time decrease for the calculation of a part can be up to three weeks [Ste95]. This time reduction, though difficult to express in terms of costs, is nevertheless an important factor in competition.

10 Examples

The use of strategies, organization, and tools of cost management can be best understood with actual examples from industrial practice, especially if the examples are not from the reader's field of knowledge. Therefore these examples were not too complex and are understandable for engineers. Besides those shown here, other examples are spread throughout the book, though often not as fully explained.

The procedures and the circumstances of the examples correspond to actual cases. The numbers and the boundary conditions have been simplified and modified because of the limited coverage, to prevent any direct connection to companies.

The following overview is meant to facilitate orientation (see Sect. 4.5.2).

Section	Object (type of design)	Lowered costs	Attained through
10.1	**Concrete mixer** (Adaptive and partially new design)	Manufacturing costs Life-cycle costs	Target Costing Collaboration Concept, shape and production change
10.2	**Centrifuge base** (Adaptive design)	Manufacturing costs	Target Costing Collaboration of design and production planning Shape and production changes
10.3	**Bearing pedestal**	Manufacturing costs	Use and comparison of different quick cost calculation procedures

In other sections:

Section	Object (type of design)	Lowered costs	Attained through
4.7	**Marking laser** (New design)	Manufacturing costs	Methodical design, Target Costing, concept changes
5.5	**Labeler**	Lifecycle costs	Modular design Breakdown costs Market research Customer contacts
7.12.6.8	**Modular design of a plant**	Manufacturing, logistics and bid quotation costs	Combination of different modular design systems, computer use
7.12.6.9	**Modular design of sports cars**	Development, manufacturing, and logistics costs	Modular design Similar parts usage
7.12.10	**Modular design of tractors**	Development, manufacturing, and logistics costs	Modular design Similar parts usage
7.13.8	**Two-stage industrial gear drives** (Adaptive and variant design)	Manufacturing costs	Target Costing, use of FVA findings from Sect. 7.13 (benchmarking), shape and production change
7.14.3	**Coffee percolator** (Adaptive design)	Disposal costs	Procedure for lowering the disposal costs

10.1 Example "Concrete mixer"

10.1.1 Goal of the example

By using the redesign of a twin-shaft concrete mixer the essential aspects of the target cost-oriented design ("target costing") are presented here. We show how by a systematic cross-departmental procedure, costs can be adequately lowered to serve the market. The example involves single-unit or small-lot production in the machine industry.

Focal points are the determination and distribution of the target costs and an early estimation of the cost reduction potentials, to realize low-cost solutions (adaptive and new design).

10.1.2 Problem description

An important customer wants to place a large order for the production of twin-shaft concrete mixers if the price is lowered by at least 20%, otherwise he wants to set up the mixer for in-house production. The manufacturing firm has already used up its maneuvering room in price setting, so that only by a manufacturing cost reduction can the price objective of the customer be achieved. Furthermore, the competition has been in the market for some time, with a more economical plate mixer. Since about 50 concrete mixers per year are built by in-house production, the expense for the cost-cutting measures should be beneficial (Sect. 4.8.2).

The twin-shaft concrete mixers have already been produced in lot sizes of four to six units in the basic design for a long time. For every customer, adaptive designs involving the connections, premix feeding, concrete outflow, and so forth are necessary. Such revisions always lead to some improvements in the mixer. A undamental revision has not been implemented up to now.

Now the pressure from a customer has led to a redesign. This is dangerous, since at times it is too late for a response and the customer can switch to the competition. It is better to not wait so long, but rather to begin the necessary revisions early, from one's own initiative (Fig. 4.6-8).

10.1.3 Description of the competing products

A functional description of the mixers (**Fig. 10.1-1**) is as follows. They are installed as stationary mixers in mixing towers at concrete factories. In each batch, 1.25 m^3 of solid concrete are mixed, at 40 batches/hour. The controlled premix of gravel, water, cement, aggregates, and chemical additives is fed from above, and the ready-mixed concrete is emptied below with an emptying slide, into concrete trucks. There are mixers operating on two different principles on the market.

Own product: Twin-shaft concrete mixer (Fig. 10.1-1, right)
With this mixing principle, two horizontal counter-rotating mixing shafts equipped with mixing arms mix the concrete, in the lower third of the trough. So that the mixing arms do not get jammed, the mixing shafts must run synchronously. This mixing principle is to be retained because of its good operation and reputation in the market. The mixer is part of a product line with mix volumes of 0.75–9 m^3. That is not pertinent to this example.

Competing product: Plate concrete mixer (Fig. 10.1-1, left)
In the plate concrete mixer, the mixing arms are attached to a vertical mixing shaft. Mixing takes place over the whole surface of the plate. The mixing principle does not entirely achieve the mix quality of the twin-shaft concrete mixer. It is, however, of lower cost due to simpler design (only one shaft, one drive, etc.). To further improve the mix quality, an additional stirrer can be installed.

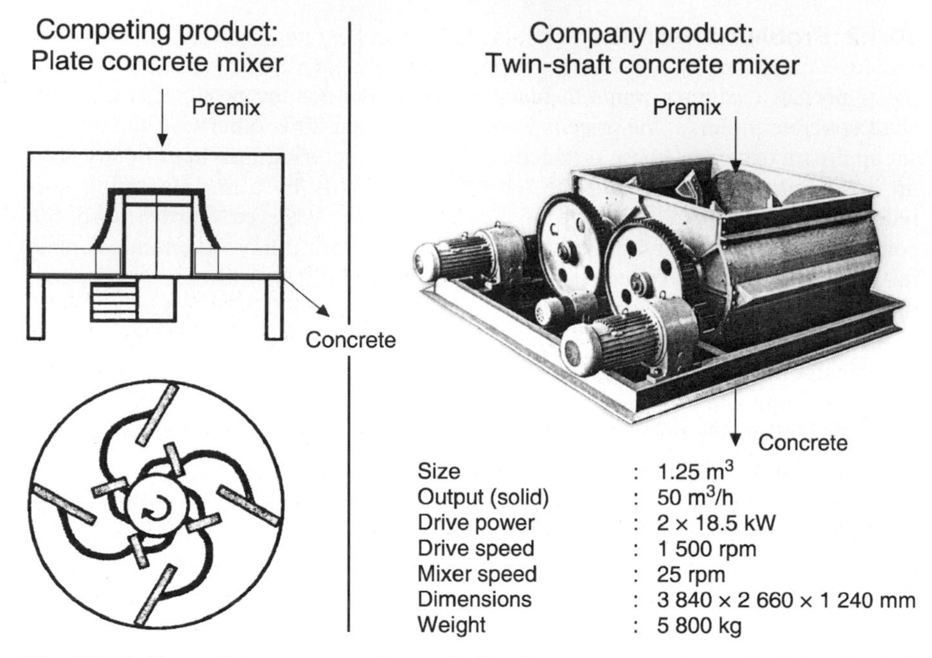

Fig. 10.1-1. Competing concrete mixers (*Left*) plate concrete mixer. (*Right*) twin-shaft concrete mixer (existing design)

10.1.4 Steps in the cost reduction project

The realization of the project took place according to the procedure cycle for cost reduction (Fig. 4.5-7). The following description is also similarly organized.

I Clarify the problem and procedure

I.0 Planning the procedure

Because of its importance, the deadline constraint, and the scope of the problem, the revision of the concrete mixer was carried out in project form. A project team was formed and they planned a project sequence with strict guidelines. The core team consisted of a design leader, a design engineer, a project engineer, a work process scheduler, a costing representative, the head of purchasing, and the assembly manager. It was complemented, as needed, by the sales manager and the works manager.

I.1 Determine the total target cost

In this specific case the customer required a lowering of the price by 20%; this could be achieved in a first approximation only by a corresponding lowering of the manufacturing costs. Because a basic revision was planned and it was desirable to have "play room" for later corrections, or that the earnings situation in this division of the company had to be improved, management required a manufacturing cost reduction of 30%, for an entire target cost of 93 100 $. Notice that the linear reduction of prices to manufacturing costs for cost reduction of products can only be vaguely hinted at because:

- Manufacturing costs are not directly related to the prices (Chap. 8).
- Processes that cause the difference between the total and manufacturing costs (about 20–50% of the total costs) cannot usually be modified through constructive measures; they are charged however, linearly, using overhead surcharges. With logical comprehensive target costing (Sect. 4.4.3 [Seid93]), the costs of hese processes must be included in the investigation and lowered just as logically.

I.1a Determination of the total target costs through order discussion

In single-unit and limited-lot production, the **order discussion** transpires with an offer from the manufacturer to the customer. It is one of the essential means of determining the customer requirements, wishes, and, of course, the price target or the target costs. The term **order discussion** means not only **one** individual discussion with a customer, but the often long process of inquiries up to the time when an order is ultimately received or missed.

In general, the customer starts with more or less comparable offers from competing manufacturers. Benchmarking, that is, a comparison of one’s own product with the best of the competition from the customer’s viewpoint, occurs at this point. However, the customer’s tactics enter here; he mixes reality and fantasy to his advantage. He combines the sum of all properties of the competing products with the respective lowest price.

The **first step** for the preparation of a **customer-oriented** requirements list is therefore the statement validation which renders the competing products technically and economically comparable.

A property checklist of the competing products, which was prepared from one’s own investigations and information from the literature and earlier order discussions (**Fig. 10.1-2**), is of help here.

In the order discussion, the customer recognized that the twin-shaft mixer has advantages over the plate mixer (better mix quality, easier maintenance and repair, better robustness and longer total service life) which can be quantified, although sometimes only with difficulty. These advantages are then particularly decisive for the purchase of the twin-shaft mixer when the higher purchase price is balanced by low wear costs. Thus, for example, after a year, breakeven point for the customer might be expected between the twin-shaft and plate mixers from the sum of (one-time) procurement and wear costs. Operating and maintenance costs, etc. are not considered further because they are similar for both types.

Properties	Plate mixer	Twin-shaft mixed (old design)
Mix quality	Good (without stirrer) (limited for coarse material)	Very good (for all grain sizes)
Mixing time	60 s	60 s
Installation space [mm]	ø 2 800, h = 1 400 (large ø, less headroom)	3 800 × 2 700 × 1 700
Energy requirements	0.6 kWh/m^3 (30 % more with stirrer)	0.6 kWh/m^3
Wear costs (speed at mixing implement)	0.9 \$/m^3 (higher with stirrer)	0.6 \$/m^3 (1.5 m/s)
Ease of maintenance and repair	For mixer good, for drive (underneath) not good	Good
Total life span	ca. 8 to 10 years	ca. 15 to 20 years
Factory price (incl. taxes)	134 000 \$ (Stirrer, 6 000-8 000 \$)	190 000 \$

Fig. 10.1-2. Property checklist for 1.25 m^3 concrete mixer

I.1b Determination of total target costs from the viewpoint of lifecycle costs

In **the second step**, the target costs from the viewpoint of the customer concerning the sales price and the wear costs of the new twin-shaft mixer were determined. This was achieved by an estimate calculation in which the interest on the capital was not considered.

Time to the breakeven point of the old twin-shaft mixer (TSM) vis-a-vis a plate mixer (PM) from viewpoint of the customer:

Difference in prices of twin-shaft and plate mixer:

$$\Delta P = 190\,000\,\$ - 134\,000\,\$ = 56\,000\,\$$$

From Fig. 10.1-2, at an expected utilization of the mixers at 40 000 m^3/year results in a difference in wear costs/year (ΔWC):

$$WC_{\text{TSM}} = 40\,000\,\text{m}^3/\text{year} \cdot 0.60\,\$/\text{m}^3 = 24\,000\,\$/\text{year}$$

$$WC_{\text{PM}} = 40\,000\,\text{m}^3/\text{year} \cdot 0.90\,\$/\text{m}^3 = 36\,000\,\$/\text{year}$$

$$\Delta WC = 12\,000\,\$/\text{year}$$

Time to breakeven point:

$$\Delta P/\Delta WC = 56\,000\,\$/12\,000\,\$/\text{year} = 4.67\text{ years}$$

To reduce this time to a year, a reduction of the manufacturing costs by 25% and a reduction of the wear costs WC by 10% to 0.5 \$/m^3 are seen as attainable. This assumes a proportionality of manufacturing costs MC and price P. The manufacturing costs are about 70% of the calculated sales price.

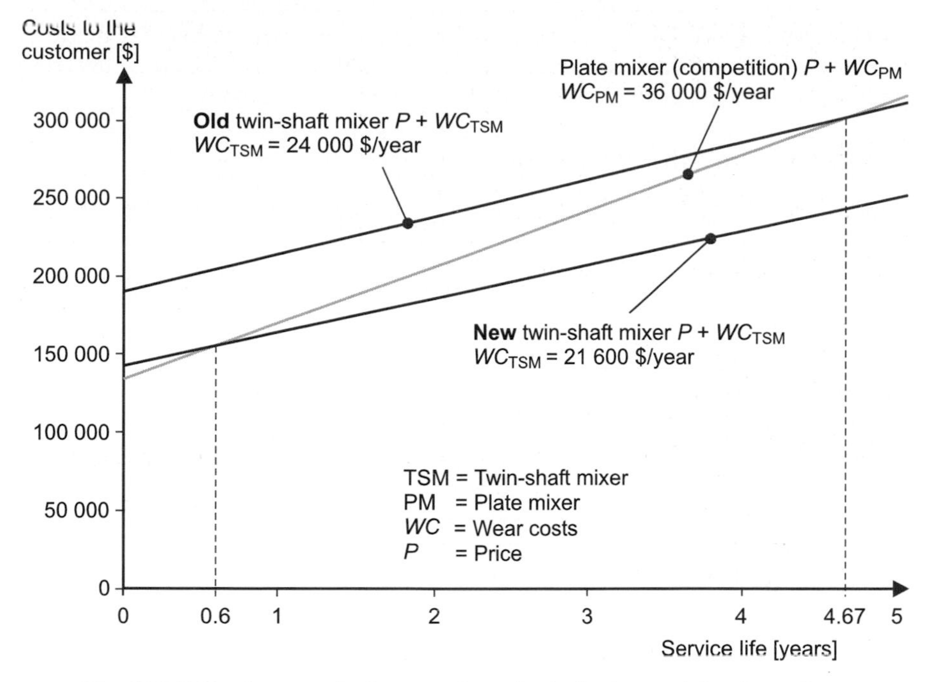

Fig. 10.1-3. Breakeven point between the twin-shaft mixer and the plate mixer

Then, the time to the **breakeven point** (**Fig. 10.1-3**) of the new twin-shaft mixer is found as follows:

With a new price for the twin-shaft mixer:

$$P_{\text{TSMnew}} = 190\,000\,\$ \cdot 0.75 \qquad = 142\,500\,\$$$

The price difference ΔP relative to the plate mixer becomes:

$$\Delta P = 142\,500\,\$ - 134\,000\,\$ \qquad = 8\,500\,\$$$

New difference in wear costs/year (ΔWC):

$$WC_{\text{TSMnew}} = 40\,000\,\text{m}^3/\text{year} \cdot 0.54\,\$/\text{m}^3 \qquad = 21\,600\,\$/\text{year}$$

$$\Delta WC = 36\,000\,\$/\text{year} - 21\,600\,\$/\text{year} \qquad = 14\,400\,\$/\text{year}$$

New time up to the breakeven point:

$$\Delta P/\Delta WC = 8\,500\,\$/14\,400\,\$/\text{year} \qquad = 0.6\ \text{years}$$

With that, the **target costs are evident from the life-cycle costs viewpoint:**

1. Reduction of manufacturing costs from 133 000 $ (190 000 $ · 0.7) by at least 25% to less than 100 000 $.
2. Reduction of the wear costs from 0.6 $/m³ by 10% to 0.54 $/m³.

The goal has been met, viz., achieve the breakeven point within a year. However, we should reckon with the fact that the competition is also active in lowering their costs. Thus the goals should rather be set too high than too low.

I.2a Distribution into sub-target costs, linearly for each assembly

After the total target costs for the complete twin-shaft concrete mixer are known, they have to be divided up into costs for the sub-targets. For that, the existing product must be organized according to its different aspects (features) and the costs split similarly. This leads to corresponding **cost structures** (Sect. 4.6.2) [Ehr85, VDI87].

Besides the distribution of target costs, **cost reduction potentials** and **change options** are also searched for during cost analysis. It is useful to study one's own product as well as similar products from one's own company. The competition products and their production should especially be included in the analysis. Since not all cost reduction potentials can be exploited in product development, for every cost reduction potential the party who is responsible for its realization must also be ascertained. In the case of design changes, it is the design group leader; for more production-related changes, it is the production manager; in the case of procurement

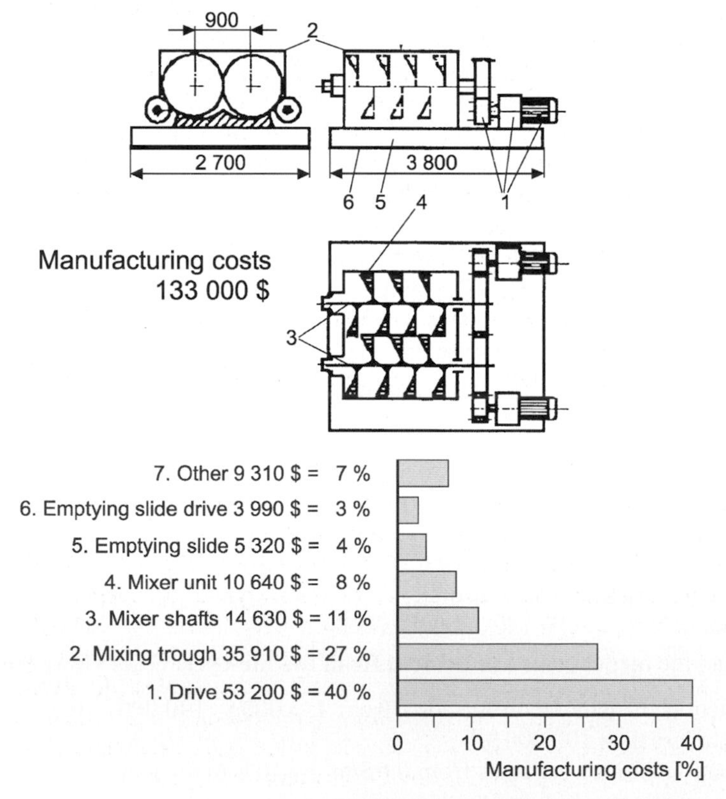

Fig. 10.1-4. Cost structure setup, according to assemblies (previous design)

changes, it is the purchasing head, etc. However, a competent project leader bears the overall responsibility for the new product including its success (Sect. 4.3.1).

At this point we represent only the cost structures of the entire twin-shaft concrete mixer according to assemblies, and of the drive unit according to its parts. **Figure 10.1-4** shows the manufacturing cost elements for the assemblies of the previous design. The essential assemblies ("A" parts) can be recognized. In the **first attempt**, we can distribute the necessary cost reduction **linearly** into the sub-target costs:

- Assembly 1 drive: 53 000 $ · 70% = 37 100 $,
- Assembly 2 mixing trough: 36 000 $ · 70% = 25 100 $, etc.

The assemblies 3 to 7 ("B" and "C" parts) are not considered here. The procedure description will be carried on here only for Assembly 1 (drive).

I.2b Distribution into sub-target costs according to customer functions (see Sect. 4.5.1.4)

The method of dividing the costs according to customer preferences, into components or groups, which was learned in target costing, is shown in this example of the concrete mixer [Seid93, Nie93, Rös96, Tan89]. During small-lot production it is not always appropriate to carry out the division as is required for target costing, in series manufacturing, and assignment of the target costs to individual properties desired by the customer (Fig. 4.5-2, part I.3). However, it should be pursued as far as possible to recognize the **properties** that are really desired and which the **customer is willing to pay for**, and to show the saving potentials. It is not a matter here of the "cent-precise" allocation of costs, but of the **recognition of central points**. Weightings are included in this division. The determination of these factors is a joint task of marketing, product development, services, etc., which requires intense discussion and several tests. It is also advisable to carry out this calculation with a simple spreadsheet.

First the most important **customer functions** are established for the concrete mixer (**Fig. 10.1-5a**).

There should be not more than ten, otherwise the calculations become too obscure and the statements too vague. The team **weights** the qualities from customer viewpoint (Fig. 4.5-4).

Comment: These customer functions are, from the methodical design viewpoint, not "functions" of a machine (in the sense of dealing with material, energy or signal flow), but rather **customer-relevant "properties"**.

Starting from the cost structure of the old design (Fig. 10.1-4), **Fig. 10.1-5b** shows which **portion every unit** has in the realization of the respective **customer function** (column total = 100%, similar to the function costs, Sect. 4.6.2).

In the table in **Fig. 10.1-5c**, the weighting of the customer functions determined in Fig. 10.1-5a is multiplied by the portions of the assemblies' involvement in the functional performance (Fig. 10.1-5b); for example, in the first field 0.26 · 20 = 5.2%. The lines are then summed in the last column and show the partial weighting for the assemblies from the customer viewpoint.

a) Weighting of the main functions from customer's viewpoint										
		Mix quality	Mixing time	Installation space	Energy costs	Wear costs	Maintenance and repair costs	Life span	Overload capacity	Sum
Functions		F1	F2	F3	F4	F5	F6	F7	F8	
Weighting		0.26	0.24	0.15	0.10	0.08	0.10	0.02	0.05	1
b) Contribution of assemblies to function fulfillment										
Drive	As1	20	20	30	50		10	20	40	
Mixing trough	As2	30	20	50	10	50	20	10	10	
Mixer shaft; bearings	As3	10	10	5	10	10	10	30	10	
Mixing unit	As4	30	30		25	30	50	30	40	
Emptying slide	As5	5	10			5	4	5		
Emptying slide drive	As6		5	5	5	5	4	5		
Other	As7	5	5	10			2			
Sum [%]		100	100	100	100	100	100	100	100	
c) Assembly significance = Weight function x Assembly portion (0.26 x 20 = 5.2, etc.)										
Weighting		0.26	0.24	0.15	0.10	0.08	0.10	0.02	0.05	1
Drive	As1	5.2	4.8	4.5	5.0	0.0	1.0	0.4	2.0	**22.9**
Mixing trough	As2	7.8	4.8	7.5	1.0	4.0	2.0	0.2	0.5	**27.8**
Mixer shaft; bearings	As3	2.6	2.4	0.8	1.0	0.8	1.0	0.6	0.5	**9.7**
Mixing unit	As4	7.8	7.2	0.0	2.5	2.4	5.0	0.6	2.0	**27.5**
Emptying slide	As5	1.3	2.4	0.0	0.0	0.4	0.4	0.1	0.0	**4.6**
Emptying slide drive	As6	0.0	1.2	0.8	0.5	0.4	0.4	0.1	0.0	**3.4**
Other	As7	1.3	1.2	1.5	0.0	0.0	0.2	0.0	0.0	**4.2**
Sum [%]		26.0	24.0	15.0	10.0	8.0	10.0	2.0	5.0	100.0

d) Determine target cost index for target cost control diagram							
Total cost target = 93 100 $		Weight portion target	Partial target costs	Actual costs [$]	Actual cost/total cost target	Target cost index	Comments
Drive	As1	**22.9**	21 320	53 200	57.1	0.40	too "expensive"
Mixing trough	As2	**27.8**	25 882	35 910	38.6	0.72	too "expensive"
Mixer shaft; bearings	As3	**9.7**	8 984	14 630	15.7	0.61	too "expensive"
Mixing unit	As4	**27.5**	25 603	10 640	11.4	2.41	too "cheap"
Emptying slide	As5	**4.6**	4 283	5 320	5.7	0.81	too "expensive"
Emptying slide drive	As6	**3.4**	3 119	3 990	4.3	0.78	too "expensive"
Other	As7	**4.2**	3 910	9 310	10.0	0.42	too "expensive"

Fig. 10.1-5. Cost split according to customer functions of the concrete mixer (page of a spreadsheet, here graphically adapted)

From the total target costs, the theoretical partial target costs can be determined with partial weighting for the assemblies and compared with the actual costs (**Fig. 10.1-5d**).

If we **divide** the **partial target costs by** the **actual costs** of the assemblies, we get the **target costs index**. Assemblies that have a target cost index of less than 1 are much too expensive from the customer's viewpoint, and there is a great need

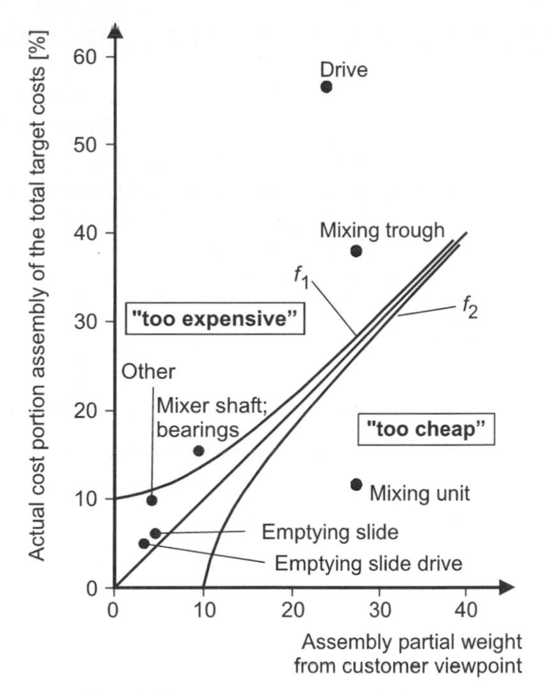

Fig. 10.1-6. Target cost control diagram

for cost reduction. If the target cost index is larger than 1, the customer estimates the value of the unit to be higher than that indicated by the actual costs. In the target cost control diagram (**Fig. 10.1-6**), the results are presented graphically. The actual cost portions of the assemblies are shown plotted against the partial weights from the customer's viewpoint. If they are above the angle bisector line, cost reduction is absolutely necessary. If they are below that line, the customer estimates the costs of the unit to be higher than they actually are.

In the target cost diagram [Rös96], two boundary curves $f_{1,2}$ are shown, defined by the functions:

$$f_{1,2} = \sqrt{x^2 \pm q^2}$$

f_1 = lower limit of the target cost curve ($-q^2$)
f_2 = upper limit of the target cost curve ($+q^2$)
x = partial weighting

q = Decision parameter for the definition of the target cost zone, set by the top management.

According to Tanaka [Tan89], they indicate the range in which the cost deviations balance out. The decision parameter q must always be determined in each case (here, 10). The **message** from the **target cost diagram** is: For components that account for a small portion of the total costs, greater deviations can be allowed between actual costs and customer weighting than with assemblies with higher cost portion. This is covered, in principle, by the statements for error compensation in Sect. 9.3.7.3. However, if we would enter the error compensation curves indicated there in the target cost diagram, a considerably narrower field would result.

In clarifying the calculation steps, it should not be forgotten that the procedure helps in comparing the most varied aspects ("apples and oranges") in a standardized form. The result cannot be understood as an exact numerical value. It is similar as the results of weighted point evaluation, the utility value analysis, and the FMEA would be. The mutual discussion in the team and the systematical comparison, as shown in Fig. 4.5-4, are important.

With the division of the target costs starting from the customer functions and the process of determining cost potentials described in I.2 above, the final partial target costs should be found. In proceeding further, it can happen that they must be modified again because the assumed cost reduction potentials could not be achieved for one assembly, or that for other assemblies greater cost reductions were possible.

➔ Attaining the total target costs is important! Therefore, the **partial target costs can and must be jointly adapted.**

I.3 Cost reduction potentials and final partial target costs for the drive

Function of the drive:

The drive transmits torque from the electric motors, with a speed change from 1 500 rpm, to the counter-rotating geared shafts running at about 25 rpm. The platform includes the drive assemblies and the mixing trough.

A precise analysis of the drive brought out the following **important points for the cost reduction** of the drive unit (**Fig. 10.1-7**, see Fig. 4.8-6 too), with the corresponding **questions from the team** for possible changes:

- The **drive platform** is expensive: Is it really needed? What other solutions are there (for example, a self-supporting design without a platform)?
- The **synchronizing gear set** is very expensive: Can purchase parts or other designs or principles (for example, other drive assemblies) replace it?
- The **drive motors** are very expensive: Are they replaceable through other purchase parts? Can the synchronizing be integrated? Are there other suppliers?
- Investigation of **production**: What are the available possibilities? What alternatives can be created by investment? Instead of small-lot production, can specific components be made by serial production?
- Change from **in-house to external production**?

We also see here that a careful analysis and discussion of the requirements motivate new solution ideas. The change options and the resulting costs were estimated so that, as partial target costs for the drive assembly a saving of at least 36% = partial target costs, 34 100 $ was found. That is more than the "customer value" and less than a linear cost reduction over all assemblies would show, but it appears realistic. The same procedure was carried out for other assemblies. To attain the partial target costs, the parties responsible were named (**Fig. 10.1-8**). For

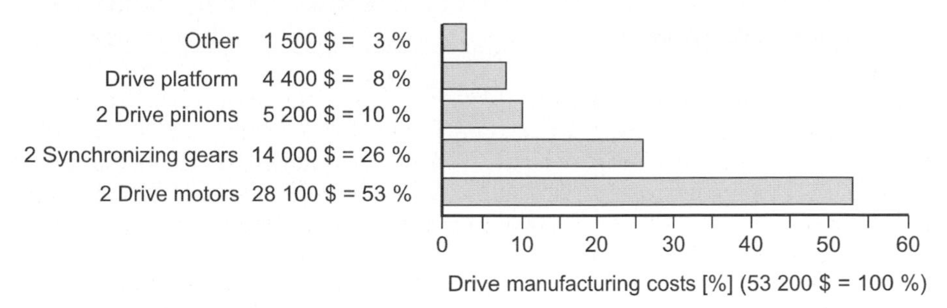

Fig. 10.1-7. Actual cost structure of the drive (see Fig. 10.1-1; 10.1-4)

the change of the drive principle, it was the chief designer; for the search for suppliers, it was the head of purchasing.

II Solution search

To attain the large cost reduction for the drive, alternative methods of implementation (new concepts, new design) had to be sought and preliminary designs created. **Figure 10.1-9** shows five possible alternatives that fulfill the demands on the drive. A self-supporting design of the mixing trough was the best choice (the drive platform is omitted, thus the mixer also becomes smaller and lighter).

It should be noted that by using worm gears their losses require an increase in the driving power of about 20%.

Parallel to the solution search, a **cost pre-checking** was realized by the close collaboration of design, production planning and cost calculation departments in the **target costing team** [Seid93]. The manufacturing costs were calculated concurrently, bids obtained, etc. Thus it was ensured that the needed information flowed into a short-loop feedback control system and the given target costs are also reached.

It is expedient for this purpose to pursue the costs in the form of a cost search table; with PCs today, this is easily done with spreadsheets. As shown in Fig. 10.1-8, the assembly or part of the product are entered, with actual and target costs. This list is complemented by the corresponding data of competition products, as far as they are available. The measures that are planned for the cost reduction are entered in another column. At the next deadline, the costs arising to that point are entered. We recognize immediately where the targets were reached and where not, and can accordingly plan further measures. This list is continually updated to the end of the project.

1. Session: Analysis

Twin-shaft mixer (TSM)	Actual costs TSM	Share TSM	Actual costs Plate mixer	Partial target costs TSM	Required cost reduction	Cost reduction potential	
Assemblies	[$]	[%]	[$]	[$]	[%]	Measures	Responsibi
Drive	53 000	39.8	32 000	34 100	-36	Other principl.	Developm
Mixing trough	36 000	27.0	27 000	21 600	-40	Welded design	Devl.+Proc
M.shaft + Brngs.	15 000	11.0	10 000	12 750	-15	Simplify	Developm
Mixing unit	11 000	8.1	8 000	9 400	-15	Prod. changes	Production
Emptying slide	5 200	3.9	5 000	4 300	-17	Simplify	Developm
Empt.sld.driv.	4 000	3.1	3 800	3 400	-15	Simplify	Developme
Other	9 000	7.0	8 000	7 550	-16	Simplify	Developme
Sum	**133 200**	100	93 800	93 100	**-30**		

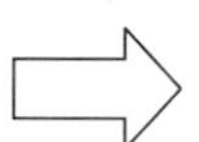

2. Session:

Stand TSM	Achieved cost reduction	New measures	
[$]	[%]		Responsibi
38 000	-28	Other suppliers	Purchasing
23 000	-36	Prod. changes	Production
11 830	-21	ok	
10 000	-9	Prod. changes	Production
3 400	-35	ok	
3 400	-15	ok	
7 800	-13	ok	
97 430	-27		

more sessions ...

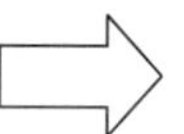

Results:

Costs TSM	Achieved cost reduction
[$]	[%]
31 700	-40
18 000	-50
11 800	-21
8 400	-24
3 400	-35
3 400	-15
7 800	-13
84 500	**-37**

For space reasons the table has been "broken up". With a spreadsheet analysis it can be adapted to requirements in practice.

Fig. 10.1-8. Search table for costs for the concrete mixer (s. Fig. 4.8-6)

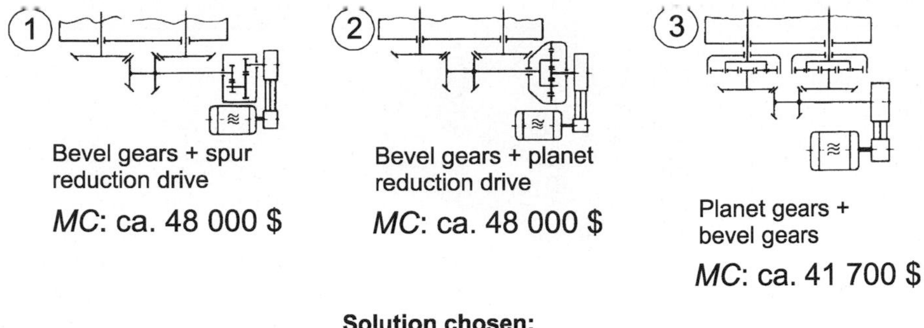

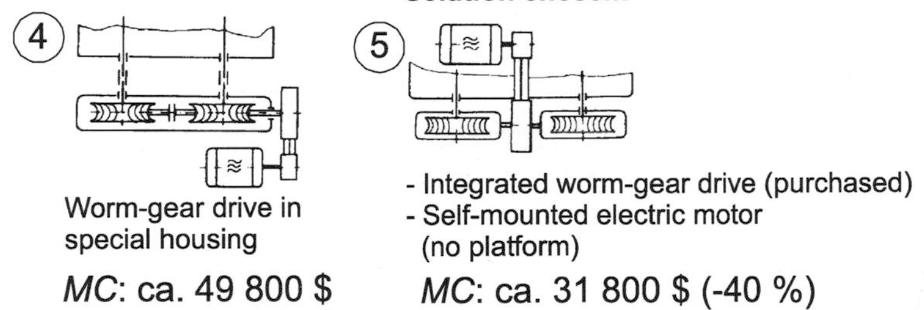

Fig. 10.1-9. Alternative drive solutions

The solution finally realized had to be selected from the different preliminary designs using cost and functional criteria. Finally, the drive choice came to solution 5 shown in Fig. 10.1-9, with the integrated worm gear drive (the drive platform is eliminated). These are driven by V-belt pulleys on the worm shafts and synchronized by a flexible coupling. The electrical motor is set directly on a clamping yoke on the mixing trough (**Fig. 10.1-10**).

For the mixing trough, a completely welded design was chosen. Machining on the boring mill for the mixer shaft bearings is eliminated by the use of welded bearing inserts.

IV Realization of the design

During the design some minor changes were still made, which barely affected the manufacturing costs. The alignment of the shafts (±0.5 mm) could be held with the welded design by the use of a fixture. The fixture costs must be included in the cost considerations.

The changes for the size considered here could be applied over the entire product line; for the larger mixers, two separate drive motors are used.

V Production and testing

During the tests, it was found that due to the shock insensitivity of the worm gears and the mixing trough, the mixer allows a 25% overloading so that a higher performance could be achieved. In addition, the higher energy costs are thus compensated for by the worm-gear drive.

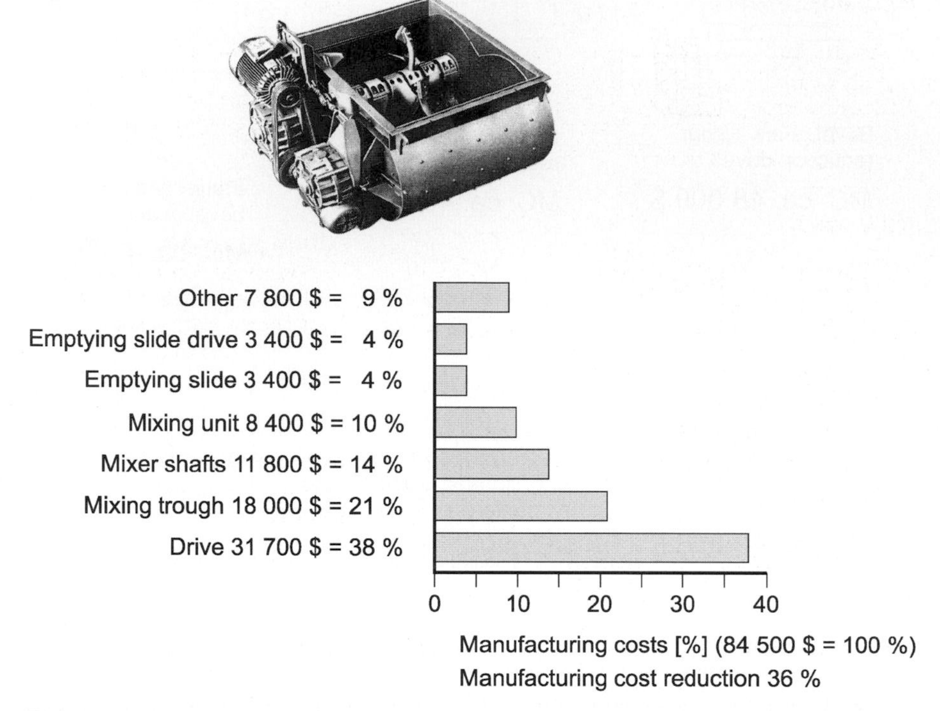

Fig. 10.1-10. Manufacturing cost structure of the twin-shaft concrete mixer (new design), by assemblies

The result of the target cost-oriented design is a smaller, lighter (by about 80%), cheaper (by 64%) mixer, as shown in **Fig. 10.1-11** for comparison. In addition it is 125% stronger, less noisy (because of the worm gears), and needs less maintenance. The saving with the basic design is so large that the costs of the entire line of products can probably be lowered by at least 20%. The constructive redesign of the concrete mixer, carried out with an interdisciplinary team and driven by target costing, was therefore a complete success. Through the intense mutual cooperation, the development time, which was usually 18 months, could be reduced to 8 months (see Fig. 6.2-2).

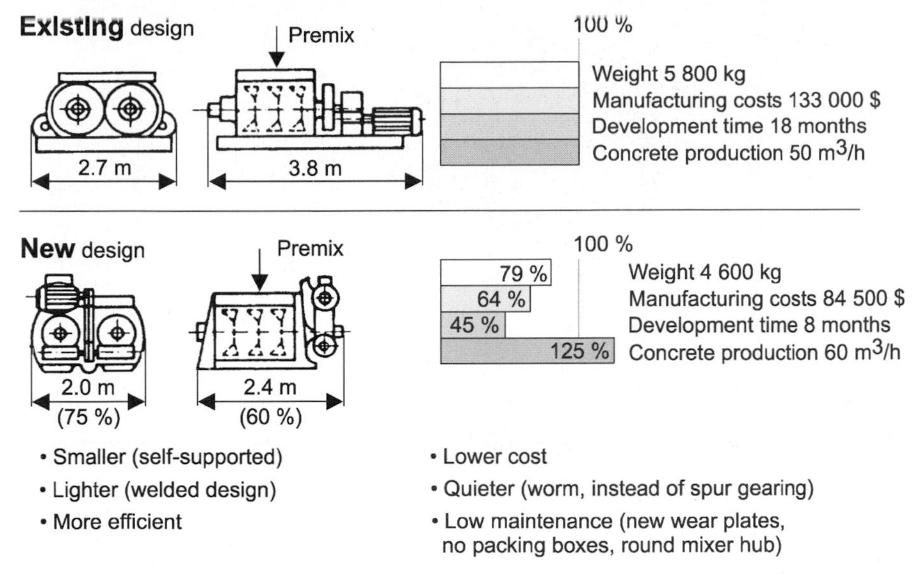

Fig. 10.1-11. Comparison of the old and new twin-shaft concrete mixers

10.1.5 Summary and conclusions of the example

Figure 10.1-12 schematically shows the process sequence of the example in a summarized form [Ehr95]. A team is formed. The target costs are split up into partial target costs. For attainment of the partial target costs of the individual assemblies, procedures are ascertained and the parties responsible are designated.

The measures are implemented by a new design. According to their magnitude, the costs of the new design for "A" and "B" parts are calculated relatively precisely, but only estimated for "C" parts (see Pareto analysis, Fig. 4.6-4). At set intervals the team meets and reports on the status of the project. If it is recognized that with specific assemblies the target is not reached (Drive A1, mixing trough A2, mixing unit B2), these are reworked until the target costs are reached or the team decides on other cost-cutting measures [Ehr87a].

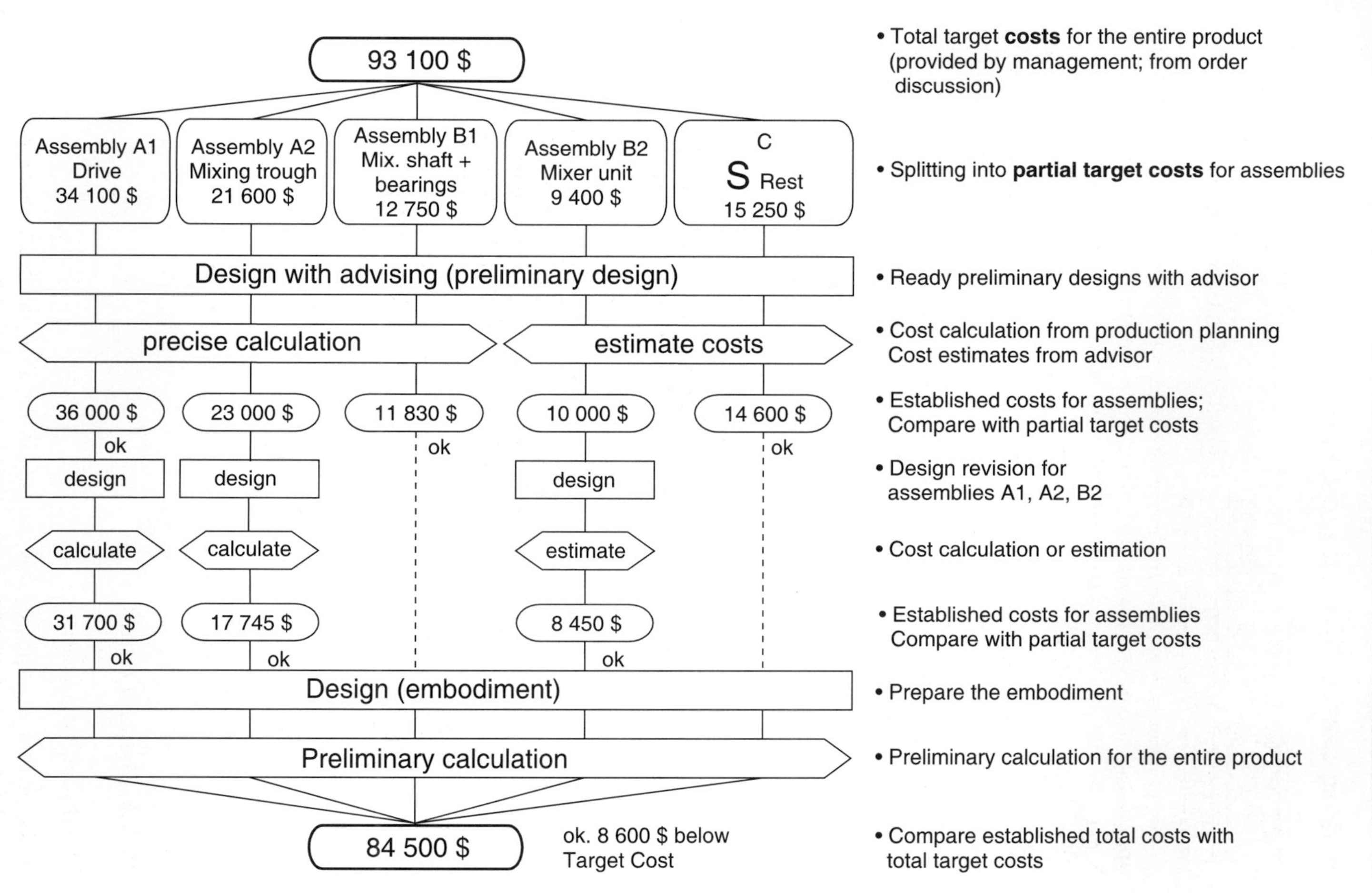

Fig. 10.1-12. Target cost oriented design [Ehr06]

10.2 Example "Centrifuge base"

10.2.1 Introduction

The procedure for low-cost design will be shown with this example of a welded centrifuge base for a process apparatus.

The product was already on the market. It is to be redesigned for the changing market conditions and is expected be produced in the company's own shops to lower the manufacturing costs, without a significant change in the function. If possible, the welded design is to be retained to use the factory's welding shop to capacity. The type of task is an adaptive design (Sect. 4.5.2).

The **function of the centrifuge base (Fig. 10.2-1)** is to accommodate the support bearings of a drum (centrifuge) in the upper openings. The lower part serves as an oil tank and holds the drive motor and the various additional parts and connections. The entire centrifuge apparatus is attached (on the right) to the centrifuge base.

Here also, the steps follow the procedure cycle (Fig. 4.5-7); the rest of the procedure is organized accordingly.

Fig. 10.2-1. Previous centrifuge base (Variant 0)

10.2.2 Clarify the task

I Find total target costs

In the present example, a large increase in the sales of these centrifuges was expected for the coming years. So that the product remains competitive on the market on a long-term basis and that the market coverage is enlarged, a manufacturing cost reduction of 10% is to be implemented on the entire product. Since the design engineer alone would be overtaxed with the job, a project team was formed, consisting of the design engineer (project leader), a process planner, a cost estimator, the supervisor of the welding shop, and, at times, the welder.

I.1 Divide up the total target costs by analysis of previous or similar machines

The distribution of the total target costs can occur based on individual cost categories, assemblies, or functional entities. For the manufacturing cost portions distribution of the total target costs can occur based on individual cost of the assemblies, 15% fell on the centrifuge base in the previous design. According to a Pareto analysis, the most expensive assembly was not the centrifuge base, but it was decided to consider its cost reduction potentials.

Assemblies with functional emphasis (for example, the centrifuge drum, the motor, etc.) are generally more difficult than are free-form assemblies (e. g., housing parts) to alter much without affecting the function; the latter were therefore preferred. So a greater cost reduction (20%) was assumed for the selected centrifuge base assembly.

I.2 Investigate focal points for cost reduction for the assembly

For the existing centrifuge base, a cost structure was set up first for individual cost elements and process cycles. Complementing that, other parameters were examined that affect cost (e. g., part quantities, weld length, etc.; see **Fig. 10.2-2**).

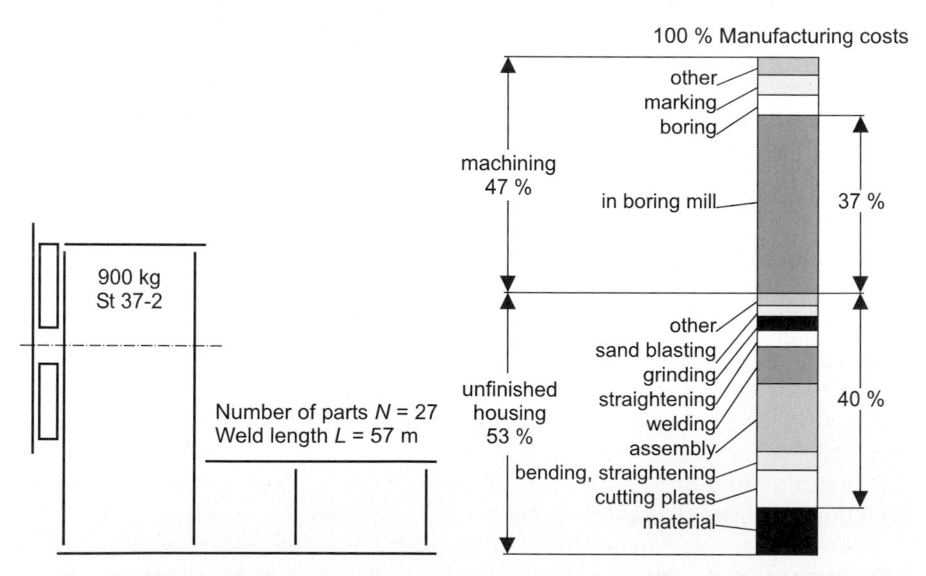

Fig. 10.2-2. Manufacturing cost structure of the existing centrifuge base (Variant 0)

The analysis of the cost structure, particularly of the high-cost parts, was very extensive here, as in most cases. The comparison was carried out with other production possibilities and the rules for the low-cost design brought out the first important findings:

- **Unfinished part material costs**: The portion of the material costs is relatively small so that purely a weight or material saving probably would not bring about any great cost savings. Rather the question here is, whether a material that is easier to process could be found (may be more expensive).
- **Unfinished part production costs**: The process cycles of assembling, welding, sheet metal cutting, and straightening constitute about 40% of the manufacturing costs. A simpler and cheaper solution was discussed in the team: to reduce the part quantities and weld lengths through more beveling, edging, bending, and the choice of some other semi-finished parts.
- **Machining** (production costs): Machining on a boring mill is very expensive. An investigation determined that the machine (boring mill) is indeed expensive and that the process controlling factors are unfavorable. For example, the worktables must be machined further (additional machining and idle times) and the boring spindle projecting length is very large due to the shape of the centrifuge base. Therefore, while machining from the right, the depths of cut must be greatly reduced. A decrease in the number of machined faces would also be helpful.

The tolerance requirements and the primary measurements could be changed very little because of the functions to be fulfilled. This point is in general very important since significant cost savings without large expenditures are often possible here. For the centrifuge base, the processing requirements could be reduced only for covers (flat parts) by the use of elastic seals.

The influence of the lot size on the manufacturing costs of the centrifuge base is small in the range of interest ($N = 4$), so a lot size increase would not significantly influence the production costs.

II Solution search

II.1 Cost reduction potentials and possibilities

Essentially the following possibilities were explored:

- **Manufacturing/production**: The machining costs of weldments can be lowered when the expensive and not absolutely necessary boring mill job is replaced by turning on a lathe (machining on vertical mill or welding of preturned bearing blocks with fixtures).
- **Material**: The use of uniform sheet metal thickness and optimization of gas cutting produce less waste. Furthermore, it is preferable to use a slightly more expensive material which is more favorable from a production viewpoint, since materials cost is a small portion of the total cost (e. g., St 52-3 provides easier processing and less delay).

- **Shape design**: Decrease of the number of parts and weld volumes (thin joints inside) leads to lower production costs of the unfinished welded part (especially important with small unfinished parts). Parts number reduction is achieved through beveling, edging, bending, and use of semi-finished section stock.

Two fundamentally different production philosophies can be pursued during the production of unfinished welded parts:

1. Produce sheet metal parts to exact size: Through higher accuracy during the production of the parts, together with more beveling, edging, and bending, there is less expense during assembling and welding.
2. Produce sheet metal parts less accurately: With higher accuracy during the production of the parts, together with higher number of parts and longer weld lengths (simple sheets), the expenditure increases enormously during assembling, straightening, and welding.

The second path was chosen in the earlier design. Now the attempt will be to go with the first choice, to reduce the costs for assembling and welding.

II.2 Propose alternative solutions

Beginning with the possibilities that were explored, three alternatives (differential design) were proposed (**Fig. 10.2-3**):

Variant 1

Here cost savings were achieved for the unfinished part by decreasing the number of parts (from 27 to 14) and weld length (from 57 to 23 m) in comparison to the initial variant 0. The smaller number of parts simplifies assembling and straightening. A beveled hood plate inserted into the forehead walls (beveling is in general cheaper than bending) replaces the outside bent and welded hood plates and uniform sheet thicknesses of steel St 52-3. Together with the optimization of the gas cutting strategy and the welding, the production costs of the unfinished part could be reduced to 70%. The weight is now about 1 000 kg instead of 900 kg.

Variant 2

Starting from variant 1, the next cost savings were in the boring mill work, achieved by separating the centrifuge base into two compound parts. The bearing holes can be made horizontally on a vertical mill and are then welded together. By that, production costs from machining reduce to 32% of the initial costs (variant 0). Due to the additional parting line (joint), the number of parts and the weld length increase, so that the production costs of the unfinished part increase from 70% to 76% of the initial value (variant 0).

Variant 3

As an alternative to welded parts a centrifuge base with a parting line and bolted joint was proposed.

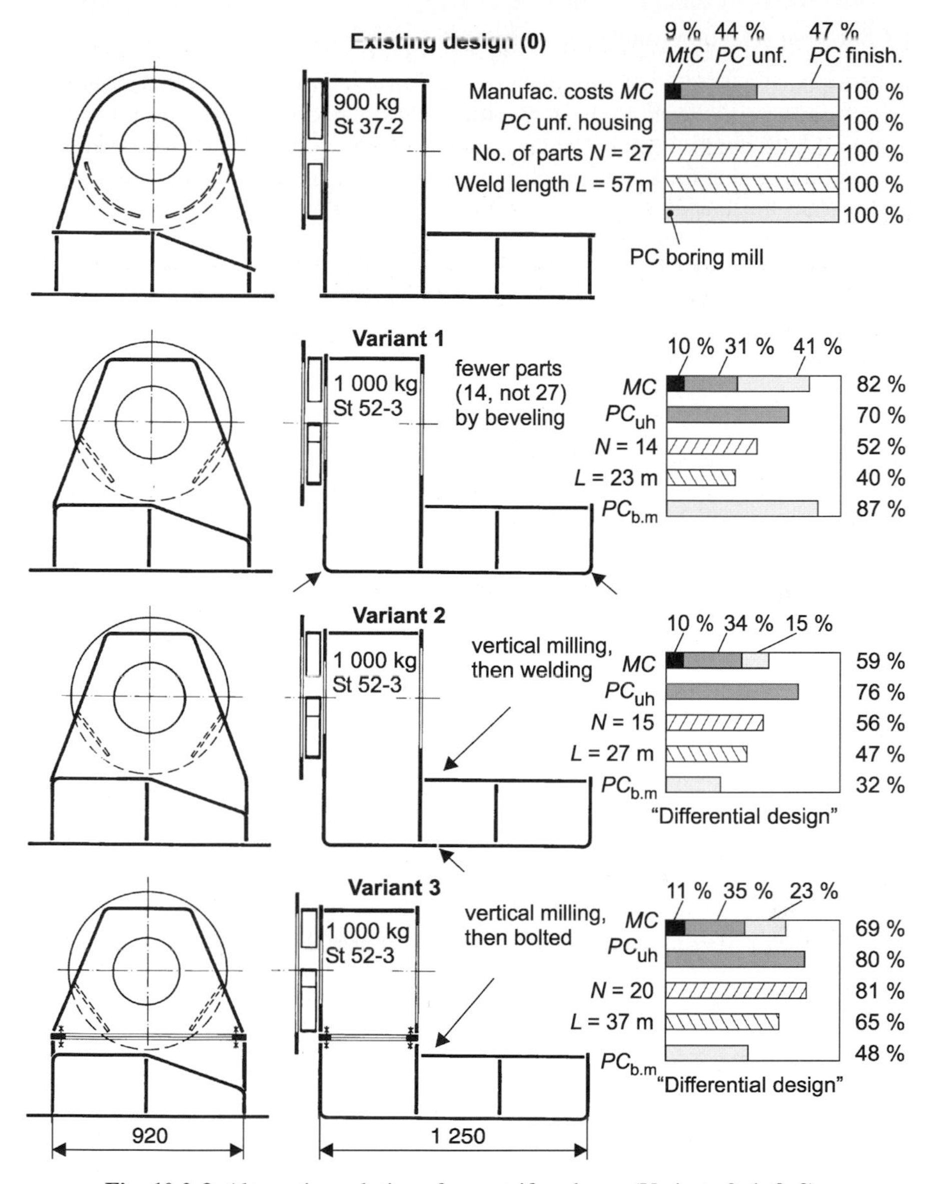

Fig. 10.2-3. Alternative solutions for centrifuge bases (Variants 0, 1, 2, 3)

The reason for this was that an unacceptable deformation was feared in the welding of the actual centrifuge base with oil tank of variant 2. Due to that, the production costs increase from 76% to 80% of the initial value (variant 0) in comparison to variant 2 because of the greater number of parts and weld length for the unfinished part. The additional machining of the joint surfaces causes the production costs of machining to grow from 32% to 48% of the initial variant 0.

III Solution selection (analysis, evaluation, selection)

Because the manufacturing firm had no capability to make quick cost calculations, the manufacturing costs of the three alternatives were calculated by production planning only approximately. In the evaluation no serious functional differences were found, so that the selection fell to variant 2 (59% of the initial manufacturing costs) from the manufacturing cost point of view. Variant 2 represents an ideal compromise for the centrifuge base due to low costs for unfinished part and the machining. **Figure 10.2-4** shows the results of the preliminary calculation.

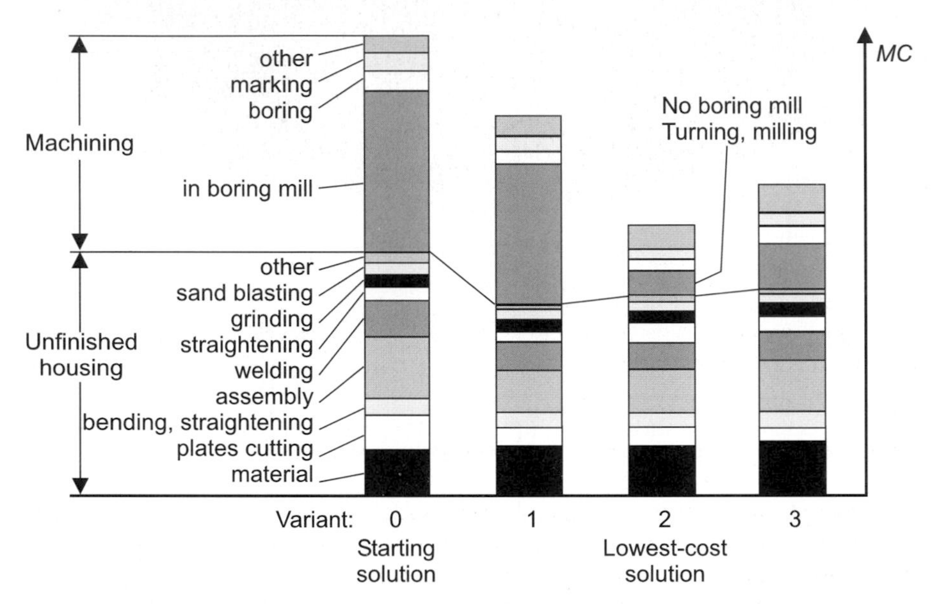

Fig. 10.2-4. Manufacturing cost structure of the centrifuge base variants 0, 1, 2, 3

IV Realization of the design

With variant 2, additional measures were taken in agreement with manufacturing specialists to reduce the deformation of the two centrifuge base parts for welding (**Fig. 10.2-5**). The initial objections because of the danger of the deformation (in manufacturing) were later found unwarranted. The tolerances could be kept.

V Production and testing

Tests showed that the centrifuge base was easy to combine with other assemblies and fulfilled its function entirely. During the follow-up calculation it came out that due to training effects during production, the manufacturing costs of the new centrifuge base were only about 50% of the initial manufacturing costs (Sect. 7.5.2b). With that, the goal of saving of 20% for the centrifuge base was clearly exceeded. The manufacturing costs of the entire machine were thus lowered by 7.5%. With

Fig. 10.2-5. New centrifuge base (variant 2)

that, the savings still expected of the other assemblies were reduced to 2.5%, which were also easily reached.

10.2.3 Summary and conclusions of the example

The example shows that a **competent team** is a prerequisite for a realistic **task clarification**. **Cost reduction potentials** can be derived from the recognized **cost focal points** (cost structure). Important elements are the teamwork through which more and better solutions are found, and the **search for several solutions**, from which the lowest cost solution can be selected. Equally important is the **concurrent calculation**, from which one recognizes whether the target costs have perhaps been reached, and which always shows the approaches for lowering the costs. Including the welder in the team was valuable because the welder drastically showed the great difficulties in the assembly of the initial design: "A housing like a house of cards". The target-oriented procedure led faster to good results.

10.3 Example for the application and comparison of quick cost calculation procedures: "Bearing pedestal"

10.3.1 Introduction

The **purpose of this example** is to show how, for a simple object, we determine the manufacturing costs with the conventional preliminary calculation, and how it compares to some quick cost calculation procedures.

The selected object was a **bearing pedestal** that supports a shaft on rolling bearings and is attached to a base with four bolts. The bearing pedestal was designed in three variants: as a **cast design** (GCI, **Fig. 10.3-1**), as a **weldment** (**Fig. 10.3-2**), and made of **solid material** (**Fig. 10.3-3**). Furthermore, the **lot sizes** (1, 5, 20 pieces) and the **dimensions** (half and double size) will be varied.

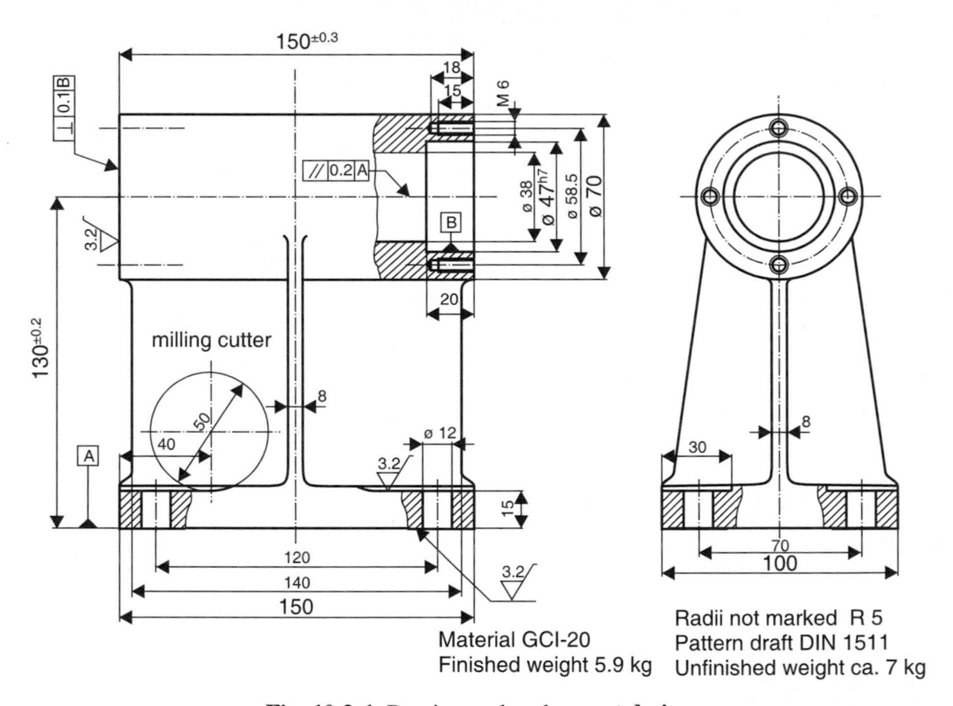

Fig. 10.3-1. Bearing pedestal as **cast design**

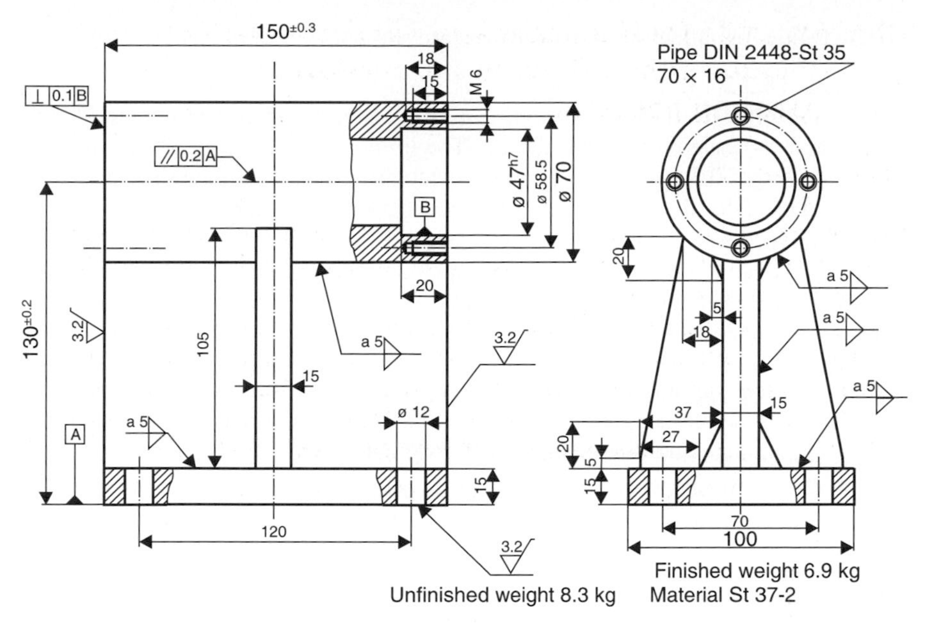

Fig. 10.3-2. Bearing pedestal as **welded design**

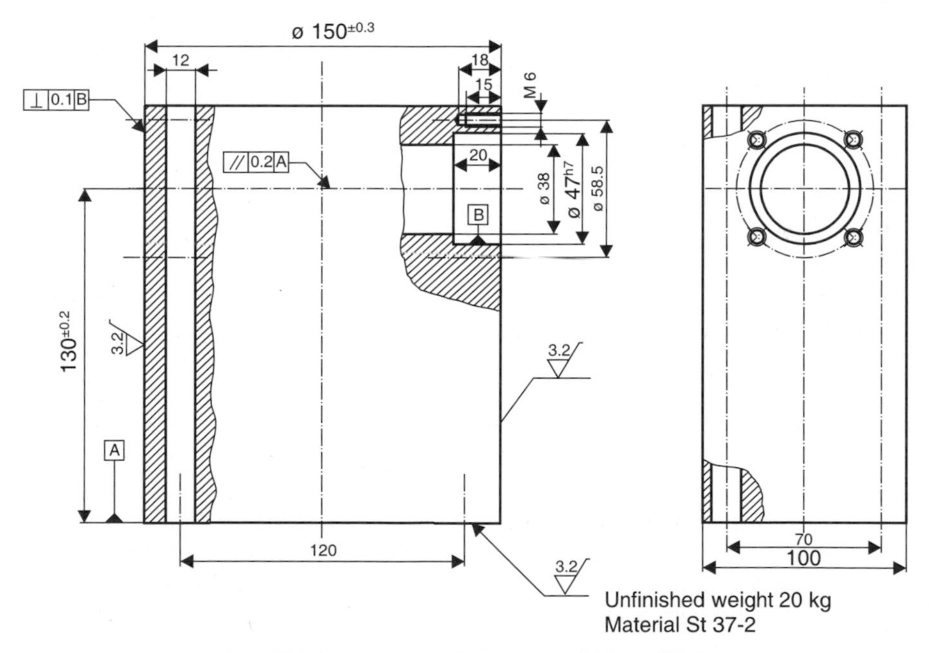

Fig. 10.3-3. Bearing pedestal from **solid material**

The following calculation procedures are shown:

- **Section 10.3.2**
  Conventional preliminary calculation of all three variants with a production plan and workstation cost calculation. The limit quantities of the design are calculated (Fig. 10.3-5). The preliminary calculation is at the same time the comparison basis for the application of the quick cost calculations.
- **Section 10.3.3**
  Only the example of the **welded design** shows how quick cost calculation procedures are used, and/or which results are obtained with that in comparison to the preliminary calculation:
  The **weight cost calculation** (Sect. 9.3.2.1), the calculation with **summary** and with **differentiated cost growth laws** (Sect. 9.3.5)

10.3.2 Determination of costs in production planning and preliminary calculation

To determine the actual manufacturing costs and the cut-off quantities between the variants, in **Fig. 10.3-4** the production plans and preliminary calculations of the bearing pedestals for lot size = 1 are combined.

With the **cast design** example (Fig. 10.3-1), the steps of work scheduling and preliminary calculation for the **production of one piece** will be explained:

- The **material direct costs** are calculated by multiplying the weight cost unit rate by 1.25 \$/kg for the cast part with the unfinished weight of 7 kg:

  $MtDC = 7 \cdot 1.25 = 8.75$ \$/piece

- The **material overhead costs** are determined by multiplying the material direct costs by the material overhead surcharge factor (*MtOCS*) of 15%:

  $MtOC = 8.75 \cdot 0.15 = 1.31$ \$/piece

- The **material costs** result from adding material direct and overhead costs:

  $MtC = 8.75 + 1.31 = 10.06$ \$/piece

- The **pattern costs** of 500 \$ must still be considered. If only one bearing pedestal is made, these pattern costs must be charged to this one pedestal. If a greater number is produced, the pattern costs are spread over the number of pieces.
- For the **preprocessing of the casting**, a setup time $t_s = 10$ min per lot and a production time $t_e = 10$ min per piece are given. This time multiplied by the workstation cost rate $WSC = 0.50$ \$/min (production wage and production overhead costs in one rate, Sect. 8.4.5) gives the production (preprocessing) costs PC_P:

  $PC_P = (10 + 10) \cdot 0.50 = 10$ \$/piece

a) Cast design				1 piece/lot		5 pieces/lot		20 pieces/lot
Pattern costs				500.00		100.00		25.00
Material direct costs *MtDC*	7 kg		1.25 $/kg	8.75		8.75		8.75
Material overh. costs *MtOC*			15 %	1.31		1.31		1.31
Material costs *MtC*				**10.06**		**10.06**		**10.06**
Production plan	t_s	t_e	*WSC* [$/min]		t_{s5}		t_{s20}	
Preparation	10	10	0.50	10.00	2	6.00	0.5	5.25
Boring mill	30	40	1.00	70.00	6	46.00	1.5	41.50
Production costs *PC*				**80.00**		**52.00**		**46.75**
Manufacturing costs *MC*				**590.06**		**162.06**		**81.81**
b) Welded design				1 piece/lot		5 pieces/lot		20 pieces/lot
Material direct costs *MtDC*	8.3 kg		1.50 $/kg	12.45		12.45		12.45
Material overh. costs *MtOC*			15 %	1.87		1.87		1.87
Material costs *MtC*				**14.32**		**14.32**		**14.32**
Production plan	t_s	t_e	*WSC* [$/min]		t_{s5}		t_{s20}	
Cut to size	15	40	0.60	33.00	3	25.80	0.8	24.45
Tack & weld	15	30	0.75	33.75	3	24.75	0.8	23.06
Deburr & straighten	15	10	0.75	18.75	3	9.75	0.8	8.06
Boring mill	30	30	1.00	60.00	6	36.00	1.5	31.50
Production costs *PC*				**145.50**		**96.30**		**87.08**
Manufacturing costs *MC*				**159.82**		**110.62**		**101.39**
c) Solid material				1 piece/lot		5 pieces/lot		20 pieces/lot
Material direct costs *MtDC*	20 kg		1.50 $/kg	30.00		30.00		30.00
Material overh. costs *MtOC*			15 %	4.50		4.50		4.50
Material costs *MtC*				**34.50**		**34.50**		**34.50**
Production plan	t_s	t_e	*WSC* [$/min]		t_{s5}		t_{s20}	
Cut to size	15	10	0.60	15.00	3	7.80	0.8	6.45
Boring mill	30	30	1.00	60.00	6	36.00	1.5	31.50
Long hole boring	10	20	1.00	30.00	2	22.00	0.5	20.50
Production costs *PC*				**105.00**		**65.80**		**58.45**
Manufacturing costs *MC*				**139.50**		**100.30**		**92.95**

Fig. 10.3-4. Work plans and preliminary calculations for the bearing pedestals

- For the **complete machining** (bearing bore, supporting surface, etc.) on a **boring mill** a setup time $t_s =$ of 30 min (per lot) and a production time $t_e = 30$ min/piece are determined. The workstation cost rate for the boring mill amounts to $WSC = 1$ \$/min. That gives the production costs PC_M:

  $PC_M = (30 + 40) \cdot 1 = 70$ \$/piece

- With that, in the case of production of only **one bearing pedestal** the manufacturing costs are the sum

  $MC_1 = 500 + 10.06 + 10 + 70 = 590.06$ \$/piece

For the production of **5** or **20 pieces**, the cost/piece decreases considerably, as indicated in Fig. 10.3-4, to

$MC_5 = 162.06$ \$/piece and $MC_{20} = 81.81$ \$/piece,

since the production costs from setup times and the pattern costs are divided by the number produced per lot. Only the production costs from production times for each piece are independent of the lot size. It is assumed that the production processes do not change with the quantity. If very large quantities were considered, then the design and the manufacturing of the bearing pedestal would be different, and thus the production time and cost would change too. This is true even with similar production processes (for example, in changing from sand casting to permanent mold casting).

The work schedules and cost calculations for the other designs (solid material and welding, Fig. 10.3-4) are carried out accordingly. The costs per piece do not reduce for the lot sizes of 5 and 20 pieces/lot as strongly as for the cast design because no pattern is needed.

If we calculate the quantities from 1 to 20 continuously, we obtain the cut-off quantities (**Fig. 10.3-5**). We recognize that the cast design is more economical from a cut-off quantity of 11 pieces and higher, than the welded design and from 14 pieces on, more economical than the design with solid material.

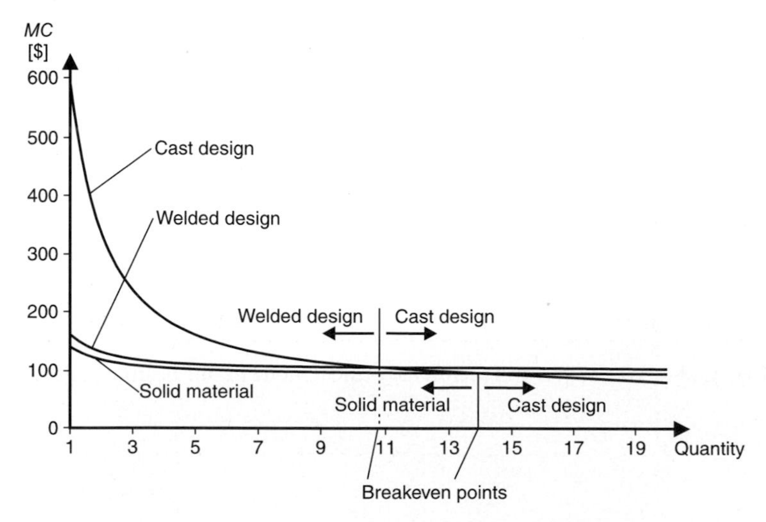

Fig. 10.3-5. Crossover quantities for bearing pedestal variants

The solid material design is, in this example, always more economical than the welded design. The curves have, however, very flat intersection points. That is, that in case of smaller changes (e. g., of the production times) the cut-off quantities change. These quantities are valid only for this example.

10.3.3 Weight cost calculation for welded design; lot size = 1

Figure 10.3-6 shows for the cited company a weight-cost curve found from available parts for MC_w of welded parts that slopes down (small parts cost relatively more than large parts with reference to the finished weight, Sect. 9.3.2.1). The weight of the welded design is 6.9 kg. For this weight, one gets from the curve in Fig. 10.3-6 a weight-cost rate of $MC_w = 19$ \$/kg and determines the manufacturing costs MC by multiplying with the weight:

$MC = 6.9 \cdot 19 = 131$ \$, i. e., $MC =$ about 130 \$.

In the preliminary calculation shown above, the manufacturing costs were found to be about 160 \$, the deviation here is −30 \$, therefore about −19% (Fig. 10.3-4). That can be sufficient for a cost estimate covering several different parts if it is assumed that in a more complex product there are also parts for which weight cost calculation sets the costs too high. The accuracy could be increased if different weight cost curves for different parts (e. g., simple versus complicated) are used. In the same manner, the welded and cast designs (with different weight cost unit rates for each) could also be processed.

If the number is increased to 5 or 20 pieces, the manufacturing costs found with weight cost calculation remain constant at 130 \$ since the quantity does not have any effect in weight cost calculation. The costs determined from the preliminary calculation are $MC_5 = 111$ \$ and $MC_{20} = 101$ \$ that errors in weight cost calculation correspond to +19 \$ (+17%) and +29 \$ (+29%).

We see that the weight cost calculation is very simple but it can only give provisional values for the costs.

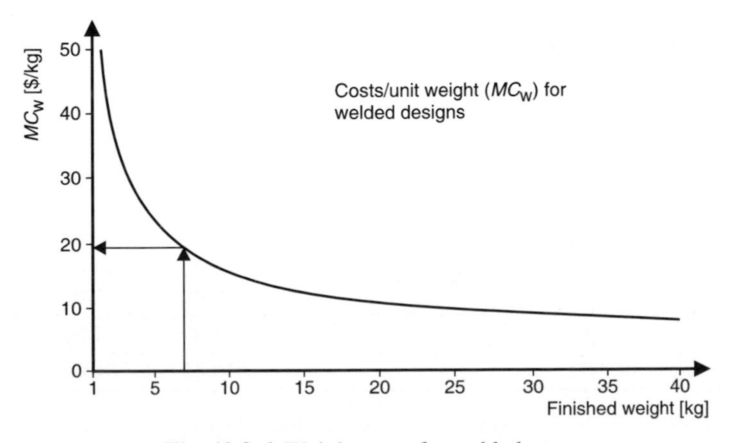

Fig. 10.3-6. Weight costs for welded parts

10.3.4 Cost calculation with Cost Growth Laws: Welded design, size ratio $\varphi_L = 0.5$ and 2

To show the procedure and the limits of the summary and differentiated cost growth laws given in Sect. 9.3.5, a comparative calculation will be made here. We start with the costs determined by preliminary calculation for welded design, (basic embodiment: ratio $\varphi_{L0} = 1$, number = 1). The costs are then determined for the ratios $\varphi_{L1} = 0.5$ and 2 (i. e., half or twice as large) and axis heights of 70 mm and 280 mm, with strictly geometrically similar reduction or enlargement.

a) Calculation for basic embodiment		$\varphi_L = 1$	1 piece/lot
Material direct costs *MtDC*	8.3 kg	1.50 $/kg	12.45
Material overh. costs *MtOC*		15 %	1.87
Material costs *MtC*			**14.32**

Production plan	t_s	t_e	*WSC* [$/min]		PC_s	PC_e
Cut to size	15	40	0.60	33.00	9.00	24.00
Tack & weld	15	30	0.75	33.75	11.25	22.50
Deburr & straighten	15	10	0.75	18.75	11.25	7.50
Boring mill	30	30	1.00	60.00	30.00	30.00
Production costs *PC*				**145.50**	**61.50**	**84.00**
Manufacturing costs *MC*				**159.82**		

b) CGL-summary		$\varphi_L = 1$	$\varphi_L = 0.5$	$\varphi_L = 2$
PCs	$PCs_1 = PCs_0 \cdot \varphi_L^{0.5}$	62.00	43.00	87.00
PCe	$PCe_1 = PCe_0 \cdot \varphi_L^{2}$	84.00	21.00	336.00
MtC	$MtC_1 = MtC_0 \cdot \varphi_L^{3}$	14.00	2.00	115.00
MC		160.00	66.00	538.00

c) CGL-differentiated		$\varphi_L = 1$	$\varphi_L = 0.5$	$\varphi_L = 2$
Cut to size	$PCs_1 = PCs_0 \cdot \varphi_L^{0.5}$	9.00	6.00	13.00
	$PCe_1 = PCe_0 \cdot \varphi_L^{1}$	24.00	12.00	48.00
Tack & weld	$PCs_1 = PCs_0 \cdot \varphi_L^{0.5}$	11.00	8.00	16.00
	$PCe_1 = PCe_0 \cdot \varphi_L^{2}$	23.00	6.00	90.00
Deburr & straighten	$PCs_1 = PCs_0 \cdot \varphi_L^{0.5}$	11.00	8.00	16.00
	$PCe_1 = PCe_0 \cdot \varphi_L^{1}$	8.00	4.00	15.00
Boring mill	$PCs_1 = PCs_0 \cdot \varphi_L^{0.5}$	30.00	21.00	42.00
	$PCe_1 = PCe_0 \cdot \varphi_L^{1}$	30.00	15.00	60.00
Material costs *MtC*	$MtC_1 = MtC_0 \cdot \varphi_L^{3}$	**14.00**	**2.00**	**115.00**
Manufacturing costs *MC*		**160.00**	**82.00**	**415.00**

Fig. 10.3-7. Application of the summary and differentiated cost growth laws (CGL) for the welded design for half ($\varphi_L = 0.5$) and double ($\varphi_L = 2$) ratios

Figure 10.3-7a shows the **cost calculation of the basic embodiment** with a division of the production costs from setup times (*PCs*) and production times (*PCe*) for every work step as a basis for further calculation. In **Fig. 10.3-7b** the costs are determined for the ratios $\varphi_{L1} = 0.5$ and 2 with **summary cost growth law** (CGL) (Eq. (7.7/1)) (rounded to full $).

With **differentiated cost growth law** (Fig. 10.3-7c) **for every individual work step** multiply the production costs from setup times *PCs* with $\varphi_L^{0.5}$ and the production costs from production times *PCe* with the size ratio φ_L and the respective exponent from Fig. 9.3-7.

Figure 10.3-8 shows the comparison of the results. The cost components of the material costs and the production costs from setup times did not distinguish between the summary and differentiated cost growth law, since there is the same calculation term (φ_L^3 or $\varphi_L^{0.5}$). On the other hand there are great differences when finding the production costs from production times. This is because in the differentiated cost growth law the part of the costs that increase with exponent 2 is small compared to the part of the costs that increases with the exponent 1.

This result does not contradict an individual quick cost calculation procedure; rather, it shows that **every procedure has its range of application and its limits**! We can assume that in this case the differentiated cost growth law provides a more precise cost estimate. If there are several similar designs, the exponent could be adapted for the production costs from production times in the summary cost growth law. Here, for example, the use of an exponent of 1.6 instead of 2 would provide a more precise result.

For the quantities 5 and 20 the cost growth laws show similar cost decreases with the conventional calculation of the limiting quantities in Fig. 10.3-4 or Fig. 10.3-5, because exactly as there, the production costs from setup times are divided by the pieces/lot.

➔ Before applying a quick cost calculation procedure, check its applicability (e. g., accuracy, area of application, etc.).

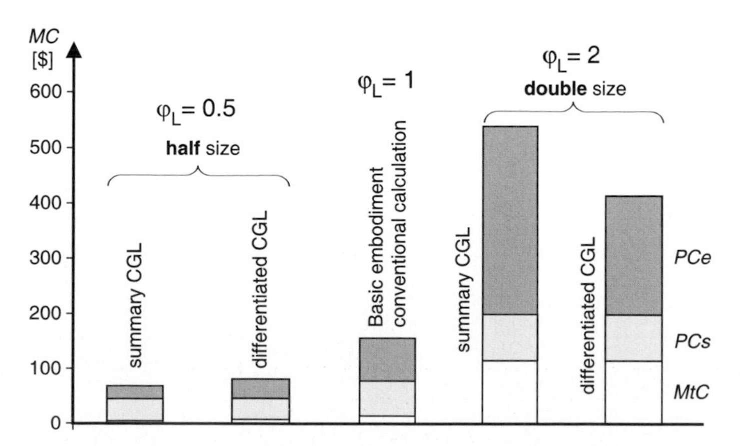

Fig. 10.3-8. Comparison of summary and differentiated cost growth laws

Appendix: Help for Cost Reduction

A 1 Overview: Introduction

Here we have put together surveys, checklists, and collections of rules for practical applications in cost reduction. This is a framework for looking at the process sequence of a cost reduction project, with references to the most important rules and figures as guidelines. The starting point is the procedure cycle for the cost reduction of products, Fig. 4.5-7.

Consider the following:

- We do not want to give the impression that this guideline is **"the recipe"** for target cost-oriented design. It is supposed to **help** you to **develop your own procedure, adapted** to your **problem**.
- **Checklists and rule collections are never complete.** Also, they are **specific** to the **company** and the **product, and are sometimes even inconsistent**. They are **supposed to stimulate your own reflection and action**. Do not work doggedly. Carry out **changes and enhancements for your product** or **your company**.
- You will recognize what is best in your case if you separate **the important from the unimportant**. For this purpose, what will help you most is **the clarification of the requirements** (Sect. 4.5.1), especially the cost target, **Pareto analyses** (Fig. 4.6-4), and **cost structures** (Figs. 4.6-5, 7.11-51, 7.13-15, 10.2-2). To set these up **before the actual beginning** of the task under time pressure is laborious, with the usually incomplete mass of data. However, **it is worth it!** Early on, ask questions of the specialists who know more about specific fields. Do not be intimidated by know-it-alls, departmental walls, killer phrases, etc.
- Cost reduction and target cost-oriented designing is systematical engineering work! Furthermore, it requires the trust and open **cooperation** of all the departments that influence and account for costs. Thus, an **interdisciplinary team** must be set up that brings down the departmental walls (Fig. 3.2-2).
- This guideline deals mainly with the **reduction of manufacturing costs**. The reduction of other costs, for example, lifecycle costs (Fig. 5.3-1) and product development costs (Fig. 6.2-1) follows the same scheme, with contents and terms appropriately adapted.

A 2 Guidelines for Cost Reduction

Starting from Fig. 4.5-7, more explanations and more references are provided here for cost reduction of products than was possible at the time of the brief presentation of the **procedure cycle** in that figure.

A 2.1 I Clarify the task and the procedure

I.0 Plan the **procedure**, form the **team**, designate the **responsible parties**.

- Before a cost reduction project is begun at all, a suitable **climate** must be created. If it does not yet exist, the climate must be created in the course of the project (Fig. 3.2-2). Genuine **support** from **management** is important (patron, steering committee; Sect. 3.2, 4.8), as is the **disclosure of cost data** (but handled confidentially) (Sect. 3.2.2). Compose a procedure plan with intermediate deadlines (Sect. 4.8.3, 6.2.2). "Sell" teamwork to the outside!
- For interdisciplinary **team formation** rules, see Sect. 4.3.1.
- **Clarify the task**:
  In the following, only the costs are dealt with. However, the normal technical and organizational clarification of the task, with preparation of the requirements list (Sect. 4.5.1.1), also go with that. Do we in fact know what the customer is most likely willing to pay? How can one improve the situation, vis-a-vis the competition?
  If possible, at the time of quotation release or order discussion, agree to few (or **few strict) demands**, functions, tolerance constraints, warranties, acceptance conditions, rules to be kept, or standards. Every demand and warranty promise is usually tied to additional costs. In discussions with the customer and suppliers, involve the design department in the determination of the customer performance specification and the requirements list.
- A revision or redesign of a product should not be limited only to cost reductions, but should always include other improvements besides. Core rule: **Increase customer utility!**

I.1 **Determine total target costs** (Sect. 4.5).
Clarify which costs are to be considered: lifecycle costs, total costs, manufacturing costs? Profit target, efficiency target?
What does the customer wish?
Do not simply accept a cost target -xx%; also scrutinize the grounds. Questions: Why are we so expensive? Why is the competition so much more economical?

I.2 Cost structures (Sect. 4.6.2, Figs. 10.1-4, 10.1-7, and 10.2-2).
Draw up and analyze cost structures of the previous and competing products (benchmarking, Sect. 7.13) according to assemblies, parts, and by qualities and functions desired by the customer.
Aid: Spreadsheet computation (Figs. 10.1-5, 10.1-8). The cost structures also help later in costing the new solution.
Goal: Clarify what the costs consist of, how they arise, and where the priorities are in the project.

I.3 Search for priorities, cost reduction potentials.
What can be changed; what is fixed?
If large cost reductions are called for, massive changes in the product and process must also surely occur (Sect. 4.8.2).
For the establishment of a difficult but attainable cost target, it must first be fully clear how it could be achieved. To determine **cost reduction potentials**, a solution search is already necessary during task clarification! It should take place in the team. In the course of the project, further cost reduction potentials sometimes become apparent, which should be pursued.

I.4 Divide up the total target costs (i. e., determine partial target costs).
(Sect. 4.5.1.4, Fig. 4.5-4, Sect. 10.1.4). Determine partial cost targets on the basis of the cost structures and the cost reduction potentials for assemblies, parts or functions, so that workable and attainable tasks are defined for individuals or groups.
Establish the measures and the parties responsible (Fig. 10.1-8) for the cost targets. Strive for short loops between the establishment and the ascertaining of the costs, to set intermediate deadlines (Figs. 4.2-2, 4.4-2).
A good task clarification, especially the target costs, and detailed procedure planning are the bases for the success of a project!

A 2.2 II Search for solutions

- Because one cannot be sure of finding the best solution immediately, and the "first-best" solution is probably not the best solution, search for several solutions!
- Split the problem into sub-problems, then search for partial solutions for each.
- Employ methods (intuitive and systematical, Sect. 4.5.2) for the solutions search.
- In the systematical solution search, the **concretization steps** (Sect. 4.5.2) go from the abstract to the concrete. The more abstract is

the level at which the solutions search begins, the greater the chance of finding a new low-cost solution. However, the expense and the risk also increase.

II.1 Function: (Sect. 4.5.1.2)

- Do we know the functions on which the customer places value?
- Fewer functions, more? Function combination?
- Have the functions of the assembly and the part been clarified?
- Is the function fulfillment unambiguous, simple, and safe?
- Can the functions be integrated into another component
- Can the functions be distributed over several components
- Are the material and the manufacturing expense justified for the function fulfillment?

II.2 Principle: (Sect. 7.3)

- Other principle (concept); possible size decrease.
  Concepts for **small** and **light design** usually lead to low-cost devices. Machines with strong physical effects (for example, mechanical and hydrostatic energy) tend to be small and light, through parallel connection of active surfaces (power branching), or increase in velocities or rotation speeds (Sect. 7.9.2.2).
  Small design (i.e., size decrease) lowers manufacturing costs, especially with large parts in small-lot production. The same is also true for small **and** large parts in series production (Sect. 7.9.2.2).
- **Concepts with simple construction and few parts** (e.g., function combination, integral design; Sect. 7.12.3.3) are more economical, with small dimensions and/or in large numbers.

II.3 Form design: (Sect. 7.4, 7.8)

- Fewer parts (integral design)?
- Company-internal standardization (Sect. 7.12), similar parts, repeating parts, part families, size ranges, modular design?

Form variants to be implemented as late as possible in the manufacturing process, for example, only during the final assembly, (Fig. A8, below).

With **parts made in large quantities**, strive for **integral design**. Approximate the final shape through primary and secondary forming procedures, such that significant material waste is avoided (Sect. 7.12.3.3).

Integral design is more economical than differential design for **small quantities** and machining from a blank, particularly with **small and medium size parts** (e.g., fixture design, Figs. 7.12-15, 7.11-42).

Differential design is more economical with **large parts** and/or with expensive material in small-lot production and with **small quantities**.

II.4 Material: (Sect. 7.9)

Less material? Less waste? More low-cost material? Standard, serial, and purchase parts (Figs. 7.9-3, A5)?

Cost of materials reduces through small designing (avoid over-dimensioning; use FEM analysis), velocity and/or rotation speed increase; use higher-strength materials (usually only slightly more expensive); use of low-cost standard materials where there are no high loads. In the case of **series production**, choose material-saving, near-net-shape production processes such as casting, forging, or deep drawing. Adapt design to quantities made. Strive for small material thicknesses. Pay attention to direct flow of force between supports (strive for axial loading, avoid bending and torsion loading).

- Can the raw material or a purchased part be procured at a lower cost?
- Can other, lower-cost material be used?
- Can standard parts (modular design) be used?
- Can the unfinished part be made from another semi-finished product?
- Can the waste be reduced by suitable form design?
- Can the unfinished part be made as a cast, forged, sintered, or sheet metal part?
- Can the semi-finished product and/or the blank be preproces sed?

II.5 Production: (Sect. 7.11)

There are many **production processes** (Figs. 7.11-4, A6)! Think not only of the standard processes for parts in one's own factory, but search for systematically different procedures in the team and possibly use specialized suppliers. Invite suppliers to the team discussions. Pass on the cost target.

Other and fewer production steps, other fixtures, equipment? Lower accuracy? Assembly variants (Figs. 7.11-52, A7)? In-house or external manufacturing (Sect. 7.10)?

- Are there in-house experts on the production technologies?
- Does the component fit into the company-specific part spectrum?
- Must the component be produced in-house?
- Are the production times justified?
- Is the sequence of work steps optimal?
- Is the production more economical on other machines?
- Are other procedures possible for material separation, for surface treatment, for joining and assembly?
- Are all machined surfaces used for function fulfillment?
- Must all active surfaces be machined?
- Are coarser tolerances and a lower surface quality possible?
- Can different measurements be unified?
- Call in a production specialist!

• **Assembly procedures** are as numerous as the part production processes (Sect. 7.11.7) and should be decided upon in the team in conjunction with the production processes and the material choices.

Measures for reduction of assembly costs (Figs. 7.11-51, A7).

A 2.3 III Select solution

- This step is not only for use at the end of a project, but to be used repeatedly in the course of the project.
  Strive for **short feedback loops** (Sect. 4.4.3). As far as possible after every solution search, (at least before agreed upon deadlines), check the compliance of the cost targets, and introduce new measures.
  Document the state that has been attained.

III.1 Analysis: (Sects. 4.5.3, 9, 10.3)
Not only the attainment of the cost target, but also all other demands are to be checked. Qualities of the solutions are to be determined for this purpose with suitable methods for early cost identification.
Concurrent cost calculation (Sect. 9, Fig. 10.1-8). Courage to estimate (Sect. 9.2)! Early determination of costs and properties is necessarily inaccurate. If carried out correctly and supported, sufficiently precise results can be achieved (balancing out of accidental errors: Sect. 9.3.7.3).
Involve production planning and costing in the process.

III.2 Selection: (Sect. 4.5.3.2)
It is not only the hard cost and technical data that go into the solution decision, but also soft factors. The boss should also be involved in the decision with the team.

A 2.4 Project follow-through, evaluation

- **Further progress on the project:**
  With the end of the design phase, the product has appeared only as a plan on paper. This plan has to be moved along! For example, the involvement of manufacturing and purchasing is helpful at this point in the design process. Furthermore, the realization of the plan must be checked regarding deadlines and cost (Fig. A4).
- **Evaluation of experiences:**
  With every project, there are positive and negative experiences. It is up to the team to learn from these experiences: "postmortem."

A 3 Important Figures and Rules

A 3.1 Cost calculation (origins of costs)

The bases for cost management are cost accounting in the company and the costs of the existing and earlier products. (Analogy from mechanics of materials: mechanics, strength values, etc.) They are company- and product-specific! It is important that from the time the target costs are ascertained, through the concurrent calculation, to the follow-up calculation, that the cost data are always calculated in the same cost structure and on the same basis. This allows comparisons, correctly recognizing deviations, and introducing countermeasures (Fig. A4).

- How are the costs calculated in your company?
- Where are the focal points?
- Which costs may be influenced through the current project and which may not?
- Who can obtain the information for you (access to data)?
- How must you read cost calculations, data master records, work schedules, etc.?
- Does cost accounting represent cost generation correctly? For example, consideration of the influence of the quantities produced (Sect. 8.4.3 b and d).
- ...

Handle data as absolutely confidential!
Figure A1 shows the scheme of differentiating overhead costing, machine hourly rate, and workstation costing. It is intended to serve as a stimulant to carry out corresponding cost analyses for your product. Bear in mind that the terms, partitioning, etc., are company-specific. If you want to know the order of magnitude of the individual quantities for comparison as mean values from VDMA, see Fig. 8.4-2.

A 3.2 Cost structures; cost targets; concurrent calculation

Cost structures are suited to finding focal points for costs and cost reduction potentials. They can be prepared from the previous product and from similar products (also, from competition products for which the costs are calculated in-house) (Sect. 4.6.2, 10.2; organizational aspects, Fig. 4.6-3).

It is advisable to continually record the partial target costs and the actual costs of the components during a project (for example, for every team session) with a spreadsheet program. This cost calculation takes place concurrently with product development (Sect. 4.8.3 and 9.1.2). Thus, it is possible to keep the whole picture in view and introduce new measures if difficulties or new possibilities for cost reduction arise. Also advisable for this purpose is to keep continuous notes regarding cost reduction potentials, parties responsible for the various steps, etc.

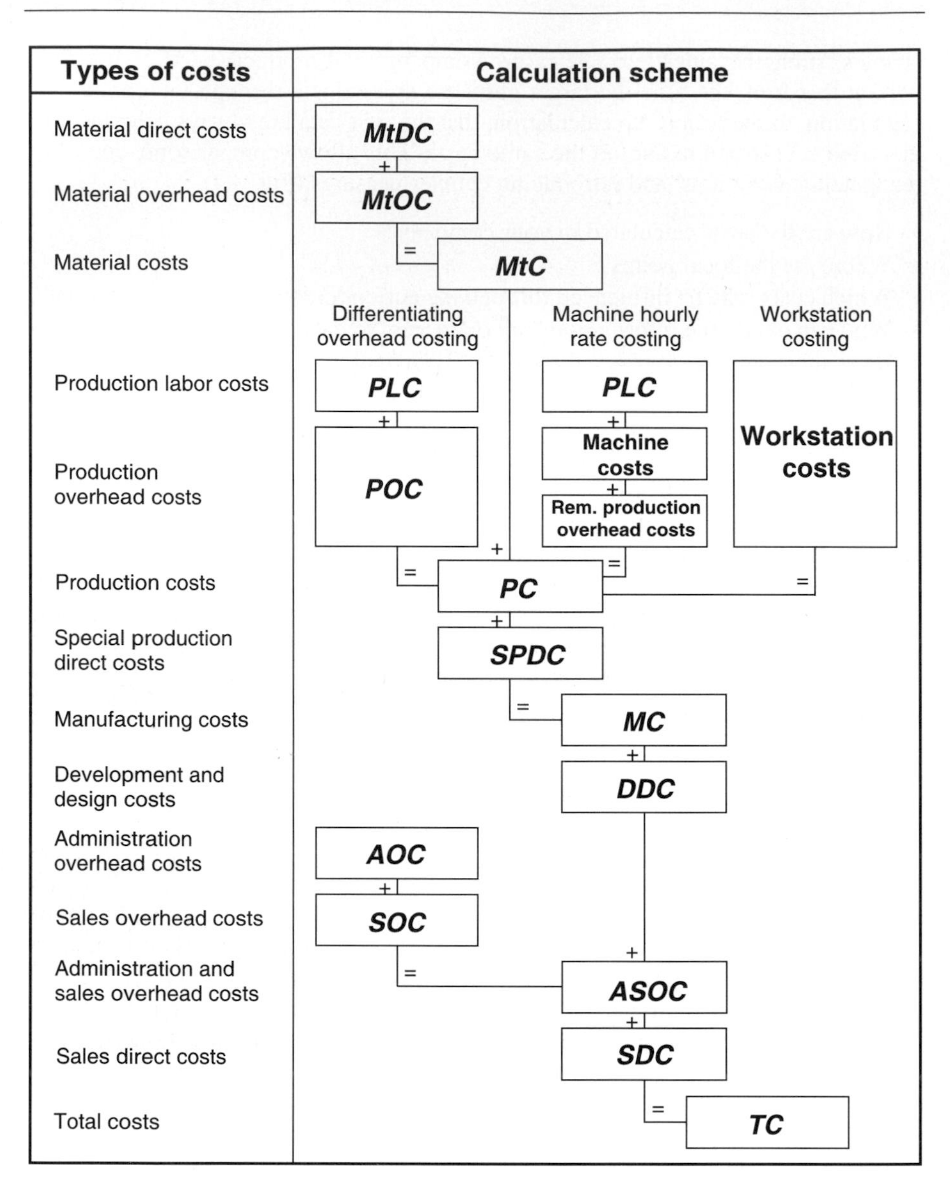

Fig. A1. Differentiating overhead costing, machine hourly rate, and workstation costing (Fig. 8.4-9)

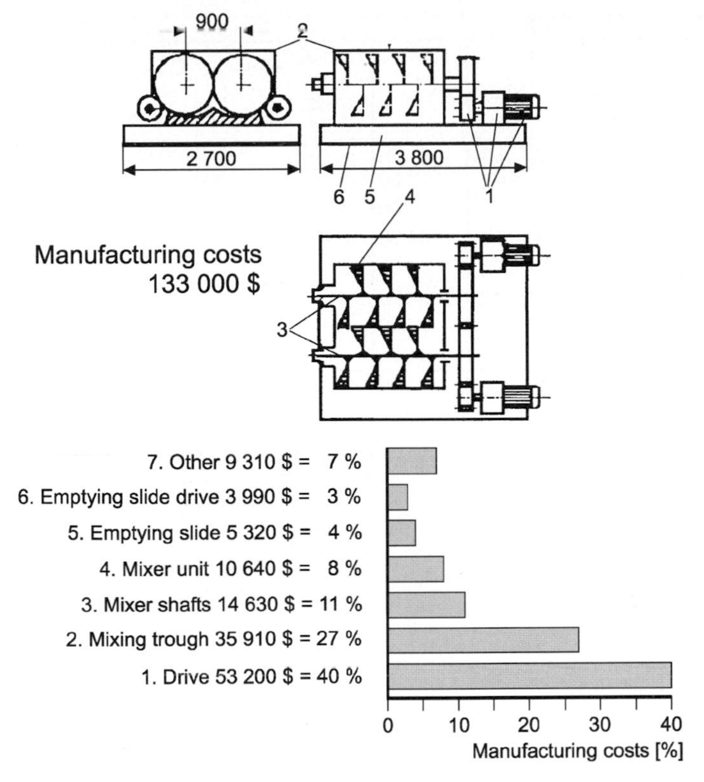

Fig. A2. Cost structure of a concrete mixer (Fig. 10.1-4)

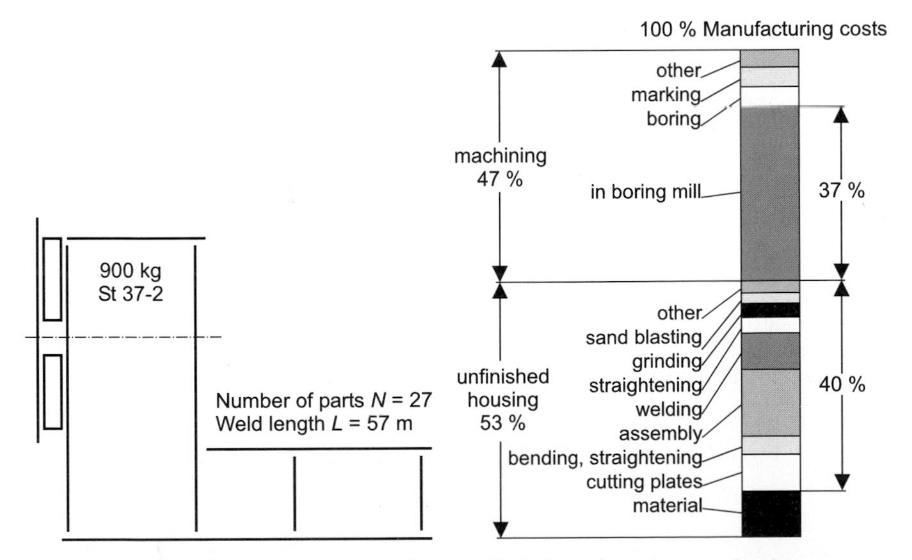

Fig. A3. Manufacturing cost structure of a centrifuge base, based on production processes and material (Fig. 10.2-2)

1. Session: Analysis

Twin-shaft mixer (TSM)	Actual costs TSM	Share TSM	Actual costs Plate mixer	Partial target costs TSM	Required cost reduction	Cost reduction potential	
Assemblies	[$]	[%]	[$]	[$]	[%]	Measures	Responsibi
Drive	53 000	39.8	32 000	34 100	-36	Other principl.	Developme
Mixing trough	36 000	27.0	27 000	21 600	-40	Welded design	Devl.+Prod
M.shaft + Brngs.	15 000	11.0	10 000	12 750	-15	Simplify	Developme
Mixing unit	11 000	8.1	8 000	9 400	-15	Prod. changes	Production
Emptying slide	5 200	3.9	5 000	4 300	-17	Simplify	Developme
Empt.sld.driv.	4 000	3.1	3 800	3 400	-15	Simplify	Developme
Other	9 000	7.0	8 000	7 550	-16	Simplify	Developme
Sum	**133 200**	100	93 800	93 100	**-30**		

2. Session:

Stand TSM	Achieved cost reduction	New measures	
[$]	[%]		Responsibi
38 000	-28	Other suppliers	Purchasing
23 000	-36	Prod. changes	Production
11 830	-21	ok	
10 000	-9	Prod. changes	Production
3 400	-35	ok	
3 400	-15	ok	
7 800	-13	ok	
97 430	-27		

more sessions ...

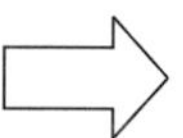

Results:

Costs TSM	Achieved cost reduction
[$]	[%]
31 700	-40
18 000	-50
11 800	-21
8 400	-24
3 400	-35
3 400	-15
7 800	-13
84 500	**-37**

For space reasons the table has been "broken up". With a spreadsheet analysis it can be adapted to requirements in practice.

Fig. A4. Search table for costs for the concrete mixer (Fig. 10.1-8)

A 3.3 Material costs

Material costs include not only the costs for the raw materials but also the costs of everything that is obtained from the suppliers. This becomes increasingly more important in the course of concentrating on core products and processes and through outsourcing. Provide the cost targets to selected suppliers. Strive for close cooperation.

From **Fig. A5** we see that costs of materials can be lowered not only by cheaper material (per volume) but also by using high-quality material, through better concepts that lead to smaller and lighter products.

A 3.4 Production costs

Figure A6 gives only a brief overview of the variety of production processes (Sect. 7.11). Only eight welding operations are shown; there are however, about 250 different ones! The choice of part production, assembly, material, and joining methods is associated closely with each other (Sect. 7.11.1, Fig. 7.11-1). They should be discussed in the team and possibly with experienced suppliers, and then the decisions made.

A 3.5 Assembly costs

The central theme here is to design as few parts as possible: What is not there does not need to be assembled (see also Fig. A8). In addition, it is necessary to first determine the assembly cost structure (Fig. 7.11-50). The recognized focal points must first be discussed in close cooperation with the assembly department (**Fig. A7**).

A 3.6 Reduction of variants

The increasing attention on fulfilling customer requirements reduces the number of similar products and increases the number of product and part variants. Both drive up costs, the manufacturing costs (Sect. 7.12) as well as the total costs (Sect. 6.3). It is a matter of recognizing, with marketing and sales, the **necessary variants** and reducing the unnecessary variants. The earlier this is begun during the program and product conception, the more effective it is.

Figure A8 gives an overview of the strategies and measures for variant reduction (see Index for further information).

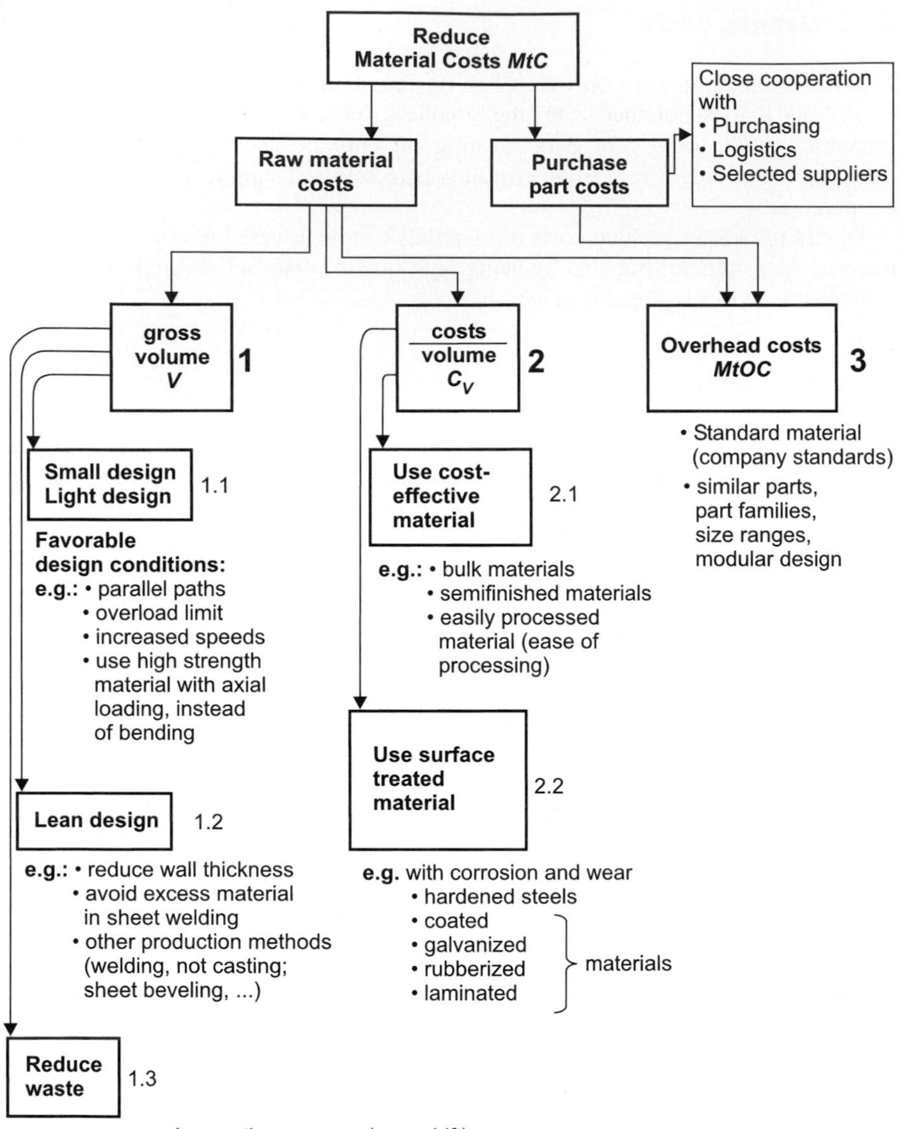

Fig. A5. Designing for material efficiency and its rules (Fig. 7.9-3)

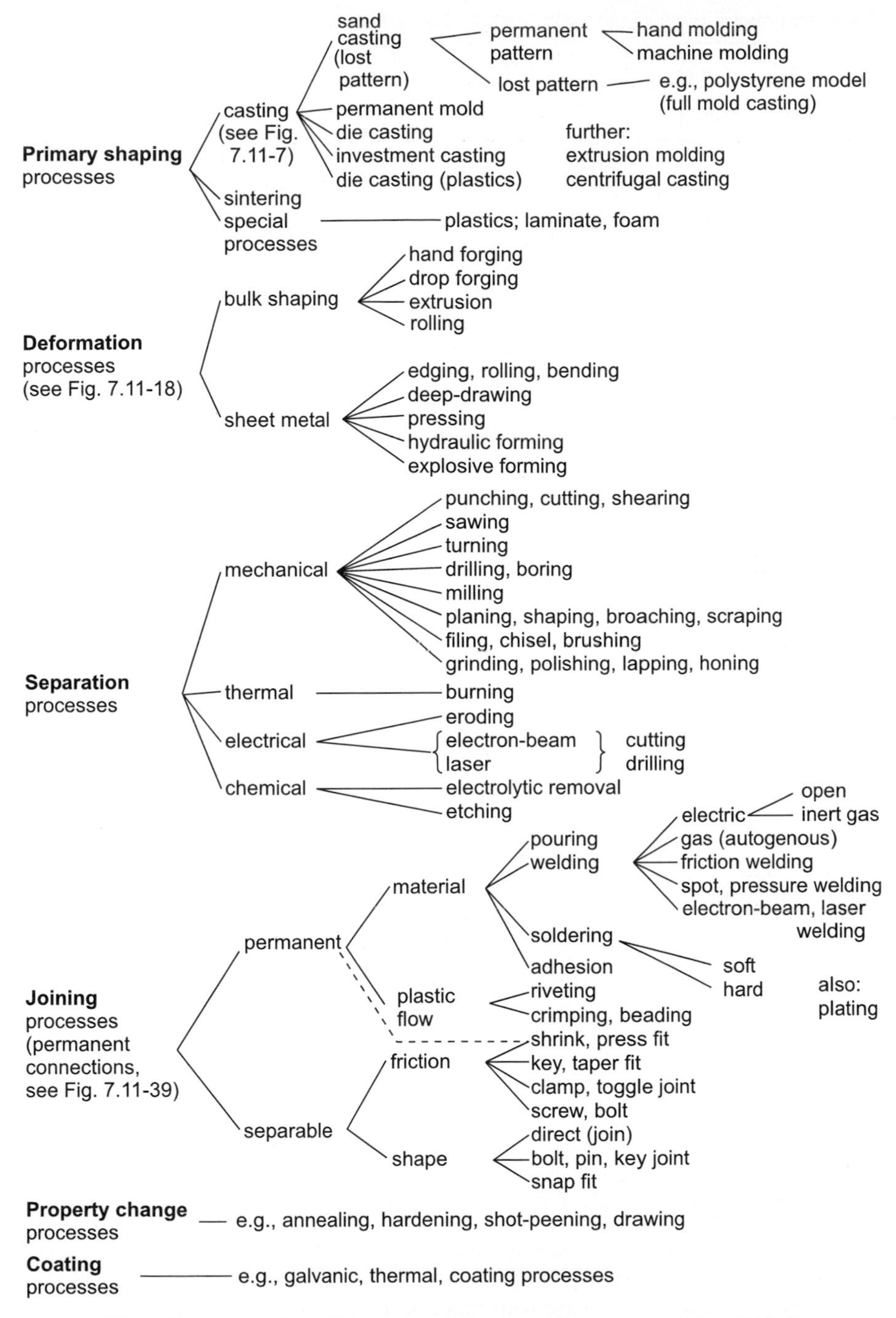

Fig. A6. An overview of the customary production processes (Fig. 7.11-4)

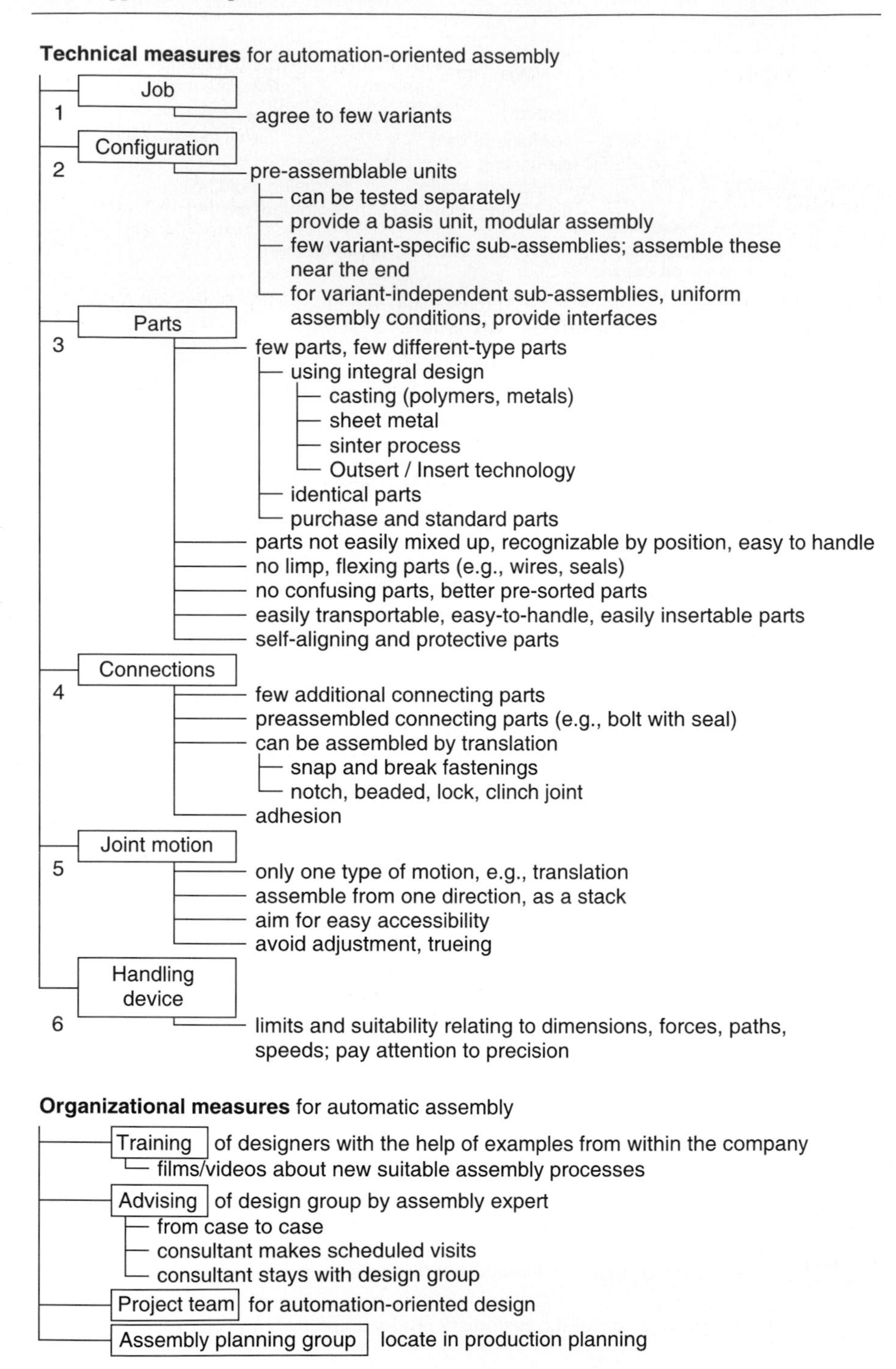

Fig. A7. Measures for low-cost assembly (Fig. 7.11-52)

1. Integral design	• Choose **primary and deformation processes** (e. g., casting, injection molding, sheet metal forming, forging, deep drawing, erosive removal): Quantities made? • Choose suitable semi-finished products.
2. Standardize shape elements	• Shape elements (part geometries, CAD-features), i. e., standardize mounting spaces, connection geometries that are needed in different variants.
3. Identical parts (occurring repeatedly in a product)	• Make as many parts as possible of a product identical and choose **integral design**. • Avoid **left-right** designs. • Also reduce **small hardware items** (e. g., screws, seals, retaining rings) "radically" and standardize: The production and assembly often becomes much simpler! • **Unify materials:** In case of doubt use the "better" material, with test report.
4. Repeating parts (often occurring with different products)	• **Use standardized parts and vendor supplied parts** (internal selection). Usually lower cost because of larger quantities. • **Symmetrical** components, which can be mounted in different positions. • Use parts of **other product families**: Arrange it to be so! Problems with changes?
5. Part family	• Standardize parts of same function. "Common tidying up action for spread-out uncontrolled growth" (Fig. 7.12-13).
6. Size ranges	• Strive for coarser step sizes.
7. Modular design	• Use **suppliers'** modular systems. • Plan for **local** modular systems (Fig. 7.12-29). • Plan for **pre-assembled** and **testable** assemblies. • **Platform strategy:** same basic building blocks for several products.
8. Concept for low number of parts	• Through choice of suitable physical principles the part count often can be remarkably reduced (Example: mechanical-to-bubble jet printer).

Fig. A8a. Strategies for variant reduction (1–8 essentially constructive)

9. Supplier setup	• Reduce the **number of suppliers** (more favorable conditions). Provide them with the cost target! • Lock in **outline contracts**. • Settle on **standardization** together.
10. Standardization degree	• Introduce as a **target** (Fig. 7.12-6): $\frac{\text{S repeating, identical, purchased, standard parts}}{\text{S different parts}}$ ≤ limiting value
11. Variant tree	• To judge whether variants occur early or late. The best: only near the end of the assembly (Fig. 7.12-7).
12. Malus	• A **cost limitation** per new part (e. g., 3 000 $) that must be met first by cost reduction; else no new part (Fig. 7.12-11)!
13. Introduce part search system	• Database with independent terms from the master data field, and with CAD-Figures. (Searches must go faster than making new parts).
14. Sell standardization!	• Make the sales documents as well as the sale such that technology and costs are transparent to the salesperson. **Pass on the cost advantage of** standardization to the customer (that needs courage.) Special wishes cost extra. (Not every order is wanted.) **Sales commission** based on profit, instead of sales.
15. Process costs	• Estimate with hourly rate of cost center accounting or from time sheets. Distinguish active (high demand) from passive (low demand) variants. Passive, "leave alone", active, at times redesign with fewer parts!
16. Problem areas due to variety	• Variant management demands reduction of variety in products, parts, suppliers, customers and orders. • Recognize the variants **necessary** for the market, reduce **unnecessary** variants (Fig. 7.12-2).

Fig. A8b. Strategies for variant reduction (9–16, more organizational) (Close coordination is necessary with cost control, production, assembly, purchasing, sales and customer service)

Literature

[Aak95] Aaker, D.; Kumar, V.; Day, G. S.: Marketing Research. Crawfordsville: Wiley& Sons 1995.

[Aka92] Akao, Y.: QFD: Quality Function Deployment. Landsberg: Moderne Industrie 1992.

[Aki94] Akiyama, K.: Funktionenanalyse: Der Schlüssel zu erfolgreichen Produkten und Dienstleistungen. Landsberg/Lech: Moderne Industrie 1994.

[Alt84] Altschuller, G. S.: Erfinden. Wege zur Lösung technischer Probleme. Berlin: VEB Verlag Technik 1984.

[Amb97] Ambrosy, St.: Methodenwerkzeuge für die integrierte Produktentwicklung. Aachen: Shaker 1997. (Konstruktionstechnik München, Band 26). Also: München: TU, Diss. 1996.

[Amf02] Amft, M. Phasenübergreifende bidirektionale Integration von Gestaltung und Berechnung. München: TU, Diss. 2002.

[And75] Andresen, U.: Die Rationalisierung der Montage beginnt im Konstruktionsbüro. Konstruktion 27 (1975), S. 478–484.

[And81] Andreasen, M.; Hubka, V.: Methodisches Konstruieren von Maschinensystemen. Zürich: Edition Heurista 1981.

[And86] Andreas, D.: Selbst fertigen oder kaufen? Rechen- und Entscheidungsschema. Frankfurt/M.: Maschinenbau-Verlag 1986.

[And87] Andreasen, M. M.; Hein, L.: Integrated Product Development. New York: Springer 1987.

[And91] Andreasen, M. M.: Design for Production – Overview of Methodology. In: Proceedings of the International Conference on Engineering Design. Zürich: 1991, S. 516–521.

[Ard88] Ardenne, M. v.; Musiol, G.; Reball, S.: Effekte der Physik und ihre Anwendungen. Berlin: VEB Deutscher Verlag der Wissenschaften 1988.

[Arn95] Arnaout, A.; Hildebrandt, J.; Werner, H.: Einsatz der Conjoint-Analyse im Target Costing. Ein Fallbeispiel des Geschäftsfeldes Personenwagen der Daimler-Benz AG. Controlling (1998) 5, S. 306–315.

[Bäß88] Bäßler, R.: Integration der montagegerechten Produktgestaltung in den Konstruktionsprozeß. Berlin: Springer 1988.

[Bau78] Baumann, G.: Systemstudie für die Erstellung eines Kostenfrüherkennungsverfahrens in der Einzelfertigung. München: TU, Lehrstuhl für Konstruktion im Maschinenbau, Unveröffentlichte Diplomarbeit 1978.

[Bau80] Bauer, C. O.: Verbindungstechnik zwischen Diaspora und Irredenta. VDI-Z 122 (1980), S. 82–96.

[Bau82] Baumann, G.: Ein Kosteninformationsystem für die Gestaltungsphase im Betriebsmittelbau. München: TU, Diss. 1982.

[Bau91] Bauer, C.; Althof, W.; Haferkamp, H.; Hamkens, J.; Kaschner, M.; Koller, R.; Stellberg, M.: Handbuch der Verbindungstechnik. München: Hanser 1991.

[Bau93] Bauer, H.; Herrmann, A.: Preisfindung durch „Nutzenkalkulation“ am Beispiel einer PKW-Kalkulation. Controlling (1993) 5, S. 236–240.

[Bec94] Becker, J.; Prischmann, M.: Konstruktionsbegleitende Kalkulation mit neuronalen Netzen – Methoden aus der Informatik zur Lösung betriebswirtschaftlicher Probleme. krp (1994) 3, S. 167–171.

[Bec96] Becker, J.: DV-Verfahren zur Unterstützung frühzeitiger Kosteneinschätzungen. krp (1996) 1, S. 81–85.

[Bee72] Beeley, P. R.: Foundry Technology. New York: Halsted Press 1972.

[Bei74] Beitz, W. u. A.: Rechnergestütztes Entwickeln und Konstruieren im Maschinenbau. Forschungskuratorium Maschinenbau, Frankfurt/M.: Maschinenbau-Verlag 1974.

[Bei77] Beitz, W.: Senkung des Konstruktions- und Fertigungsaufwands durch Einsatz bereits konstruierter Teile und Baugruppen – Kennzeichnende Merkmale für wiederkehrende Baugruppen. DIN-Mitt. 56 (1977), S. 351–359.

[Bei82a] Beitz, W.; Klasmeier, U.: Kostenfrüherkennung bei komplexen Schweißgruppen. Düsseldorf: VDI-Verlag 1982, S. 115–127. (VDI-Berichte 457)

[Bei82b] Beitz, W.: Was ist unter „normungsfähig“ zu verstehen? – Ein Standpunkt aus der Sicht der Konstruktionstechnik. DIN-Mitt. 61 (1982), S. 518–522.

[Bei97] Beitz, W.; Helbig, D.: Neue Wege zur Produktentwicklung – Berufsfähigkeit und Weiterbildung. In: Beitz, W. (Hrsg.): Schriftenreihe Konstruktionstechnik, Nr. 37. Berlin: TU 1997.

[Ben90] Benthake, H.: Baukastensysteme – grundsätzliche Möglichkeiten, Optimierung am Beispiel einer Industriegetriebereihe. 7. Konstrukteurstagung Dresden 1990, Berlin: Kammer der Technik 1990, S. 217–238.

[Ben93] Benecke, R.: Beitrag zum räumlichen Laserstrahlschweißen von Stahlwerkstoffen im Feinblechbereich mit CO2-Hochleistungslasern. Düsseldorf: VDI-Verlag 1993. (Fortschrittsberichte VDI Reihe 5, Nr. 301)

[Ber95] Berkau, C.: Vernetztes Prozeßkostenmanagement – Konzeption und Realisierung mit einem Blackboardsystem. Wiesbaden: Gabler 1995.

[Bie71] Biegert, H.: Die Baukastenbauweise als technisches und wirtschaftliches Getaltungsprinzip Karlsruhe: TH, Diss. 1971.

[Bin97] Binder, M.: Technisch-wirtschaftlich integrierte Steuerung von Produktkosten in den Phasen Entwicklung und Konstruktion. Stuttgart: Universität, Diss. 1997.

[Bir92] Birkhofer, H.: Erfolgreiche Produktentwicklung mit Zulieferkomponenten. In: Praxiserprobte Methoden erfolgreicher Produktentwicklung, Mannheim. Düsseldorf: VDI-Verlag 1992, S. 155–170. (VDI-Berichte 953)

[Bir93] Birkhofer, H.; Reinemuth, J.: Zulieferer im Wettbewerb. Düsseldorf: VDI-Verlag 1993, S. 1–22. (VDI-Berichte 1089)

[Boo89] Boothroyd, G.; Dewhurst, P.: Product Design for Assembly. Wakerfield: Boothroyd Dewhurst 1989.

[Boo94] Boothroyd, G.; Dewhurst, P.; Knight, W. A.: Product Design for Manufacturing. New York: Marcel Dekker 1994.

[Bor61] Borowski, K.-H.: Das Baukastensystem in der Technik. Hannover: TH, Diss. 1961.

[Bor96] Born, M; Eiselin, S.: Teams – Chancen und Gefahren: Grundlagen; Anwendung am Beispiel von Lean Management. Bern: Hans Huber 1996. (Psychologie-Praxis Organisation, Arbeit, Wirtschaft)

[Bor97] Beck-Bornholdt, H.-P.; Dubben, H.-H.: Der Hund, der Eier legt. Erkennen von Fehlinformationen durch Querdenken. Reinbek bei Hamburg: Rowohlt 1997.

[Bos91] Bossert, J. L.: Quality Function Deployment – A Practitioner's Approach. Milwaukee, WI: ASQC Quality Press 1991, S. 45.

[Bot96] Botta, V.: Mitlaufende Kalkulationen für ein frühzeitiges Kostenmanagement. krp (1996) 1, S. 39–52.

[Bra86] Bralla, J. G.: Handbook of Product Design for Manufacturing. New York: McGraw-Hill 1986.

[Bre93] Breeing, A.; Flemming, M.: Theorie und Methoden des Konstruierens. Berlin: Springer 1993.

[Bri95] Brinkmann, T. et al: Umwelt- und recyclinggerechte Produktentwicklung. Augsburg: WEKA 1995.

[Bro66a] Bronner, A.: Zukunft und Entwicklung der Betriebe im Zwang der Kostengesetze. Werkstattstechnik 56 (1966), S. 80–89.

[Bro66b] Bronner, A.: Kostenrechnung als Mittel der Arbeitsgestaltung und Investitionsplanung. Ind.-Anz. 88 (1966), S. 482–486.

[Bro68a] Bronner, A.: Wertanalyse als integrierte Rationalisierung. Wt-Z. d. ind. Fertig. 58 (1968) 1, S. 16–21.

[Bro68b] Bronner, A.: Wertanalyse als Grundlage der Erzeugnisplanung. VDI-Z. 110 (1968), S. 1583–1588.

[Bro89] Bronner, A.: Einsatz der Wertanalyse in Fertigungsbetrieben. Köln: RKW-Verlag, Verlag TÜV Rheinland 1989.

[Bro96] Bronner, A.: Angebots- und Projektkalkulationen – Leitfaden für Praktiker. Berlin: Springer 1996.

[Bro98] Brokemper, A.; Gleich, R.: Benchmarking von Arbeitsvorbereitungsprozessen in der Maschinenbaubranche. krp 42 (1998) 1, S. 16–25.

[Bru93] Bruckner, J.; Ehrlenspiel, K.: Kosteninformationen für den Konstrukteur. Entwicklung von Hilfsmitteln in Zusammenarbeit von Hochschule und Industrie. VDI-Z 135 (1993) 11/12, S. 107–112.

[Bru94] Bruckner, J.: Kostengünstige Wärmebehandlung durch Entscheidungsunterstützung in Konstruktion und Härterei. München: Hanser 1994. (Konstruktionstechnik München, Band 16). Also: München: TU, Diss. 1993.

[Buc99] Buchner, H.; Mauer, C.: Schlankes Prozeßkostenmanagement für den Mittelstand am Beispiel des Maschinen-Zulieferers Ringspann GmbH. Controlling (1999) 2, S. 81–86.

[Büt95] Büttner, K.; Kohlhase, N.; Birkhofer, H.: Rechnerunterstütztes Kalkulieren komplexer Produkte mit neuronalen Netzen. Konstruktion 47 (1995), S. 295–301.

[Bug95] Buggert, W.; Wielpütz, A.: Target Costing. Grundlagen und Umsetzung des Zielkostenmanagements. München: Hanser 1995.

[Bul95] Bullinger, H.-J.: Kostenorientierte Entwicklung mit innovativen CAD-Systemen. MoKoKo – durch Montage-, Fertigungs- und Kostengerechte Konstruktion schneller zu neuen Produkten. Stuttgart: IRB-Verlag 1995.

[Bul98] Bullinger, H.-J.; Warschat, J.; Frech, J.: Kostengerechte Produktentwicklung durch Target Costing und Wertanalyse. Müssen wir markt- und kostengerechte Produktentwicklung wieder von Japan lernen? Internetdokumente des IAO: http://www.rdm.iao.fhg.de/engineering/kpe/TCWA.html v. 17.12.98.

[Bun92] Bundesministerium für Umwelt Naturschutz und Reaktorsicherheit (Hrsg.): Elektronikschrottverordnung. Arbeitspapier 1992.

[Bun94] Bundestag-Drucksache 12/8084, vom 08.07.94.

[Bur93] Burghardt, M.: Projektmanagement – Leitfaden für die Planung, Überwachung und Steuerung von Entwicklungsprojekten. 2. überarb. Aufl. Berlin, München: Siemens AG, 1993.

[Bus83] Busch, W.: Relativkostenermittlung von Schraubenverbindungen. In: Schraubenverbindungen heute. Düsseldorf: VDI-Verlag 1983, S. 137–142. (VDI-Berichte 478)

[Cae91] Caesar, C.: „Kostenorientierte Gestaltungsmethodik für variantenreiche Serienprodukte", Variant Mode and Effects Analysis (VMEA). Düsseldorf: VDI-Verlag 1991. (Fortschrittsberichte VDI Reihe 2, Nr. 218)

[Cam94] Camp, R. C.: Benchmarking. München: Hanser 1994.

[Cho78] Chow, W.: Cost reduction in product design. New York: Van Nostrand Reinhold Company 1978.

[Cla91] Clark, K. B.; Fujimoto, T.: Product Development Performance: Strategy, Organization and Management in the World Auto Industry. Boston, Mass.: Harvard Business School Press 1991.

[Cla92] Clark, K. B.; Fujimoto, T.: Automobilentwicklung mit System. Strategie, Organisation und Management in Europa, Japan und USA. Frankfurt/Main: Campus 1992.

[Coe94] Coenenberg, A. G.; Fischer, T.; Schmitz, J.: Target Costing und Product Life Costing als Instrumente des Kostenmanagements. Zeitschrift für Planung (1994) 5, S. 1–38.

[Con98] Conrat, J. I.: Änderungskosten in der Produktentwicklung. München: TU, Diss. 1998.

[Coo97] Cooper, R; Slagmulder, R.: TARGET COSTING and VALUE ENGINEERING. IMA Publication Number 240-97319 (1997).

[Cre90] Creese, R. C.; Moore, T.: Cost Modelling for Concurrent Engineering , in: Cost Engineering, 32. Jg. (1990) H. 6, S. 23–27.

[Dae95] Daenzer, W. F.; Huber, F.: Systems Engineering. 9. Aufl. Zürich: Verl. Industr. Org. 1995.

[Dan96] Danner, S.: Ganzheitliches Anforderungsmanagement mit QFD – ein Beitrag zur Optimierung marktorientierter Entwicklungsprozesse. Aachen: Shaker 1996. (Konstruktionstechnik München, Band 24). Also: München: TU, Diss. 1996.

[Das88] Daschbach, J. M.; Apgar, H.: Design Analysis through Techniques of Parametric Cost Estimation. In: Engineering Costs and Production Economics, 14 (1988) 2, S. 87–93.

[Deb98] Debuschewitz, M.: Integrierte Methodik und Werkzeuge zur herstellkostenorientierten Produktentwicklung. Berlin: Springer Verlag 1998. (Reihe Forschungsberichte: Berichte aus dem Institut für Werkzeugmaschinen und Betriebswissenschaften der Technischen Universität München Bd. 118). Also: München: TU, Diss. 1998.

[DeJ56] De Jong, J. R.: Fertigkeit, Stückzahl und benötigte Zeit. REFA-Nachr., Sonderheft, Darmstadt 1956.

[Dem86] Deming, W. E.: Out of the crisis. 2nd ed. Cambridge: 1986.

[Der71] Derndinger, H. O.: Einfluß der Massenfertigung auf die konstruktive Gestaltung. Wt–Z. ind. Fertig. 61 (1971), S. 284–287.

[Der95] Derhake, T.: Konstruktionsmanagement im Sondermaschinenbau. In: Proceedings of the International Conference on Engineering Design 1995, Vol. 4, 1995, S. 400ff. (Schriftenreihe WDK)

[Die88] Diels, H.: Kostenwachstumsfunktionen als Hilfsmittel zur Kostenfrüherkennung – Ihre Erstellung und Anwendung. Düsseldorf: VDI-Verlag 1988. (Fortschrittsberichte VDI Reihe 1, Nr. 157)

[Die89] Diels, O. A.: Systematischer Aufbau von Methodenbanken für die Arbeitsplanung dargestellt am Beispiel der Arbeitsplanerstellung und NC-Programmierung. Aachen: RWTH, Diss. 1989.

[DIN74] DIN 323: Normzahlreihen. Berlin: Beuth 1974.

[DIN77] DIN 199, Teil 2: Begriffe im Zeichnungs- und Stücklistenwesen. Berlin: Beuth 1977.

[DIN87a] DIN 32 991, Teil 1: Kosteninformationen. Kosteninformations-Unterlagen Gestaltungsgrundsätze Berlin: Beuth 1987.

[DIN87b] DIN 32 992, Teil 3: Kosteninformationen, Berechnungsgrundlagen, Ermittlung von Relativkostenzahlen. Berlin: Beuth 1987.

[DIN87c] DIN-Manuskriptdruck: Kosteninformationen zur Kostenfrüherkennung. Berlin: Beuth 1987.

[DIN87d] DIN 69 910: Wertanalyse. Berlin: Beuth 1987.

[DIN89a] DIN 32 990, Teil 1: Kosteninformationen, Begriffe zu Kosteninformationen in der Maschinenindustrie. Berlin: Beuth 1989.

[DIN89b] DIN 32 992, Teil 1: Kosteninformationen, Berechnungsgrundlagen, Kalkulationsarten und -verfahren. Berlin: Beuth 1989.

[DIN93] DIN 32 992, Teil 2: Kosteninformationen, Berechnungsgrundlagen, Verfahren der Kurzkalkulation. Berlin: Beuth 1993.

[DIN94] DIN EN ISO 9 004-4: Qualitätsmanagement und Elemente eines Qualitätsmanagementsystems, Teil 4. Berlin: Beuth 1994, S. 42–92.

[Dom85] Domin, A; Maskow, J.: Design to Cost – eine Methode zur Kostenreduzierung in der Produktentwicklungsphase. Konstruktion 37 (1985) 19, S. 395–399.

[Dun82] Dunkler, H.; Hedrich, P.: Fertigungstechnische Beratung durch die Fertigungsvorbereitung. Düsseldorf: VDI-Verlag 1982, S. 43–52. (VDI-Berichte 457)

[Dud92] Dudick, T. S.: MANUFAKTURING COST CONTROLS. Prentice-Hall 1992.

[Dut93] Dutschke, W.: Fertigungsmeßtechnik. Stuttgart: Teubner 1993.

[Dyl91] Dylla, N.: Denk- und Handlungsabläufe beim Konstruieren. München: Hanser 1991. (Konstruktionstechnik München, Band 5). Also: München: TU, Diss. 1990.

[Eck77] Eckert, F.: Erstellung eines Material-Relativkosten-Kataloges. München: TU, Lehrstuhl für Konstruktion im Maschinenbau, Unveröffentlichte Studienarbeit 1977.

[Ehr73] Ehrlenspiel, K.; Dehner, E.: Planetengetriebe für Schiffsantriebe. Hansa 110 (1973) 4, S. 289–295.

[Ehr77] Ehrlenspiel, K.: Betriebsanforderungen an Maschinen – Bedeutung und Einteilung. Konstruktion 29 (1977), S. 29–35.

[Ehr78] Ehrlenspiel, K.; Balken, J.: Auswertung von Wertanalysen zur Ermittlung von Kosteneinflüssen und Hilfsmitteln zum kostenarmen Konstruieren. München: TU, Lehrstuhl für Konstruktion im Maschinenbau, DFG-Ber. zu Projekt Eh 46/6, 1978.

[Ehr79] Ehrlenspiel, K.; Kiewert, A.; Lindemann, U.: Kostenfrüherkennung im Konstruktionsprozeß. Düsseldorf: VDI-Verlag 1979, S. 133–142. (VDI-Berichte 347)

[Ehr80a] Ehrlenspiel, K.: Möglichkeiten zum Senken der Produktkosten – Erkenntnisse aus einer Auswertung von Wertanalysen. Konstruktion 32 (1980) 5, S. 173–178.

[Ehr80b] Ehrlenspiel, K.: Genauigkeit, Gültigkeitsgrenzen, Aktualisierung der Erkenntnisse und Hilfsmittel zum kostengünstigen Konstruieren. Konstruktion 32 (1980) 12, S. 487–492.

[Ehr82a] Ehrlenspiel, K.; Fischer, D.: Relativkosten von Stirnrädern in Einzel- und Kleinserienfertigung. Abschlußbericht Teil I und II. FVA Forschungsvorhaben Nr. 61. Heft 116 u. 146. Frankfurt/M.: Forschungsverein. Antriebstech. e. V. 1982 u. 1983.

[Ehr82b] Ehrlenspiel, K.; Fischer, D.: Kostenanalyse von Zahnrädern – Ergebnisse eines FVA-Forschungsvorhabens. Düsseldorf: VDI-Verlag 1982, S. 143–154. (VDI-Berichte 457)

[Ehr83] Ehrlenspiel, K.; Hillebrand, A.; Rutz, A.: Kostenvergleichssystem Gußteile. In: Abschlußbericht KfK-PFT 48 Kosteninformationssystem, S. 77–132, Karlsruhe 1983.

[Ehr85] Ehrlenspiel, K.: Kostengünstig Konstruieren. Berlin: Springer 1985.

[Ehr87a] Ehrlenspiel, K.; Sauermann, H.-J.: Produktkostenplanung in Teamarbeit. Harvard manager (1987) 4, S. 113–118.

[Ehr87b] Ehrlenspiel, K.: Methoden zur Kostenfrüherkennung. In: Herstellkosten im Griff? Düsseldorf: VDI-Verlag 1982, S. 13–32. (VDI-Berichte 651)

[Ehr91] Ehrlenspiel, K.: Kittsteiner, H.-J.: Auswahl und Gestaltung kostengünstiger Welle-Nabe-Verbindungen. Antriebstechnik 30 (1991) 10, S. 73–80.

[Ehr93] Ehrlenspiel, K.: Industrieprobleme in Entwicklung und Konstruktion sowie Folgerungen gemäß einer Umfrage. Konstruktion 45 (1993), S. 389–396.

[Ehr93a] Ehrlenspiel, K.; Milberg, J.; Schuster, G.; Wach, J.: Rechnerintegrierte Produktkonstruktion und Montageplanung. CIM Management (1993) 2, S. 23–28.

[Ehr93b] Ehrlenspiel, K.; Seidenschwarz, W.: Target Costing, ein Rahmen für kostenzielorientiertes Konstruieren – eine Praxisdarstellung. In: VDI-Gesellschaft EKV (Hrsg.): Konstrukteure gestalten Kosten. Düsseldorf: VDI-Verlag 1993, S. 167–187.

[Ehr96] Ehrlenspiel, K.; Lindemann, U.; Kiewert, A.; Steiner, M.: Konstruktionsbegleitende Kalkulation in der integrierten Produktentwicklung. Kostenrechnungspraxis (1996) Sonderheft 1, S. 69–76.

[Ehr97] Ehrlenspiel, K.: Knowlege-explosion and its consequences. In: Riitahuhta, A. (Ed.): Proceedings of the 11th International Conference on Engineering Design 1997, Vol. 2, Tampere, Finnland 1997. Tampere: Univ. of Technology 1997, S. 474–484. (Schriftenreihe WDK 25)

[Ehr98] Ehrlenspiel, K.; Bernard, R.; Mußbach, G.: Methodische Auswahl und Gestaltung von Kunststoffverbindungen in der Praxis. In: Verbindungstechniken bei Kunststoff-Bauteilen in der Serienfertigung, 1./2.4.1998, Würzburg. Würzburg: Süddeutsches Kunststoff-Zentrum 1998, S. A1–A18.

[Ehr06] Ehrlenspiel, K.: Integrierte Produktentwicklung: Methoden für die Prozeßorganisation, Produkterstellung und Konstruktion. München: Hanser 1995, 3. Aufl. 2006.

[Eil98] Eiletz, R.: Zielkonfliktmanagement in der Entwicklung komplexer Produkte. München: TU, Diss. 1998.

[End00] Endebrock, K.: Ein Kosteninformationsmodell für die frühzeitige Kostenbeurteilung in der Produktentwicklung. Aachen: Shaker Verlag 2000. Also: Bochum: Univ., Diss. 2000.

[Eng93] Engemann, B.: Schneiden mit Laserstrahlung und Wasserstrahl. Ehningen bei Böblingen: expert 1993. (Kontakt und Studium, Band 362)

[EUR95] European Commission (Hrsg.): Value Management Handbook. Luxemburg: 1995.

[Eve77] Eversheim, W.; Minolla, W.; Fischer, W.: Angebotskalkulation mit Kostenfunktionen in der Einzel- und Kleinserienfertigung. Berlin: Beuth 1977.

[Eve88] Eversheim, W.; Schuh, G; Caesar, C.: Variantenvielfalt in der Serienproduktion. VDI-Z 130 (1988) 12, S. 45–49.

[Eve90] Eversheim, W.; Caesar, C.: Kostenmodell zur Bewertung von Produktvarianten – Das PC-Programmsystem KOMO. VDI-Z 132 (1990) 6, S. 75–79.

[Eve92] Eversheim, W.; Böhmer, D.; Kümper, R.: Die Variantenvielfalt beherrschen. Entwicklung geeigneter Organisationsformen – Praxisbeispiel Automobilindustrie. VDI-Z 134 (1992) 4, S. 47–53.

[Eve95] Eversheim, W.; Bochtler. W.; Laufenberg, L.: Simultaneous Engineering. Berlin: Springer 1995.

[Eve97a] Eversheim, W.; Kümper, R.: Prozeß- und ressourcenorientierte Vorkalkulation in den Phase der Produktentstehung. In: Männel, W. (Hrsg): Frühzeitiges Kostenmanagement: Kalkulationsmethoden und DV-Unterstützung. Wiesbaden: Gabler 1997, S. 91–107.

[Eve97b] Eversheim, W.; Warnke, L.; Schröder, T.: Änderungsmanagement in Entwicklungskooperationen. VDI-Z 139 (1997) 3, S. 60–63.

[Fer87] Ferreirinha, P.: Rechnerunterstützte Vorkalkulation im Maschinenbau für den Konstrukteur und den Arbeitsvorbereiter mit dem HKB-Programm. Düsseldorf: VDI-Verlag 1987, S. 343–350. (VDI-Berichte 651)

[Fer92] Ferlemann, F.: Schleifen mit höchsten Schnittgeschwindigkeiten. Düsseldorf: VDI-Verlag 1992. (Fortschrittsberichte VDI Reihe 2, Nr. 280)

[Fig88] Figel, K.: Optimieren beim Konstruieren: Einsatz von Optimierungsverfahren, CAD und Expertensystemen. München: Hanser 1988. Also: München: TU, Diss. 1988.

[Fis83] Fischer, D.: Kostenanalyse von Stirnzahnrädern. Erarbeitung und Vergleich von Hilfsmitteln zur Kostenfrüherkennung. München: TU, Diss. 1983.

[Fis93] Fischer, J.; Koch, R.; Schmidt-Faber, B.; Hauschulte, K.-B.: Gemeinkostenvermeiden durch entwicklungsbegleitende Prozeßkostenkalkulation. Ein Ansatz zur prozeßsynchronen Prognose von Lebenszykluskosten. In: Horváth, P. (Hrsg.): Marktnähe und Kosteneffizienz schaffen. Stuttgart: 1993, S. 259–274.

[Fis94] Fischer, J.; Koch, R.; Hauschulte, K.-B.; Jakuschona, K.: Lebenszyklusorientierte Prozeßkostenplanung in frühen Konstruktionsphasen. ZwF (1994) 11, S. 566–568.

[Fra82] Franke, H.-J.: Berücksichtigung der Produkt-Gesamtkosten bei Konzeption und Gestaltung von Maschinen und Geräten. Düsseldorf: VDI-Verlag 1982, S. 71–77. (VDI-Berichte 457)

[Fra87] Franke, H.-J.; Schill, J.: Kosten senken durch Einsparen von Teilen. Düsseldorf: VDI-Verlag 1987, S. 139–152. (VDI-Berichte 651)

[Fra93] Franz, K.-P.: Target Costing. Konzept und kritische Bereiche. Controlling (1993) 3, S. 124–130.

[Fra97] Franz, K.-P.; Kajüter, P. (Hrsg): Kostenmanagement: Wettbewerbsvorteile durch systematische Kostensteuerung. Stuttgart: Schäffer-Poeschel 1997.

[Fra02] Franz, K.-P.; Kajüter, P. (Hrsg): Kostenmanagement: Wertsteigerung durch systematische Kostensteuerung. Stuttgart: Schäffer-Poeschel 2002.

[Fra98] Franken, T.: Modellbasierte Beherrschung von Konstruktionsabläufen. Düsseldorf: VDI-Verlag 1998.

[Fran97] Frankenberger, E.: Arbeitsteilige Produktentwicklung. Empirische Untersuchung und Empfehlungen zur Gruppenarbeit in der Konstruktion. Darmstadt: TU, Diss. 1997.

[Fre78] Fremgens, G. J.: Kostensenkung durch Sortenverminderung. DIN-Mitt. 57 (1978), S. 337–339.

[Fre98] Frech, J. T.: Ein Verfahren zur integrierten, prozessbegleitenden Vorkalkulation für die kostengerechte Konstruktion. Berlin: Springer 1998. Also: Stuttgart: Univ., Diss. 1998.

[Frö94] Fröhling, O.: Zielkostenspaltung als Schnittstelle zwischen Target Costing und Target Cost Management. krp (1994) 6, S. 421–425.

[Gai81] Gairola, A.: Montagegerechtes Konstruieren – Ein Beitrag zur Konstruktionsmethodik. Darmstadt: TU, Diss. 1981.

[Gam96] Gamber, P.: Ideen finden, Probleme lösen – Methoden, Tips und Übungen für einzelne und Gruppen. Weinheim: Beltz 1996.

[Gau98] Gauer, C.; Skriba, J.: Die Standortlüge – Abrechnung mit einem Mythos. Frankfurt: Fischer 1998.

[Gau00] Gausemeier, J., Lindemann, U. Reinhart, G. Wiendahl, H.-P.: Kooperatives Produktengineering: Ein neues Selbstverständnis des ingenieurmäßigen Wirkens. Paderborn: HNI, 2000 (HNI-Verlagschriftenreihe; Bd. 79).

[Gem95] Gemmerich, M.: Technische Produktänderungen: Betriebswirtschaftliche und empirische Modellanalyse. Wiesbaden: Gabler 1995. Also: Passau: Univ., Diss. 1995.

[Gem98] Gembrys, S.-N.: Ein Modell zur Reduzierung der Variantenvielfalt in Produktionsunternehmen. Berlin: IPK 1998.

[Ger94] Gerhard, E.: Kostenbewußtes Entwickeln und Konstruieren. Renningen-Malmsheim: expert 1994.

[Ger02] Gerst, M.: Strategische Produktentscheidungen in der integrierten Produktentwicklung. München: TU, Diss. 2002.

[Geu96] Geupel, H.: Konstruktionslehre. Berlin: Springer 1996.

[Gie92] Giere, R.: Einflüsse des Laserstrahlschneidens auf die mechanisch/technologischen Eigenschaften von Feinkornbaustählen und Aluminiumlegierungen. Düsseldorf: VDI-Verlag 1992. (Fortschrittsberichte VDI Reihe 5, Nr. 267)

[Gle96] Gleich, R.: Target Costing für die montierende Industrie. München: Vahlen 1996. Also: Stuttgart: Univ., Diss. 1996.

[Gle97] Gleich, R.; Brokemper, A.: Gemeinkostenanalyse und Prozeßkostenrechnung in der Antriebstechnikbranche. Antriebstechnik 36 (1997) 2, S. 57–61.

[Göt78] Goetze, H.: Kostenplanung technischer Systeme am Beispiel der Werkzeugmaschine. Berlin: TU, Diss. 1978.

[Gor61] Gordon, W. J. J.: Synectics, the Development of Creative Capacity. New York: Harper 1961.

[Gra97a] Grabowski, H.; Geiger, K. (Hrsg.): Neue Wege zur Produktentwicklung. Stuttgart: Raabe 1997.

[Gra97b] Graßhoff J.; Gräfe, C.: Kostenmanagement in der Produktentwicklung. Controlling (1997) 1, S. 14–23.

[Grä98] Gräfe, C. Kostenmanagement im Entstehungszyklus eines Serienerzeugnisses. Verlag Dr. Kovac, Hamburg 1998.

[Gre88] Green, P. E.; Tull, D. S.; Albaum, G.: Research for Marketing Decisions. New Jersey: Prentice Hall 1988.

[Grö91] Gröner, L.: Entwicklungsbegleitende Vorkalkulation. Berlin: Springer 1991.

[Gro04] Grob, H. L.; Lahme, N.: Total Cost of Ownership-Analyse mit vollständigen Finanzplänen. Controlling H3 2004, S157–164.

[Gut98] Gutzmer, P.: Management und Umsetzung der Gleichteilstrategie bei den neuen Porschefahrzeugen Boxster und 911. In: 2. Fachkonferenz Automobilentwicklung & Management `98. 29. und 30. Juni 1998 in Wiesbaden.

[Hab96] Habenicht, W.: Teilefamilienbildung. In: Kern, W.; Schröder, H. H.; Weber, J. (Hrsg.): Handwörterbuch der Produktionswirtschaft. Stuttgart: Schäffer-Poeschel 1996, S. 2040–2050.

[Hab97] Habenicht, G.: Kleben. 3., völlig neubearb. u. erw. Aufl. Berlin: Springer 1997.

[Haf87] Hafner, J.: Entscheidungshilfen für das kostengünstige Konstruieren von Schweiß- und Gußgehäusen. München: TU, Diss. 1987.

[Hal93] Hales, C.: Managing Engineering Design. Harlow: Longman 1993.

[Ham93] Hamelmann, S: Rechnerunterstützte Arbeitsplanung – was gibt der Markt her? AV 30 (1993) 2, S. 51–55.

[Hei91] Heinen, E.: Industriebetriebslehre. Entscheidungen im Industriebetrieb. Wiesbaden: Gabler 1991.

[Hei93a] Heil, H.-G.: Kosten senken in der Konstruktion. Analyse von Kostensenkungspotentialen. Frankfurt/M.: Maschinenbau-Verlag 1993.

[Hei93b] Heidutzek, R.: Wissensbasierte Angebotskalkulation und Fertigungsplanung. ZWF CIM 88 (1993) 11. Sonderdruck.

[Hei94] Heider, P.: Lasergerechte Konstruktion und lasergerechte Fertigungsmittel zum Schweißen großformatiger Aluminium-Strukturbauteile. Düsseldorf: VDI-Verlag 1994. (Fortschrittsberichte VDI Reihe 2, Nr. 326)

[Hei95] Heine, A.: Entwicklungsbegleitendes Produktkostenmanagement. Gestaltung des Führungssystems am Beispiel der Automobilindustrie. Wiesbaden: Gabler 1995.

[Hein95] Heine, A.: Entwicklungsbegleitendes Produktkostenmanagement: Gestaltung des Führungssystems am Beispiel der Automobilindustrie. Wiesbaden: Gabler 1995. Also: Koblenz: Wiss. Hochschule für Unternehmensführung, Diss. 1994.

[Hel78] Hellfritz, H. (Hrsg.): Innovation via Galeriemethode. Königsstein/Taunus: Eigenverlag 1978.

[Hen84] Henderson, B.: Die Erfahrungskurve in der Unternehmensstrategie. Frankfurt/M.: Campus 1984.

[Hen93] Henzler, H. H.; Späth, L.: Sind die Deutschen noch zu retten? München: Bertelsmann: 1993.

[Her00] Herb, R.; Kohnhauser, V.: TRIZ. Der systematische Weg zur Innovation. Landsberg: Moderne Industrie 2000.

[Hic85] Hichert, R.: Probleme der Vielfalt, Teil 1 „Was kostet eine Variante?“ Wt-Z. 75 (1985), S. 235–237.

[Hic86] Hichert, R.: Probleme der Vielfalt, Teil 2 „Was kostet eine Variante?“ Wt-Z. 76 (1986), S. 141–145.

[Hil86] Hillebrand, A.; Ehrlenspiel, K.: Suchkalkulation – ein Hilfsmittel zum kostengünstigen Konstruieren. CAD-CAM REPORT (1986) 1, S. 53–57.

[Hil90] Hillebrand, A.: Ein Kosteninformationssystem für die Neukonstruktion. München: Hanser 1991. (Konstruktionstechnik München, Band 4). Also: München: TU, Diss. 1990.

[Hom97] Homburg, C; Daum, D.: Wege aus der Komplexitätskostenfalle. ZWF 92 (1997) 7-8, S. 333–337.

[Hor91] Horváth, P.; Mayer, R.: Prozeßkostenrechnung – Der neue Weg zu mehr Kostentransparenz und wirkungsvolleren Unternehmensstrategien. Controlling 1 (1989) 4, S. 214–219.

[Hor93] Horváth, P. (Hrsg.): Target Costing – Marktorientierte Zielkosten in der deutschen Praxis. Stuttgart: Schäffer-Poeschel 1993.

[Hor96] Horváth, P.; Gleich, R.; Scholl, K.: Vergleichende Betrachtung der bekanntesten Kalkulationsmethoden für das kostengünstige Konstruieren. krp Sonderheft 1/1996, S. 53–61.

[Hor97] Horváth, P; Scholl, K.; v. Wangenheim, S.: Controlling in der Produktentwicklung. Controlling (1997) Jahrbuch, S. 1–11.

[Hua96] Huang, G. Q.: Design for X. London: Chapman Hall 1996.

[Hub92] Hubka, V.; Eder, E. W.: Einführung in die Konstruktionswissenschaft. Berlin: Springer 1992.

[Hub95a] Huber, Th.: Senken von Montagezeiten und -kosten im Getriebebau; München: Hanser 1995. (Konstruktionstechnik München, Band 20). Also: München: TU, Diss. 1994.

[Hub95b] Huber, Th.; Ehrlenspiel, K.: Kostensenken bei der Getriebemontage Antriebstechnik 34 (1995) 10, S. 61–63 (Teil I) u. 34 (1995) 11, S. 63–65 (Teil II).

[Hum04] Humphreys, K.: Projekt und Cost Engineers` Handbook. New York: Marcel Decker 2004.

[Hun97] Hundal, M. S.: Systematic Mechanical Designing. New York: ASME Press 1997.

[Ima93] Imai, M.: Kaizen – der Schlüssel zum Erfolg der Japaner im Wettbewerb. 10. Aufl. München: Lingen Müller/Herbig 1993.

[Jes96] Jeschke, A.: Beitrag zur wirtschaftlichen Bewertung von Standardisierungsmaßnahmen in der Einzel- und Kleinserienfertigung durch die Konstruktion. Braunschweig: TU, Diss. 1996.

[Käs74] Käser, A.: Wo stehen wir auf dem Gebiet der Vorbestimmung der Schnittdaten und der Arbeitszeit in der spanenden Fertigung? REFA-Nachr. 27 (1974) 1, S. 21–32.

[Kan93] Kannheiser, W.; Hormel, R.; Aichner, R.: Planung im Projektteam. München: Hampp 1993.

[Kap96] Kaplan, S.: An Introduction to TRIZ: The Russian Theory of Inventive Problem Solving. Southfield, Michigan: Ideation International 1996.

[Kap98] Kaplan, R; Cooper, R.: Cost & Effects. Boston 1998.

[Ker96] Kern, W. (Hrsg.): Handwörterbuch der Produktionswirtschaft. Stuttgart: Schäffer-Poeschel 1996.

[Ker99] Kersten, W.: Wirksames Variantenmanagement durch Einbindung in den Controlling- und Führungsprozeß im Unternehmen. In: Plattformkonzepte auch für Kleinserien und Anlagen. Düsseldorf: VDI-Verlag 1999, S. 155–175.

[Kes54] Kesselring, F.: Technische Kompositionslehre. Berlin: Springer 1954.

[Ket71] Kettner, H.; Klingenschmitt, V.: Die morphologische Methode und das Lösen konstruktiver Aufgaben. 1. Teil: Wt-Z. d. ind. Fertig. 61 (1971), S. 737–741. 2. Teil: Wt-Z. d. ind. Fertig. 63 (1973), S. 357–363.

[Kie79] Kiewert, A.: Systematische Erarbeitung von Hilfsmitteln zum kostenarmen Konstruieren. München: TU, Diss. 1979.

[Kie82] Kiewert, A.: Kurzkalkulation und die Beurteilung ihrer Genauigkeit. VDI-Z 124 (1982) 12, S. 443–446.

[Kie88] Kiewert, A.: Wirtschaftlichkeitsbetrachtungen zum kostengünstigen Konstruieren. Konstruktion 40 (1988), S. 301–307.

[Kin89] King, B.: Better Designs in Half the Time. 3. Aufl. Methuen, MA: GOAL/QPC 1989, S. E-1.

[Kin94a] Kinzel, A.: Beitrag zum Laserstrahlschweißen von Feinblechen. Düsseldorf: VDI-Verlag 1994. (Fortschrittsberichte VDI Reihe 5, Nr. 347)

[Kin94b] King, B.: Doppelt so schnell wie die Konkurrenz – Quality Function Deployment. St. Gallen: gmft 1994.

[Kit90] Kittsteiner, H.-J.: Die Auswahl und Gestaltung von kostengunstigen Welle-Nabe-Verbindungen. München: Hanser 1990. (Konstruktionstechnik München, Band 3). Also: München: TU, Diss. 1996.

[Kle98] Kleedörfer, R.: Prozeß- und Änderungsmanagement der Integrierten Produktentwicklung. München: TU, Diss. 1998.

[Klu94] Kluge, J. u. A.: Wachstum durch Verzicht. Stuttgart: Schäffer-Poeschel 1994.

[Koc94] Koch, R.; Fischer, J.; Jakuschona, K.; Kou-I Szu; Hauschulte, K.-B.: Konstruktionsbegleitende Kalkulation auf der Basis eines Prozeßkostenansatzes. Konstruktion 46 (1994), S. 427–433.

[Kön95] König, T.: Konstruktionsbegleitende Kalkulation auf der Basis von Ähnlichkeitsvergleichen. Bergisch Gladbach: Eul, 1995.

[Koh96] Kohlhase, N.: Marktgerechte Baukastenentwicklung – Strategien, Methoden und Instrumente. Darmstadt: TH, Diss. 1996.

[Koh97] Kohlhase, N.: Integrated Variant Reduction in Practice. In: Riitahuhta, A. (Ed.): Proceedings of the 11th International Conference on Engineering Design 1997, Vol. 3, Tampere, Finnland 1997. Tampere: Univ. of Technology 1997, S. 175–180. (Schriftenreihe WDK 25)

[Koh98] Kohlhase, N.; Schnorr, R.; Schlücker, E.: Reduzierung der Variantenvielfalt in der Einzel- und Kleinserienfertigung. Konstruktion 50 (1998) 6, S. 15–21.

[Koh99] Kohlhase, N.: Beherrschung der Variantenvielfalt. Seminarunterlagen, TA Esslingen 1999.

[Kok99] Kokes, M.; Neff, T.; Schulz, M.: Zielkostenmanagement in unternehmensübergreifenden Wertschöpfungsketten durch Front Load Costing. DAIMLER-CHRYSLER Technische Notiz Nr. TN-FT"/K-1999-005.

[Kol94] Koller, R.: Konstruktionslehre für den Maschinenbau. Berlin: Springer 1994.

[Kre81] Krehl, H.: Erfolgreiche Produkte durch Value Management. In: Proceedings of the International Conference on Engineering Design 1981. Zürich: Edition Heurista 1981, S. 246–253. (Schriftenreihe WDK 20)

[Kre97] Kreuz, W.: Kostenbenchmarking. Konzept und Praxisbeispiel. In: Franz, K. P.; Kajüter, P. (Hrsg.): Kostenmanagement. Stuttgart: Schäffer-Poeschel 1997.

[Kru98] Kruschewitz, R.: Plattformstrategie und Gleichteileverwendung im Volkswagenkonzern. In: 2. Fachkonferenz Automobilentwicklung & Management `98. 29. und 30. Juni 1998 in Wiesbaden.

[Küh86] Kühborth, W.: Baureihen industrieller Erzeugnisse zur optimalen Nutzung von Kostendepressionen. Mannheim: Univ., Diss. 1986.

[Lak93] Laker, M.: Target Costing nicht ohne Target Pricing: Was darf ein Produkt kosten? Gablers Magazin (1993) 3, S. 61–63.

[Lan72] Langheinrich, G.: Rationelle und sichere Vorkalkulation nach Rechenprogrammen mit elektronischen Tischrechnern. Werkst. u. Betr. 105 (1972), S. 21–27.

[Lan74] Langheinrich, G.: Die neuzeitliche Vorkalkulation der spanenden Fertigung im Maschinenbau. Berlin: Technischer Verlag Herbert Cramm 1974.

[Lan96] Lange, J.-U.; Schauer, B. D.: Ausgestaltung und Rechnungszwecke mittelständischer Kostenrechnung. krp 40 (1996) 4, S. 202–208.

[Las00] Lashin, G.: Baukastensystem für modulare Straßenbahnfahrzeuge. Konstruktion (2000) 1/2, S. 61–66.

[Lei97] Leidich, E.; Schumann, F. J.: Target Cost Design based on Working Structure Cost Models. In: Riitahuhta, A. (Ed.): Proceedings of the 11th International Conference on Engineering Design 1997, Vol 2, Tampere, Finnland 1997. Tampere: Univ. of Technology 1997, S. 499–503. (Schriftenreihe WDK 25).

[Lei01] Leidich, E.; Jurklies. I.; Schumann, F. J. : Fuzzybasierte Kostenprognose in der Konzeptphase. Konstruktion (2001) 3, S. 83–87.

[Les64] Leslie, H. L. G.: Design guide to value. Prod. Eng. 28.10.1963 u. 2.3.1964.

[Lin80] Lindemann, U.: Systemtechnische Betrachtung des Konstruktionsprozesses unter besonderer Berücksichtigung der Herstellkostenbeeinflussung beim Festlegen der Gestalt. Düsseldorf: VDI-Verlag 1980.

[Lin84] Lindemann, U.: CAD-Körpermodelle ermöglichen neue Zahnradgetriebe-Konstruktion. CAMP-Bericht. Berlin: 1984.

[Lin90] Lindemann, U.; Peiker, S.: Integration von Konstruktion und Fertigung im Getriebebau. In: Rechnerunterstützte Produktentwicklung, Bad Soden, 01./02.03.1990. Düsseldorf: VDI 1990. (VDI Berichte 812).

[Lin92] Lindemann, U.: Zeit- und Kostenmanagement – Herausforderung und Hilfe für den Konstrukteur. In: Praxiserprobte Methoden erfolgreicher Produktentwicklung. Düsseldorf: VDI-Verlag 1992. (VDI-Berichte 953)

[Lin93a] Linde, H.; Hill, B.: Erfolgreich erfinden – Widerspruchsorientierte Innovationsstrategie. Darmstadt: Hoppenstedt 1993.

[Lin93b] Linner, S.: Konzept einer integrierten Produktentwicklung. München: TU, Diss. 1993.

[Lin98a] Lindemann, U.; Birkhofer, H.; Amft, M.; Assmann, G.; Wallmeier, S.; Wulf, J.: Rechnerunterstützung für frühe Phasen der Entwicklung. F&M 106 (1998), S. 123–127.

[Lin98b] Lindemann, U.; Reichwald, R. (Hrsg.): Integriertes Änderungsmanagement. Berlin: Springer 1998.

[Lor76] Lorenzen, H.: Wirtschaftliche Produktgestaltung. In: Leistungssteigerung von Entwicklung und Forschung im Maschinenbau. Frankfurt/M.: VDMA 1976.

[Lin01] Lindemann, U.; Stricker, H.; Gramann, J.; Pulm, U.: Kosteneinsparungen in Wertanalysen – Eine Systematik zur Wirkungskontrolle. ZWF 96 (2001) 10, S. 543–546.

[Lin05] Lindemann, U.: Methodische Entwicklung technischer Produkte. Berlin: Springer 2005.

[Lot86] Lotter, B.: Wirtschaftliche Montage. Ein Handbuch für Elektrogerätebau und Feinwerktechnik. Düsseldorf: VDI-Verlag 1986.

[Lüt76] Lüttgert, H.: Fortschritte bei Bauelementen des Maschinenbaus und der Elektrotechnik. Düsseldorf: VDI-Verlag 1976, S. 121–129. (VDI-Berichte 229)

[Mag82] Mager, H.: Moderne Regressionsanalyse. Frankfurt/M.: Sauerländer 1982.

[Mai01] MAI, A.: Ein Marketingplan macht noch keine Strategie aus. Absatzwirtschaft (2001) 3, S. 134–135.

[Mal93] Malhotra, N. K.: Marketing Research. New Jersey: Prentice Hall 1993.

[Män90] Männel, W.: Wahl zwischen Eigenfertigung und Fremdbezug in der Praxis. Wiesbaden: Gabler 1990.

[Män97] Männel, W. (Hrsg.): Frühzeitiges Kostenmanagement: Kalkulationsmethoden und DV-Unterstützung. Wiesbaden: Gabler 1997.

[Mas01] www.maschinenmarkt.de/german/framedef/service/software/winkalk/fachartikel

[Mat57] Matousek, R.: Konstruktionslehre des allgemeinen Maschinenbaues. Berlin: Springer 1957.

[Mat98] Matt, D.: Objektorientierte Prozeß- und Strukturinnovation – Methode und Leitfaden zur Steigerung der Produktivität indirekter Leistungsprozesse. Karlsruhe: TU, Diss. 1998. (Forschungsberichte aus dem Institut für Werkzeugmaschinen und Betriebstechnik der Universität Karlsruhe, Bd. 80)

[Mau01] Mauerer, H.; Renaud. M.: Hoch integrierte Lösungen – Konstruktive Möglichkeiten mit Blasformteilen aus technischen Kunststoffen. Der Konstrukteur (2001) 5, S. 26–28.

[Mee89] Meerkamm, H.; Finkenwirth, K.: Konstruktionssystem Fertigungsgerecht – ein Expertensystem für den Konstrukteur? Düsseldorf: VDI-Verlag 1989, S. 99–115. (VDI-Berichte 775)

[Mei77] Meinl, F.: Kosten der Einführung eines neuen Bauteils. DIN-Mitt. 56 (1977), S. 469–476.

[Mei92] Meinecke, C.: Konstruktionsbegleitende Kostenermittlung für Erzeugnisse im Rahmen einer computerintegrierten Produktion. Hannover: Univ., Diss. 1992.

[Mel92] Melchert, M.: Entwicklung einer Methode zur systematischen Planung von Make-or-Buy-Entscheidungen. Aachen: Shaker 1992.

[Mer94] Mertins, K.; Kempf, S.; Siebert, G.: Benchmarking, ein Management-Werkzeug. ZwF 89 (1994) 7/8, S. 359–361.

[Mer96] Merat, P.: Rechnergestützte Auftragsabwicklung an einem Praxisbeispiel. Aachen: Shaker 1996. (Konstruktionstechnik München, Band 25). Also: München: TU, Diss. 1996.

[Mic89] Michaels, J. V.; William, P. W.: Design to Cost. New York: Wiley 1989.

[Mil87] Miles, L. D.: Value Engineering. Wertanalyse, die praktische Methode zur Kostensenkung. 2. Aufl. München: Moderne Industrie 1987.

[Mir91] MIRAKON Vertriebsunterlagen: Kurzbeschreibung von HKB Rev. 6.0. Hittnau: 1991.

[Mon89] Monden, Y.: Total cost management system in Japanese automobile corporations. In: Monden, Y.; Sakurai, M. (Ed.): Japanese Management Accounting. A World Class Approach to Profit Management. Cambridge, Mass.: Productivity Press 1989, S. 15–34.

[Mon99] Monden, Y.; Hoque, M.: Target Costing based on QFD. Controlling (1999) 11, S. 525–534.

[Mör02] Mörtl, M.: Entwicklungsmanagement für langlebige, upgradinggerechte Produkte München: Dr. Hut 2002. (Produktentwicklung München, Band 51). Also: München: TU, Diss. 2002.

[Mül91] Müller, R.: Datenbankgestützte Teileverwaltung und Wiederholteilsuche. München: Hanser 1991. (Konstruktionstechnik München, Band 6). Also: München: TU, Diss. 1990.

[Mül93] Müller, H.: Prozeßkonforme Grenzplankostenrechnung. Wiesbaden: Gabler 1993.

[Mül94] Müller, R.; Pickel, H.: Ein neues Verfahren zur Klassifizierung von Teilen. CIM Management 10 (1994) 2, S. 31–35.

[Nef00] Neff, T.; Kokes, M.; Mathes, H. D.; Hertel, G.; Virt, W.: Front LoadCosting – Produktkostenmanagement auf Basis unvollkommener Information. Krp-Kostenrechnungspraxis 44 (2000), S. 15–24.

[Nie83] Niemann, G.; Winter, H.: Maschinenelemente II. Berlin: Springer 1983.

[Nie93] Niemand, S.: Target Costing im Anlagenbau. krp (1993) 5, S. 327–332.

[Opi66] Opitz, H.: Werkstückbeschreibendes Klassifizierungssystem. Essen: Giradet 1966.

[OPL00] Optimierung der Produktlebensdauer zur nachhaltigen Abfallreduzierung. Forschungsprojekt für den Bayerischen Forschungsverbund Abfallforschung u. Reststoffverwertung (BayFORREST). Laufzeit: 11/1999–12/2001.

[Osb57] Osborn, A.: Applied Imagination, Principles and Procedures of Creative Thinking. New York: Charles Cribner´s Sons 1957.

[Pac76] Pacyna, H.; Traub, K.: Arbeitsvorbereitungen in Gießereien (Zweckmäßige Arbeitsplanung und treffsichere Kalkulation). Düsseldorf: Gießerei-Verlag 1976.

[Pac80] Pacyna, H.: Einfluß der konstruktiven Gestaltung auf die Herstellkosten von Gußstücken. Düsseldorf: VDI-Verlag 1980, S. 97–111. (VDI-Berichte 362)

[Pac82a] Pacyna, H.; Hillebrand, A.; Rutz, A.: Kostenfrüherkennung für Gußteile. Düsseldorf: VDI-Verlag 1982, S. 103–114. (VDI-Berichte 457)

[Pac82b] Pacyna, H.: Die Kosten vor dem Gießen berechnet. VDI-Nachrichten 36 (1982) 16.

[Pät77] Pätzold, H.: Senkung des Konstruktions- und Fertigungsaufwandes durch Einsatz bereits konstruierter Teile und Baugruppen – praktische Anwendung von Sachmerkmalen bei Wiederholteilen. DIN-Mitt. 56 (1977), S. 341–350.

[Pah74] Pahl, G.; Beitz, W.: Baureihenentwicklung. Konstruktion 26 (1974) 2, S. 71–79, u. 26 (1974) 3, S. 113–118.

[Pah79] Pahl, G.; Beelich, K. H.: Ermittlung von Herstellkosten für ähnliche Bauteile. Düsseldorf: VDI-Verlag 1979, S. 155–164. (VDI-Berichte 347)

[Pah82] Pahl, G.; Beelich, K. H.: Kostenwachstumsgesetze nach Ähnlichkeitsbeziehungen für Baureihen. Düsseldorf: VDI-Verlag 1982, S. 69–69. (VDI-Berichte 457)

[Pah84] Pahl, G.; Rieg, F.: Kostenwachstumsgesetze für Baureihen. München: Hanser 1984.

[Pah04] Pahl, G.; Beitz, W.: Konstruktionslehre. Berlin: Springer 2004.

[Pah05] Pahl, G.; Beitz, W.: Engineering Design – A Systematic Approach. London: Springer 2005.

[Pat82] Patzak, G.: Systemtechnik – Planung komplexer, innovativer Systeme. Berlin: Springer 1982.

[Pat97] Paterak, J.: Vom theoretischen Minimum zum praktischen Optimum. VDI-Nachrichten (1997) 7, v. 14.2.97.

[Pau78] Paul, J.: Überwindung von Engpässen im Konstruktionsbereich durch Planung und Steuerung. In: Das Konstruktionsbüro – Arbeitsmittel und Organisation, Stuttgart. Düsseldorf: VDI-Verlag 1978, S. 113–120. (VDI-Berichte 311)

[Paw05] Pawellek, G.; O´Shea, M.; Schramm, A.: Logistikgerechte Produktentwicklung. Konstruktion 3 (2005) S. 71–76.

[Pee93] Peemöller, V. H.: Zielkostenrechnung für die frühzeitige Kostenbeeinflussung. krp (1993) 6, S. 375–380.

[Pei67] Peithmann, L.: Das Baukastensystem – eine Gestaltungsaufgabe bei Hebezeugen. Wirtschaftliche Aspekte für Hersteller und Benutzer. Vortrag anl. Fachtagung Fördertechnik; DIM Hannover 1967.

[Pfe93] Pfeiffer, T.: Qualitätsmanagement – Strategien, Methoden, Techniken. München: Hanser 1993.

[Pfl79] Pflicht, W.: Sachmerkmalssystem – genormtes Verfahren zur Wiederholteilfindung. Konstruktion 31 (1979) 12, S. 475–477.

[Phl97] Phleps, U.; Meier-Staude, R.: Methoden und Hilfsmittel zur Senkung der Entsorgungskosten als Teil der Produkt-Gesamtkosten. Abschlussbericht für das Vorhaben Nr. 189 Entsorgungskostensenkung AiF-Nr. 9847. Frankfurt/M.: Forschungskuratorium Maschinenbau e.V. 1997.

[Phl99] Phleps, U.: Recyclinggerechte Produktdefinition – Methodische Unterstützung für Upgrading und Verwertung. Aachen: Shaker 1999. (Konstruktionstechnik München, Band 34). Also: München: TU, Diss. 1999.

[Pic89] Pickel, H.: Kostenmodelle als Hilfsmittel zum kostengünstigen Konstruieren. München: Hanser 1989. (Konstruktionstechnik München, Band 2). Also: München: TU, Diss. 1988.

[Pic96] Picot, A.; Reichwald, R.; Wigand, R. T.: Die grenzenlose Unternehmung: Information, Organisation und Management; Lehrbuch zur Unternehmensführung im Informationszeitalter. Wiesbaden: Gabler 1996.

[Pie95] Pieske, R.: Benchmarking – Lernen von den Besten. VDI-Z 137 (1995) 1/2, S. 80–83.

[Pil98] Pilller, F. Th.: Kundenindividuelle Massenproduktion: Die Wettbewerbsstrategie der Zukunft. München: Hanser 1998.

[Pil99] Pilller, F. Th.; Waringer, D.: Modularisierung in der Automobilidustrie – neue Formen und Prinzipien. Aachen: Shaker 1999.

[Pok74] Pokorny, F.: Die Kosten der Normung. DIN-Mitt. 53 (1974), S. 227–230.

[Pro96] ProMeKreis – Verbundprojekt: Methoden zur Verbesserung der Kreislauffähigkeit von Geräten unter Berücksichtigung der Recyclingkosten und unter Einbeziehung innovativer Verwertungstechniken für die Produktentwicklung. Im Rahmen von Produktion 2000, Projektträger des BMBF für Fertigungstechnik und Qualitätssicherung: Forschungszentrum Karlsruhe. Laufzeit 12/96–11/99.

[Puc89] Pucher, H.: Entwicklungstendenzen bei Großdieselmotoren und ihre Auswirkungen auf deren Konstruktion. Konstruktion 41 (1989), S. 221–228.

[Rad84] Radke, M.: Kosten senkt man heute so. 222 praxiserprobte Maßnahmen zur Kostensenkung. Landsberg: Moderne Industrie 1984.

[Rau78] Rauschenbach, T.: Kostenoptimierung konstruktiver Lösungen – Möglichkeiten für die Einzel- und Kleinserienproduktion. Düsseldorf: VDI-Verlag 1978. (VDI-Taschenbuch 31)

[Rec97] Rechberg, U. v.: Systemgestützte Kostenschätzung – Eine Controlling-Perspektive. Wiesbaden: Gabler 1997. Also: Koblenz: WHU, Diss. 1997.

[REF71] REFA: Methodenlehre des Arbeitsstudiums. Teil 2 Datenermittlung. München: Hanser 1971.

[Reh81] Rehm, S.: Reduzierung der Tätigkeiten in der Konstruktion und Arbeitsplanung durch Teile und Fertigungsfamilien. REFA-/AKIE-Tagung, München 1981.

[Reh87] Rehm, S.: Einfluß auf das kostenbewußte Konstruieren durch Zusammenarbeit der Entwicklung und anderer Bereiche. Düsseldorf: VDI-Verlag 1987, S. 75–90. (VDI-Berichte 651)

[Rei93] Reinemuth J.; Birkhofer, H.: Elektronische Katalogsysteme für Zulieferkomponenten. In: Wettbewerbsfähiger mit Zulieferkomponenten. Düsseldorf: VDI-Verlag 1993, S. 157–172. (VDI-Berichte 1098)

[Rei95] Reinhart, G.; Lindermaier, R.; Kuba, R.: Im Schwitzkasten von Kosten und Zeit. Produktion 38 (1995), S. 3.

[Rei96] Reinhart, G.; Lindemann, U.; Heinzl, J.: Qualitätsmanagement. Berlin: Springer 1996.

[Rei96a] Reischl, C.: Implementierung des Kosteninformationssystems XKIS in der Zahnradfabrik Passau GmbH. krp 40 (1996) 4, S. 215–221.

[Rei97] Reischl, C.; Stößer, R.; Lindemann, U.: Design Concurrent Calculation – A Tool for the Target Costing Oriented Design Process. In: Riitahuhta, A. (Hrsg.): Proceedings of ICED 97, Tampere, Finnland. Tampere: University of Technology 1997 Band 2, S. 505–510. (WDK Schriftenreihe 25)

[Ren97] Renner, A.; Sauter, R.: Target Manager – Erste Standardsoftware zur Unterstützung des gesamten Target Costing-Prozesses. Controlling 1997, H. 1.

[Ren99] Renius, K. T.: Two Axle Tractors. In: CIGR Handbook of Agricultural Engineering, Vol.III, S.115–184. American Society of Agricultural Engineers, St. Joseph, MI, USA 1999. ISBN 1-892769-02-6.

[Ren02a] Renius, K. T.: Persönliche Mitteilungen 2002.

[Ren02b] Renius, K. T.: Global Tractor Development: Product Families and Technology Levels. Plenarvortrag XXX. Internat. Symposium: "Actual Tasks on Agricultural Engineering". 12–15.3.2002. Opatija, Kroatien. Proceedings S. 87–95. ISSN 1333-2651.

[Ren02c] Renius, K. T.: Gesamtentwicklung Traktoren. In: Jahrbuch Agrartechnik 14 (2002), S. 41–48, 232, 233. Münster: Landwirtschaftsverlag 2002. ISBN 3-7843-3133-5.

[Rep94] Repota, P.: Welche Technik ist die Beste? PM 4 (1994), S. 84–94.

[Rie82] Rieg, F.: Kostenwachstumsgesetze für Baureihen. Darmstadt: TH, Diss. 1982.

[Rie85] Riebel, P.: Einzelkosten und Deckungsbeitragsrechnung. 5. Aufl. Wiesbaden: Gabler 1985, S. 177.

[Rie93] Rieker, S.: Kundenorientierung als tragender Erfolgsfaktor des Key Account Managements. In: Hofmaier, R. (Hrsg.): Investitionsgüter- und High-Tech-Marketing (ITM). 2. Aufl. Landsberg/Lech: Moderne Industrie 1993, S. 355–378.

[RKW90] Relativkosten für den Konstrukteur: Nutzen und Risiken; Bedarf, Erwartungen, Leistungsvermögen und Kosten bezüglich Relativkostenblätter und andere Methoden der Kostenbeeinflussung des Konstruktionsprozesses. Eschborn: RKW-Verlag; Köln: TÜV-Rheinland 1990.

[Rod91] Rodenacker, W. G.: Methodisches Konstruieren – Grundlagen, Methodik, praktische Beispiele. 4., überarb. Aufl. Berlin: Springer 1991.

[Rös96] Rösler, F.: Target Costing für die Automobilindustrie. Wiesbaden: Gabler 1996. Also: Koblenz: Wiss. Hochsch. für Unternehmensführung, Diss. 1996.

[Roh69] Rohrbach, B.: Kreativ nach Regeln – Methode 635, eine neue Technik zum Lösen von Problemen. Absatzwirtschaft 12 (1969), S. 73–75.

[Rom95] Rommel, G.; Brück, F.; Diederichs, R.; Kempis, R.-D.; Kaas, H.-W.; Fuhry, G.; Kurfess, V.: Qualität gewinnt. Stuttgart: Schäffer-Poeschel 1995.

[Roma93] Romanow, P.: Konstruktionsbegleitende Kalkulation von Werkzeugmaschinen. Berlin: Springer 1994. Also: München: TU, Diss. 1993.

[Romm93] Rommel, G.; Brück, F.; Diederichs, R.; Kempis, R.-D.; Kluge, J.: Einfach überlegen. Stuttgart: Schäffer-Poeschel 1993.

[Roo95] Roozenburg, N.; Eekels, J.: Product design. Chichester: Wiley 1995.

[Ros96] Rosenberg, O.: Variantenfertigung. In: Kern. W.; Schröder, H. H.; Weber, J.: Handwörterbuch der Produktionswirtschaft. 2. Aufl. Stuttgart: Schäffer-Poeschel 1996.

[Ros02] Rosenberg, O.: Kostensenkung durch Komplexitätsmanagement. In: Franz, K.-P.; Kajüter, P. (Hrsg): Kostenmanagement: Wertsteigerung durch systematische Kostensteuerung. Stuttgart: Schäffer-Poeschel 2002.

[Rot86] Rothenbücher, J.: Ein Beitrag zur Kostenprognose und Analyse der Effizienz im Betriebsmittelbau. Aachen: RWTH, Diss. 1986.

[Rot94] Roth, K.: Konstruieren mit Konstruktionskatalogen. Band 1: Konstruktionslehre; Band 2: Kataloge. Berlin: Springer 1994.

[Rot96] Roth, K.: Konstruieren mit Konstruktionskatalogen. Band 3: Verbindungen und Verschlüsse, Lösungsfindung. Berlin: Springer 1996.

[RTO00] RTO: Design for Low Cost Operation und Support. RTO MEETING PROCEEDINGS 37 (2000).

[Ruc82] Ruckes, J.: Analytische Angebotskalkulation im Stahl- und Apparatebau mit empirischen Werten. Berlin: Springer 1982.

[Sab97] Sabisch, H.; Tintelot, C.: Integriertes Benchmarking. Berlin: Springer 1997.

[Sak89a] Sakurai, M.: Target Costing and how to use it. Journal of Cost Management for the manufacturing industry (1989) 3, S. 39–50.

[Sak89b] Sakurai, M.; Huang, P. Y.: A Japanese Survey of Factory Automation and its Impact on Management Control Systems. In: Monden, Y.; Sakurai, M. (Ed.): Japanese Management Accounting. A World Class Approach to Profit Management. Cambridge, Mass.: Productivity Press 1989, S. 261–279.

[Sak94] Sakurai, M.; Keating, P. J.: Target Costing und Activity-Based Costing. Controlling (1994) 2, S. 84–91.

[Sau86] Sauermann, H. J.: Eine Produktkostenplanung für Unternehmen des Maschinenbaus. München: TU, Diss. 1986.

[Say84] Saynisch, M.: Konfigurationsmanagement – fachlich-inhaltliche Entwurfssteuerung, Dokumentation und Änderungswesen im ganzheitlichen Projektmanagement. Köln: TÜV Rheinland, 1984.

[Sch92] Schaal, S.: Integrierte Wissensverarbeitung mit CAD am Beispiel der konstruktionsbegleitenden Kalkulation. München: Hanser 1992. (Konstruktionstechnik München, Band 8). Also: München: TU, Diss. 1992.

[Sch95] Schultz, V.: Projektkostenschätzung. Kostenermittlung in den frühen Phasen von technischen Auftragsprojekten. Wiesbaden: Gabler 1995.

[Sch96] Schmidt, A.: Kostenrechnung. Grundlagen der Vollkosten-, Deckungsbeitrags-, Plankosten- und Prozeßkostenrechnung. Stuttgart: Kohlhammer 1996.

[Sch98a] Schmalzl, B.; Schröder, J.: Managementkonzepte im Wettstreit. München: Beck 1998.

[Sch98b] Scholl, K.: Konstruktionsbegleitende Kalkulation. Computergestützte Anwendung von Prozeßkostenrechnung und Kostentableaus. München: Vahlen 1998.

[Sch01] Schumann, F. J.: Methoden und Werkzeuge zur Integration der kundengerechten Wertgestaltung in die Konzeptphase des Produktentwicklungsprozesses. Chemnitz: TU, Dis.2001.

[Sche90] Scheer, A.-W.; Bock, M.; Bock, R.: Konstruktionsbegleitende Kalkulation mit Expertensystem-Unterstützung. ZwF 85 (1990) 11, S. 576–579.

[Sche93] Scheucher, F.: FTA-Fehlerbaumanalyse, Qualitätsmanagement. Augsburg: WEKA Fachverlag 1993.

[Sche97] Scheffels, G.: Welches (Fahrzeug-)Teil darf es sein? „Ford Power Products" – ein Komponenten-Baukasten mit über 100000 Teilen. Der Zuliefermarkt 11 (1997), S. 19–20.

[Schh89] Schuh, G.: Gestaltung und Bewertung von Produktvarianten. Aachen: TH, Diss. 1989.

[Schh93] Schuh, G.; Steinfatt, E.: Konstruktionsbegleitende Prozeßkostenrechnung. ZwF 88 (1993), S. 344–346.

[Schl89] Schultz, H.: Elektronenstrahlschweißen. Düsseldorf: DVS-Verlag 1989. (Fachbuchreihe Schweißtechnik, Band 93)

[Schm96] Schmidt, F.: Gemeinkostensenkung durch kostengünstiges Konstruieren. Wiesbaden: Dt. Univ.-Verlag; Wiesbaden: Gabler 1996. (Gabler Edition Wissenschaft). Also: Kaiserslautern: Univ., Diss. 1996.

[Schu89] Schuh, G.: Gestaltung und Bewertung von Produktvarianten. Aachen: TH, Diss. 1989.

[Schu96] Schulz, H.: Hochgeschwindigkeitsbearbeitung. München: Hanser 1996.

[Schu97] Schuh, G. u. A.: Zielkostenmanagement im Anlagenbau. ZWF 92 (1997) 11, S. 583–587.

[Schw01] Schwankl, L.: Analyse und Dokumentation in den frühen Phasen der Produktentwicklung. München: TU, Diss. 2001.

[Sei90] Seicht, G.: Moderne Kosten- und Leistungsrechnung. Wien: 1990.

[Seid93] Seidenschwarz, W.: Target Costing: Marktorientiertes Zielkostenmanagement. München: Vahlen 1992. Also: Stuttgart: Univ., Diss. 1992.

[Seid97] Seidenschwarz, W.: Nie wieder zu teuer! 10 Schritte zum marktorientierten Kostenmanagement. Stuttgart: Schäffer-Poeschel 1997.

[Sim97] Simon, H.: Hidden Champions: Wie deutsche Unternehmen Weltspitze werden. Münchner Koll. 1997. Landsberg: Moderne Industrie 1997.

[Smi98] Smith, P.G.; Reinertsen, D.G.: Developing products in half the time. New York. J. Wiley 1998.

[Som92] Sommer, P.: Kosten- und Leistungsrechnung in Härtereien. Bericht der Dr. Sommer Werkstofftechnik, Issum-Seveler 1992.

[Spr95] Sprenger, R.: Das Prinzip Selbstverantwortung. Wege zur Motivation. Frankfurt/M.: Campus 1995.

[Spu81] Spur, G.; Stöferle, T.: Handbuch der Fertigungstechnik. Band 1. München: Hanser 1981.

[Spu82] Spur, G.; Kreisfeld, P. : Hilfsmittel zur Kostenprognose abgespanter Einzelteile. VDI-Verlag 1982, S. 91–102. (VDI-Berichte 457)

[Sta03] Stang, S.; Warneck G.: Plattform Engineering – neue Wege zu neuartigen Konzepten. Konstruktion (2003) 3, S. 72–75.

[Ste92] Steitz, T.: Methodik zur markorientierten Entwicklung von Werkzeugmaschinen mit Integration funktionsbasierter Strukturierung und Kostenschätzung. Karlsruhe: Forschungsberichte aus dem Institut für Werkzeugmaschinen u. Betriebstechnik 1992.

[Ste93a] Steiner, M.; Ehrlenspiel, K.; Schnitzlein, W.: Erfahrungen mit der Einführung wissensbasierter Erweiterungen eines CAD-Systems zur konstruktionsbegleitenden Kalkulation. Düsseldorf: VDI-Verlag 1993. (VDI-Berichte 1079)

[Ste93b] Steinhilper, R.; Hudelmaier, U.: Erfolgreiches Produktrecycling zur erneuten Verwendung oder Verwertung. Eschborn: RKW 1993.

[Ste95] Steiner, M. Rechnergestütztes Kostensenken im praktischen Einsatz. Aachen: Shaker 1996. (Konstruktionstechnik München, Band 22). Also: München: TU, Diss. 1995.

[Ste98] Steinmeier, E.: Realisierung eines systemtechnischen Produktmodells – Einsatz in der Pkw-Entwicklung. Aachen: Shaker 1998. (Konstruktionstechnik München, Band 28). Also:: München: TU, Diss. 1998.

[Sti98] Stippel, N.; Reichmann, Th.: Target Costing und Wertanalyse. Konkurrierende oder sich ergänzende Instrumente für die marktorientierte Gestaltung neuer Produkte? Controlling (1998) 2, S. 98–105.

[Stö75] Stöferle, Th.; Dilling, H. J.; Rauschenbach, Th.: Rationelle Montage – Herausforderung an den Ingenieur. VDI-Z 117 (1975), S. 715–719.

[Stö97] Stößer, R.: Managing Target Costs – A Computer Based Tool to Support the Designer in a Design to Cost Process. In: Riitahuhta, A. (Hrsg.): Proceedings of ICED 97, Tampere, Finnland. Tampere: University of Technology 1997 Band 2, S. 521–524. (WDK Schriftenreihe 25)

[Stö98] Stößer, R.; Lindemann, U.: Conception of a Cost Information System for a Target-Costing-Oriented Product Development Process. In: Proceedings of IV International Congress of Project Engineering. Cordoba, Spain 1998. Cordoba: Universidad de Córdoba, Área de Proyectos de Ingeniería, 1998.

[Stö99] Stößer, R.: Zielkostenmanagement in integrierten Produkterstellungsprossen. Aachen: Shaker 1999. (Konstruktionstechnik München, Band 33). Also: München: TU, Diss. 1999.

[Sto87] Stoeck, N.: Made in Germany als Qualitätsbegriff ist durchaus noch entwicklungsfähig. MEGA (12) 1987, S. 88–92.

[Sto94] Stoll, G.: Montagegerechte Produkte mit featurebasiertem CAD. München: Hanser 1994. (Konstruktionstechnik München, Band 21). Also: München: TU, Diss. 1994.

[Sto99] Stoi, R.: Prozeßkostenmanagement in Deutschland. Ergebnisse einer empirischen Untersuchung. Controlling (1999) 2, S. 53–59.

[Stu94] Stuffer, R.: Planung und Steuerung der integrierte Produktentwicklung. München: Hanser 1994. (Konstruktionstechnik München, Band 13). Also: München: TU, Diss. 1994.

[Stu97] Stuffer, R.; Kleedörfer, R.: Prozeßmanagement im Wandel – Zeitgerechte Ansätze zur Prozeßplanung und -steuerung. EDM-Report (1997) 2, S. 42–47.

[Tan89] Tanaka, M.: Cost Planning and Control Systems in the Design Phase of a New Product. In: Moden, Y.; Sakurai, M. (Ed.): Japanese Management Accounting. A World Class Approach to Profit Management. Cambridge, Mass.: Productivity Press 1989, S. 49–71.

[Ter97] Terninko, J.; Zusman, A.; Zlotin, B.: Step-by-Step TRIZ: Creating Innovative Solution Concepts. Nottingham, New Hampshire: Responsible Management 1997.

[The89] Theuerkauff, P.; Groß, A.: Praxis des Klebens. Heidelberg: Springer 1989.

[Tre78] Trechsel, F.: Produkt/Markt-Strategie ... Kernstück jeder Unternehmensstrategie. Management-Z. 47 (1978), S. 383–387.

[TRU96] TRUMPF: Faszination Blech – Flexible Bearbeitung eines vielseitigen Werkstoffs. Stuttgart: Raabe 1996. – Firmenschrift

[Ull92] Ullman, D. G.: The Mechanical Design Process. New York: McGraw-Hill 1992.

[Ulr95] Ulrich, K. T.; Eppinger, S. D.: Product Design and Development. New York: McGraw-Hill 1995.

[VDG66] VDG: Konstruieren mit Gußwerkstoffen. Düsseldorf: Gießerei-Verlag 1966.

[VDI77] VDI-Richtlinie 2225: Technisch-wirtschaftliches Konstruieren. Blatt 2. Düsseldorf: VDI-Verlag 1977.

[VDI78] VDI-Richtlinie 3381: Schaumstoffmodelle; Konstruktionshinweise für Gußteile. Düsseldorf: VDI-Verlag 1978.

[VDI79] VDI-Richtlinie 2006: Gestalten von Spritzgußteilen aus thermoplastischen Kunststoffen. Düsseldorf: VDI-Verlag 1979.

[VDI82] VDI-Richtlinie 2222, Bl. 2: Konstruktionsmethodik. Erstellung und Anwendung von Konstruktionskatalogen. Düsseldorf: VDI-Verlag 1982.

[VDI87] VDI-Richtlinie 2235: Wirtschaftliche Entscheidungen beim Konstruieren – Methoden und Hilfen. Düsseldorf: VDI-Verlag 1987.

[VDI90] VDI-Richtlinie 2234: Wirtschaftliche Grundlagen für den Konstrukteur. Düsseldorf: VDI-Verlag 1990.

[VDI93a] VDI-Richtlinie 2221: Methodik zum Entwickeln und Konstruieren technischer Systeme und Produkte. Düsseldorf: VDI-Verlag 1993.

[VDI93b] VDI-Richtlinie 2243: Konstruieren recyclinggerechter technischer Produkte. Grundlagen und Gestaltungsregeln. Düsseldorf: VDI-Verlag 1993.

[VDI95] VDI-Zentrum Wertanalyse (Hrsg.): Wertanalyse; Idee-Methode-System. 4. Aufl. Düsseldorf: VDI-Verlag 1995.

[VDI97] VDI-Richtlinie 2225: Technisch-wirtschaftliches Konstruieren. Blatt 1 u. 4. Düsseldorf: VDI-Verlag 1997.

[VDI98a] VDI-Richtlinie 2223: Methodisches Entwerfen technischer Produkte. Düsseldorf: VDI-Verlag 1998.

[VDI98b] VDI-Berichte 1434: Effektive Entwicklung und Auftragsabwicklung variantenreicher Produkte.: Düsseldorf: VDI-Verlag 1998.

[VDI99] VDI (Hrsg.): pm-praxis. 1. praxisorientierter Anwendertag zum Projektmanagement, München, 6./7.5.99; Düsseldorf: VDI-Verlag 1999. (VDI-Berichte 1490)

[VDM78a] VDMA: Selbst fertigen oder kaufen? Rechen- und Entscheidungsschema. Bw V182. Frankfurt/M.: Maschinenbau-Verlag 1978.

[VDM78b] VDMA: Kennzahlen-Kompaß. Frankfurt: Maschinenbau-Verlag 1978.

[VDM06] VDMA: Kennzahlenkompaß 2002. Frankfurt/M.: Maschinenbau-Verlag 2006.

[VDM97] VDMA: Total Profit Management. Frankfurt/M.: Maschinenbau-Verlag 1997.

[VDM98] VDMA: Studie „Schneller oder besser?" Frankfurt/M.: Maschinenbau-Verlag 1998.

[Vec86] Vecemik, P.: Bei schrumpfendem Absatz kann eine Preisreduktion rasch den Ruin bedeuten. Management-Z. 55 (1986) 4, S. 189–191.

[Ver94] Verlemann, E.: Prozeßgestaltung beim Hochgeschwindigkeitsaußenrundschleifen von Ingenieurkeramik. Aachen: Shaker 1994. (Berichte aus der Produktionstechnik, Band 1994,17). Also: Aachen: TH, Diss. 1994.

[Wal87] Wallace, K; Hales, C.: Detailed Analysis of an Engineering Design Project. In: Proceedings of the International Conference on Engineering Design 1987. New York: 1987. (Schriftenreihe WDK 13)

[War80a] Warnecke, H. J.; Bullinger, H.-J.; Hichert, R.: Wirtschaftlichkeitsrechnung für Ingenieure. München: Hanser 1980.

[War80b] Ward, S. A.: Cost control in design and construction. McGraw-Hill 1980.

[War90] Warnecke, H. J.; Bullinger, H.-J.; Hichert, R.: Kostenrechnung für Ingenieure. München: Hanser 1990.

[War92] Warnecke, H. J.: Die fraktale Fabrik. Revolution der Unternehmenskultur. Berlin: Springer 1992.

[Web97] Webhofer, M.: Zielkostenorientierte Umkonstruktion eines Nd-YAG-Laserresonators. München: TU, Lehrstuhl für Konstruktion im Maschinenbau, Unveröffentlichte Semesterarbeit 1997.

[Wel74] Welschof, G.: Entwicklungslinien im Schlepperbau. Kriterien für die heutige und zu-künftige Entwicklung. Grundlagen der Landtechnik 24 (1974) H.1, S. 6–13.

[Wel97] Welg, M. K.; Amshoff, B: Neuordnung der Kostenrechnung zur Unterstützung der strategischen Planung. In: Franz, K. P.; Kajüter, P. (Hrsg.): Kostenmanagement. Stuttgart: Schäffer-Poeschel 1997.

[Wel98] Welp, E. G.; Endebrock, K.; Albrecht, K.: Entwicklungs- und konstruktionsbegleitende Kostenbeurteilung – Ergebnisse einer Befragung von Konstruktionsleitern. krp 42 (1998) 5, S. 257–265.

[Wer97] Werner, W.: Schmäh, M.: Gemeinsam zum Erfolg – Die Zukunft in (West-) Europa gehört der „intelligenten Zulieferung". Der Zuliefermarkt 11 (1997), S. 19–20.

[Wes02] Westekemper, M.: Methodik zur Angebotspreisfindung – am Beispiel des Werkzeug- und Formenbaus. Dis. RWTH Aachen 2002.

[Weu99] Weule, H. Trender, L.: Entwicklungsbegleitende Kostenkalkulation – prozeßorientierte Bestimmung der Produktlebenszykluskosten unter Beachtung von Zielkosten. Konstruktion (1999) 9, S. 35–39.

[Wie72] Wiendahl, H.-P.; Grabowski, H.: Systematische Erfassung von Konstruktionstätigkeiten – Voraussetzung für eine Rationalisierung im Konstruktionsbereich. Konstr. 24 (1972), S. 175–180.

[Wie90] Wierda, L.: Cost Information Tools for Designers. Delft: Univ., Diss. 1990.

[Wil94] Wildemann, H.: Die modulare Fabrik: Kundennahe Produktion durch Fertigungssegmentierung. 4. Aufl. München: TCW Transfer-Centrum 1994.

[Wil96] Wildemann, H.: Kernkompetenzen – Leitfaden zur Ermittlung und Entwicklung von Kernfähigkeiten in Produktion, Entwicklung und Logistik. München: TCW Transfer-Centrum 1996.

[Wil98] Wildemann, H.: Produktklinik – Wertgestaltung von Produkten und Prozessen. München: TCW Transfer-Centrum 1998. (TCW-report Nr. 2)

[Wil99] Wildemann, H.: Produktklinik. Wertgestaltung von Produkten und Prozessen Methoden und Fallbeispiele. München: TCW Transfer-Centrum 1999.

[Wil02] Wildemann, H.: Komplexitätsmanagement. Leitfaden. 3. Aufl. München 2002.

[Wip81] Wippersteg, H. H.: Kosten der Teilevielfalt – Methoden zu ihrer Erfassung und Reduzierung. Pers. überlass. Manuskr., Fa. Claas, Harsewinkel 1981.

[Wit84] Witte, K.-W.: Konstruktion senkt Montagekosten. VDI-Z. 126 (1984), S. 835–840.

[Wöh96] Wöhe, G.: Einführung in die allgemeine Betriebswirtschaftslehre. 16. Aufl. München: Vahlen 1996.

[Wol94] Wolfram, M.: Feature-basiertes Konstruieren und Kalkulieren. München: Hanser 1994. (Konstruktionstechnik München, Band 19). Also: München: TU, Diss. 1994.

[Wom91] Womack, J. P.; Jones, D. T.; Roos, D.: Die zweite Revolution in der Automobilindustrie. Frankfurt/M.: Campus 1991.

[Wom98] Womack, J. P.; Jones, D. T.: Auf dem Weg zum perfekten Unternehmen (Lean Thinking). München: Heyne 1998 u. Frankfurt: Campus 1998.

[Zan70] Zangemeister, C.: Nutzwertanalyse in der Systemtechnik. München: Wirtmannsche Buchhandlung 1970.

[Zol96] Zoll, G.: Entwicklung von Automobilteilen aus Kunststoff mit systemtechnische Methoden. Berlin: TU, Diss. 1996.

[ZWA92] ZWA: Hrsg.: Bundeswirtschaftskammer (Gruppe Technik und Betriebwirtschaft) Wien; ZWA Bericht 3: „Erfolg mit Methode“, 1992.

[Zwi66] Zwicky, F.: Entdecken, Erfinden, Forschen im morphologischen Weltbild. Zürich: Droemer Knaur 1966.

Index

Page numbers in bold indicate the primary discussion of a topic.

M

N

O

P

S

T

U

V

W